Achieving sustainable production of poultry meat

Volume 1: Safety, quality and sustainability

It is widely recognised that agriculture is a significant contributor to global warming and climate change. Agriculture needs to reduce its environmental impact and adapt to current climate change whilst still feeding a growing population, i.e. become more 'climate-smart'. Burleigh Dodds Science Publishing is playing its part in achieving this by bringing together key research on making the production of the world's most important crops and livestock products more sustainable.

Based on extensive research, our publications specifically target the challenge of climate-smart agriculture. In this way we are using 'smart publishing' to help achieve climate-smart agriculture.

Burleigh Dodds Science Publishing is an independent and innovative publisher delivering high quality customer-focused agricultural science content in both print and online formats for the academic and research communities. Our aim is to build a foundation of knowledge on which researchers can build to meet the challenge of climate-smart agriculture.

For more information about Burleigh Dodds Science Publishing simply call us on +44 (0) 1223 839365, email info@bdspublishing.com or alternatively please visit our website at www.bdspublishing.com.

Related titles:

Achieving sustainable production of poultry meat Volume 2: Breeding and nutrition
Print (ISBN 978-1-78676-068-5); Online (ISBN 978-1-78676-070-8, 978-1-78676-071-5)

Achieving sustainable production of poultry meat Volume 3: Health and welfare
Print (ISBN 978-1-78676-072-2); Online (ISBN 978-1-78676-074-6, 978-1-78676-075-3)

Achieving sustainable production of eggs Volume 1: Safety and quality
Print (ISBN 978-1-78676-076-0); Online (ISBN 978-1-78676-078-4, 978-1-78676-079-1)

Achieving sustainable production of eggs Volume 2: Animal welfare and sustainability
Print (ISBN 978-1-78676-080-7); Online (ISBN 978-1-78676-082-1, 978-1-78676-083-8)

Chapters are available individually from our online bookshop: https://shop.bdspublishing.com

climate•SMART•publishing
IN AGRICULTURAL SCIENCE

BURLEIGH DODDS SERIES IN AGRICULTURAL SCIENCE

NUMBER 13

Achieving sustainable production of poultry meat

Volume 1: Safety, quality and sustainability

Edited by Professor Steven C. Ricke, University of Arkansas, USA

burleigh dodds SCIENCE PUBLISHING

Published by Burleigh Dodds Science Publishing Limited
82 High Street, Sawston, Cambridge CB22 3HJ, UK
www.bdspublishing.com

Burleigh Dodds Science Publishing, 1518 Walnut Street, Suite 900, Philadelphia, PA 19102-3406, USA

First published 2017 by Burleigh Dodds Science Publishing Limited

Library of Congress Control Number: 2016956458

British Library Cataloguing in Publication Data
A catalogue record for this book is available from the British Library

ISBN 978-1-78676-064-7 (print)
ISBN 978-1-78676-066-1 (online)
ISBN 978-1-78676-067-8 (online)
ISSN 2059-6936 (print)
ISSN 2059-6944 (online)

Typeset by Deanta Global Publishing Services, Chennai, India
Printed by Lightning Source

Contents

Part 3 Sustainability

Series list

Title	Series Number
Achieving sustainable cultivation of maize - Vol 1 From improved varieties to local applications Edited by: Dr Dave Watson, CGIAR Maize Research Program Manager, CIMMYT, Mexico	001
Achieving sustainable cultivation of maize - Vol 2 Cultivation techniques, pest and disease control Edited by: Dr Dave Watson, CGIAR Maize Research Program Manager, CIMMYT, Mexico	002
Achieving sustainable cultivation of rice - Vol 1 Breeding for higher yield and quality Edited by: Prof. Takuji Sasaki, Tokyo University of Agriculture, Japan	003
Achieving sustainable cultivation of rice - Vol 2 Cultivation, pest and disease management Edited by: Prof. Takuji Sasaki, Tokyo University of Agriculture, Japan	004
Achieving sustainable cultivation of wheat - Vol 1 Breeding, quality traits, pests and diseases Edited by: Prof. Peter Langridge, The University of Adelaide, Australia	005
Achieving sustainable cultivation of wheat - Vol 2 Cultivation techniques Edited by: Prof. Peter Langridge, The University of Adelaide, Australia	006
Achieving sustainable cultivation of tomatoes Edited by: Dr. Autar Mattoo, USDA-ARS, USA & Prof. Avtar Handa, Purdue University, USA	007
Achieving sustainable production of milk - Vol 1 Milk composition, genetics and breeding Edited by: Dr Nico van Belzen, International Dairy Federation (IDF), Belgium	008
Achieving sustainable production of milk - Vol 2 Safety, quality and sustainability Edited by: Dr Nico van Belzen, International Dairy Federation (IDF), Belgium	009
Achieving sustainable production of milk - Vol 3 Dairy herd management and welfare Edited by: Prof. John Webster, University of Bristol, UK	010
Ensuring safety and quality in the production of beef - Vol 1 Safety Edited by: Prof. Gary Acuff, Texas A&M University, USA & Prof.James Dickson, Iowa State University, USA	011
Ensuring safety and quality in the production of beef - Vol 2 Quality Edited by: Prof. Michael Dikeman, Kansas State University, USA	012
Achieving sustainable production of poultry meat - Vol 1 Safety, quality and sustainability Edited by: Prof. Steven C. Ricke, University of Arkansas, USA	013
Achieving sustainable production of poultry meat - Vol 2 Breeding and nutrition Edited by: Prof. Todd Applegate, University of Georgia, USA	014
Achieving sustainable production of poultry meat - Vol 3 Health and welfare Edited by: Prof. Todd Applegate, University of Georgia, USA	015

Acknowledgements

We wish to acknowledge the following for their help in reviewing particular chapters:

- Chapter 5: *Dr Rolf Joeger, University of Delaware, USA; and Dr Michael Hume, USDA-ARS, USA*
- Chapter 14: *Dr Anand Mohan, University of Georgia, USA*
- Chapter 20: *Dr Josh Payne, Oklahoma State University, USA; and Dr Claudia Dunckley, University of Georgia, USA*

Introduction

Poultry meat production faces a range of challenges. To meet growing demand, the FAO has estimated that the world poultry production needs to grow by 2.5% per year up to 2030, and by as much as 3.4% in developing countries. Much of the increase in output already achieved has been as a result of improvements in commercial breeds combined with rearing in more intensive production systems. However, these systems face a number of problems in achieving further sustainable increases in production. More intensive systems and complex supply chains have increased the risk of rapid transmission of zoonoses, which remains a significant challenge (including the risk of pathogens developing antibiotic resistance). At the same time consumers' expectations of sensory and nutritional quality have never been higher. Consumers' attitudes now also encompass wider concerns such as the environmental impact and sustainability of poultry production. These challenges are addressed in this collection: *Achieving sustainable production of poultry meat Volume 1: Safety, quality and sustainability.*

Poultry meat safety

The first part of the book reviews continuing challenges in ensuring the safety of poultry meat. Chapter 1 begins by reviewing current research on *Campylobacter*, which is emerging as the leading cause of bacterial foodborne illness related to poultry products in many countries. Despite stringent biosecurity on farm and control measures during processing, the number of cases of *Campylobacter* has continued to rise each year in counties such as the United Kingdom. Unlike *Salmonella*, currently no effective vaccine for *Campylobacter* spp. has been discovered for either the human or the poultry populations. As the chapter shows, farm interventions including biosecurity, altering the diet, use of prebiotics and probiotics, and additives have not been found to be consistently effective. Other techniques during processing such as rapid chilling, surface washes and irradiation have also had limited success in controlling the pathogen without affecting quality. In addition, having traditionally been seen as a commensal of broiler chickens, *Campylobacter* is increasingly regarded as potentially pathogenic to poultry. As the chapter points out, there is a need for more research on *Campylobacter* pathogenesis to develop more effective interventions. There is also a need to explore the biology of *Campylobacter* in chicken to better understand the effect it has on the health, welfare and performance of broilers.

Chapter 1 is complemented by Chapter 10, which discusses in more detail where we are in achieving effective control of *Campylobacter*. The chapter reviews ways by which *Campylobacter* can spread around farms and how effective biosecurity measures are in reducing horizontal transmission. It also discusses key findings in the use of antimicrobial therapy, including the relative ineffectiveness of probiotics against *Campylobacter*, as well as studies of the potential use of botanicals. Finally, it summarises progress in developing vaccines, including DNA vaccines. Like Chapter 1, it underlines the need for more research on this pathogen in the context of poultry meat production.

Chapter 2 provides an authoritative review of the state of the art in control of *Salmonella*, which remains the major cause of poultry-related foodborne illness in countries such as the United States, including emerging risks associated with new serovars. The chapter summarises our current understanding of the invasion and transmission mechanisms of *Salmonella*. As it indicates, further research is needed to understand the complete roles of humoral and cell-mediated immunity.

The chapter also reviews the wide range of methodologies for the sampling and bacteriological detection of *Salmonella* in poultry flocks, their weaknesses and strengths, and the development of new techniques, including polymerase chain reaction-based techniques with greater sensitivity, specificity and reduced assay time. Finally, the chapter discusses the wealth of techniques used for control of *Salmonella* on the farm. Although vaccination has been the primary cause of reduction of *Salmonella* incidence all over the world, the efficacy of vaccines depends on several factors and they are not universally effective. This has placed the emphasis on other strategies based on diet modulation and feed supplementation.

The chapter reviews the diverse dietary strategies that have been investigated to reduce *Salmonella* in the gastrointestinal (GI) tract, such as feeding whole cereals or mash diet to improve intestinal resistance, the use of exogenous enzymes to reduce the amount of substrate remaining in the gut for pathogenic microorganisms or the use of bioactive compounds with immunomodulatory properties. Other techniques to boost immune response include feeding maternal antibodies to hatchlings, the use of antimicrobial peptides, the potential acidification of feed or water by short-chain fatty acids to reduce *Salmonella* colonisation of broilers, essential oils and competitive exclusion treatments such as the use of probiotics, prebiotics and synbiotics. The chapter also discusses newer techniques such as bacteriophages and the increasing interest in ways of combining of different dietary treatments allowing potential addition, complementarity, or even synergism, of protagonist effects directly at the gut level.

Chapter 2 is complemented by Chapter 9 on the control programme for *Salmonella* in Sweden, which has focused on pre-harvest and harvest interventions and which has largely eliminated *Salmonella* from fresh poultry meat. This remarkable achievement is based on the following key steps: preventing contamination of the breeding pyramid (by compulsory sampling of grandparent and parent flocks, including hatcheries), strict biosecurity on farm sites; preventing *Salmonella* contamination in feed given to broiler flocks (e.g. through HACCP systems based on heat treatment), detailed surveillance of the production chain at critical control points to detect *Salmonella* contamination and immediate action to remove *Salmonella* from the food chain if detected. Given the decision not to use vaccination as a control measure, the chapter reviews the effectiveness of other approaches such as competitive exclusion treatments as well as biosecurity arrangements. The chapter also discusses the various cost–benefit analyses of the *Salmonella* control programme, including the balance of responsibilities and costs faced by farmers and others such as feed manufacturers in achieving a very high level of safety whilst sustaining a viable business.

Chapter 2 refers to increasing concern over the past 30 years regarding the worldwide emergence of multidrug-resistant phenotypes among *Salmonella* serotypes such as *S*. Typhimurium, *S*. Enteritidis and *S*. Newport. This problem is discussed in Chapters 4 and 5. Chapter 4 highlights public health concerns associated with a strong link between antibiotic use in food production and the rise of antibiotic-resistant pathogens that affect public health, for example ciprofloxacin-resistant and erythromycin-resistant *Campylobacter*. Antibiotics have been conventionally used in food animals to treat clinical disease (therapy), to prevent and control common disease events (prophylaxis and metaphylaxis) and to enhance animal growth (growth promoter), though the latter use has now been banned in many countries. The chapter discusses the range of initiatives to reduce antibiotic use, exemplified by the World Organisation for Animal Health (OIE), which has identified the following targets: tracking rates of veterinary antibiotic use, resistance and residues through a nationwide surveillance and monitoring system; changing incentives to discourage unnecessary antibiotic use in animals; educating farmers, veterinarians and consumers on the dangers of antibiotic resistance; and phasing out the sub-therapeutic use of antibiotics in animals. The chapter also highlights the shift to methods that reduce reliance on antibiotics, for example, efficient biosecurity and hygiene protocols, the use of prebiotics and probiotic bacteria for competitive exclusion, bacteriophages, additives such as organic acids, genetic resistance, vaccination, and organic and antibiotic-free farming. It also highlights recent reductions in antimicrobial-resistant *Salmonella* associated with *Salmonella* control programmes adopted by the poultry industry, linking back to the kinds of intervention reviewed in Chapter 2.

Building on the overview in Chapter 2, Chapter 5 reviews in detail the range of prebiotics and related compounds used in competitive exclusion treatments. These include non-digestible carbohydrates, fructooligosaccharides, yeast-derived components and mannan derivatives, galactooligosaccharides and guar gum. The chapter reviews the latest research into their mode of action and effectiveness in control of pathogens such as *Salmonella*. It highlights where further research is needed, for example in how well these treatments work on less-studied pathogens such as *Campylobacter* as well as the growing interest in combining probiotics and prebiotics into synbiotics which offers the opportunity to enhance the growth of beneficial bacteria in a more controlled way. It highlights the need for more research to understand the GI tract microbiome and its mechanisms of action in dealing with pathogens, host animal responses, pathogen resistance and variability among strains within a pathogen species. There also needs to be more research in such areas as multifunctional carriers (e.g. nanoparticles) to optimise delivery of compounds.

Chapters 3, 6, 7 and 8 review safety control measures at different stages in the poultry meat supply chain. Echoing Chapter 9, Chapter 3 discusses farm safety management. As it points out, producing and maintaining a healthy flock requires good husbandry practices, backed by effective procedures, training, monitoring and record keeping. These practices start with the selection, housing and management of healthy chicks coming into a flock. They then cover quarantine and other biosecurity measures such as pest control, housing, stocking and diet as well as waste management and health monitoring.

Chapter 6 discusses processing operations from live transportation of birds and slaughter to post-chill processing, and considers the effects of these operations on carcass contamination. The chapter starts with transportation and handling of live birds, including physical separation between flocks, keeping the birds clean and dry during transport and unloading, reducing cross-contamination risks and minimising stress associated with feed withdrawal and confinement during transport. It then reviews current research on the effects of processing steps such as scalding and evisceration on pathogen survival and spread, as well as the role of prerequisite and HACCP systems for effective management of slaughterhouse operations. Finally, it reviews the effectiveness of the range of antimicrobial treatments: chemical, physical (such as washing, chilling, the use of ozone or irradiation) and biological treatments such as bacteriophages. Throughout it highlights the importance of a pre-harvest pathogen control approach is crucial as a key part of an overall food safety programme.

Chapter 7 looks at good practice in official inspection of slaughterhouses, highlighting the need for more research on ways of optimising compliance. As chilling is critical in storage, transport and sale of fresh poultry meat, Chapter 8 reviews research on its effectiveness in the context of the range of decontamination techniques available, as well as emerging techniques such as the use of natural antimicrobials and edible coatings in preservation.

The chapters in Part 1 show what can be achieved in controlling pathogens such as *Salmonella* in poultry, including appropriate interventions at each point in the supply chain that systematically reduce the pathogen load. They also highlight the continuing challenges faced in dealing with other pathogens such as *Campylobacter*. These include a deeper understanding of pathogen behaviour, further improving detection techniques, a better knowledge of the GI tract microbiome, and the use of that foundation to better understand and extend the range of methods to control pathogens in the GI tract, as well as ways of combining techniques to create a series of hurdles which, in combination, keep the pathogen under effective control.

Poultry meat quality

Poultry meat quality can be defined in many different ways. Chapter 11 provides an overview of these different aspects of quality, which range from safety (discussed in Part 1) to nutritional value and from physiological and processing qualities such as water-holding capacity and shelf life to sensory properties such as colour, juiciness, tenderness and flavour. Consumption quality is based on an assessment of subjective sensory qualities and nutritive value. In contrast, processing quality is mainly based on objective physical and/or chemical properties. This chapter highlights recent research on defining, measuring and enhancing poultry meat quality attributes, including the use of indicators of meat maturation such as pH or conductivity and ways of enhancing nutritional value.

Building on the overview in Chapter 11, Chapter 12 reviews current research specifically on enhancing the nutritional quality of poultry meat. It discusses current research on understanding lipid metabolism and carcass lipid deposition. The chapter then summaries methods for the nutritional enrichment of poultry meat to increase polyunsaturated fatty acid content, particularly long chain *n-3*, using differing feed ingredients such as marine, plant and, more recently, algal ingredients. However, it has been found that such enrichment can be accompanied by the potential for increased oxidation of muscle lipids, the major contributor to the decline in meat quality during storage. The chapter therefore also reviews ways of reducing tissue lipid oxidation, for example, through use of antioxidants.

Flavour is an essential attribute of poultry meat quality and is discussed in Chapter 13. It reviews the wealth of research on the complex chemistry of flavour compounds and precursors underpinning poultry meat flavour as well as the process of flavour formation during cooking. The chapter also discusses the relative importance of factors influencing the formation of flavour and off-flavour in poultry meat, including breed/strain of chicken, diet and type of production system, and what this means for strategies to preserve or enhance the flavour of poultry meat.

The colour of poultry meat is often the first thing used by consumers to assess meat quality, and so it plays a vital role in purchasing decisions. Chapter 14 begins by surveying the extensive literature on meat pigments, including the types and chemistry of haeme pigments. It then considers what determines the colour of different types of poultry meat, starting with factors affecting the colour of fresh meat such as sex, diet, bird management and pre-slaughter stress as well the mechanisms of discolouration in poultry meat, and how discolouration can be prevented. It also reviews the colour of cooked, cured and irradiated poultry meat. Finally, recent developments in methods for objective colour measurement of meat products are reviewed.

Chapter 15 discusses how the characteristics of texture and tenderness are affected by the post-slaughter processing of poultry carcasses. It reviews what we know about the changes that occur in poultry muscle after slaughter, including pH decline and the development of rigor mortis. The chapter shows how the quality of poultry meat is influenced by chilling type, temperature and time, and how processing techniques can improve the quality of the meat and reduce the incidence of defects such as pale, soft and exudative meat.

Freshly processed poultry meat may be contaminated by microorganisms carried into the processing facility by live birds and by microorganisms that contaminate the carcass during processing. Chapter 16 focusses on the range of spoilage bacteria, conditions favouring their growth and how they can be controlled through techniques such as modified atmosphere packaging.

All these chapters in Part 2 highlight both how far back in the supply chain quality traits can be shaped and how critical a holistic approach is in seeking to optimise quality at each step in the process. Chapters in Part 2 also show that there is still much to learn, for example, about the genetic basis of individual quality attributes, the role of dietary interventions and the ways managing birds up to the point of slaughter affects final meat quality.

Sustainability

Increasingly quality for consumers in not just about sensory properties but broader issues such as the environmental impact of production, which is the focus of the final part of the book. As Chapter 17 shows, intensive poultry production systems face a number of environmental challenges. These systems are very much dependent on external feed production, especially protein sources such as soya, the production of which has raised environmental concerns. Another issue is nitrogen emissions, for example in the forms of ammonia emissions to air, nitrous oxide emissions that contribute to global warming and nitrate leaching contributing to eutrophication. This chapter reviews recent studies applying life cycle assessment modelling to quantify the contribution of different sub-systems of poultry production (production of feed, housing emissions, manure management, etc.) to the overall environmental impact of poultry production and to suggest measures which could make the intensive poultry production more environmentally friendly. These include improving feed efficiency, for example by using additives such as enzymes in feed or alternatively improving the efficiency of the birds through genetic selection, use of alternative feed ingredients, such as locally grown protein crops and agricultural by-products, and other management options such as improving poultry housing and new strategies for manure management.

Given its importance in the overall environmental impact of poultry production, Chapter 18 focuses on poultry feed. As it shows, the potential to reduce the adverse environmental impact of intensive poultry production through precise feed formulation and manufacturing is extensive, as new additives and supplements are developed to increase nutrient retention and decrease nutrient excretion, enhance performance and prevent diseases. The chapter starts by reviewing research on processing issues such as improving nutrient uniformity, particle size and pelleting as well as ways of optimising nutrient density and the use of multiphase feeding. It shows how formulating diets with low dietary crude protein and supplementation with highly digestible synthetic amino acids can reduce the excretion of excess nitrogen. The addition of exogenous enzymes, probiotics and prebiotics has also been shown to increase nutrient bioavailability, reduce nutrient variation of ingredients, enable the use of potentially less expensive ingredients and reduce the excretion of excess nitrogen and phosphorus. However, more needs to be discovered about the mechanism of action of these new additives given that the response of poultry to these products is highly variable, depending on the substrate, additive source, concentration and chemical relationship with other feed components.

Chapters 19 and 20 review recent research on measuring and reducing energy and water use, as well as the management of by-products and waste, in conventional poultry processing operations, focussing on processes such as bleed out, scalding, defeathering and evisceration. The following chapter reviews the growing sector of organic poultry production, with sections covering housing and nutrition requirements for organically reared poultry, the management of farm resources and bird health, ensuring food safety and the oversight of poultry meat processing and product labelling. It identifies the need to develop new organic ingredients able to provide required levels of non-essential amino acids as well as further research into appropriate use of probiotics, prebiotics, essential oils and organic acids.

As the final chapter in the book, Chapter 22 looks at sustainability in the context of the small-scale family poultry farming practiced by a large proportion of smallholder families in the rural and peri-urban areas of most of the world's developing countries. It shows that by enabling improvements in disease control, breeding and genetics, bird management and nutrition through smallholder poultry development projects, we have the potential to impact very meaningfully on poverty alleviation, household nutrition and food security, the health and education of children and the empowerment of women as the principal poultry keepers. This is shown via case studies from Myanmar, Tanzania and Zambia. The chapter suggests how the complex nature of the mixed farming systems and value chains incorporating family poultry production will increasingly benefit from inter- and transdisciplinary research in such areas as identifying the composition and nutritional value of alternative feedstuffs as well as better feed formulation.

Summary

This collection shows how interdependent questions of safety, quality and sustainability are. As an example, it demonstrates the central role of diet in such areas as food safety (e.g. in the control of *Salmonella*), aspects of quality (such as nutritional value, colour and flavour) and sustainability, whether in reducing the environmental impact of intensive systems or supporting smallholder poultry production. It also demonstrates the importance of an interdisciplinary approach in addressing issues of safety, quality and sustainability in an integrated way. The book shows the benefits of researchers collaborating to ensure synergies across the supply chain, for example by combining enhanced biosecurity with other interventions to combat pathogens in a way that is less reliant on the use of antibiotics. It is hoped that this collection will help support such an integrated approach.

Part 1

Poultry meat safety

Chapter 1

Zoonoses affecting poultry: the case of *Campylobacter*

Tom J. Humphrey and Lisa K. Williams, Swansea University, UK

1 Introduction

There are several bacterial zoonoses that affect, and/or are associated with, poultry worldwide, including those caused by *Campylobacter*, *Salmonella*, *Escherichia coli* and *Listeria* spp. In the United Kingdom, during the 1980s, an outbreak of *Salmonella* linked to eggs made headlines; this led to most laying hens in the United Kingdom, reared under the Lion Code, to be vaccinated against *Salmonella*. This control measure led to a reduction in the number of cases in the human population seen in the United Kingdom. *Campylobacter* is now the leading cause of bacterial foodborne illness, with the majority of cases being attributed to the consumption of poultry and poultry products (Wilson et al. 2010; Hermans et al. 2011; Levin 2007; Williams 2014), although there have been cases associated with pigs (Denis et al. 2008; Horrocks et al. 2009; Zweifel et al. 2008; Little et al. 2008; Oporto et al. 2007), cows (Gilpin et al. 2008; Kwan et al. 2008; Ingilis et al. 2004; Busato et al. 1999), sheep (Sahin et al. 2008; Oporto et al. 2007; Zweifel et al. 2004; Açik and Cetinkaya 2006), companion animals (Kemp et al. 2005; Brown et al. 2004; Andrzejewska et al. 2013) and wild birds (Horrocks et al. 2009; Kwan et al. 2008; Meerburg et al. 2006).

Despite stringent biosecurity on farm and control measures during processing, the number of cases of *Campylobacter* continues to rise each year. In the United Kingdom, it is thought that 82% of hospital admissions with a diagnosis of food poisoning can be attributed to a *Campylobacter* infection (Adak et al. 2002). In humans, the infectious dose is believed to be low, at around 500 cells in an adult male (Robinson 1981) and an incubation

http://dx.doi.org/10.19103/AS.2016.0010.01

period of up to 10 days, with most people exhibiting symptoms by day four. Symptoms include diarrhoea, which may be bloody, particularly in children; acute abdominal pain; and fever. Most infections are self-limiting and those who suffer an infection usually recover quickly after resting and maintenance of fluid levels; a fatal outcome is rare and would usually occur in the elderly or those already suffering from another serious illness (Skirrow and Blaser 2000). Despite treatment with antibiotics being rare, resistance to clinically important antimicrobials including macrolides and fluoroquinolones has been reported (Humphrey et al. 2007). Approximately 1% of cases develop long-term sequelae, such as inflammatory bowel syndrome or reactive arthritis.

2 *Campylobacter* in poultry

To date there are 26 species, 2 provisional species and 9 subspecies (Kaakoush et al. 2015) in the *Campylobacter* family. Of these species, *C. jejuni* and *C. coli* are most often found in the human cases, and are also commonly found in poultry meat (Humphrey et al. 2007).

Campylobacter has a long-standing association with poultry; it is well suited to 42°C, the body temperature of chickens, and is commonly found in the gastrointestinal tract of these birds (Hermans et al. 2011; Williams et al. 2014). *Campylobacter* was commonly referred to as a commensal of poultry (Dhillion et al. 2006; Macpherson and Uhr 2004) despite studies examining the impact of infection on the health and welfare of broiler chickens. In the field, symptoms we now know can be caused by *Campylobacter*, such as diarrhoea, were believed to be the action of other poultry pathogens. There have been studies based on laboratory and field data, going back to 1981, which have demonstrated that *Campylobacter* can and does have an effect on the health, welfare and performance of broiler chickens; this will be discussed in detail later.

There are two main routes of transmission from poultry to humans: first, via cross-contamination during processing, and secondly, via the spread of the bacterium from the intestines to other organs. Internationally, undercooked chicken muscle and liver are important vehicles for human infection. The chicken gut can be colonised with high numbers of *Campylobacter* spp. During processing, which is highly automated, these bacteria cross-contaminate the external surface of the carcass. *Campylobacter* can also be aerosolised, contaminating the carcass that way (Shane 1992; Gruntar et al. 2015; Elvers et al. 2011).

More recently, it has been shown that *Campylobacter* has the ability to leave the gut and infect other organs, the liver being the predominant one. Extra-intestinal spread is a major public health issue: several outbreaks of human campylobacteriosis have been linked to the consumption of chicken livers (Fernandez and Pison 1996; Mederiros et al. 2008; Cox et al. 2006; Whyte et al. 2006; O'Leary et al. 2009; Simaluiza et al. 2015), indicating that perhaps extra-intestinal spread is more common than previously thought. The ability of *Campylobacter* to leave the gut is poorly understood, requiring further investigation. *Campylobacter* spp. are isolated throughout poultry production, including rearing and at slaughter (Nebola et al. 2006), and their occurrence is well documented (Wagenaar et al. 2006; Humphrey 2006; Denis et al. 2001; Stern et al. 2001). However, little is known about the epidemiology of *Campylobacter* in poultry flocks (Bull et al. 2006), making control measures too difficult to implement and monitor.

3 Control measures in poultry

There is no doubt that controlling *Campylobacter* in poultry flocks would have an impact on the number of cases seen in humans; a vaccination strategy in the United Kingdom in the 1980s of laying flocks led to a reduction in the number of cases of *Salmonella* observed in the human population. Unfortunately, there is no effective vaccine for *Campylobacter* spp. for either the human or poultry populations, and this is mainly due to the complicated antigenic diversity *C. jejuni* has, a lack of small animal models and a poor understanding of *Campylobacter* pathogenesis (Riddle and Guerry 2016). In the absence of an effective vaccine, alternative control measures have been used during rearing on the farm and during processing with limited success.

During rearing, at the farm interventions including biosecurity, altering the diet, use of pre- and probiotics and additives (Meunier et al. 2015) have not been effective either individually or in combination. Other inventions during processing such as rapid chilling and surface washes have not been effective in reducing *Campylobacter* spp. to a level sufficient to lower the infection risk. *Campylobacter* has the ability to survive the harsh environments that it is exposed to during rearing and processing, and as a consequence of this, 71% of chicken carcasses on retail sale in the United Kingdom (Food Standards Agency 2015) contain *Campylobacter*. This is a problem not just in the United Kingdom; across Europe approximately 75% of carcasses on retail sale contain *Campylobacter* spp. Despite the interventions during rearing, processing and media campaigns aimed at raising awareness to consumers, the number of cases in the human population in the United Kingdom increases each year. There is no 'silver bullet' that will eradicate *Campylobacter* from poultry flocks, and individually, no single method appears to have a long-term effect on the bacteria. Combinations of control measures have been tried, including the use of antimicrobial feed additives containing fatty acids; pro- and prebiotics have been used together and combined with improved biosecurity (Meunier et al. 2015). However, in some cases these treatments predisposed birds to disease and increased the number of *Campylobacter*. Control measures that have proven to be effective in one country do not necessarily work in another. A good example of this is the use of fly screen on poultry houses in Denmark that led to a reduction in the level of *Campylobacter* in the flock (Hald et al. 2007, 2008), but when fly screens were used on poultry houses elsewhere, little effect was observed.

Other studies (Janez and Loc Carrillo 2013; Loc Carrillo et al. 2005; Atterbury et al. 2005) have examined the use of bacteriophage against *Campylobacter*. Although this has been shown in some experiments to be effective in controlling *Campylobacter*, its use is limited as consumers need convincing that these are 'good viruses' (Janez and Loc Carrillo 2013).

There are other control measures that have undergone testing, including rapid chilling, irradiation and use of elevated chlorine levels with carcasses. These have been tested with limited success (Berrang et al. 2011; James et al. 2007; Collins et al. 1996). Rapid chilling and irradiation have one major drawback: they affect carcass appearance and quality. Interventions are only useful if they are effective in controlling *Campylobacter* and are reported to have no impact on carcass appearance or meat quality. There are also some concerns that consumers do not find irradiation acceptable. There are strict controls in the United Kingdom regulating the amount of chlorine that is permitted to be used in potable water; and so, although chlorine has been shown to be effective (Berrang et al. 2011) in reducing *Campylobacter* levels during processing, the level used cannot be greater than is permitted by regulations.

4 *Campylobacter* as a pathogen rather than a commensal of poultry

Whether *Campylobacter* is a pathogen or a commensal of poultry is a topic of debate in the wider *Campylobacter* research community. Historically, *Campylobacter* was referred to as a commensal of poultry, as it was believed to cause no obvious signs of disease or have an effect on the health and welfare of broiler chickens. *Campylobacter* is considered part of the caecal microbiome of broiler chickens, because it is found in nearly every flock reared in the United Kingdom. It is generally accepted that every broiler chicken will be colonised with or exposed to *Campylobacter* during their lifetime irrespective of breed and rearing system used. Birds usually become colonised by the time they are three weeks old, and due to the rapid growth and short lifespan of the birds, the immune system is underdeveloped, which may make them more vulnerable to infection. However, chickens do mount innate and adaptive immune responses to infection with *C. jejuni* (Humphrey et al. 2014), but this does not result in clearance and may have the objective of confining the bacteria to the gut.

In many ways *Campylobacter* in broiler chickens can meet the criteria to be considered a commensal, although this can depend on the strain of the bacterium and broiler type (Humphrey et al. 2014). However, the reason for this is that, historically, signs of disease, which we now know can be associated with *Campylobacter*, were thought to be due to other bacteria or diseases. There have been changes in the performance and welfare of broiler chickens that are associated with other conditions and with carcass rejections at slaughter, which are examined thoroughly. For more than 30 years *Campylobacter* had been considered part of the normal gut microbiota of a chicken, despite early evidence indicating otherwise (Neil et al. 1984; Ruiz-Palacious et al. 1981; Sanyal et al. 1984). Certainly within some bird types and with certain strains of *Campylobacter* the behaviour of the bacterium is more like a commensal, but this is not universal.

It was only recently accepted that chickens mount an immune response to *Campylobacter*. Prior to this, chickens were thought not to do so in the same way as they would not for other commensal pathogens such as *Lactobacillus*. As we develop a deeper understanding of the chicken immune system, we begin to realise that it is more complex than previously thought and that *Campylobacter* is dealt with by the chicken immune system in a similar way to other pathogens. Many still consider *Campylobacter* to be a commensal of chickens, but there is an increasing body of evidence from several research groups (Awad et al. 2014; Williams et al. 2013, 2014; Wigley and Humphrey 2014; Humphrey et al. 2014), suggesting that it can no longer be considered in this way.

Several studies have reported the ability of *Campylobacter* to leave the chicken gut and infect other internal organs, including the liver (Knudsen et al. 2006; Van Deun et al. 2007; Sanyal et al. 1984; Jennings et al. 2011; O'Leary et al. 2009). Muscle contamination has also been reported in some studies (Berntson et al. 1992; Scherer et al. 2006; Luber and Bartlett 2007). The source of this contamination has previously been debated, and cross-contamination from the intestine during processing is not the only contributing factor to infected meat (Williams et al. 2014). Studies dating back to 1984 (Sanyal et al.1984; Knudsen et al. 2006; O'Leary et al. 2009; Jennings et al. 2011) have isolated *Campylobacter* from organs other than the gut; if *Campylobacter* was a true gut commensal of chickens, it should only be found there.

It is well understood that, within the poultry gut, *Campylobacter* adheres to epithelial cells and that this adherence is essential for colonisation (Hermans et al. 2011). Extra-intestinal spread does not always indicate disease or that an infectious process is occurring (Williams et al. 2014), but the invasive behaviour of *Campylobacter* in chickens suggests that it should be regarded as either a true or an opportunistic pathogen. It is increasingly being recognised that *Campylobacter* does alter the state of the host (Williams et al. 2014; Humphrey et al. 2014).

Campylobacter can also have an effect on the health, welfare and performance of broiler chickens, and there has been an association between certain welfare-associated conditions such as hock marks and pododermatitis. These foot and leg lesions are associated with diarrhoea in the birds and are considered risk factors for *Campylobacter* infection in commercial flocks (Neil et al. 1984; Bull et al. 2008; Rushton et al. 2009). Williams et al. (2013) demonstrated that the incidence and severity of hock marks and pododermatitis increased when *Campylobacter* is present in artificially infected birds. These leg and foot conditions are caused by prolonged contact with wet litter, and this study strongly suggests that *Campylobacter* is causing these leg lesions indirectly by increasing the looseness of the faeces, which increases the wetness of the litter (Williams et al. 2013, 2014). Similar results were obtained by Humphrey et al. (2014), who showed that diarrhoea in certain broiler breeds was associated with high levels of inflammation and much damage to gut mucosa (Humphrey et al. 2014; Fig. 1). Figure 1 is taken from Humphrey et al. (2014) and shows the breed with high levels of diarrhoea and raised levels of pododermatitis, which exhibited profound damage to the mucosa of the ileum, whereas in the breed with essentially normal faeces and no pododermatitis, the damage was much less (Fig. 1).

Damage to gut mucosa has been seen in other studies using artificial infection of the rapid growing breeds currently used commercially (Awad et al. 2014 and 2015). These studies provide evidence to suggest that *Campylobacter* is having a direct impact on the health and welfare of broiler chickens. Awad et al. (2014, 2015) showed that birds infected with *Campylobacter* performed significantly less well in the laboratory. Ruiz-Palacios

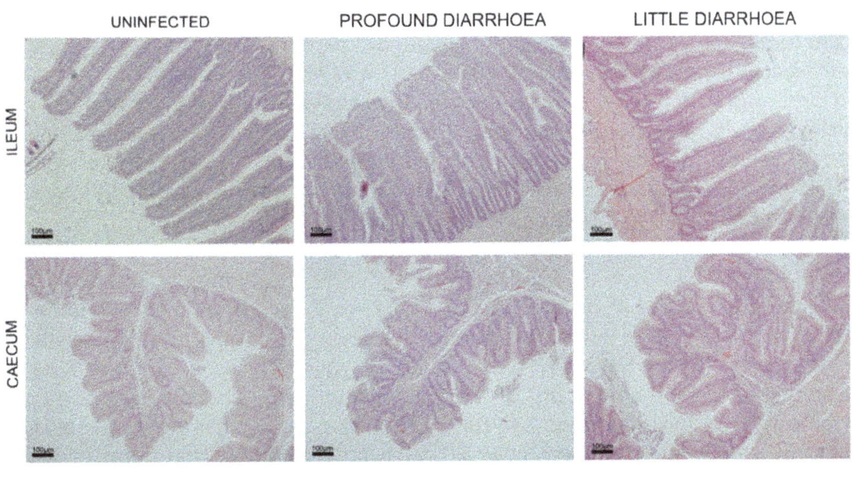

Figure 1 In some broilers *C.jejuni* damages the gut epithelia.

et al. (1981) showed that, in chickens infected with *C. jejuni* isolates from human cases of diarrhoea, over 80% of the birds got diarrhoea and around 40% died; and there was around a 40% drop in growth rate in the survivors, making this the first documented evidence suggesting that *Campylobacter* has an effect on the health and welfare of broiler chickens. This has previously been overlooked, allowing *Campylobacter* to be referred to as a commensal when the evidence indicates otherwise. A recent paper from India (Singh et al. 2012) found that the presence of *Campylobacter* in the gut microbiota of chickens was significantly ($p < 0.0004$) associated with poorer feed conversion ratios (FCRs).

Recent fieldwork in the United Kingdom and Ireland (Sparks and Whyte pers. comm.) has found that *Campylobacter*-positive commercial broiler flocks had significantly reduced performance than negative ones, although there were bacterial strain-to-strain differences in the impact of *Campylobacter*. These studies suggest that there are both economic and welfare reasons for better *Campylobacter* control on farm; these bacteria threaten the economic sustainability of chicken production, and the infection of chickens is likely to cost the international poultry industry considerable a lot each year.

Campylobacter being linked to disease in the chicken is not new; in 1954, a link was made between vibrio-like organisms, which were later identified as *C. jejuni*, and vibrionic hepatitis (Tudor 1954). Other authors have found an association between avian vibrio hepatitis (AVH) and the presence of *Campylobacter* (Lukas 1955; Hofstad et al. 1958; Moore 1958; Sevolan et al. 1958; Whenham et al. 1961). This disease, which causes focal lesions that are greyish-white in colour and 1–2 mm in size (Burch 2005; Crawshaw and Young 2003), persisted until 1965 and re-emerged in the 2000s (Crawshaw and Young 2003; VLA surveillance report). *Campylobacter* has not been conclusively linked with the focal lesions (Jennings et al. 2011) despite being initially linked with the disease (Peckham 1958). Jennings et al. (2011) found that significantly fewer healthy livers contained *Campylobacter* than those with greyish white spots, and using FISH they demonstrated that *Campylobacter* was present in higher numbers in spotty livers than in livers without spots, which were deemed healthy. *Campylobacter* has been found previously in the livers of apparently healthy birds (Cox et al. 2006, 2007, 2009). A further study (Richardson et al. 2011) identified *Campylobacter* in the bloodstream of broiler chickens with blood taken from naturally infected birds at slaughter, as did Sanyal et al. (1984), using artificially infected birds.

As the bacteria may be present in the blood of birds without apparently causing harm to these birds, and thus no rejection at slaughter, they could lodge in the small blood vessels in chicken muscle when the birds are bled. Undercooked chicken meat is an internationally important vehicle for human infection. In their experimental studies, Ruiz-Palacios et al. (1981) found that around 15% of birds had *Campylobacter* in circulating blood, and Richardson et al. (2011) found approximately 12%.

Damage to gut mucosa has been reported in some studies (Gharib et al. 2012; Awad et al. 2014; Ruiz-Palacios et al. 1981; Sanyal et al. 1984; Humphrey et al. 2014), but not all (Dhillion et al. 2006; Larson et al. 2008). There were also differences in whether birds suffered diarrhoea. Some authors reported this (Humphrey et al. 2014; Sanyal 1984, Ruiz-Palacios et al.1981; Sang 1989), particularly in young chickens. The bird type could have an impact on the effects of *Campylobacter*, but bacterial strain is also important. The same type of bird has been used in several studies, but has been colonised with different *C. jejuni* strains (Williams et al. 2013; Awad et al. 2014; Humphrey et al. 2014), and diarrhoea was not reported in all. Differences between strains in their ability to cause diarrhoea in chickens have been reported by others (Humphrey et al. 2014). It is clear that certain *Campylobacter* strains cause disease in different chicken types, although few papers have looked beyond caecal colonisation.

The chicken immune system is highly developed, but in comparison with mammals, it is considered smaller and simpler, although the functions are the same. The immune response to *Campylobacter* colonisation is complex, and little is known about the interaction between *Campylobacter* and the chicken immune system (Wigley and Humphrey 2014). *Campylobacter* is recognised by pattern-recognition receptors (PRRs), which lead to the activation of an immune response. Toll-like receptors (TLRs), in particular, in terms of *Campylobacter* colonisation, TLR 4 and TLR 21, are activated in the gut, although it has been reported that the increase in the expression of these TLRs in response to *Campylobacter* colonisation is short-lived (Shaughnessy et al. 2009). Following PRR activation, there is an increase in the pro-inflammatory cytokines, IL-6 and IL-1β, along with a pro-inflammatory chicken chemokine IL-8 (chIL-8) (Larson 2008; Shaughnessy 2000). After this initial pro-inflammatory response, cytokine levels decrease, suggesting that, although initial colonisation is dealt with by the host immune system as an attack, it reaches a level of tolerance (Williams et al. 2014; Smith et al. 2008; Hermans et al. 2012). Previous studies (Poh 2008) have shown that both hetrophils and monocytes are attracted by IL-8 to areas of inflammation. However, there is conflicting evidence whether this happens in IL-8 produced with *Campylobacter* colonisation (Williams et al. 2014). Several studies have shown no heterophilia (Meade 2009; Van Deun 2008), but Smith et al. (2008) observed an increase in hetrophils in the caeca and ileum of experimentally infected birds at two weeks of age. TLR5 recognises flagellin and there is a strong inflammatory response to invasion, mainly mediated by the recognition of flagellin through TLR5, leading to the production of CXCL chemokines and pro-inflammatory cytokines including IL-6 and IL-1β (Iqbal et al. 2005; Wigley 2013). This leads to inflammatory damage, which is considered to be mild in some chicken types (Humphrey et al. 2014), and can also lead to immune activation and largely restricts infection to the gut (Henderson et al. 1999; Withanage et al. 2004; Wigley 2013). Certain strains of *Salmonella* have adapted to cause systemic infection in the chicken, namely the serovars Gallinarum and Pullorum lack flagella and so evade recognition by the host immune system, allowing systemic infection to occur (Chappell et al. 2009). In a similar way to *Salmonella* Gallinarum and Pullorum, *Campylobacter* can also evade TLR-5 recognition by glycosylation of its flagella (Howard et al. 2009), but it is recognised by other receptors including TLR-21 (Meade et al. 2009). Several studies (Humphrey et al. 2014; Meade et al. 2009; Shaughnesaay et al. 2009; Smith et al. 2008) have shown that there is an inflammatory response to *Campylobacter* in the intestine. Although the bacterium was believed to be poorly invasive, it has been found in other organs, including the liver (Knudsen et al. 2006; Van Deun et al. 2007; Sanyal et al. 1984; Jennings et al. 2011; O'Leary et al. 2009). Infection of the liver can also be influenced by the *Campylobacter* strain (Medeiros et al. 2008; Cox et al. 2006; Whyte et al. 2006; O'Leary et al. 2009).

As well as mounting an innate immune response to *Campylobacter*, chickens also mount an adaptive one, producing antibodies against a number of *Campylobacter* proteins, including flagellin (Cawthraw et al. 1994; Sahin et al. 2001, 2002; Smith et al. 2008). Our understanding of the adaptive response of chicken to *Campylobacter* is rudimentary (Wigley and Humphrey 2014). Chickens have been shown to produce a humoral response against T-cell-dependent and T-cell-independent type 1 antigens; they produce a low response against T-cell-independent type 2 antigens (Jeurissen et al. 1998). The polysaccharide component of the *Campylobacter* capsule is a type 2 antigen and this has been suggested to contribute to the lack of clearance of *Campylobacter* from the chicken gut (Williams et al. 2014).

There is no doubt that chickens mount a response to infection; although an initial one is to be expected to any commensal organism, reports of pathology including heterophilia indicate that *Campylobacter* may be causing a diseased-like state, and is behaving more like a pathogen than a commensal (Williams et al. 2014; Humphrey et al. 2015).

5 Conclusions

There is strong evidence that indicates that *Campylobacter* from broiler chickens causes the majority of cases of human bacterial foodborne illness. Despite interventions on the farm during rearing and during processing the incidence of *Campylobacter* does not seem to be decreasing, indicating a need for an alternative approach. There is a need to understand the biology of *Campylobacter* in the chicken to determine the effect it is having on the health, welfare and performance of broiler chickens. Historically, *Campylobacter* has been referred to as a commensal of broiler chickens, but there is increasing evidence to challenge this viewpoint. Several studies, from different research groups, have shown that *Campylobacter* is capable of invading the gut cells, causing extra-intestinal infection of other internal organs and muscle. In addition, it has been shown that chickens mount innate and adaptive immune responses to *Campylobacter* higher than they do to commensal bacteria. The evidence suggests that, under the right conditions, *Campylobacter* behaves more like a pathogen, which can have a direct effect on the health and welfare of chickens; therefore, it should not always be considered a commensal of chickens.

6 Future trends

Campylobacter is an ongoing problem to the food industry. Chicken is the main reservoir for human infections, and the number of cases in the human population has largely increased despite interventions on the farm and during processing. There is no 'silver bullet' to tackle *Campylobacter*; it is likely to be a combination of factors that will lead to significant reductions in the number of cases observed in the human population. Ongoing work is looking at the viability of a vaccine for *Campylobacter* and work in this area will continue. There is a need to understand the effect *Campylobacter* has on a broiler chicken; for so long it has been considered a harmless commensal despite evidence to challenge this view. It is likely that the high incidence of human infection by *Campylobacter* will continue in most countries worldwide until we fully understand the infection biology of different strains of these bacteria in various broiler chicken breeds.

7 Where to look for further information

A good review on *Campylobacter* being more than a commensal in chickens is Williams et al. (2014) Conditional Commensalism of *Campylobacter* in chickens, which is in the book *Campylobacter* Ecology and Evolution Edited by S. K. Sheppard. There are also chapters in this book on the chicken immune system and responses to *Campylobacter*.

The World Health Organization is a useful resource for detailed information on *Campylobacter*, http://www.who.int/topics/campylobacter/en/

The European Food Standards Agency regularly publishes advice, trends and information about *Campylobacter* and poultry, https://www.efsa.europa.eu/en/topics/topic/campylobacter

Individual countries also publish their own guidance, trends and information. In the UK the Food Standards Agency, https://www.food.gov.uk/science/microbiology/campylobacterevidenceprogramme and Public Health England, https://www.gov.uk/government/collections/campylobacter-guidance-data-and-analysis are good sources of information, particularly about human campylobacterosis.

In the US the Centers for Disease Control and Prevention (CDC), http://www.cdc.gov/foodsafety/diseases/campylobacter/index.html and the US Food and Drug Administration http://www.fda.gov/default.htm contain valuable information.

Details on how UK funding bodies have joined together to tackle *Campylobacter* in the food chain can be found here, http://www.bbsrc.ac.uk/innovation/collaboration/collaborative-programmes/tackling-campylobacter/

Campylobacter in poultry information can be found here http://www.thepoultrysite.com/diseaseinfo/22/campylobacter-infection/

8 References

Açik, M. N. and Cetinkaya, B. (2006). Random amplified polymorphic DNA analysis of *Campylobacter jejuni* and *Campylobacter coli* isolated from healthy cattle and sheep. *J. Med. Microbiol.* 55(Pt 3): 331–4.

Adak, G. K., Long, S. M. and O'Brien, S. J. (2002). Trends in indigenous foodborne disease and deaths, England and Wales: 1992–2002. *Gut* 51: 832–41.

Andrzejewska, M., Szczepanska, B., Klawe, J. J., Spica, D. and Chudzinska, M. (2013). Prevalence of *Campylobacter jejuni* and *Campylobacter coli* species in cats and dogs from Bydgoszcz (Poland) region. *Polish J. Vet. Sci.* 1: 115–20.

Atterbury, R. J., Dillon, E., Swift, C., Connerton, P. L., Frost, J. A., Dodd, C. E. R., Rees, C. E. D. and Connerton, I. F. (2005). Correlation of *Campylobacter* bacteriophage with reduced presence of hosts in broiler chicken ceca. *Appl. Environ. Microbiol.* 71: 4885–7.

Awad, W. A., Aschenbach, J. R., Chareeb, K., Khayal, B., Hess, C. and Hess, M. (2014). *Campylobacter jejuni* influences the expression of nutrient transporter genes in the intestines of chickens. *Vet. Microbiol.* 172: 195–201.

Awad, W. A., Molnár, A., Aschenbach, J. R., Chareeb, K., Khayal, B., Hess, C., Liebhart, D., Dublecz, K. and Hess, M. (2015). *Campylobacter infection* in chickens modulates the intestinal epithelial barrier function. *Innate Immun.* 19: 1–10.

Berndtson, E., Tiverno, M. and Engvall, A. (1992). Distribution and numbers of *Campylobacter* in newly slaughtered broiler chickens and hens. *Int. J. Food Microbiol.* 15: 45–50.

Berrang, M. E., Windham, W. R. and Meinersmann, R. J. (2011). *Campylobacter*, *Salmonella*, and *Escherichia coli* on broiler carcasses subjected to a high pH scald and low pH postpick chlorine dip. *Poult. Sci.* 90: 896–900.

Brown, P. E., Christensen, O. F., Clough, H. E., Diggle, P. J., Hart, C. A., Hazel, S., Kemp, R., Leatherbarrow, A. J. H., Moore, A., Sutherst, J., Turner, J., Williams, N. J., Wright, E. J. and

French, N. P. (2004). Frequency and spatial distribution of environmental *Campylobacter* spp. *Appl. Environ. Microbiol.* 70: 6501–11.

Bull, S. A., Thomas, A., Humphrey, T., Ellis-Iversen, J., Cook, A. J., Lovell, R. and Jorgensen, F. (2008). Flock health indicators and *Campylobacter* spp. in commercial housed broilers reared in Great Britain. *Appl. Environ. Microbiol.* 74: 5408–13.

Burch, D. (2005). Avian vibrionic hepatitis in laying hens. *Vet. Rec.* 157: 528.

Busato, A., Hofer, D., Lentze, T., Gaillard, C. and Burnens, A. (1999). Prevalence and infection risks of zoonotic enteropathogenic bacteria in Swiss cow-calf farms. *Vet. Microbiol.* 69: 251–63.

Cawthraw, S., Ayling, R., Nuijten, P., Wassenaar, T. and Newell, D. G. (1994). Isotype, specificity, and kinetics of systemic and mucosal antibodies to *Campylobacter jejuni* antigens, including flagellin, during experimental oral infections of chickens. *Avian. Dis.* 38: 341–9.

Chappell, L., Kaiser, P., Barrow, P., Jones, M. A., Johnston, C. and Wigley, P. (2009). The immunobiology of avian systemic salmonellosis. *Vet. Immunol. Immunopathol.* 128: 53–9.

Collins, C. L., Murano, E. A. and Wesley, I. V. (1996). Survival of *Arcobacter butzleri* and *Campylobacter jejuni* after irradiation treatment in vacuum-packaged ground pork. *J. Food Protect.* 11: 1153–247.

Cox, N. A., Richardson, L. J., Buhr, J. R., Fedorka-Cray, J. P., Bailey, J. S., Wilson, J. L. and Hiett, K. L. (2006). Natural presence of *Campylobacter* spp. in various internal organs of commercial broiler breeder hens. *Avian. Dis.* 50: 450–3.

Cox, N. A., Richardson, L. J., Buhr, R. J. and Fedorka-Cray, P. J. (2009). *Campylobacter* species occurrence within internal organs and tissues of commercial caged Leghorn laying hens. *Poult. Sci.* 88: 2449–56.

Cox, N. A., Richardson, L. J., Buhr, R. J., Northcutt, J. K., Bailey, J. S. and Fedorka-Cray, P. J. (2007). Recovery of *Campylobacter* and *Salmonella* serovars from the spleen, liver/gallbladder, and ceca of six and eight week old commercial broilers. *J. Appl. Poult. Res.* 16: 477–80.

Crawshaw, T. and Young, S. (2003). Increased mortality on a free-range layer site. *Vet. Rec.* 153: 664.

Denis, M., Refregier-Petton, J., Laisney, M. J., Ermel, G. and Salvat, G. (2001). *Campylobacter* contamination in French chicken production from farm to consumers. Use of a PCR assay for detection and identification of *C. jejuni* and *C. coli*. *J. App. Microbio.* 91: 255–67.

Denis, M., Rose, V., Huneau-Salaün, A., Balaine, L. and Salvat, G. (2008). Diversity of pulsed-field gel electrophoresis profiles of *Campylobacter jejuni* and *Campylobacter coli* from broiler chickens in France. *Poult. Sci.* 87: 1662–71.

Dhillon, A. S., Shivaprasad, H. L., Schaberg, D., Wier, F., Weber, S. and Bandli, D. (2006). *Campylobacter jejuni* infection in broiler chickens. *Avian Dis.* 50: 55–8.

Elvers, K. T., Morris, V. K., Newell, D. G. and Allen, V. M. (2011). Molecular tracking, through processing, of *Campylobacter* strains colonizing broiler flocks. *Appl. Envion. Microbiol.* 77: 5722–9.

Fernández, H. and Pisón, V. (1996). Isolation of enteropathogenic thermotolerant species of *Campylobacter* from commercial chicken livers. *Int. J. Food Microbiol.* 29: 75–80.

Food Standards Agency (2015). *Campylobacter* retail survey. Available from, http://www.food.gov.uk/news-updates/campaigns/campylobacter/actnow/act-e-newsletter/campylobacter-retail-survey last (accessed 04 April 2016).

Gharib Naseri, K., Rahimi, S. and Khaki, P. (2012). Comparison of the effects of probiotic, organic acid and medicinal plant on *Campylobacter jejuni* challenged broiler chickens. *J. Agr. Sci. Tech.* 14: 1485–96.

Gilpin, B. J., Thorrold, B., Scholes, P., Longhurst, R. D., Devane, M., Nicol, C., Walker, S., Robson, B. and Savill, M. (2008). Comparison of *Campylobacter jejuni* genotypes from dairy cattle and human sources from the Matamata-Piako District of New Zealand. *J. Appl. Microbiol.* 105: 1354–60.

Gruntar, I., Biasizzo, M., Kusar, D., Pate, M. and Ocepek, M. (2015). *Campylobacter jejuni* contamination of broiler carcasses: population dynamics and genetic profiles at slaughterhouse level. *Food Microbiol.* 50: 97–101.

Hald, B., Skovgård, H., Pedersen K. and Bunkenborg, H. (2008). Influxed insects as vectors for *Campylobacter jejuni* and *Campylobacter coli* in Danish broiler houses. *Poult. Sci.* 87: 1428–34.
Hald, B., Sommer, H. M. and Skovgård, H. (2007). Use of fly screens to reduce *Campylobacter* spp. introduction in broiler houses. *Emer. Infect. Dis.* 13: 1951–3.
Hermans, D., Pasmans, F., Heyndrickx, M., Van Immerseel, F., Martel, A., Van Deun, H. and Haesebrouck, F. (2012). A tolerogenic mucosal immune response leads to persistent *Campylobacter jejuni* colonization in the chicken gut. *Crit. Rev. Microbiol.* 38: 17–29.
Hermans, D., Van Deun, K., Martel, A., Van Immerseel, F., Messens, W., Heyndrickx, M., Haesebrouck, F. and Pasmans, F. (2011). Colonization factors of *Campylobacter jejuni* in the chicken gut. *Vet. Res.* 42: 82.
Hofstad, M. S., McGehee, E. H. and Bennett, P. C. (1958). Avian infectious hepatitis. *Avian Dis.* 2: 358.
Horrocks, S. M., Anderson, R. C., Nisbet, D. J. and Ricke, S. C. (2009). Incidence and ecology of *Campylobacter jejuni* and *coli* in animals. *Anaerobe* 15: 18–25.
Humphrey, S., Chaloner, G., Kemmett, K., Davidson, N., Williams, N., Kipar, A., Humphrey, T. and Wigley, P. (2014). *Campylobacter jejuni* is not merely a commensal in commercial broiler chickens and affects bird welfare. *MBio* 5: 01364.
Humphrey, S., Lacharme-Lora, L., Chaloner, G., Gibbs, K., Humphrey, T., Williams, N. and Wigley, P. (2015). Heterogeneity in the infection biology of *Campylobacter jejuni* isolates in three infection models reveals an invasive and virulent phenotype in a ST21 isolate from poultry. *PLoS ONE.* 2310: e0141182.
Humphrey, T. (2006). Are happy chickens safer chickens? Poultry welfare and disease susceptibility. *Br. Poult. Sci.* 47: 379–91.
Humphrey, T. J., O'Brien, S. and Madsen, M. (2007). Campylobacters as zoonotic pathogens: a food production perspective. *Int. J. Food Microbiol.* 117: 237–57.
Inglis, G. D., Kalischuk, L. D. and Busz, H. W. (2004). Chronic shedding of *Campylobacter* species in beef cattle. *J. Appl. Microbiol.* 97: 410–20.
Iqbal, M., Philbin, V. J., Withanage, G. S., Wigley, P., Beal, R. K., Goodchild, M. J., Barrow, P., McConnell, I., Maskell, D. J., Young, J., Bumstead, N., Boyd, Y. and Smith, A. L. (2005). Identification and functional characterization of chicken toll like receptor 5 reveals a fundamental role in the biology of infection with *Salmonella enterica* serovar *typhimurium*. *Infect. Immun.* 73: 2344–50.
James, C., James, S. J., Hannay, N., Purnell, G., Barbedo-Pinto, C., Yaman, H., Araujo, M., Luisa Gonzalex, M., Calvo, J., Howell, M. and Corry, J. E. L. (2007). Decontamination of poultry carcasses using steam or hot water in combination with rapid cooling, chilling or freezing of carcass surfaces. *Int. J. Food. Microbiol.* 114: 195–203.
Janez, N. and Loc-Carrillo, C. (2013). Use of phages to control *Campylobacter* spp. *J. Micro. Methods* 95: 68–75.
Jennings, J. L., Sait, L. C., Perrett, C. A., Foster, C., Williams, L. K., Humphrey, T. J. and Cogan, T. A. (2011). *Campylobacter jejuni* is associated with, but not sufficient to cause vibrionic hepatitis in chickens. *Vet. Microbiol.* 149: 193–9.
Jeurissen, S. H., Janse, E. M., van Rooijen, N. and Claassen, E. (1998). Inadequate anti-polysaccharide antibody responses in the chicken. *Immunobiology* 198: 385–95.
Kaakoush, N. O., Castano-Rodriguez, N., Mitchell, H. M. and Man, S. I. M. (2015). Global epidemiology of *Campylobacter* infection. *Clin. Microbiol. Rev.* 28: 687–720.
Kemp, R., Leatherbarrow, A. J. H., Williams, N. J., Hart, C. A., Clough, H. E., Turner, J., Wright, E. J. and French, N. P. (2005). Prevalence and genetic diversity of *Campylobacter* spp. in environmental water samples from a 100-square-kilometer predominantly dairy farming area. *Appl. Environ. Microbiol* 71: 1876–82.
Knudsen, K. N., Bang, D. D., Andresen, L. O. and Madsen, M. (2006). *Campylobacter jejuni* strains of human and chicken origin are invasive in chickens after oral challenge. *Avian Dis.* 50: 10–14.
Kwan, P. S., Barrigas, M., Bolton, F. J., French, N. P., Gowland, P., Kemp, R., Leatherbarrow, H., Upton, M. and Fox, A. J. (2008). Molecular epidemiology of *Campylobacter jejuni* populations in

dairy cattle, wildlife, and the environment in a farmland area. *Appl. Environmen. Microbiol.* 74: 5130–8.

Larson, C. L., Shah, D. H., Dhillon, A. S., Call, D. R., Ahn, S., Haldorson, G. J., Davitt, C. and Konkel, M. E. (2008). *Campylobacter jejuni* invade chicken LMH cells inefficiently and stimulate differential expression of the chicken CXCLi1 and CXCLi2 cytokines. *Microbiology* 154: 3835–47.

Levin, R. E. (2007). *Campylobacter jejuni*: a review of its characteristics, pathogenicity, ecology, distribution, subspecies characterization and molecular methods of detection. *Food Biotechnol.* 21: 271–347.

Little, C. L., Richardson, J. F., Owen, R. J., de Pinna, E. and Threlfall, E. J. (2008). Prevalence, characterisation and antimicrobial resistance of *Campylobacter* and *Salmonella* in raw poultrymeat in the UK, 2003–5. *Int. J. Environ. Health Res.* 18: 403–14.

Loc Carillo, C., Atterbury, R. J., El-Shibiny, A., Connerton, P. L., Dillon, E., Scott, A. and Connerton, I. F. (2005). Bacteriophage therapy to reduce *Campylobacter jejuni* colonization of broiler chickens. *Appl. Environ. Microbiol.* 71: 6554–63.

Luber, P. and Bartelt, E. (2007). Enumeration of *Campylobacter* spp. on the surface and within chicken breast fillets. *J. Appl. Microbiol.* 102: 313–8.

Lukas, G. N. (1955). Avian infectious hepatitis-a preliminary report. *J. Am. Vet. Med. Assoc.* 126: 402.

Macpherson, A. J. and Uhr, T. (2004). Compartmentalization of the mucosal immune responses to commensal intestinal bacteria. *Ann. NY Acad. Sci.* 1029: 36–43.

Meade, K. G., Narciandi, F., Cahalane, S., Reiman, C., Allan, B. and O'Farrelly, C. (2009). Comparative in vivo infection models yield insights on early host immune response to *Campylobacter* in chickens. *Immunogenetics* 61: 101–10.

Medeiros, D. T., Sattar, S. A., Farber, J. M. and Carrillo, C. D. (2008) Occurrence of *Campylobacter* spp. in raw and ready-to-eat foods and in a Canadian food service operation. *J. Food Prot.* 71: 2087–93.

Meerburg, B. G., Jacobs-Reitsma, W. F, Wagenaar, J. A. and Kijlstra, A. (2006). Presence of *Salmonella* and *Campylobacter* spp. in wild small mammals on organic farms. *Appl. Environ. Microbiol.* 72: 960–2.

Meunier, M., Guyard-Nicodème, M., Dory, D. and Chemaly, M. (2015). Control strategies against *Campylobacter* at the poultry production level: biosecurity measures, feed additives and vaccination. *J. Appl. Microbiol.* (Epub ahead of print).

Moore, R. W. (1958). Studies of an agent causing hepatitis in chickens. *Avian Dis.* 2: 39.

Nebola, M. and Steinhauserova, I. (2006). PFGE and PCR/RFLP typing of *Campylobacter jejuni* strains from poultry. *Br. Poult. Sci.* 47: 456–61.

Neil, S. D., Campbell, J. N. and Greene, J. A. (1984). *Campylobacter* species in broiler chickens. *Avian Pathol.* 13: 777–85.

O'Leary, M. C., Harding, O., Fisher, L. and Cowden, J. (2009). A continuous common-source outbreak of campylobacteriosis associated with changes to the preparation of chicken liver pâté. *Epidemiol. Infect.* 137: 383–8.

Oporto, B., Esteban, J. I., Aduriz, G., Juste, R. A. and Hurtado, A. (2007). Prevalence and strain diversity of thermophilic *Campylobacters* in cattle, sheep and swine farms. *J. Appl. Microbiol.* 103: 977–84.

Peckham, M. C. (1958). Avian vibrionic hepatitis. *Avian Dis.* 2: 348.

Poh, T. Y., Pease, J., Young, J. R., Bumstead, N. and Kaiser, P. (2008). Re-evaluation of chicken CXCR1 determines the true gene structure: CXCLi1 (K60) and CXCLi2 (CAF/interleukin-8) are ligands for this receptor. *J. Biol. Chem.* 283: 16408–15.

Richardson, L. J., Cox, N. A., Buhr, R. J. and Harrison, M. A. (2011). Isolation of *Campylobacter* from circulating blood of commercial broilers. *Avian Dis.* 55(3): 375–8.

Riddle, M. S. and Guerry, P. (2016). Status of vaccine research and development for *Campylobacter jejuni*. *Vaccine* (in press).

Robinson, D. A. (1981). Infective dose of *Campylobacter jejuni* in milk. *Br. Med. J.* 282: 1584.

Ruiz-Palacios, G. M, Escamilla, E. and Torres, N. (1981). Experimental *Campylobacter* diarrhea in chickens. *Infect. Immun.* 34: 250–5.

Rushton, S. P., Humphrey, T. J., Shirley, M. B., Bull, S. and Jorgensen, F. (2009). *Campylobacter* in housed broiler chickens: a longitudinal study of risk factors. *Epidemiol. Infect.* 37: 1099–110.

Sahin, O., Zhang, Q., Meitzler, J. C., Harr, B. S., Morishita, T. Y. and Mohan, R. (2001). Prevalence, antigenic specificity, and bactericidal activity of poultry anti-*Campylobacter* maternal antibodies. *Appl. Environ. Microbiol.* 67: 3951–7.

Sahin, O., Morishita, T. Y. and Zhang, Q. (2002). *Campylobacter* colonization in poultry: sources of infection and modes of transmission. *Anim. Health Res. Rev.* 3: 95–105.

Sahin, O., Plummer, P. J., Jordan, D. M., Sulaj, K., Pereira, S., Robbe-Austerman, S., Wang, L., Yaeger, M. J., Hoffman, L. J. and Zhang, Q. (2008). Emergence of a tetracycline-resistant *Campylobacter jejuni* clone associated with outbreaks of ovine abortion in the United States. *J. Clin. Microbiol.* 46: 1663–71.

Sang, F. C., Shane, S. M., Yogasundram, K., Hagstad, H. V. and Kearney, M. T. (1989). Enhancement of *Campylobacter jejuni* virulence by serial passage in chicks. *Avian Dis.* 33: 425–30.

Sanyal, S. C., Islam, K. M., Neogy, P. K., Islam, M., Speelman, P. and Huq, M. I. (1984). *Campylobacter jejuni* diarrhea model in infant chickens. *Infect. Immun.* 43: 931–6.

Scherer, K., Bartelt, E., Sommerfield, C. and Hildebrandt, G. (2006). Quantification of *Campylobacter* on the surface and in the muscle of chicken legs at retail. *J. Food Prot.* 69: 757–61.

Sevolan, M. C., Winterfield, R. W. and Goldman, C. O. (1958). Avian infectious hepatitis; I. Clinical and pathological manifestations. *Avian Dis.* 2: 3.

Shane, S. M. (1992). The significance of *Campylobacter jejuni* infection in poultry: a review. *Avian Pathol.* 21: 189–213.

Shaughnessy, R. G, Meade, K. G, Cahalane, S., Allan, B., Reiman, C., Callanan, J. J. and O'Farrelly, C. (2009). Innate immune gene expression differentiates the early avian intestinal response between *Salmonella* and *Campylobacter*. *Vet. Immunol. Immunopathol.* 132: 191–8.

Singh, K. M., Shah, T., Deshpande, S., Jakhesara, S. J., Koringa, P. G., Rank, D. N. and Joshi, C. G. (2012). High through put 16s rRNA gene-based pyrosequencing analysis of the fecal microbiota of high FCR and low FCR broiler growers. *Mol. Biol. Rep.* 39: 10595–602.

Simaluiza, R. J., Toledo, Z., Ochoa, S. and Fernández, H. (2015). The prevalence and antimicrobial resistance of *Campylobacter jejuni* and *Campylobacter coli* in chicken livers used for human consumption in Ecuador. *J. Anim. Vet. Adv.* 14: 6–9.

Skirrow, M. B. and Blaser, M. J. (2002). Clinical aspects of *Campylobacter* infection. In *Campylobacter* 2nd Edition, Edited by I Nachamkin and M Blaser, ASM Press, Washington, DC.

Smith, C. K., Abuoun, M., Cawthraw, S. A., Humphrey, T. J., Rothwell, L., Kaiser, P., Barrow, P. A. and Jones, M. A. (2008). *Campylobacter* colonization of the chicken induces a proinflammatory response in mucosal tissues. *FEMS Immunol. Med. Microbiol.* 54: 114–21.

Stern, N. J., Cox, N. A., Bailey, J. S., Berrang, M. E. and Musgrove, M. T. (2001). Comparison of mucosal competitive exclusion and competitive exclusion treatment to reduce *Salmonella* and *Campylobacter* spp. colonization in broiler chickens. *Poult. Sci.* 80: 156–60.

Tudor, D. C. (1954). A liver degeneration of unknown origin in chickens. *J. Am. Vet. Med. Assoc.* 125: 219–20.

Van Deun, K., Haesebrouck, F., Heyndrickx, M., Favoreel, H., Dewulf, J., Ceelen, L., Dumez, L., Messens, W., Leleu, S., Van Immerseel, F., Ducatelle, R. and Pasmans, F. (2007). Virulence properties of *Campylobacter jejuni* isolates of poultry and human origin. *J. Med. Microbiol.* 56: 1284–9.

Van Deun, K., Pasmans, F., Ducatelle, R., Flahou, B., Vissenberg, K., Martel, A., Van den Broeck, W., Van Immerseel, F. and Haesebrouck, F. (2008). Colonization strategy of *Campylobacter jejuni* results in persistent infection of the chicken gut. *Vet. Microbiol.* 130: 285–97.

Veterinary Laboratories Agency (2005). VLA surveillance report. *Vet. Rec.* 157: 399–402.

Wagenaar, J. A., van Bergen, M. A., Newell, D. G., Grogono-Thomas, R. and Duim, B. (2001). Comparative study using amplified fragment length polymorphism fingerprinting, polymerase chain reaction genotyping and phenotyping to differentiate *Campylobacter fetus* strains from animals. *J. Clin. Microbiol.* 39: 2283–6.

Whenham, G. R., Carlson, H. C. and Aksel, A. (1961). Avian vibrionic hepatitis in Alberta. *Can. Vet. J.* 2: 3–7.

Whyte, R., Hudson, J. A. and Graham, C. (2006) *Campylobacter* in chicken livers and their destruction by pan frying. *Lett. Appl. Microbiol.* 43: 591–5.

Wigley, P. (2013). Immunity to bacterial infection in the chicken. *Dev. Comp. Immunol.* 41: 413–17.

Wigley, P. and Humphrey, S. (2014). The long road ahead: unravelling the immune response to *Campylobacter* in the chicken. In S. K. Sheppard (Ed.), *Campylobacter Ecology and Evolution*. Caister Academic Press, London.

Williams, L. K., Sait, L. C., Trantham, E. K., Cogan, T. A. and Humphrey, T. J. (2013). *Campylobacter* infection has different outcomes in fast and slow growing broiler chickens. *Avian Dis.* 57: 238–41.

Williams, L. K., Trantham, E. K. and Cogan, T. A. (2014). Conditional commensalism of *Campylobacter* in poultry. In S. K. Sheppard (Ed.), Campylobacter *Ecology and Evolution*. Caister Academic Press, London.

Wilson, D. L., Rathinam, V. A. K., Qi, W., Wick, L. M., Landgraf, J., Bell, J. A., Plovanich-Jones, A., Parrish, J., Finley, R. L., Mansfield, L. S. and Linz, J. E. (2010). Genetic diversity in *Campylobacter jejuni* is associated with differential colonization of broiler chickens and C57BL/6J IL10-deficient mice. *Microbiology* 156: 2046–48.

Withange, G. S., Wigley, P., Kaiser, P., Mastroeni, P., Brooks, H., Powers, C., Beal, R., Barrow, P., Maskell, D. and McConnell, I. (2005). Cytokine and chemokine responses associated with clearance of a primary *Salmonella enterica* serovar Typhimurium infection in the chicken and in protective immunity to rechallenge. *Infect. Immunity.* 73: 5173–82.

Zweifel, C., Zychowska, M. A. and Stephan, R. (2004). Prevalence and characteristics of Shiga toxin-producing *Escherichia coli, Salmonella* spp. and *Campylobacter* spp. isolated from slaughtered sheep. *Int. J. Food Microbiol.* 92: 45–53.

Chapter 2

Zoonoses affecting poultry: the case of *Salmonella*

Sabrina Vandeplas, Adisseo France SAS, France

1 Introduction

During all the stages of production, poultry face the risk of infection by several types of bacteria, either pathogens, which lead to diseases and eventually flock's morbidity, or mortality, or bacteria, being carried asymptomatically by poultry but are pathogenic for humans. The spread of disease-causing microorganisms or diseases from animals to humans is called zoonosis. Several zoonotic agents potentially infect poultry and humans, as recorded by Huang et al. (2016). The United States Foodborne Diseases Active Surveillance Network (FoodNet), which monitors the incidence of laboratory-confirmed infections caused by nine zoonotic pathogens, showed that the pathogen exhibiting the highest incidence is *Salmonella* (Table 1).

Salmonella is a major cause of gastroenteritis, known as salmonellosis, in human worldwide. The estimated total cost for salmonellosis is in excess of $14 billion/year in the United States alone, with almost 41 930 cases of non-typhoidal foodborne salmonellosis confirmed annually and an estimated total number of 1 million cases of foodborne salmonellosis not reported (Cosby et al., 2015). In the European Union (EU), a total of

http://dx.doi.org/10.19103/AS.2016.0010.26

Table 1 Number of cases and incidence of confirmed infections*, hospitalizations and deaths by pathogen – Foodborne Diseases Active Surveillance Network, United States, 2015 (Huang et al., 2016)

	Cases			Hospitalizations	Deaths
Pathogen	No.	Incidence§	Objective‡	No. (%)	No. (%)
Bacteria					
Campylobacter	6,309	12.97	8.5	1,065 (17)	11 (0.2)
Listeria	116	0.24	0.2	111 (96)	15 (12.9)
Salmonella	7,728	15.89	11.4	2,074 (27)	32 (0.4)
Shigella	2,688	5.53	—**	619 (23)	1 (0.0)
STEC O157	463	0.95	0.6	180 (39)	3 (0.6)
STEC non-O157	796	1.64	—**	126 (16)	1 (0.1)
Vibrio	192	0.39	0.2	47 (24)	5 (2.6)
Yersinia	139	0.29	0.3	37 (27)	1 (0.7)
Parasites					
Cryptosporidium	1,612	3.31	—**	268 (17)	8 (0.5)
Cyclospora	64	0.13	—**	4 (6)	0 (0.0)
Total	20,107	NA	NA	4,531	77

NA = not applicable; STEC = Shiga toxin-producing *Escherichia coli*.

*Confirmed infections are defined as culture-confirmed bacterial infections and laboratory-confirmed parasitic infections.

**No national health objective exists for these pathogens.

†Data for 2015 are preliminary.

§Per 100 000 population.

‡*Healthy People 2020* objective targets for incidence of *Campylobacter*, *Listeria*, *Salmonella*, STEC O157, *Vibrio* and *Yersinia* infections per 100 000 population.

88 715 confirmed salmonellosis cases were reported in 2014, resulting in incidence rate of 23.4 cases per 100 000 population, increasing by 15.3% compared to 2013 (EFSA and ECDC, 2015). The European Food Safety Authority has estimated that the overall economic burden of human salmonellosis could be as high as €3 billion a year, with a range between €560 million and €2.8 billion (Korsak et al., 2004).

It has been shown that human infections emanate from consumption of a variety of foodstuffs covering food of both animal and plant origin. However, the food that is most often incriminated is raw or undercooked poultry products (EFSA, 2007; Guo et al., 2011). Outbreaks of human salmonellosis associated with live poultry contact have been reported since 1955 in the United States. *Salmonella* is considered a threat to the poultry industry, especially because of the serotypes that are carried most of the time asymptomatically in birds that can then spread via the slaughter process to raw or finished products (Mead, 2004).

It is because of the huge impact these bacteria have on public health that several countries have implemented specific regulations to reduce *Salmonella* prevalence, especially in eggs and chicken carcasses. In the United States, contamination of eggs by *Salmonella*

Enteritidis has been a prominent cause of human illness for several decades and is the focus of a recently implemented national regulatory plan for egg-producing flocks. The Food Safety Modernization Act of 2011 gives FDA additional authority to regulate food facilities, establish standards for safe produce, recall contaminated foods and oversee imported foods; it also calls on CDC to strengthen surveillance and outbreak response (Crim et al., 2014). Similarly, the EU has established a European regulation mandating its member states to set up surveys and control programmes focused on *Salmonella* and other pathogenic agents in poultry. As carcass disinfectants are prohibited in the EU and decontaminating fresh table eggs is difficult, the European control programmes mostly focus on integrated approach along the poultry production chain (Snijders et al., 2002), based on prevention and on monitoring/eradication during the living phase, transport, at the slaughterhouse and the transformation stage.

As a consequence, the salmonellosis incidence tends to decrease worldwide, but vigilance must not be relaxed, as several measures to control *Salmonella* contamination in poultry production may be only partly effective. Control strategies have to be continuously improved and implemented in order to ensure sustainable and reliable poultry production. An efficient management programme needs to (a) clearly understand the mode of transmission of this pathogen through the poultry chain, (b) to better assess the virulence and the mechanism of infection of the birds and (c) to develop accurate and relevant methods of detection and quantification of *Salmonella* at the farm and processing levels. The chapter also aims to suggest future developments in the area of *Salmonella* management in poultry, which may be of value to the industry in the coming years.

2 *Salmonella* microorganisms and their invasion mechanism

2.1 A description of *Salmonella*

Salmonella spp. are Gram-negative and oxidase-negative bacteria belonging to the Enterobacteriaceae family. *Salmonellae* are motile, facultative anaerobic, rods measuring 0.3–1.5 by 1.0–2.5 μm in size. The small rod-shaped, straight-sided and non-spore-forming cells are mostly motile with peritrichous flagellae. *Salmonella* can reduce nitrate to nitrite, ferment glucose and grow under aerobic or anaerobic conditions. They are mesophilic organisms with an optimum for growth between 35°C and 43°C, close to the body temperature of warm-blooded animals, while they are able to grow over a wide temperature range (5–46°C). Values of water activity (aw) lower than 0.94 and pH greater than 9.5 or lower than 3.8 also inhibit the growth of salmonellae. All these conditions are met inside the gastrointestinal tract of mammals, allowing *Salmonella* species to easily colonize and proliferate in the proximal and distal part of the intestine.

2.2 *Salmonella* serotypes

The genus *Salmonella* lies within the kingdom Eubacteria, class Gammaproteobacteria, order Enterobacteriales and family Enterobacteriaceae. The genus *Salmonella* is divided into two species (*Salmonella bongori* and *S. enterica*), the first being usually isolated from cold-blooded animals whereas the second being mainly associated with warm-blooded vertebrates (Chan et al., 2003). The species *S. enterica* is further divided into 6 subspecies (*arizonae, diarizonae, enterica, houtanae, indica* and *salamae*) and over 2 500 serovars based on their flagellar (H), envelope (Vi) and somatic (O) antigen structures (Wisner et al., 2012). The *Salmonella* strains mainly associated to zoonotic contamination in poultry production are members of *Salmonella enterica* subspecies *enterica*. Of a total of 2 541 serotypes officially recognized, 1 504 are of *S. enterica* subspecies *enterica*, which show different virulence, species specificity and prevalence all over the world (Popoff et al., 2003).

The host-adapted serovars *Salmonella enterica* serovar Gallinarum and serovar Pullorum are important pathogens of chicken worldwide causing systemic disease, often with high levels of morbidity and mortality, that is the fowl typhoid and the pullorum disease, respectively (Wigley, 2014). These diseases have been eradicated in many developed countries, but they remain responsible for economic losses in the poultry industry in developing countries. By contrast, non-host-specific *Salmonella* are commensal in poultry and can persist in the gastrointestinal tract. They are mainly asymptomatic but are associated with widespread human illness (Revolledo and Ferreira, 2012). Non-host-specific *Salmonella* Typhimurium and *Salmonella* Enteritidis are generally recognized as the most prevalent serotypes in poultry, the first serotype being mostly associated with poultry meat whereas the second is mainly recovered from eggs, although *S.* Typhimurium has been sometimes associated with outbreaks involving contamination of egg content (Wales and Davies, 2011). *S.* Enteritidis and *S.* Typhimurium, represented 44.4% and 17.4%, respectively, of all reported serovars confirmed in human cases in the EU in 2014 (EFSA and CFDC, 2015). Moreover, other strains may be frequently recovered at the poultry farm. Besides *S.* Typhimurium and *S.* Enteritidis, *S.* Heidelberg, *S.* Kentucky and *S.* Senftenberg are the most prevalent serovars in the US poultry (He et al., 2012). In Europe, *S.* Hadar, *S.* Virchow and *S.* Infantis have also been often reported (EFSA, 2007).

The ability to colonize the intestine and the adaptive characteristics of the pathogen is dependent on the *Salmonella* serovars (Wallis and Barrow, 2005). He et al. (2012) showed a lower rate of systemic infection in birds challenged with *Salmonella* for *S.* Senftenberg compared with *S.* Typhimurium, *S.* Enteritidis, *S.* Heidelberg and *S.* Kentucky (14% vs 86–98%). Nevertheless, many similarities for different chicken breeds and tissues, and for different *Salmonella* serovars measured at different times post infection, were identified regarding the mechanisms for chicken host response to *Salmonella* infection (Pas et al., 2012).

2.3 Other proposed mechanisms of infection

Even though some experimental and epidemiological evidence has indicated the respiratory route to be a potential portal of entry for salmonellae in poultry (Kallapura et al., 2014), this pathogen usually infects chickens via the faecal–oral route with spread from the intestinal tract primarily at the distal ileum and caeca of the bird (1). A feature of avian salmonellosis is persistent infection or carrier state, which may occur for several months following infection with broad-host-range serovars. This persistence may be related to a substantial immune response requiring evasion or modulation of the pathogen's virulence.

2.3.1 Invasion

The intestines, especially the caeca have been reported to be the primary sites of colonization for *Salmonella* in live chickens (Desmidt et al., 1997). Ingested organisms easily overcome the acidic conditions of the proventriculus, because of the rapid adaptation of the bacteria to lower pH (Kwon and Ricke, 1998) and their tendency to colonize the small intestine, but also the underlying lymph tissue where it can proliferate. As the number of organisms increases, *Salmonella* can further infect essentially all tissues of the body, including the crop, spleen, liver, gall bladder, heart, ovaries and oviducts (Desmidt et al., 1997). In order for an infection to occur, an infective *Salmonella* must first survive the rather hostile lower gastrointestinal environment and peristalsis (Slauch et al., 1997), then be able to attach themselves to and invade the intestinal epithelia and Peyer's patches, and finally to counteract the host immune response while entering the systemic system.

Different virulence factors are necessary for the pathogen to promote colonization and adhesion, including flagella, fimbriae and *Salmonella* pathogenicity islands (Wallis and Barrow, 2005; Wisner et al., 2012). As each fimbria is typically specific to a given host receptor, the differences in fimbrial distribution among serovars may contribute to host specificity. So far, 21 *Salmonella* pathogenicity islands have been identified, which were described by Wisner et al. (2012). Bacterial surface adhesins, haemagglutinins, siderophores for the accumulation of sufficient environmental iron, and epithelial cell induction of bacterial polypeptides are other mechanisms contributing to colonization and invasion of the host. The length of the O side chains of lipopolysaccharide molecules will contribute to the resistance of *Salmonella* to lytic action of complement (D'Aoust, 1991). Moreover, virulence plasmids are associated with the ability to spread after colonization, and to invade internal organs, while inducing general suppression of the host immune response (Slauch et al., 1997). Virulence plasmids have been confirmed in *S.* Enteritidis, *S.* Typhimurium, *S.* Dublin and in the host-adapted *S.* Gallinarum-pullorum. Once pathogen invasion has occurred, *Salmonella* invade and persist within macrophages and dendritic cells and the progression of infection is to a large extent dependent on the susceptibility of the animal (Chappell et al., 2009). In resistant animals, infection is relatively controlled by macrophages and eventually cleared via adaptive responses, whereas infection may rapidly disseminate leading to septicaemia in susceptible chickens (Withanage et al., 2005).

2.3.2 Interaction with the immune system

Even though invasion by non-host-specific serotypes is sometimes accompanied by the establishment of systemic infection while avoiding activation of immunity (Wigley, 2014), *Salmonella* is normally initially recognized in chicken intestine by TLR receptors via its flagellin. Ramasamy et al. (2014) highlighted differential expression levels of TLRs induced by *S.* Pullorum during the early stages of infection in chicks, shedding light on tailored innate immune response following infection. Toll-like receptors (TLRs) play a pivotal role in early recognition of pathogen as well as in the initiation of robust and specific immune response. This recognition by TLRs is followed by induction of chemokines, pro-inflammatory cytokines and many effector genes (Yazdani et al., 2015), especially at the earliest stage of infection (Crhanova et al., 2011) and reduction of anti-inflammatory interleukins as observed by Bai et al. (2015) for IL-4 and IL-10 in the caecal tonsils. However, it is noteworthy that the expression of chicken β-defensins (i.e. gallinacins 1, 2, 4 and 6), which normally protect the caecum for the first three days of life, appear to not respond to *S.* Enteritidis colonization (Crhanova et al., 2011). The consequence of activation of innate immunity is an influx of heterophils (Shah et al., 2013), macrophages, B and T lymphocytes, and changes in total gene expression of several protein and enzymes in the caecal lamina propria (Rychlik et al., 2014). Whereas these can lead to inflammatory

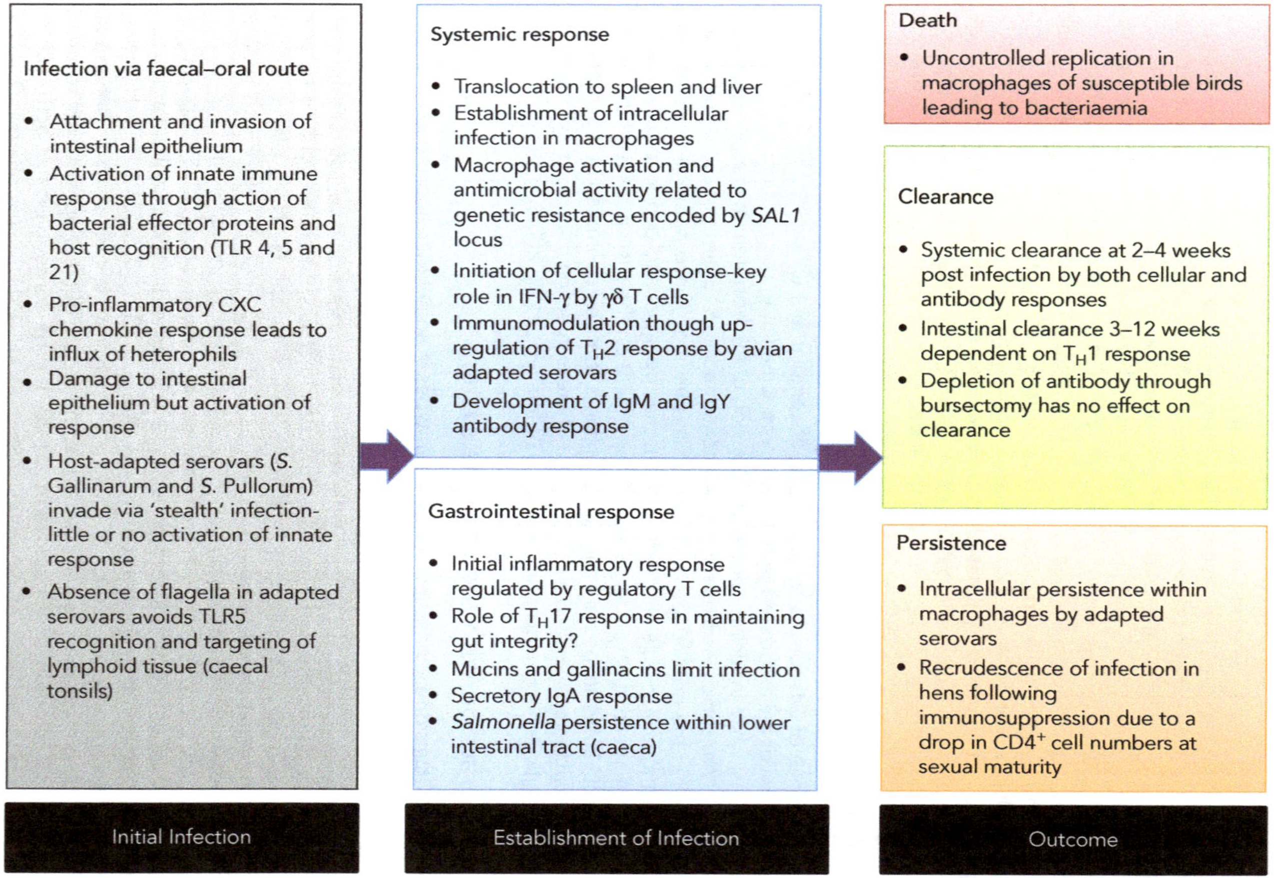

Figure 1 Major interactions between *Salmonella enterica* and chicken immune system (Wigley, 2014).

damage, they also largely limit invasive disease. Once the infection is under control, which happens approximately 2 weeks after infection, the adaptative immune system is activated and expression of IgY and IgA increases to facilitate *Salmonella* elimination from the gut lumen. Salisbury et al. (2014) determined that colonization of the gastrointestinal tract and the spleen, elicited an increase in specific IgA, IgG and IgM antibodies after *S.* Virchow infection. The authors also observed quantitative changes in several leukocyte populations in the gastrointestinal tract, as well as increased expression of pro-inflammatory cytokines IL-1β and IL-6 and of the chemokine CXCLi2.

The stages of infection in avian salmonellosis and interactions with the immune system are summarized in Fig. 1. Although progress has been made in understanding immune responses against *Salmonella* infections, further research is needed to understand the complete roles of humoral and cell-mediated immunity. Chickens infected with enteric *Salmonella* serovars show high levels of specific antibodies, a T-cell response, cytokines and chemokines, whose function would be further discriminated by the presence of cellular determinants, such as CD4+ (T-helper cells) and CD8+ (T-cytotoxic cells).

2.3.3 Effect on metabolism

Salmonella infection in the intestine can have widespread systemic effects on the metabolism of chickens without disease-like symptoms. Arsenault et al. (2013) observed key metabolic changes that affected fatty acid and glucose metabolism through the 5′-adenosine monophosphate-activated protein kinase (AMPK) and the insulin/mammalian target of rapamycin (mTOR) signalling pathway, in skeletal muscle of *S.* Typhimurium-infected chickens. Altered glucose metabolism resulted in a strong inhibition of glycogenesis and AMPK deactivation would lead to an anabolic response, which appears biased towards fatty acid metabolism. *Salmonella* infection could then elevate lipid synthesis and reduce lipid transportation in chicken hepatocytes. These changes may lead to excessive lipid accumulation in liver and slower lipoprotein deposition in the yolk (Wang et al., 2016). Coble et al. (2013) actually showed that *S.* Enteritidis infection influenced the liver of broilers with a pronounced downregulation of genes that function in liver metabolic pathways, inflammation and mitochondria-mediated apoptosis.

2.3.4 Interaction with the microbiota

It was also shown that infection of poultry with *Salmonella* may cause changes in the intestinal microbiota. Mon et al. (2015) observed that *S.* Enteritidis infection in young chicks significantly reduced the overall diversity of the microbiota population with expansion of Enterobacteriaceae family. Juricova et al. (2013) and Videnska et al. (2013) observed a similar increase of Lactobacillaceae by using terminal restriction fragment length polymorphism combined to real-time PCR and pyrosequencing the V3/V4 variable regions of genes coding for 16S rRNA, respectively. These authors suggested that the increase of Lactobacillaceae population may correlate with their microaerophilic growth characteristic compared with the obligate anaerobes from other families. Significant inverse correlation between Enterobacteriaceae and Lachnospiraceae family in both non-infected and infected groups was observed in the study of Mon et al. (2015), whereas a similar inverse trend was observed with Clostridiales, Lactobacillales and Bifidobacteriales by Juricova et al. (2013). These observations suggest possible antagonistic interaction among members of different taxa, which could potentially influence the overall microbial population in the gut. However, the microbiota changes following *Salmonella* infection, as measured in all these studies, were considered as minor in nature and should be understood more as an indirect consequence of infection and inflammation rather than a positively selected evolutionary trait, as stated by Videnska et al. (2013).

2.4 Antibiotic resistance

There has been an increasing concern over the past 30 years regarding the worldwide emergence of multidrug-resistant phenotypes among *Salmonella* serotypes such as *S.* Typhimurium, *S.* Enteritidis and *S.* Newport. Of special concern is the emergence of resistance to quinolones, fluoroquinolones or extended-spectrum cephalosporins such as ceftiofur and ceftriaxone. Antimicrobial resistance in 30 *S.* Typhimurium isolates of chicken and human origin in Egypt showed high resistance to chloramphenicol and trimethoprim–sulphamethoxazole (73.3%, each), followed by streptomycin (56.7%), tetracycline and ampicillin (53.3%, each), and gentamicin (30%) (Heba et al., 2016). Besides *S.* Typhimurium DT104, Parsons et al. (2013) pointed out the emergence of new variants in recent years, such as antimicrobial-resistant *S.* Typhimurium DT193 and specifically the monophasic *S.* Typhimurium-like variants of this phage type, serotypes 4,12:i:- and 4,5,12:i:-, which have become an increasing risk to public health in Europe and the United States. Usually associated with pigs, these strains are capable of colonizing chicken and represent a significant and potential emerging threat to poultry production from 'spillover' of these isolates from the pig industry or other sources. Moreover, the efficacy of vaccination against monophasic variants, which possess only phase 1 flagella unlike *S.* Typhimurium that possesses two forms of flagella, is not yet known, but may differ given the changes to the flagellar antigenic profile and may result in reduced protection in broiler breeder flocks. The consequences of antimicrobial resistance can be an accompanying change in the virulence of the organism, or poorer response to treatment with antimicrobial to which the organism is resistant (Travers and Barza, 2002). An increase in virulence could result from linkage of resistance factors to other virulence genes, such as those for adherence, invasion and toxin production. A study by the US Centers for Disease Control and Prevention (CDC) (Lee et al., 1994) revealed that subjects with infections caused by antimicrobial-resistant *Salmonella* were significantly more likely to be hospitalized than those with antimicrobial-susceptible infections (35% vs 27%), reflecting increased virulence of the infecting organism rather than inappropriate choice of treatment. The increased virulence of antimicrobial-resistant strains would cause more prolonged or more severe illness than do antimicrobial-susceptible strains (Travers and Barza, 2002). The reader who is interested in the mechanisms of resistance in *Salmonella* by class, transmission of antimicrobial resistance and in the global implications of antimicrobial resistance are invited to read the recent review by Cosby et al. (2015).

3 *Salmonella* transmission in poultry

3.1 Factors affecting poultry colonization

Numerous factors can affect the susceptibility of poultry to *Salmonella* colonization, and can be environmental (pathogen dose and serotype) (Foley et al., 2011), physiological (age, genetics, health status) (Quinteiro-Filho et al., 2012), social (housing, transport) (Gomes et al., 2014; Hahn, 2014) or nutritional (interaction with mycotoxin) (Grenier and Applegate, 2013). Young birds are more susceptible to *Salmonella* infections as shown by Martelli et al. (2014) that can lead to increased early mortality. On the other hand, adult birds can harbour *Salmonella* in their intestinal tract without showing clinical signs. Infective dose is consequently generally lower for chicks than for adults (Desmidt et al., 1997; Milner and Shaffer, 1952). It was shown that 50% of day-old birds became infected with *S.* Typhimurium after the oral challenge of only 10 cells, while even a challenge dose of a million cells infected only 10% of birds at 14 days of age. Besides broilers and layers, other poultry species may be asymptomatic carrier for *Salmonella*, as observed for turkeys (Morris et al., 2015), although the predominant serotypes may be different. Morris et al. (2015) isolated only persistent *S.* Newport contamination in commercial turkey flocks from placement at 1 d old until slaughter.

3.2 Poultry flock prevalence

In the United States, the Pathogen Reduction/HACCP Programme for broiler establishments was launched in 1988. In 2002, this programme could demonstrate a reduction of contaminated ground chicken from a baseline of 44.6% to 14.4% (FDA, 2002). The per cent of *Salmonella*-positive broiler flocks increased again between 2002 and 2005 from 11.5% to 16.3%, but decreased to 11.4% in 2006, presumably as a result of the USDA-FSIS's redirection of the sampling programme and new initiatives to achieve and maintain consistent process control (USDA-FSIS, 2007). *S.* Enteritidis has remained a significant problem in commercial poultry, but its prevalence has also declined in chickens since the mid-1990s. Coinciding with this decrease, *S.* Heidelberg and *S.* Kentucky have emerged as the predominant serovars in commercial broilers (Foley et al., 2011). In 2014, the EU-level prevalence of *Salmonella* target serovar-positive poultry flocks was very low (<1%), for breeding flocks, for laying hen flocks, broiler flocks as well as for flocks of breeding and of fattening turkeys (EFSA and ECDC, 2015), due to the implementation of regulatory management along the whole production chain. However, the prevalence in poultry flock is dependent on many other factors than just the regulatory context. Lake et al. (2004) compiled studies of the prevalence of *Salmonella* in poultry and raw poultry products from 12 countries worldwide showing a variation from 0% to >50%. In northern Thailand, the prevalence of *Salmonella* in chickens at the slaughterhouse was 9% (Padungtod and Kaneene, 2006). In New Zealand, 1.2% of 1 318 samples taken on chicken carcasses after chilling, draining and dripping were positive for *Salmonella* (Lake et al., 2004), while in Vietnam 7.9% of faecal or intestinal samples from chicken were positive for *Salmonella* (Tran et al., 2004). In the province of Quebec, Canada, the prevalence of *Salmonella*-positive flocks was shown to be 50% for chickens and 54% for turkeys, based on 81 broiler chicken and 59 turkey flocks with about 30 birds sampled per flock from 24 April 2003 to 23 February 2004, in the four largest commercial processing plants in Quebec (Arsenault et al., 2007).

3.3 Mode of transmission

In order to implement effective measures for *Salmonella* control, it is first necessary to clearly identify the sources of flock infection, the means of transmission and the risk factors along the production chain. Several epidemiological studies were carried out all over the world to help answer that question. Pre- and post-harvest transmission route was reviewed by Ricke (2014). At the farm level, two types of transmission of *Salmonella* in poultry flocks were pointed out, that is, vertical or horizontal transmission. They were also detailed in the review cited above and in the papers by Vandeplas et al. (2010), and will be only briefly summarized below.

3.3.1 Vertical transmission

Vertical transmission originate from an infection of the reproductive organs of the breeder via systemic infection, leading to an ascending infection from the cloaca to the vagina and lower regions of the oviduct in hens (Rasschaert, 2007), or contamination of semen in roosters (Russell, 2010), which may pass the pathogen to the female during mating. Only through vertical transmission, the yolk, the yolk membrane or the albumen surrounding it are directly contaminated before the eggs are covered by the shell, with a potential final transfer to the progeny. However, some authors did not find any relationship between the numbers of *Salmonella* that colonize reproductive organs and the associated frequency of egg contamination (Gast et al., 2011). According to Heyndrickx et al. (2002), vertical transmission is nowadays of less importance mainly due to the vaccination of breeder flocks, so that horizontal transmission is the main determinative factor for colonization of broiler flocks.

3.3.2 Horizontal transmission

Salmonella is able to survive and maybe even multiply in the external environment (Winfield and Groisman, 2003), which represents a risk of propagation into the poultry house. Many environmental factors and management practices were identified as risk factors of flock infection. In the study by Denagamage et al. (2015) and Van Hoorebeke et al. (2010), the presence of previous *Salmonella* infection, absence of cleaning and disinfection, presence of rodents, induced moulting, larger flock size (>30000 hens), multiage management, cage housing systems, in-line egg processing, rearing pullets on the floor, pests with access to feed before movement to the feed trough, visitors allowed in the layer houses, trucks near farms and air inlets and sampling in winter were identified as risk factors for *Salmonella* contamination of laying hen premises. Another important factor is the quality of the feed and drinking water. Several analyses of commercially manufactured feeds have linked the occurrence of *Salmonella* in poultry and its presence in feed (Heyndrickx et al., 2002; Davies and Wray, 1996). Moreover, the inclusion of a feed ingredient contaminated with mycotoxin was correlated with higher susceptibility of poultry to infection (Fukata et al., 1996), and increased translocation of *Salmonella* over the intestinal cells (Vandenbroucke et al., 2011).

However, it must be kept in mind that for some factors inside the house, such as rodents, insects, feed or litter, it is generally difficult to determine if the contamination follows or precedes the flock contamination. As an example, Heyndrickx et al. (2002) showed that water contamination usually follows flock colonization rather than preceding it so that water source is actually considered as a low-risk factor for flock colonization.

4 Detection methods

The detection of *Salmonella* in primary poultry production is an issue of great concern as control of this zoonotic disease is in part based on the reduction of the prevalence at the farm level. The national control plans set up all over the world are likely to require harmonized sampling and analysis of production flocks, in addition to sensitive monitoring of breeding flocks and hatcheries. A wide range of methodologies for the sampling and bacteriological detection of *Salmonella* in poultry flocks is available, each with weaknesses and strengths (Carrique-Mas and Davies, 2008). The selection of an appropriate method will depend on the final application which will make mandatory to define an acceptable compromise between improved sensitivity, speed, ease of use and economic considerations for routine monitoring.

4.1 The importance of sampling

Success of detection is likely to be highly dependent on the choice of an adequate sampling procedure combined with a sensitive culture method. The analysis of *Salmonella* in poultry at the primary production stage may involve either the collection of material from the environment in the poultry house (e.g. faeces, litter, dust) or the collection of samples from the birds themselves (e.g. blood, cloacal swabs, post-mortem samples of the small intestine and caecum). It is at least theoretically possible that positive environmental samples may not reflect established infection in the flock, so that collection of intestinal samples from the animal (faeces) is important to have a consistent overview of the flock sanitary status. Carrique-Mas and Davies (2008) reviewed the advantages and limits of each sampling methods. As *Salmonella* has been reported to survive in poultry houses for at least 53 weeks in dust, and as fresh faeces provide an indication of current infection, these authors recommend sampling of both faeces/litter and dust for *Salmonella* monitoring programmes, with boot swabs as the preferred method for the collection of faeces in floor system. Cloacal swabs have been the preferred method in many *Salmonella* monitoring schemes in poultry, but this method is likely to be relatively insensitive compared with the culture of more voluminous faecal material. At the slaughterhouse, microbiological sampling of poultry carcasses may be conducted by whole-carcass rinsing, tissue excision or skin swabbing. In general, neck skin or carcass rinse sampling is regarded as more sensitive than skin swabbing, even if results of the former method may vary depending on when the sample is taken in the process (i.e. pre-chill, post-chill).

4.2 Detection methods

A wide range of laboratory tests exist, which can be broadly classified into those that aim to detect the presence of viable *Salmonella* organisms (culture, immunological detection of antigens following culture), specific components of these organisms (i.e. molecular DNA amplification techniques) or a response of the host to infection (i.e. serological testing). Traditional culture methods for *Salmonella* detection in foods consist of a series of steps that include non-selective enrichment, selective enrichment, and selective/differential plating, and finally biochemical and serological confirmation, such as the ISO 6579:2002 method, which is the standard for poultry environmental samples in the EU.

The traditional biological method for *Salmonella* isolation is labour-intensive and requires a minimum of four days to produce a negative result and up to seven days to produce a confirmed positive result. As monitoring programmes aiming to reduce *Salmonella* prevalence in human are based on efficient monitoring and control of bacterial contamination of poultry product before they enter the retail market, there has been a need to develop and validate faster screening and detection methods for this pathogen.

Existing bacteriological methods are consequently progressively challenged by new molecular-based technologies that may deliver results quicker or more efficiently. These rapid methods should deliver reliable results in time to make it possible to steer the finished products in poultry slaughterhouses into the direction of the fresh poultry market or into the direction of industrial treatment. The most important requirements are the detection limit, the time of analysis (within hours up to a maximum of 24 h), the sensitivity and specificity, and the method validation (Eijkelkamp et al., 2009). Attention is now being focused on molecular-based detection methods, such as polymerase chain reaction (PCR)-based techniques, because of their sensitivity, specificity and reduced assay time. This method allows 10^7-fold amplification of the target DNA from as little as one copy in 2–3 h (Bailey and Cosby, 2003). Several recent studies have compared PCR methods with conventional plating protocols, and showed similar or higher performance of PCR (Jinu et al., 2014; Langkabel et al., 2014). Jinu et al. (2014) tested the two methods for detection of *Salmonella* from field poultry samples and showed that the PCR employed for detection of *Salmonella* was found to be 100% sensitive for poultry blood but its sensitivity was lower (77.77%) for faeces samples as compared with the culture method. However, PCR was 100% specific with regard to faeces samples, and 97.89% specific for blood samples, demonstrating the usefulness of genus-specific PCR for detection of *Salmonella* in poultry clinical samples. Besides developing molecular approaches for pathogen detection on live birds, methods for analysing final poultry product (Ferretti et al., 2001) have also been investigated, in order to perform fast diagnosis of food poisoning. In China, Yan et al. (2011) developed a PCR approach based on *Salmonella*-specific primers (PsalF-PsalR), which was showed to be sensitive and specific, as 30 strains of *Salmonella* standard and isolates were detected positive and 18 non-*Salmonella* strains were negative. PCR detection limit for artificially contaminated samples after enriching for 12 h was 1 CFU/25 g sample, and positive rate out of total 16 market-sold samples was 43.75% (7/16) for enrichment-PCR, while the rate was 31.25% (5/16) for the traditional method. A selection of commercially available rapid detection methods for testing of *Salmonella* in poultry meat is given in the review by Eijkelkamp et al. (2009).

4.3 Method improvement

In the case of faecal and environmental samples of poultry, alternative methods to traditional culture techniques are still in the early stages of development, and there is considerable work to be done before they can be consistently applied. Worldwide, several research teams are studying different ways to improve existing methodologies in order to allow consistent and monitoring of poultry production.

Real-time multiplex PCR assays were developed for simultaneous detection of several *Salmonella* serovars, such as the protocol set up for poultry carcasses during a research project completed in 2015 at the University of Arkansas and sponsored by the US Poultry & Egg Association (Park and Ricke, 2015). The multiplex PCR method developed by O'Regan et al. (2008) for poultry-associated serotypes Enteritidis, Gallinarum, Typhimurium, Kentucky and Dublin was as sensitive as the traditional cultural method with artificially contaminated chicken samples resulting in a detection limit of between 1 and 10 CFU per 25 g sample for both methods. A total of 63 naturally contaminated chicken samples were investigated by both methods and relative accuracy, relative sensitivity and relative specificity of the real-time PCR method were determined to be 89%, 94% and 87%, respectively.

Other derived techniques such as the BAX system described by Bailey and Cosby (2003) or PCR linked to immunological assay (Eyigor et al., 2007) were also investigated. By analysing naturally occurring and spiked *Salmonella* in 183 chicken carcass rinses and 90 chicken hot dogs, Bailey and Cosby (2003) were able to detect *S.* Typhimurium, *S.* Heidelberg, *S.* Montevideo and *S.* Enteritidis at 1 to 250 cells per ml of rinse or 1 to 250 cells per g of meat, with similar or higher accuracy for the automated PCR compared with the culture procedures. The automated PCR system provided reliable *Salmonella* screening of chicken product samples within 24 h, using a detection method combining the unique properties of a fluorescent DNA intercalating dye, SybrGreen, and the thermal denaturation characteristics of a specific DNA fragment. Eyigor et al. (2007) combined PCR and ELISA to detect *Salmonella*-specific DNA from chicken intestine, to be used particularly in high sample throughput laboratories. *Salmonella* genus-specific PCR product was produced using *invA*-specific primers and digoxigenin by commercial PCR-ELISA DIG Labelling and ELISA DIG Detection kits. PCR-ELISA detection limits with *S.* Enteritidis 64K DNA, pure culture, and with the intestinal homogenate artificially contaminated were found as 0.0079 μg ml^{-1}, 50 cfu ml^{-1} and 70 cfu ml^{-1}, respectively. Sensitivity and specificity of the assay was determined as 100% with the tests performed with 41 *S. enterica* serovar DNAs, with non-*Salmonella* strains and with 124 tetrathionate broth enrichment cultures of chicken intestinal samples. Immunological techniques can also be used alone, as showed by Schneid et al. (2006). The authors used indirect ELISA based on a monoclonal antibody (MAb) specific for an outer membrane protein of *S.* Enteritidis for detection of *Salmonella* in 154 samples of chicken meat, and showed an ELISA sensitivity and specificity of 94%, and positive and negative predictive values of 82% and 98%, respectively.

Finally, the immunomagnetic separation (IMS) technique is being used to increase sensitivity and specificity for detection of *Salmonella* in foods by conventional plating in selective agar (IMSplating). Conceição et al. (2008) sensitized protein A-coated microspheres with polyclonal antibodies against lipopolysaccharide and flagella from salmonellae and tested samples of chicken meat experimentally contaminated with *S.* Enteritidis immediately after contamination and after 24 h of refrigeration. The method yielded 100% sensitivity and 94% specificity, and was able to diminish by at least one day the time required for detection of *Salmonella* in chicken products by the conventional culture method.

5 Strategies to control *Salmonella* in poultry: overview of pre- and post-harvest strategies

Following the incidence of zoonotic agents like *Salmonella* and their public health impact, protecting humans against zoonoses has become of capital importance worldwide, and poultry producers are facing increasingly demanding national regulation aiming to reduce the prevalence of *Salmonella* in animal populations at the level of primary production and, where necessary, at other appropriate stages of the food chain. In the United States, one goal of the federal Healthy People 2020 (HP2020) initiative is to reduce foodborne illnesses that are caused by *Salmonella* by 25% by the year 2020. In 2010, the European Community Regulation 2073/2005 stated that samples of 25 g raw or 10 g cooked meat products must be *Salmonella* free or shall not be placed on the market without any industrial treatment able to eliminate *Salmonella* (EFSA, 2007). It should be noted that predicting the final contamination of poultry products is a difficult exercise as the infective dose of salmonellae varies according to the serotype, the strain, the immunocompetence of individuals and the type of food involved. The methods developed to limit *Salmonella* contamination of the poultry production chain are all part of a holistic animal management approach needed to significantly reduce the overall burden of salmonellosis on human health, which also includes stringent control of antimicrobial agents in the livestock industry, early clinical and microbiological diagnosis, appropriate treatment and implementation of strict sanitary standards in the food industry (Hur et al., 2012).

In this section, we provide a brief overview of pre- and post-harvest control strategies. Sections 6 and 7 then consider several pre-harvest control strategies in detail.

5.1 Post-harvest control strategies

The easier method to ensure food safety is the decontamination of poultry product during processing. *Salmonella* spp. can enter the human food chain through the contamination of chicken carcasses by either infected faecal material or organs from contaminated birds at slaughter. Therefore the level of contamination in chicken carcasses can be reduced during processing, either by immersion or by spraying of the carcass with processing water containing chemical disinfectants at the abattoir or by ionizing radiation. Several disinfectants are in common use in the processing plants, such as inorganic and organic chlorine-based treatments, acidified sodium chlorite, trisodium phosphate, quaternary ammonium compounds or bacteriophages among others (FSIS, 2015). In the United States, the *Salmonella* control is in the hands of the food business operator and that is why decontamination by treatment of carcasses in the processing plant is considered as the most economical option, and is included into the HACCP system at the slaughterhouse, in combination with other best practices described throughout the Food Safety and Inspection Service's compliance guide.

5.2 Pre-harvest control strategies

In some other parts of the world, as in Europe, sanitation of poultry product are not allowed by legislation, making it mandatory to only focus on the control of the pathogen in poultry flocks before processing. The directive 2003/99/CE and the Regulation (EC) n° 2160/2003 force the member states to implement programmes to control *Salmonella* in poultry through an integrated approach, which includes control measures applied throughout the poultry production chain.

At the primary production level, the introduction of salmonellae is primarily prevented by implementing stringent biosecurity. However, it sounds easy but is difficult to achieve in detail, and biosecurity measures are only partly effective to control pathogen infection. Sub-therapeutic antibiotics, which were used as growth promoters were also the first strategy to prevent *Salmonella* contamination, but are increasingly questioned due to recurring antimicrobial resistances. They have been banned in the EU since January 2006 and other parts of the world are also considering their reduction or elimination in livestock production. Vaccination was shown to be efficient, and was the primary cause of reduction of *Salmonella* incidence all over the world, especially in laying hens, but is less easily applicable in meat-producing birds. Other strategies have therefore been investigated. They are mostly based on diet modulation and feed supplementation with different types of additives other than feedstuffs.

6 Strategies to control *Salmonella* in poultry: managing the environment, diet and immune responses of poultry

6.1 Keeping the environment *Salmonella* free

Complementary biosecurity measures have to be taken to prevent the introduction and spread of infectious agents to poultry flocks. The starting point is to ensure that the flock environment is kept *Salmonella* free, by treatment and sanitation of the feed, the water (Doyle and Erickson, 2006) and the litter. Appropriate cleansing and disinfection procedure have to be defined and restriction of visits are recommended as well (EFSA, 2004). Finally, litter reuse should be avoided in order to reduce the risk of infection of the following flock. In hatcheries, the most common approaches to control cross-contamination from contaminated eggs to uncontaminated eggs during hatching involve disinfection of the surfaces in the incubator and hatching cabinets and disinfecting the hatching eggs, especially through electrostatic application of sanitizer (Russel, 2010).

6.2 Optimizing diet formulation

As stated by Klasing (1998), dietary characteristics can modulate the bird's susceptibility to infectious pathogens and the level or type of ingredients may be of critical importance. Diverse dietary strategies have been investigated to reduce *Salmonella* in the gastrointestinal tract, such as feeding whole cereals or mash diet to improve the intestinal resistance (Bjerrum et al., 2005; Huang et al., 2006), optimizing the level of protein content, as well as the nature of the protein source, or feeding plant protein-based feeds rich in dietary fibres, which help to maintain normal microbial populations and low pH (Dunkley et al., 2009). For example, alfalfa is one of the most extensively studied high-fibre dietary sources that increase lactic acid production and show potential inhibitory effect on *Salmonella* (Donalson et al., 2008), but soluble fibres from more exotic sources such as edible plantain banana (*Musa* spp.) have also been shown to protect poultry from *Salmonella* (Parsons et al., 2014). The level of other nutrients, such as minerals (Bullen et al., 2005) or amino acids (Hume et al., 1997), are known to play an important role in maintaining GIT integrity and natural resistance to colonization, highlighting the need for proper diet formulation and optimal nutrient balance in the poultry diet.

Increasing the digestion rate and absorption of nutrients by the use of exogenous enzymes would also reduce the amount of substrate remaining in the GIT for pathogenic microorganisms. Santos (2005) demonstrated that supplementing a corn-based diet with an enzyme product containing endoxylanase and alpha-amylase activities significantly reduced *Salmonella* prevalence and caecal population. The early feeding, and even *in ovo* feeding, may initiate early immune development and increase digestive capacity, consequently improving the resistance of chicks to disease (Tako et al., 2004; Smirnov et al., 2006). Finally, bioactive compounds with immunomodulatory properties included in the diet, such as β-glucans (Lowry et al., 2005) or antioxidants (Chew, 1993), is another important part of nutrition strategies in order to maintain animal health. These compounds may originate from ingredients, such as cereals or microbial sources of proteins, or from specific additives such as vitamins or phytobiotics.

6.3 Improving the immune response

As protective adaptive immune responses can be elicited in poultry after *Salmonella* infection, different types of vaccines are available for use in breeder flocks and in young chickens and young turkeys and are administered orally or by injection. They have been designed with inactivated or live attenuated *Salmonella* strains developed by non-reverting mutations of genes involved in metabolic functions or in virulence factors. Killed vaccines exert a poorer protective immunity than live vaccine because they stimulate mainly antibody production (Nagaraja and Rajashekara, 1999), are rapidly destroyed and eliminated from the host system and may require multiple doses or take longer to produce the immune benefits, limiting but not excluding their usefulness in chicks due to the short grow-out period. Finally, subunit vaccines, such as outer membrane proteins (Meenakshi et al., 1999) or toxoids (Mishra and Sharma, 2001), have also been used in poultry. The reader is invited to learn more about the different classes of *Salmonella* vaccines in the detailed reviews by Desin et al. (2013), which provides an insight into the future of *Salmonella* vaccines and the platforms that can be used for delivery. However, the efficacy of vaccines depends on several factors, such as the targeted pathogen serotype (Salisbury et al., 2014) or the bird's age (Johnston et al., 2012). In some cases, they are very effective as demonstrated by Dorea et al. (2010), but in others they do not have any effect, highlighting the importance of properly identifying the targeted serotypes of *Salmonella*, and the need for safer, well-defined *Salmonella* vaccines.

As maternal antibodies transferred from the yolk to the chicks were shown to prevent the chick's colonization by *Salmonella* (Hassan and Curtiss, 1996), the strategy of feeding these antibodies to provide passive immune protection to the hatchling was studied as complementary approach and has yielded promising results, as observed by Rahimi et al. (2007). Several studies, reviewed by Schade et al. (2005) and Chalghoumi et al. (2009), have investigated how hyper-immunizing hens could produce pathogen-specific antibodies in large quantities from eggs. Furthermore, it has been found that non-immunized egg yolk powder was also effective in eliminating and preventing *Salmonella* colonization in poultry (Kassaify and Mine, 2004). Egg yolk contains anti-infectious/adhesive factors besides IgY, such as ovotransferrin, which may involve the pathogens' agglutination to the egg yolk, competition for adhesion sites or the stimulation of the immune system. Finally, lymphokines may be used for passive immunity, as tested by McGruder et al. (1995). These authors administered *S.* Enteritidis-immune lymphokines *in ovo*, and after challenging the day-of-hatch chicks with 5×10^4 CFU *S.* Enteritidis, they showed that organ invasion was significantly decreased in two-day-old chicks with lymphokine treatment.

Another immune approach involves antimicrobial peptides whose structure usually contains elements that facilitate the interaction with the negatively charged membrane of target pathogen organisms. The antimicrobial activity of different cationic antimicrobial peptides against *Salmonella* has been described in some *in vitro* studies (Evans et al., 1995; Milona et al., 2007; van Dijk et al., 2007). *In vivo*, Kogut et al. (2013) suggested the usefulness of small cationic peptides produced by a Gram-positive soil bacterium, *Brevibacillus texasporus*, as local immune modulators in neonatal poultry. Peptide-containing plant material could also be added to animal feed (Joerger, 2003). However, the potential use of peptides as feed additives faces some important economic and ecological issues that need to be resolved. For example, chemical modifications or encapsulation methods should be developed to make them more resistant to proteolysis in animals. Some of the peptides such as defensins or bactencin exhibit toxic effects

(Rademacher et al., 1993; Kagan et al., 1994). Further aspects about properties and applications of antimicrobial peptides have been reviewed by van't Hof et al. (2001).

7 Strategies to control *Salmonella* in poultry: controlling the environment inside chicken

7.1 Modifying chicken environment

Volatile fatty acids (VFA), and short-chain fatty acids (SCFA) are produced by the normal anaerobic intestinal flora as end products of their metabolism and exert bacteriostatic or bactericidal effects against Gram-negative bacteria, by entering the bacterial cell and decreasing the intracellular pH. Consequently, many studies have investigated the potential acidification of feed or water by these SCFA to reduce *Salmonella* colonization of broilers. These studies, reviewed by Doyle and Erickson (2006) and Van Immerseel et al. (2006), yield conflicting results. It appears that the administration mode, the infection pressure, the type of acid and its concentration are very important. Researchers have also developed microencapsulated organic acids in order to transport them further down in the GIT, which should prevent the absorption of acids in the upper tract and ensure their release further down in the GIT. Most commercial products that are actually applied in the field consist of propionic and formic acids, either in powder form or encapsulated in silica beads. Van Immerseel et al. (2004) tested the effect of feed supplementation with microencapsulated SCFA on *S.* Enteritidis invasion of chickens challenged at five and six days post-hatch and showed that butyric acid-impregnated microbeads resulted in a significantly decreased colonization by *S.* Enteritidis in the caeca. However, some *in vitro* studies suggested that VFA might also have an undesirable effect by promoting the invasiveness and virulence of *Salmonella* (Durant et al., 2000; Van Immerseel et al., 2002, 2006), or enhancing *Salmonella's* resistance to acid upon exposure to VFA (Kwon and Ricke, 1998). Medium-chain fatty acids, from 6 to 12 carbons (caproic, caprylic, capric and lauric acid), appear to be much more effective against *Salmonella* than the SCFA. As little as 25 mM C6 to C10 acids were bacteriostatic to a *S.* Enteritidis strain, whereas the same strain tolerated 100 mM of SCFA (Van Immerseel et al., 2003). When *S.* Enteritidis and Typhimurium were incubated with 5 mM monocaprin combined with an emulsifier, the bacteria did not survive (Thormar et al., 2006). However, selecting the type of organic acids used in commercial products has been rather empirical until now, and little information is available concerning the best combination of acids in these products. Complementary studies are needed in order to analyse the effects of each VFA on *Salmonella* virulence.

Another type of acidified poultry diet is fermented liquid feed (FLF). The fermentation may happen spontaneously, by inoculating the moistened feed at a certain temperature before feeding the birds, or it may be induced. The induced fermentation can be achieved either by back slopping or inoculating with a lactic acid bacteria (LAB) strain. Consequently FLF is characterized by high numbers of LAB, high concentration of lactic acid and low pH of 3.5 to 4.5 (Plumed-Ferrer C. and von Wright, 2009). Several studies showed the potential of FLF to decrease *Salmonella* population in the intestinal tract of orally challenged chickens (Heres et al., 2003a,b; Savvidou et al., 2009). However, several variables in the fermentation process, such as the composition of the feed varying with the production stage, may affect the advantages of such FLF and should be further studied.

Besides directly feeding the bird with antibacterial acids, the treatment of its environment would be more applicable in the practice for the producer to prevent or reduce *Salmonella* contamination in broiler flocks. The effect of organic acids, formalin, sodium bisulphate, aluminium sulphate or sulphuric acid, applied on the flock's litter, on *Salmonella* and other pathogens has been described (Pope and Cherry, 2000; Ivanov, 2001; Line, 2002). Poultry Guard® litter amendment is a product based on 40%–50% sulphuric acid, used to control ammonia volatilization by converting litter ammonium to ammonium sulphate, thereby decreasing litter pH and resulting in pathogen reduction in the litter and consequently in the bird's caeca (Vicente et al., 2007).

Moreover, recently, the use of both essential oils has gained attention in the context of antibiotics ban as a potential natural method for reducing pathogens in the poultry gastrointestinal tract. Essential oils are complex mixtures of plant metabolites consisting of low-boiling phenylpropenes and terpenes (Brenes and Roura, 2010). Essential oils can be extracted from plant materials such as flowers, buds, seeds, leaves, twigs, bark, herbs, wood, fruits and roots (Brenes and Roura, 2010). They were especially investigated for their antimicrobial properties (Burt, 2004; Van Immerseel et al., 2006), although they can also have hypolipidemic, antioxidant, digestive stimulant, antiviral, antimycotic, antiparasitic and insecticidal activities (Hahn, 2014). Antimicrobial activity of essential oils is mostly due to their phenolic compounds (Burt, 2004), and can be attributed to their lipophilic character (Applegate et al., 2010). They may be able to suppress pathogenic bacteria by penetrating into the cell or by disintegrating the cell membrane. Although essential oils have shown promising results in reducing *Salmonella in vitro*, few studies have investigated their potential in live poultry and showed inconsistent results (Amerah et al., 2012; Hahn, 2014). The combination of cinnamaldehyde and thymol has been shown to be effective in preventing horizontal transmission of *S.* Heidelberg (Amerah et al., 2012). More research needs to be conducted examining the efficacy of different combinations of essential oils and their components against pathogenic bacteria *in vivo*.

7.2 Modulating the microflora

In modern poultry production, therapeutic antibiotics are used to control economically important infections such as bacterial enteritis caused by *Salmonella*. Several types of antibiotic agents are available with different modes of action based on inhibition or alteration of bacterial metabolic functions (Gaskins et al., 2002). Next to their curative use, antibiotics have also been given to poultry in feed for their growth-promoting properties as they act by reducing mainly Gram-positive bacteria, which are known to depress animal growth, either directly or indirectly, through their metabolic activities (Gaskins et al., 2002). The recent increase in the relative proportion of *Salmonella* strains with reduced sensitivity to antibiotics, as well as the extension of their spectrum of antibiotic resistance, lately including fluoroquinolone resistance, have already led to their ban in some parts of the world as prophylactic agent, and may actually impede the antibiotic therapy efficiency. Some studies also pointed out the other 'side effects' of antibiotics in live birds, such as depletion of the clostridial bacteria that are most important in preventing expansion of commensal *Escherichia coli* and maintaining 'balance' in the microbiota. The assumptions on antibiotic drawbacks have, however, led to conflicting conclusions and are still debated (Casewell et al., 2003; Phillips et al., 2004), as the suppression of growth-promoting antibiotics led to deteriorating animal welfare and increasing infections, which in turn substantially increased the use of therapeutic antibiotics. Anyway, antibiotic use does not seem to be interesting in view of the burden of antibiotic resistance in the food chain and the harmful consequences for the public health, so that development of alternative measures have thus become popular.

As antibiotic alternative, the administration of probiotic strains actually induces changes in the bacteria community structure (Netherwood et al., 1999) and stimulates the immune system as well as the indigenous enzymes, which may lead to a beneficial nutritional and growth-promoting effect (Patterson and Burkholder, 2003) as Jin et al. (1998) observed. A probiotic is defined as 'a live microbial feed supplement which beneficially affects the host animal by improving its intestinal microbial balance' (Fuller, 1989). Many studies have actually shown that inoculating cultures of one or several probiotic strains to broiler chickens may inhibit the bird's contamination by *Salmonella* (Audisio et al., 2000; Higgins et al., 2007; Van Coillie et al., 2007), based on different proposed mechanisms such as lactic acid production, competition for receptor sites, competition for limiting nutrients and production of antimicrobial metabolites (bacteriocins, VFA, hydrogen peroxide), which may inhibit certain enteric pathogens (Servin, 2004; Tsai et al., 2005). Many different microorganism species are being used as probiotics incorporated in animal feed. There are 57 products on the market, from 16 microbial species (FAO, 2016). The most products (17) are being found in the category *Lactobacillus* but bacterial or fungal strains belonging to the genera *Enterococcus*, *Pediococcus*, *Bacillus* or *Saccharomyces* have also been investigated.

Prebiotics are feed ingredients that are not digested by the host, not or little used and/or metabolized as they pass through the upper portion of the intestinal tract, so they can interact with the microbiota of the large intestine (Gibson and Roberfroid, 1995). In principle, not only non-digestible, fermentable feed carbohydrates, especially di- and oligosaccharides (Mussatto and Mancilha, 2007), but also sugar alcohols may be considered as prebiotics. Diverse types of prebiotics were listed by Šušković et al. (2001). It is generally recognized that prebiotics exert beneficial health effect on animal and human, especially by inhibiting intestinal colonization by enteric pathogens such as *Salmonella* (Grizard and Barthomeuf, 1999). This effect can be direct by binding to pathogens and therefore blocking the adhesion of the bacteria to the epithelial cells (Spring et al., 2000) or indirect, as they may constitute a substrate for the intestinal flora, which may then exert their antibacterial activity via competitive exclusion and production of antibacterial metabolites, such as lactic acid, VFA and bacteriocins (Šušković et al., 2001). However, the capacity of prebiotics to control the colonization of enteropathogens showed inconsistent results, which are detailed in the review by Rehman et al. (2009). Fructooligosaccharides (FOS) and mannanoligosaccharides (MOS) are the most extensively studied oligosaccharides in chickens with respect to their prebiotic effect and their activity against *Salmonella* (Chambers et al., 1997; Spring et al., 2000). FOS are usually produced commercially by hydrolysis of inulin or by enzymatic synthesis from sucrose or lactose (Mussatto and Mancilha, 2007), while MOS occur naturally in many products such as yeast cell walls and gums (Spring et al., 2000). More recently, other oligomers, such as bacterial isomaltooligosaccharides produced from *Leuconostoc mesenteroides* fermentation (Chung and Day, 2004) or xylo-oligosaccharides produced by hydrolysis of arabinoxylans from cereal (Eeckhaut et al., 2008), have also been studied for their potential in *Salmonella* inhibition. Combination of several prebiotics has also been designed. In the study by Murate et al. (2015), *Salmonella*-free 1-day-old layers and broiler chicks were inoculated with *S.* Enteritidis and were given 0.15 g/kg prebiotic containing inulin, fructooligosaccharide, mannanoligosaccharide and oligosaccharides that had a protective effect during the first week post-infection.

A further step has consisted in developing synbiotics in which probiotic strains and prebiotic substrate were combined (Collins and Gibson, 1999; Schrezenmeir and de Vrese, 2001). This term is reserved for products in which the prebiotic compound selectively favours the probiotic strain, as observed *in vitro* by Kontula et al. (1998), providing a synergistic effect for the modulation of the faecal microbiota. Most data available for poultry reported the effect of different synbiotics on growth performance and intestinal metabolism (Awad et al., 2008; Buteikis et al., 2008) but the low number of studies available to date do not allow conclusions about the efficiency of that kind of combination. Fasina et al. (2008) failed to show any beneficial effect from the combination of pectin and a commercially available probiotic on caecal contamination of chickens infected at 4 days of age with 10^6 CFU of *Salmonella* Typhimurium, highlighting the need to properly select probiotic strains able to efficiently metabolize a specific prebiotic and providing a specific competitive advantage.

In intensive production systems, newly hatched chicks are very sensitive to enteric infections as they have little opportunity to develop a normal intestinal microflora quickly due to the clean housing conditions in which the chicks are reared. Nurmi and Rantala (1973) then introduced the concept of competitive exclusion (CE) by stating that a suspension of alimentary tract contents from adult birds administrated orally to newly hatched chicks to establish an adult-type microbiota early in the chicks, which protects from *Salmonella* colonization. This treatment is generally used as a prophylactic measure aiming at increasing the resistance of young chicks to *Salmonella* (Mead, 2000), but it can also be used after a curative antibiotic therapy applied to eliminate an existing infection; even if the efficacy of such an approach gives an inconsistent result, its composition and early administration is critical for its efficacy (Varmuzova et al., 2016). At present, different mixed cultures derived from the caecal contents and/or the gut wall of domestic fowl (Schneitz, 2005) are commercially available, for example AviFree, Aviguard, Broilact, MSC, Preempt and CF-3, giving variable results against overwhelming colonization by salmonellae. The effect on *Salmonella* in the caeca is suggested to be mainly bacteriostatic rather than bactericidal (Mead, 2000) and may be partly explained by the VFA production by the bacteria (Hume et al., 1996; Schneitz, 2005). Other mechanisms were also proposed to explain the antagonistic effect of CE, such as competition for receptor sites in the intestinal tract and competition between pathogens and native microbiota for nutrients (Mead, 2000).

7.3 Recently developed strategies

Bacteriophages were also tested for their inhibitory effect against *Salmonella*. They represent a group of viruses that specifically infect and replicate in bacteria. Bacteriophage treatments were first applied to decontaminate poultry carcasses and products (Goode et al., 2003; Higgins et al., 2005). Subsequently, Andreatti Filho et al. (2007) applied such treatments *in vivo* by treating day-of-hatch chicks experimentally infected with *S.* Enteritidis via oral gavage with cocktails of 4 or 45 bacteriophages at 10^8 PFU/chick, and showed significantly reduced *S.* Enteritidis recovery from caecal tonsils 24 h post-treatment but not after 48 h. The actual efficiency of bacteriophages in reducing *Salmonella* in poultry for a long period would need further investigation, but would probably face serious regulatory obstacles, due to microbial resistance concerns.

Combination of different approaches, as was initiated by synbiotics, appear to be an interesting approach and many recent studies have aimed to define the best combinations (Tellez et al., 2001; Braukmann et al., 2016). In the study by Braukmann et al. (2016), they combined the use of *Salmonella* live vaccine and CE culture, and this combination resulted in an additive protective effect and prevented completely the systemic dissemination of the *Salmonella* challenge strain. The authors proposed to further improve live *Salmonella* vaccines in order to make them still capable of inducing both intestinal colonization and invasion-inhibition effects against *Salmonella* exposure, in order to exploit the potential of combined use of CE and vaccination. Table 2 summarizes different combinations of strategies studied to control *Salmonella* infection in poultry.

Table 2 Example of combination of different strategies developed to control *Salmonella* in live poultry

Tested solution	Poultry species	Observations	Reference
Probiotic[b] + NSP-enzyme[c]	Chicken	At slaughter age, *Salmonella* contamination was reduced by 2.00 and 1.85 log colony-forming units for the enzyme and combination treatment, respectively	Vandeplas et al. (2009)
NSP-enzyme[j] + essential oils[k]	Chicken	Enzyme and essential oil reduced significantly the incidence of horizontal transmission of *Salmonella* infection by 61 and 77%, respectively Combination numerically improved growth performance but no additional advantage on *Salmonella*	Amerah et al. (2012)
Salmonella vaccine + prebiotic[f] or mix[g]	Laying hen	No effect on *Salmonella* combination of additive gave the highest egg weight	Hahn (2014)
Prebiotic[h] + probiotic[i]	Pullet	No effect on *Salmonella* faecal count, decrease in *Enterobacteriaceae*	Hahn (2014)
Probiotic[d] + prebiotic[e]	Laying hen and chicken	The probiotic and synbiotic additives did not influence the *Salmonella* infection in laying hens and broilers	Murate et al. (2015)
Salmonella vaccine + CE[a] culture	Chicken	Combination of vaccine and CE prevented completely the dissemination of *Salmonella*	Braukmann et al. (2016)

[a] CE: competitive exclusion.
[b] 10^6 cfu/g *Lactobacillus plantarum*.
[c] 10 IU/kg xylanase – Belfeed B1100MP (Puratos NV).
[d] *Bacillus subtilis, Lactobacillus acidophilus, L. casei, Enterococcus faecium, Bifidobacterium longum* at the same quantitative proportion.
[e] Inulin, fructooligosaccharide, mannanoligosaccharide and oligosaccharide at the same quantitative proportion.
[f] MOS – BioSecure® MOS (BioMatrix International).
[g] Combination of additives – BioSecure®Biolex (chicory root, lactic acid, glucomannonprotein complex, *Saccharomyces cerevisiae*, spices and essential oils).
[h] 24% MOS + 25% beta-glucan – AgriMOS® (Lallemand Animal Nutrition).
[i] 1×10^{10} *Pediococcus acidilactici* – Bactocell® (Lallemand Animal Nutrition) or 2×10^{10} *Saccharomyces cerevisiae* – Levucell®, (Lallemand Animal Nutrition).
[j] 2,000 U/kg Danisco Xylanase (Danisco Animal Nutrition).
[k] Enviva EO 101 (Danisco Animal Nutrition).

8 Concluding remarks

Knowledge of infection mechanisms in the gastrointestinal tract of poultry host and mechanisms of immunity against the pathogen has first led to the development of vaccination programmes. Nevertheless, *Salmonella* control should not rely on vaccination alone because of different virulence properties according to the serotypes, modification of bird immunity with time in layer or shorter life span for broiler, and should be considered as part of an integrated control strategy including biosecurity and improved animal management practices. The use of feed additives is then more and more accepted as a valuable way to combat *Salmonella* infection in poultry standard production.

All those control strategies are based on improvement of the nutritional value of the poultry diet, modulation of the beneficial microbiota, direct elimination of *Salmonella* or modification of the intestinal environment to render it hostile for the pathogen. However, none of these treatments are actually effective to eliminate *Salmonella* contamination from poultry flocks completely. Combining several measures seems therefore the only efficient strategy to reduce the prevalence of this pathogen at farm level. An attractive developing approach is the combination of different dietary treatments allowing potential addition, complementarity, or even synergism, of the protagonist effects directly at the gut level. Whatever the defined strategy, the most important criterion to select appropriate control measures is their efficacy, which depends on many breeding parameters, but economical costs and phenomena of *Salmonella* resistance, as observed with some organic acids, have also to be taken into account.

9 Where to look for further information

More information on the epidemiology of *Salmonella*, the description of the public health problem and the impact along the food chain can be found in the Codex Committee on Food Hygiene (2007) and the review by Eng et al. (2015). The reader who wishes to have more details regarding detection method is invited to read the review by Carrique-Mas and Davies (2008), which discuss in detail the range of sampling and culture methodologies for *Salmonella* in poultry farms and carcasses. For specific information regarding the application of molecular approaches to analyse *Salmonella* at farm level, Ricke (2014) offered a general description of functional genomic technologies and their specific application to poultry vaccine developments. The antimicrobial resistance was discussed in details in the review by Cosby et al. (2015), who described the mechanisms and transmission of bacterial resistance and its global implications. The pathogenesis of *Salmonella* and immune responses to infection is briefly discussed in Revolledo and Ferreira (2012). Regarding interaction between the chicken immune system and *Salmonella*, Wigley (2014) discussed more in details the role of the immune system and its modulation in persistent infection and infection of the reproductive tract of poultry.

10 References

Amerah, A. M., Mathis, G. and Hofacre, C. L. (2012), Effect of xylanase and a blend of essential oils on performance and *Salmonella* colonization of broiler chickens challenged with *Salmonella* Heidelberg. *Poult. Sci.*, 91:94–47.

Andreatti Filho, R. L., Higgins, J. P., Higgins, S. E., Gaona, G., Wolfenden, A. D., Tellez, G. and Hargis, B. M. (2007), Ability of bacteriophages isolated from different sources to reduce *Salmonella enterica* serovar Enteritidis *in vitro* and *in vivo*. *Poult. Sci.*, 86:1904–9.

Applegate, T. J., Klose, V., Steiner, T., Ganner, A. and Schatzmayr, G. (2010), Probiotics and phytogenics for poultry: Myth or reality? *J. Appl. Poult. Res.*, 19:194–210.

Arsenault, J., Letellier, A., Quessy, S., Normand, V. and Boulianne, M. (2007), Prevalence and risk factors for *Salmonella* spp. and *Campylobacter* spp. caecal colonization in broiler chicken and turkey flocks slaughtered in Quebec, Canada. *Prev. Vet. Med.*, 81:250–64.

Arsenault, R. J., Napper, S. and Kogut, M. H. (2013), *Salmonella enterica* Typhimurium infection causes metabolic changes in chicken muscle involving AMPK, fatty acid and insulin/mTOR signaling. *Vet. Res.*, 44:35.

Audisio, M. C., Oliver, G. and Apella, M. C. (2000), Protective effect of *Enterococcus faecium* J96, a potential probiotic strain, on chicks infected with *Salmonella* Pullorum. *J. Food Prot.*, 63:1333–7.

Awad, W., Ghareeb, K. and Böhm, J. (2008), Intestinal structure and function of broiler chickens on diets supplemented with a symbiotic containing *Enterococcus faecium* and oligosaccharides. *Int. J. Mol. Sci.*, 9:2205–16.

Bai, S. P., Huang, Y., Luo, Y. H., Wang, L. L., Ding, X. M., Wang, J. P., Zeng, Q. F. and Zhang, K. Y. (2015), Effect of dietary nonphytate phosphorus content on ileal lymphocyte subpopulations and cytokine expression in the cecal tonsils and spleen of laying hens that were or were not orally inoculated with *Salmonella* Typhimurium. *Am. J. Vet. Res.*, 76(8):710–18.

Bailey, J. S. and Cosby, D. E. (2003), Detection of *Salmonella* from chicken rinses and chicken hot dogs with the automated BAX PCR system. *J. Food Prot.*, 66(11):2138–40.

Bjerrum, J., Pedersen, K. and Engberg, R. M. (2005), The influence of whole wheat feeding on *Salmonella* infection and gut flora composition in broilers. *Avian Dis.*, 49:9–15.

Braukmann, M., Barrow, P. A., Berndt, A. and Methner, U. (2016), Combination of competitive exclusion and immunisation with a live *Salmonella* vaccine in newly hatched chickens: Immunological and microbiological effects. *Res. Vet. Sci.*, 107:34–41.

Brenes, A. and Roura, E. (2010), Essential oils in poultry nutrition: main effects and modes of actions. *Anim. Feed Sci. Tech.*, 158(2):1–14.

Bullen, J. J., Rogers, H. J., Spalding, P. B. and Ward, C. G. (2005), Iron and infection: the heart of the matter. *FEMS Immunol. Med. Mic.*, 43:325–30.

Burt, S. (2004), Essential oils: their antibacterial properties and potential applications in foods-a review. *Int. J. Food Microbiol.*, 94(3):223–53.

Buteikis, G., Matusevičius, P., Januškevičius, A., Jankowski, J., Mikulski, D., Blok, J. and Kozłowski, K. (2008), Use of synbiotic preparations in turkey diets and their effect on growth performance. *Veterinarija Ir Zootechnika*, 43:14–19.

Carrique-Mas, J. J. and Davies, R. H. (2008), Sampling and bacteriological detection of *Salmonella* in poultry and poultry premises: a review. *Rev. Sci. Tech. Int. Epiz.*, 27(3):665–77.

Casewell, M., Friis, C., Marco, E., McMullin, P. and Phillips, I. (2003), The European ban on growth-promoting antibiotics and emerging consequences for human and animal health. *J. Antimicrob. Chemoth.*, 52:159–61.

Chalghoumi, R., Beckers, Y., Portetelle, D. and Théwis, A. (2009), Hen egg yolk antibodies (IgY), production and use for passive immunization against bacterial enteric infections in chicken: a review. *Biotechnol. Agron. Soc. Environ.*, 13:295–308.

Chambers, J. R., Spencer, J. L. and Modler, H. W. (1997), The influence of complex carbohydrates on *Salmonella typhimurium* colonization, pH, and density of broiler ceca. *Poult. Sci.*, 76:445–51.

Chan, K., Baker, S., Kim, C. C., Detweiler, C. S., Dougan, G. and Falkow, S. (2003). Genomic comparison of *Salmonella* enterica serovars and *Salmonella bongori* by use of a *S. enterica* serovar Typhimurium DNA microarray. *J. Bacteriol.*, 185(2):553–63.

Chappell, L., Kaiser, P., Barrow, P., Jones, M. A., Johnston, C. and Wigley, P. (2009), The immunobiology of avian systemic salmonellosis. *Vet. Immunol. Immunopathol.*, 128:53–9.

Chew, B. P. (1993), Role of the carotenoids in the immune response. *J. Dairy Sci.*, 76:2804–11.

Chung, C.-H. and Day, D. F. (2004), Efficacy of *Leuconostoc mesenteroides* (ATCC 13146) isomaltooligosaccharides as a poultry prebiotic. *Poult. Sci.*, 83:1302–6.

Coble, D. J., Sandford, E. E., JiT., Abernathy, J., Fleming, D., Zhou, H. and Lamont, S. J. (2013), Impacts of *Salmonella enteritidis* infection on liver transcriptome in broilers. *Genesis*, 51:357–64.

Codex Committee on Food Hygiene (2007), Food Safety Risk Profile for *Salmonella* species in broiler (young) chickens. CCFH Working Group on Guidelines for control of *Campylobacter* and *Salmonella* spp. in broiler (young bird) chicken meat, p. 30.

Collins, D. M. and Gibson, G. R. (1999), Probiotics, prebiotics, and synbiotics: approaches for modulating the microbial ecology of the gut. *Am. J. Clin. Nutr.*, 69:1052S–57S.

Conceição, R. C. S., Moreira, A. N., Ramos, R. J., Goularte, F. L., Carvalhal, J. B. and Aleixo, J. A. G. (2008), Detection of *Salmonella* sp. in chicken cuts using immunomagnetic separation. *Braz. J. Microbiol.*, 39:173–7.

Cosby, D. E., Cox, N. A., Harrison, M. A., Wilson, J. L., Buhr, R. J. and Fedorka-Cray, P. J. (2015), *Salmonella* and antimicrobial resistance in broilers: a review. *J. Appl. Poult. Res.*, 24:408–26.

Crhanova, M., Hradecka, H., Faldynova, M., Matulova, M., Havlickova, H., Sisak, F. and Rychlik, I. (2011), Immune response of chicken gut to natural colonization by gut microflora and to *Salmonella enterica* serovar Enteritidis infection. *Infect. Immun.*, 79(7):2755–63.

Crim, S. M., Iwamoto, M., Huang, J. Y., Griffin, P. M., Gilliss, D., Cronquist, A. B., Cartter, M., Tobin-D'Angelo, M., Blythe, D., Smith, K., Lathrop, S., Zansky, S., Cieslak, P. R., Dunn, J., Holt, K. G., Lance, S., Tauxe, R. and Henao, O. L. (2014), Incidence and trends of infection with pathogens transmitted commonly through food – Foodborne Diseases Actives Surveillance Network, 10 U. S. Sites, 2006–2013. *Morb. Mortal. Wkly Rep.*, 63(15):321–24.

D'Aoust, J. Y. (1991), Pathogenicity of foodborne *Salmonella*. *Int. J. Food Microbiol.*, 12:17–40.

Davies, R. H. and Wray, C. (1996), Persistence of *Salmonella enteritidis* in poultry units and poultry food. *Brit. Poult. Sci.*, 37:589–96.

Denagamage, T., Jayarao, B., Patterson, P., Wallner-Pendleton, E. and Kariyawasam, S. (2015), Risk factors associated with *Salmonella* in laying hen farms: systematic review of observational studies. *Avian Dis.*, 59(2):291–302.

Desin, T. S., Köster, W. and Potter, A. A. (2013), *Salmonella* vaccines in poultry: past, present and future. *Expert Rev. Vaccines*, 12(1):87–96.

Desmidt, M., Ducatelle, R. and Haesebrouck, F. (1997), Pathogenesis of *Salmonella* Enteritidis phage type four after experimental infection of young chickens. *Vet. Microbiol.*, 56: 99–109.

Donalson, L. M., McReynolds, J. L., Kim, W. K., Chalova, V. I., Woodward, C. L., Kubena, L. F., Nisbet, D. J. and Ricke, S. C. (2008), The influence of a fructooligosaccharide prebiotic combined with alfalfa molt diets on the gastrointestinal tract fermentation, *Salmonella* Enteritidis infection and intestinal shedding in laying hens. *Poult. Sci.*, 87:1253–62.

Dorea, F. C., Cole, D. J., Hofacre, C., Zamperini, K., Mathis, D., Doyle, M. P., Lee, M. D. and Maurer, J. J. (2010), Effect of *Salmonella* vaccination of breeder chickens on contamination of broiler chicken carcasses in integrated poultry operations. *Appl. Environ. Microbiol.*, 76(23):7820–5.

Doyle, M. P. and Erickson, M. C. (2006), Reducing the carriage of foodborne pathogens in livestock and poultry. *Poult. Sci.*, 85:960–73.

Dunkley, K. D., Callaway, T. R., Chalova, V. I., McReynolds, J. L., Hume, M. E., Dunkley, C. S., Kubena, L. F., Nisbet, D. J. and Ricke, S. C. (2009), Foodborne *Salmonella* ecology in the avian gastrointestinal tract. *Anaerobe*, 15:26–35.

Durant, J. A., Corrier, D. E. and Ricke, S. C. (2000), Short-chain volatile fatty acids modulate the expression of the *hilA* and *invF* genes of *Salmonella* Typhimurium. *J. Food Prot.*, 63:573–8.

Eeckhaut, V., Van Immerseel, F., Dewulf, J., Pasmans, F., Haesebrouck, F., Ducatelle, R., Courtin, C. M., Delcour, J. A. and Broekaert, W. F. (2008), Arabinoxylooligosaccharides from wheat bran inhibit *Salmonella* colonization in broiler chickens. *Poult. Sci.*, 87:2329–34.

EFSA (European Food Safety Authority) (2004), Opinion of the scientific panel on biological hazards on the request from the Commission related to the use of antimicrobials for the control of Salmonella in poultry. *EFSA J.*, 115:1–76.

EFSA (European Food Safety Authority) (2007), The community summary report on trends and sources of zoonoses, zoonotic agents, antimicrobial resistance and foodborne outbreaks in the European Union in 2006. *EFSA J.*, 130:1–352.

EFSA (European Food Safety Authority) and ECDC (European Centre for Disease Prevention and Control) (2015), The European Union summary report on trends and sources of zoonoses, zoonotic agents and food-borne outbreaks in 2014. *EFSA J.*, 13(12):4329.

Eijkelkamp, J. M., Aarts, H. J. M. and van der Fels-Klerx, H. J. (2009), Suitability of rapid detection methods for *Salmonella* in poultry slaughterhouses. *Food Anal. Methods*, 2:1–13.

Eng, S.-K., Pusparajah, P., Ab Mutalib, N.-S., Ser, H.-L., Chan, K.-G. and Lee, L.-H. (2015), Salmonella: A review on pathogenesis, epidemiology and antibiotic resistance. *Front. Life Sci.*, 8(3):284–93.

Evans, E. W., Beach, F. G., Moore, K. M., Jackwood, M. W., Glisson, J. R. and Harmon, B. G. (1995), Antimicrobial activity of chicken and turkey heterophil peptides CHP1, CHP2, THP1, and THP3. *Vet. Microbiol.*, 47:295–303.

Eyigör, A., Goncagül, G. and Carli, K. T. (2007), A PCR-ELISA for the detection of Salmonella from chicken intestine. *J. Biol. Environ. Sci.*, 1(1):45–9.

FAO (2016), Probiotics in animal nutrition – Production, impact and regulation by Yadav S. Bajagai, Athol V. Klieve, Peter J. Dart and Wayne L. Bryden. Editor Harinder P. S. Makkar. FAO Animal Production and Health Paper No. 179. Rome.

Fasina, Y. O., Bowers, J. W. J. and Mckee, S. R. (2008), Effect of pectin-Protexin®symbiotic therapy on the concentration of *Salmonella* Typhimurium in the ceca of broiler chicks. PSA Annual Meeting, Niagara Falls, Ontario, Canada, 20–23 July 2008, p. 147.

FDA (Food and Drug Administration, USA) (2002), FSIS Pathogen Reduction/HACCP. Available at http://www.fda.gov/ohrms/dockets/ac/02/slides/3816s2_07_Masters.ppt.

Ferretti, R., Mannazzu, I., Cocolin, L., Comi, G. and Clementi, F. (2001), Twelve-hour PCR-based method for detection of *Salmonella* spp. in food. *Appl. Environ. Microbiol.*, 67(2):977–8.

Foley, S. L., Nayak, R., Hanning, I. B., Johnson, T. J., Han, J. and Ricke, S. C. (2011), Population dynamics of *Salmonella enterica* serotypes in commercial egg and poultry production. *Appl. Environ. Microbiol.*, 77(13):4273–9.

Food Safety and Inspection Service (FSIS) (2015), Draft FSIS Compliance Guideline For Controlling Salmonella and Campylobacter in Raw Poultry. Available at http://www.fsis.usda.gov/wps/wcm/connect/6732c082-af40-415e-9b57-90533ea4c252/Controlling-Salmonella-Campylobacter-Poultry-2015.pdf?MOD=AJPERES. Accessed 5 August 2016.

Fukata, T., Sasai, K., Baba, E. and Arakawa, A. (1996), Effect of ochratoxin A on *Salmonella* Typhimurium-challenged layer chickens. *Avian Dis.*, 40(4):924–6.

Fuller, R. (1989), Probiotics in man and animals. *J. Appl. Bacteriol.*, 66:365–78.

Gaskins, H. R., Collier, C. T. and Anderson, D. B. (2002), Antibiotics as growth promotants: mode of action. *Anim. Biotechnol.*, 13:29–42.

Gast, R. K., Guraya, R., Guard, J. and Holt, P. S. (2011), The relationship between the numbers of *Salmonella* Enteritidis, *Salmonella* Heidelberg, or *Salmonella* Hadar colonizing reproductive tissues of experimentally infected laying hens and deposition inside eggs. *Avian Dis.*, 55(2):243–7.

Gibson, G. R. and Roberfroid, M. B. (1995), Dietary modulation of the human colonic microbiota: introducing the concept of prebiotics. *J. Nutr.*, 125:1401–2.

Gomes, A. V. S., Quinteiro-Filho, W. M., Ribeiro, A., Ferraz-de-Paula, V., Pinheiro, M. L., Baskeville, E., Akamine, A. T., Astolfi-Ferreira, C. S., Ferreira, A. J. P. and Palermo-Neto, J. (2014), Overcrowding stress decreases macrophage activity and increases *Salmonella* Enteritidis invasion in broiler chickens. *Avian Pathol.*, 43(1):82–90.

Goode, D., Allen, V. M. and Barrow, P. A. (2003), Reduction of experimental *Salmonella* and *Campylobacter* contamination of chicken skin by application of lytic bacteriophage. *Appl. Environ. Microbiol.*, 69:5032–6.

Grenier, B. and Applegate, T. J. (2013), Modulation of intestinal functions following mycotoxin ingestion: Meta-analysis of published experiments in animals. *Toxins*, 5:396–430.

Grizard, D. and Barthomeuf, C. (1999), Non-digestible oligosaccharides used as prebiotic agents: mode of production and beneficial effects on animal and human health. *Reprod. Nutr. Dev.*, 39:563–88.

Guo, C., Hoekstra, R. M., Schroeder, C. M., Pires, S. M., Ong, K. L., Hartnett, E., Naugle, A., Harman, J., Bennett, P., Cieslak, P., Scallan, E., Rose, B., Holt, K. G., Kissler, B., Mbandi, E., Roodsari, R., Angulo, F. J. and Cole, D. (2011), Application of Bayesian techniques to model the burden of human salmonellosis attributable to U.S. food commodities at the point of processing: adaptation of a Danish model. *Foodborne Pathog. Dis.*, 8(4):509–16.

Hahn, D. (2014), The effects of feed additives, housing systems, and stress on *Salmonella* shedding in single comb white and brown laying hens. PhD Dissertation. University of Nebraska-Lincoln.

Hassan, J. O. and Curtiss III, R. (1996), Effect of vaccination of hens with an avirulent strain of *Salmonella* Typhimurium on immunity of progeny challenged with wild-type *Salmonella* strains. *Infect. Immun.*, 64:938–44.

He, H., Genovese, K. J., Swaggerty, C. L., Nisbet, D. J. and Kogut, M. H. (2012), A comparative study on invasion, survival, modulation of oxidative burst, and nitric oxide responses of macrophages (HD11), and systemic infection in chickens by prevalent poultry *Salmonella* serovars. *Foodborne Pathog. Dis.*, 9(12):1104–10.

Heba, A. A., El-Hofy Fatma, I., Shafik Saleh, M., Abdelrahman Mahmoud, A. and Elsaid Gamilat, A. (2016), Characterization of virulence-associated genes, antimicrobial resistance genes, and class 1 integrons in *Salmonella enterica* serovar Typhimurium isolates from chicken meat and humans in Egypt. *Foodborne Pathog. Dis.*, 13(6):281–8.

Heres, L., Engel, B., van Knapen, F., de Jong, M. C. M., Wagenaar, J. A. and Urlings, H. A. P. (2003a), Fermented liquid feed reduces susceptibility of broilers for *Salmonella* Enteritidis. *Poult. Sci.*, 82:603–11.

Heres, L., Wagenaar, J. A., van Knapen, F. and Urlings, B. A. P. (2003b), Passage of *Salmonella* through the crop and gizzard of broiler chickens fed with fermented liquid feed. *Avian Pathol.*, 32:173–81.

Heyndrickx, M., Vandekerchove, D., Herman, L., Rollier, I., Grijspeerdt, K. and De Zutter, L. (2002), Routes for *Salmonella* contamination of poultry meat: epidemiological study from hatchery to slaughterhouse. *Epidemiol. Infect.*, 129:253–65.

Higgins, J. P., Higgins, S. E., Guenther, K. L., Huff, W., Donoghue, A. M., Donoghue, D. J., Hargis, B. M. (2005) Use of specific bacteriophage treatment to reduce *Salmonella* in poultry products. *Poultry Sci.*, 84:1141–5.

Higgins, J. P., Higgins, S. E., Vicente, J. L., Wolfenden, A. D., Tellez, G. and Hargis, B. M. (2007), Temporal effects of lactic acid bacteria probiotic culture on *Salmonella* in neonatal broilers. *Poult. Sci.*, 86:1662–6.

Huang, D. S., Li, D. F., Xing, J. J., Ma, Y. X., Li, Z. J. and Lv, S. Q. (2006), Effects of feed particle size and feed form on survival of *Salmonella* Typhimurium in the alimentary tract and cecal *S.* Typhimurium reduction in growing broilers. *Poult. Sci.*, 85:831–6.

Huang, J. Y., Henao, O. L., Griffin, P. M., Vugia, D. J., Cronquist, A. B., Hurd, S., Tobin-D'Angelo, M., Ryan, P., Smith, K., Lathrop, S., Zansky, S., Cieslak, P. R., Dunn, J., Holt, K. G., Wolpert, B. J. and Patrick, M. E. (2016), Infection with pathogens transmitted commonly through food and the effect of increasing use of culture-independent diagnostic tests on surveillance – Foodborne Diseases Active Surveillance Network, 10 U. S. Sites, 2012–2015. MMWR Morb. *Mortal. Wkly Rep.*, 65:368–71.

Hume, M. E., Corrier, D. E., Nisbet, D. J. and DeLoach, J. R. (1996), Reduction of *Salmonella* crop and cecal colonization by a characterized competitive exclusion culture in broilers during grow-out. *J. Food Prot.*, 59:688–93.

Hume, M. E., Nisbet, D. J. and DeLoach, J. R. (1997), In vitro ^{14}C-amino acid fermentation by CF3TM, a characterized continuous-flow competitive exclusion culture of caecal bacteria. *J. Appl. Microbiol.*, 83:236–42.

Hur, J., Jawale, C. and Lee, J. H. (2012), Antimicrobial resistance of *Salmonella* isolated from food animals: a review. *Food Res. Int.*, 45(2):819–30.

Ivanov, I. E. (2001), Treatment of broiler litter with organic acid. *Res. Vet. Sci.*, 70:169–73.

Jin, L. Z., Ho, Y. W., Abdullah, N., Ali, M. A. and Jalaludin, S. (1998), Effects of adherent *Lactobacillus* cultures on growth, weight of organs and intestinal microflora and volatile fatty acids in broilers. *Anim. Feed Sci. Technol.*, 70:197–209.

Jinu, M., Agarwal, R. K., Sailo, B., Wani, M. A., Kumar, A., Dhama, K. and Singh, M. K. (2014), Comparison of PCR and conventional cultural method for detection of *Salmonella* from poultry blood and faeces. *Asian J. Anim. Vet. Adv.*, 9(11):690–701.

Joerger, R. D. (2003), Alternatives to antibiotics: bacteriocins, antimicrobial peptides and bacteriophages. *Poult. Sci.*, 82:640–7.

Johnston, C. E., Hartley, C., Salisbury, A.-M. and Wigley, P. (2012), Immunological changes at point-of-lay increase susceptibility to *Salmonella enterica* serovar Enteritidis infection in vaccinated chickens. *PLoS ONE*, 7(10):e48195.

Juricova, H., Videnska, P., Lukac, M., Faldynova, M., Babak, V., Havlickova, H., Sisak, F. and Rychlik, I. (2013), Influence of *Salmonella enterica* serovar Enteritidis infection on the development of the cecum microbiota in newly hatched chicks. *Appl. Environ. Microbiol.*, 79(2):745–7.

Kagan, B. L., Ganz, T. and Lehrer, R. I. (1994), Defensins: a family of antimicrobial and cytotoxic peptides. *Toxicology*, 87:131–49.

Kallapura, G., Kogut, M. H., Morgan, M. J., Pumford, N. R., Bielke, L. R., Wolfenden, A. D., Faulkner, O. B., Latorre, J. D., Menconi, A., Hernandez-Velasco, X., Kuttappan, V. A., Hargis, B. M. and Tellez, G. (2014), Fate of *Salmonella* Senftenberg in broiler chickens evaluated by challenge experiments. *Avian Pathol.*, 43(4):305–9.

Kassaify, Z. G. and Mine, Y. (2004), Nonimmunized egg yolk powder can suppress the colonization of *Salmonella* Typhimurium, *Escherichia coli* O157:H7, and *Campylobacter jejuni* in laying hens. *Poult. Sci.*, 83:1497–506.

Klasing, K. C. (1998), Nutritional modulation of resistance to infectious diseases. *Poult. Sci.*, 77:1119–25.

Kogut, M. H., Genovese, K. J., He, H., Swaggerty, C. L. and Jiang, Y. (2013), Modulation of chicken intestinal immune gene expression by small cationic peptides as feed additives during the first week posthatch. *Clin. Vaccine Immunol.*, 20(9):1440–8.

Kontula, P., von Wright, A. and Mattila-Sandholm, T. (1998), Oat bran β-gluco- and xylooligosaccharides as fermentative substrates for lactic acid bacteria. *Int. J. Food Microbiol.*, 45:163–9.

Korsak, N., Clinquart, A. and Daube, G. (2004), *Salmonella* spp. dans les denrées alimentaires d'origine animale: un réel problème de santé publique? *Ann. Méd. Vét.*, 148:174–93.

Kwon, Y. M. and Ricke, S. C. (1998), Induction of acid resistance of *Salmonella* Typhimurium by exposure to short-chain fatty acids. *Appl. Environ. Microbiol.*, 64:3458–63.

Lake, R., Hudson, A., Cressey, P., Wong, T. L. and Gilbert, S. (2004), Risk profile: *Salmonella* (non typhoidal) in poultry (whole and pieces). Inst. Environ. Sci. Res. Ltd, Christchurch Science Centre, Christchurch, New Zealand.

Langkabel, N., Klose, A., Irsigler, H., Jaeger, D., Bräutigam, L., Hafez, H. M. and Fries, R. (2014), Comparison of methods for the detection of *Salmonella* in poultry. *J. Appl. Poult. Res.*, 23(3):403–8.

Lee, L. A., Puhr, N. D., Maloney, E. K., Bean, N. H. and Tauxe, R. V. (1994), Increase in antimicrobial-resistant *Salmonella* infections in the United States, 1989–1990. *J. Infect. Dis.*, 170:128–34.

Line, J. E. (2002), *Campylobacter* and *Salmonella* populations associated with chickens raised on acidified litter. *Poult. Sci.*, 81:1473–7.

Lowry, V. K., Farnell, M. B., Ferro, P. J., Swaggerty, C. L., Bahl, A. and Kogut, M. H. (2005), Purified β-glucan as an abiotic feed additive up-regulates the innate immune response in immature chickens against *Salmonella enterica* serovar Enteritidis. *Int. J. Food Microbiol.*, 98:309–18.

Martelli, F., Gosling, R., Kennedy, E., Rabie, A., Reeves, H., Clifton-Hadley, F., Davies, R. and La Ragione, R. (2014), Characterization of the invasiveness of monophasic and aphasic *Salmonella* Typhimurium strains in 1-day-old and point-of-lay chickens. *Avian Pathol.*, 43(3):269–75.

McGruder, E. D., Ramirez, G. A., Kogut, M. H., Moore, R. W., Corrier, D. E., Deloach, J. R. and Hargis, B. M. (1995), *In ovo* administration of *Salmonella* Enteritidis-immune lymphokines confers protection to neonatal chicks against *Salmonella* Enteritidis organ infectivity. *Poult. Sci.*, 74:18–25.

Mead, G. C. (2000), Prospects for competitive exclusion treatment to control *Salmonella* and other foodborne pathogens in poultry. *Vet. J.*, 159:111–23.

Mead, G. C. (2004), Current trends in the microbiological safety of poultry meat. *World Poult. Sci. J.*, 60:112–18.

Meenakshi, M., Bakshi, C. S., Butchaiah, G., Bansal, M. P., Siddiqui, M. Z. and Singh, V. P. (1999), Adjuvanted outer membrane protein vaccine protects poultry against infection with *Salmonella* Enteritidis. *Vet. Res. Commun.*, 23:81–90.

Milona, P., Townes, C. L., Bevan, R. M. and Hall, J. (2007), The chicken host peptides, gallinacins 4, 7, and 9 have antimicrobial activity against *Salmonella* serovars. *Biochem. Bioph. Res. Co.*, 356:169–74.

Mishra, R. S. and Sharma, V. D. (2001), Comparative efficacy of various toxoids against salmonellosis in poultry. *Vet. Res. Commun.*, 25:337–44.

Mon, K. K., Saelao, P., Halstead, M. M., Chanthavixay, G., Chang, H. C., Garas, L., Maga, E. A. and Zhou, H. (2015), *Salmonella enterica* serovars Enteritidis infection alters the indigenous microbiota diversity in young layer chicks. *Front. Vet. Sci.*, 2:61.

Morris, V. K., Carrique-Mas, J. J., Mueller-Doblies, D., Davies, R. H., Wales, A. D. and Allen, V. M. (2015), A longitudinal observational study of *Salmonella* shedding patterns by commercial turkeys during rearing and fattening, showing limitations of some control measures, *Br. Poult. Sci.*, 56(1):48–57.

Murate, L. S., Paião, F. G., de Almeida, A. M., Berchieri Jr., A. and Shimokomaki, M. (2015), Efficacy of prebiotics, probiotics, and synbiotics on laying hens and broilers challenged with *Salmonella* Enteritidis. *J. Poult. Sci.*, 52:52–6.

Mussatto, S. I. and Mancilha, I. M. (2007), Non-digestible oligosaccharides: a review. *Carbohyd. Polym.*, 68:587–97.

Nagaraja, K. V. and Rajashekara, G. (1999), Vaccination against *Salmonella enterica* serovar Enteritidis infection: dilemma and realities. In A. M. Saeed, R. K. Gast, M. E. Potter and P. G. Wall (eds), *Salmonella Enterica Serovar Enteritidis in Human and Animals*. Iowa State University Press, Ames, USA, pp. 397–404.

Netherwood, T., Gilbert, H. J., Parker, D. S. and O'Donnell, A. (1999), Probiotics shown to change bacterial community structure in the avian gastrointestinal tract. *Appl. Environ. Microbiol.*, 65:5134–8.

Nurmi, E. and Rantala, M. (1973), New aspects of *Salmonella* infection in broiler production. *Nature*, 241:210–11.

O'Regan, E., McCabe, E., Burgess, C., McGuinness, S., Barry, T., Duffy, G., Whyte, P. and Fanning, S. (2008), Development of a real-time multiplex PCR assay for the detection of multiple *Salmonella* serotypes in chicken samples. *BMC Microbiol.*, 8:156.

Padungtod, P. and Kaneene, J. B. (2006), *Salmonella* in food animals and humans in northern Thailand. *Int. J. Food Microbiol.*, 108:346–54.

Park, S. H. and Ricke, S. C. (2015), Development of multiplex PCR assay for simultaneous detection of *Salmonella* genus, *Salmonella* subspecies I, *Salm.* Enteritidis, *Salm.* Heidelberg and *Salm.* Typhimurium. *J. Appl. Microbiol.*, 118(1):152–60.

Parsons, B. N., Crayford, G., Humphrey, T. J. and Wigley, P. (2013), Infection of chickens with antimicrobial-resistant *Salmonella enterica* Typhimurium DT193 and monophasic *Salmonella* Typhimurium-like variants: an emerging risk to the poultry industry? *Avian Pathol.*, 42(5):443–6.

Parsons, B. N., Wigley, P., Simpson, H. L., Williams, J. M., Humphrey, S., Salisbury, A.-M., Watson, A. J. M., Fry, S. C., O'Brien, D., Roberts, C. L., O'Kennedy, N., Keita, Å. V., Söderholm, J. D., Rhodes, J. M. and Campbell, B. J. (2014), Dietary supplementation with soluble plantain non-starch polysaccharides inhibits intestinal invasion of *Salmonella* Typhimurium in the chicken. *PLoS ONE*, 9(2):e87658.

Pas, M. F. W., Hulsegge, I., Schokker, D., Smits, M. A., Fife, M., Zoorob, R., Endale, M.-L. and Rebel, J. M. J. (2012), Meta-analysis of chicken – *Salmonella* infection experiments. *BMC Genom.*, 13:146.

Patterson, J. A. and Burkholder, K. M. (2003), Application of prebiotics and probiotics in poultry production. *Poult. Sci.*, 82:627–31.

Phillips, I., Casewell, M., Cox, T., De Groot, B., Friis, C., Jones, R., Nightingale, C., Preston, R. and Waddell, J. (2004), Does the use of antibiotics in food animals pose a risk to human health? A critical review of published data. *J. Antimicrob. Chemoth.*, 53:28–52.

Plumed-Ferrer, C. and von Wright, A. (2009), Fermented pig liquid feed: nutritional, safety and regulatory aspects. *J. Appl. Microbiol.*, 106:351–68.

Pope, M. J. and Cherry, T. E. (2000), An evaluation of the presence of pathogens on broilers raised on Poultry Litter Treatment®-treated litter. *Poult. Sci.* 79:1351–5.

Popoff, M. Y., Bockemuhl, J. and Gheesling, L. L. (2003), Supplement 2001 (no. 45) to the Kauffmann-White scheme. *Res. Microbiol.*, 154:173–4.

Quinteiro-Filho, W. M., Gomes, A. V. S., Pinheiro, M. L., Ribeiro, A., Ferrazde-Paula, V., Astolfi-Ferreira, C. S., Ferreira, A. J. P. and Palermo-Neto, J. (2012), Heat stress impairs performance and induces intestinal inflammation in broiler chickens infected with *Salmonella* Enteritidis. *Avian Pathol.*, 41(5):421–7.

Rademacher, S. W., Schoop, V. M. and Schluesener, H. J. (1993), Bactencin, a leucocytic antimicrobial peptide, is cytotoxic to neuronal and glial cells. *J. Neurosci. Res.*, 36:657–62.

Rahimi, S., Shiraz, Z. M., Salehi, T. Z., Torshizi, M. A. K. and Grimes, J. L. (2007), Prevention of *Salmonella* infection in poultry by specific egg-derived antibody. *Int. J. Poult. Sci.*, 6:230–5.

Ramasamy, K. T., Verma, P. and Reddy, M. R. (2014), Toll-like receptors gene expression in the gastrointestinal tract of *Salmonella* serovar Pullorum-infected broiler chicken. *Appl. Biochem. Biotechnol.*, 173(2):356–64.

Rasschaert, G., Houf, K. and De Zutter, L. (2007), External contamination of *Campylobacter*-free flocks after transport in cleaned and disinfected containers. *J. Food Prot.*, 70(1):40–7.

Rehman, H., Vahjen, W., Kohl-Parisini, A., Ijaz, A. and Zentek, J. (2009), Influence of fermentable carbohydrates on the intestinal bacteria and enteropathogens in broilers. *World Poult. Sci. J.*, 65:75–90.

Revolledo, L. and Ferreira, A. J. P. (2012), Current perspectives in avian salmonellosis: vaccines and immune mechanisms of protection. *J. Appl. Poult. Res.*, 21:418–31.

Ricke, S. C. (2014), Application of molecular approaches for understanding foodborne *Salmonella* establishment in poultry production. *Adv. Biol.*, Article ID 813275, p. 25.

Russell, S. M. (2010), *Salmonella* intervention strategies at the farm. Georgia Poultry Conference, Athens, Georgia, 28–29 September 2010.

Rychlik, I., Elsheimer-Matulova, M. and Kyrova, K. (2014), Gene expression in the chicken caecum in response to infections with non-typhoid *Salmonella*. *Vet. Res.*, 45(1):119.

Salisbury, A. M., Leeming, G., Nikolaou, G., Kipar, A. and Wigley, P. (2014), *Salmonella* Virchow infection of the chicken elicits cellular and humoral systemic and mucosal responses, but limited protection to homologous or heterologous re-challenge. *Front. Vet. Sci.*, 1:6.

Santos, A. A (2005), Poultry intestinal health through diet formulation and exogenous enzyme supplementation. PhD Thesis, North Carolina State University, Raleigh, NC, USA.

Savvidou, S. E., Beal, J. D., Brooks, P. H. and La Ragione, R. M. (2009), Liquid feed fermented with *Lactobacillus salivarius* reduces susceptibility of broiler chickens to *Salmonella enterica* Typhimurium Sal 1344 nal. *Br. Poult. Abstracts*, 5:43–4.

Schade, R., Gutierrez Calzado, E., Sarmiento, R., Chacana, P. A., Porankiewicz-Asplund, J. and Terzolo, H. R. (2005), Chicken egg yolk antibodies (IGY-technology): a review of progress in production and use in research and human and veterinary medicine. *Atla*, 33:129–54.

Schneid, A. S., Rodrigues, K. L., Chemello, D., Tondo, E. C., Ayub, M. A. Z. and Aleixo, J. A. G. (2006), Evaluation of an indirect ELISA for the detection of *Salmonella* in chicken meat. *Braz. J. Microbiol.*, 37:350–5.

Schneitz, C. (2005), Competitive exclusion in poultry – 30 years of research. *Food Control*, 16:657–67.

Schrezenmeir, J. and de Vrese, M. (2001), Probiotics, prebiotics, and synbiotics approaching a definition. *Am. J. Clin. Nutr.*, 73(Suppl. 2):361S–4S.

Servin, A. L. (2004), Antagonistic activities of lactobacilli and bifidobacteria against microbial pathogens. *FEMS Microbiol. Rev.*, 28:405–40.

Shah, S. N., Kamil, S. A., Darzi, M. M., Mir, M. S. and Bhat, S. A. (2013), Haematological and some biochemical changes in experimental fowl typhoid infection in broiler chickens. *Comp. Clin. Pathol.*, 22(1):83–91.

Slauch, J., Taylor, R. and Maloy, S. (1997), Survival in a cruel world: how *Vibrio cholerae* and *Salmonella* respond to an unwilling host. *Genes Dev.*, 11:1761–74.

Smirnov, A., Tako, E., Ferket, P. R. and Uni, Z. (2006), Mucin gene expression and mucin content in the chicken intestinal goblet cells are affected by in ovo feeding of carbohydrates. *Poult. Sci.*, 5:669–73.

Snijders, J. M. A. and van Knapen, F. (2002), Prevention of human diseases by an integrated quality control system. *Livest. Prod. Sci.*, 76:203–6.

Spring, P., Wenk, C., Dawson, K. A. and Newman, K. E. (2000), The effects of dietary mannanoligosaccharides on cecal parameters and the concentrations of enteric bacteria in the ceca of *Salmonella*-challenged broiler chicks. *Poult. Sci.*, 79:205–11.

Šušković, J., Kos, B., Goreta, J. and Matošić, S. (2001), Role of lactic acid bacteria and bifidobacteria in symbiotic effect. *Food Technol. Biotechnol.*, 39:227–35.

Tako, E., Ferket, P. R. and Uni, Z. (2004), Effects of *in ovo* feeding of carbohydrates and beta-hydroxy-beta-methylbutyrate on the development of chicken intestine. *Poult. Sci.*, 83:2023–8.

Tellez, G., Petrone, V. M., Escorcia, M., Morishita, T. Y., Cobb, C. W. and Villaseñor, L. (2001), Evaluation of avian-specific probiotic and *Salmonella* Enteritidis-, *Salmonella* Typhimurium-, and *Salmonella* Heidelberg-specific antibodies on cecal colonization and organ invasion of *Salmonella* Enteritidis in broilers. *J. Food Prot.*, 64:287–91.

Thormar, H., Hilmarsson, H. and Bersson, G. (2006), Stable concentrated emulsions of the 1-monoglyceride of capric acid (monocaprin) with microbicidal activities against the food-borne bacteria *Campylobacter jejuni*, *Salmonella* spp., and *Escherichia coli*. *Appl. Environ. Microbiol.*, 72:522–6.

Tran, T. P., Ly, T L., Nguyen, T. T., Akiba, M., Ogasawara, N., Shinoda, D., Okatani, T. A. and Hayashidani, H. (2004), Prevalence of *Salmonella* spp. in pigs, chickens and ducks in the Mekong Delta, Vietnam. *J. Vet. Med. Sci.*, 66(8):1011–14.

Travers, K. and Barza, M. (2002), Morbidity of infections caused by antimicrobial-resistant bacteria. *Clin. Infect. Dis.*, 34:131–4.

Tsai, C.-C., Hsih, H.-Y., Chiu, H.-H., Lai, Y.-Y., Liu, J.-H., Yu, B. and Tsen, H.-Y. (2005), Antagonistic activity against *Salmonella* infection *in vitro* and *in vivo* for two *Lactobacillus* strains from swine and poultry. *Int. J. Food Microbiol.*, 102:185–94.

USDA-FSIS (2007), Compliance guideline for controlling *Salmonella* and *Campylobacter* in poultry, Second Edition, Draft, May 2007.

van't Hof, W., Veernan, E. C. I., Helmerhorst, E. J. and Amerongen, A. V. N. (2001), Antimicrobial peptides: properties and applications. *Biol. Chem.*, 382:597–619.

Van Coillie, E., Goris, J., Cleenwerck, I., Grijspeerdt, K., Botteldoorn, N., Van Immerseel, F., De Buck, J., Vancanneyt, M., Swings, J., Herman, L. and Heyndrickx, M. (2007), Identification of lactobacilli isolated from the cloaca and vagina of laying hens and characterization for potential use as probiotics to control *Salmonella* Enteritidis. *J. Appl. Microbiol.*, 102:1095–106.

van Dijk, A., Veldhuizen, E. J. A., Kalkhove, S. I. C., Tjeerdsma-van Bokhoven, J. L. M., Romijn, R. A. and Haagsman, H. P. (2007), The β-defensin gallinacin-6 is expressed in the chicken digestive tract and has antimicrobial activity against food-borne pathogens. *Antimicrob. Agents Ch.*, 51:912–22.

Van Hoorebeke, S., Van Immerseel, F., Schulz, J., Hartung, J., Harisberger, M., Barco, L., Ricci, A., Theodoropoulos, G., Xylouri, E., De Vylder, J., Ducatelle, R., Haesebrouck, F., Pasmans, F., de Kruif, A. and Dewulf, J. (2010), Determination of the within and between flock prevalence and identification of risk factors for *Salmonella* infections in laying hen flocks housed in conventional and alternative systems. *Prev Vet Med.*, 94(1–2):94–100.

Van Immerseel, F., Cauwerts, K., Devriese, L. A., Haesebrouck, F. and Ducatelle, R. (2002), Feed additives to control *Salmonella* in poultry. *World Poult. Sci. J.*, 58:501–13.

Van Immerseel, F., De Buck, J., Pasmans, F., Velge, P., Bottreau, E., Fievez, V., Haesebrouck, F. and Ducatelle, R. (2003), Invasion of Salmonella enteritidis in avian intestinal epithelial cells in vitro is influenced by short-chain fatty acids. *Int. J. Food Microbiol.* 85:237–48.

Van Immerseel, F., Fievez, V., de Buck, J., Pasmans, F., Martel, A., Haesebrouck, F. and Ducatelle, R. (2004), Microencapsulated short-chain fatty acids in feed modify colonization and invasion early after infection with *Salmonella* Enteritidis in young chickens. *Poult. Sci.*, 83:69–74.

Van Immerseel, F., Russell, J. B., Flythe, M. D., Gantois, I., Timbermont, L., Pasmans, F., Haesebrouck, F. and Ducatelle, R. (2006), The use of organic acids to combat *Salmonella* in poultry: a mechanistic explanation of the efficacy. *Avian Pathol.*, 35:182–8.

Vandenbroucke, V., Croubels, S., Martel, A., Verbrugghe, E., Goossens, J., Van Deun, K., Boyen, F., Thompson, A., Shearer, N., De Backer, P., Haesebrouck, F. and Pasmans, F. (2011), The mycotoxin deoxynivalenol potentiates intestinal inflammation by *Salmonella* Typhimurium in porcine ileal loops. *PLoS ONE*, 6(8):e23871.

Vandeplas, S., Dubois Dauphin, R., Beckers, Y., Thonart, P. and Théwis, A. (2010). *Salmonella* in chicken: current and developing strategies to reduce contamination at farm level. *J. Food Protect.*, 73:774–85

Vandeplas, S., Dubois Dauphin, R., Thiry, C., Beckers, Y., Welling, G. W., Thonart, P. and Théwis, A. (2009), Efficiency of a *Lactobacillus plantarum*-xylanase combination on growth performance, microflora populations, and nutrient digestibilities of broilers infected with *Salmonella* Typhimurium. *Poult. Sci.*, 88:1643–54.

Varmuzova, K., Kubasova, T., Davidova-Gerzova, L., Sisak, F., Havlickova, H., Sebkova, A., Faldynova, M. and Rychlik, I. (2016), Composition of gut microbiota influences resistance of newly hatched chickens to *Salmonella* Enteritidis infection. *Front Microbiol.*, 7:957.

Vicente, J. L., Higgins, S. E., Hargis, B. M. and Tellez, G. (2007), Effect of poultry guard litter amendment on horizontal transmission of *Salmonella* Enteritidis in broiler chicks. *Int. J. Poult. Sci.*, 6:314–17.

Videnska, P., Sisak, F., Havlickova, H., Faldynova, M. and Rychlik, I. (2013), Influence of *Salmonella enterica* serovar Enteritidis infection on the composition of chicken cecal microbiota. *BMC Vet. Res.*, 9:140.

Wales, A. D. and Davies, R. H. (2011), A critical review of *Salmonella* Typhimurium infection in laying hens. *Avian Pathol.*, 40(5):429–36.

Wallis, T. S. and Barrow, P. A. (2005), *Salmonella* epidemiology and pathogenesis in food-producing animals. *EcoSal Plus* 2005.

Wang, C. L., Fan, Y. C., Wang, C., Tsai, H. J. and Chou, C. H. (2016), The impact of *Salmonella* Enteritidis on lipid accumulation in chicken hepatocytes. *Avian Pathol.*, 9:1–38.

Wigley, P. (2014), *Salmonella enterica* in the chicken: how it has helped our understanding of immunology in a non-biomedical model species. *Front. Immunol.*, 5. Article 482.

Winfield, M. D. and Groisman, E. A. (2003), Role of nonhost environments in the lifestyles of *Salmonella* and *Escherichia coli*. *Appl. Environ. Microbiol.*, 69:3687–94.

Wisner, A., Desin, T., White, A., Potter, A. and Köster, W. (2012), The *Salmonella* Pathogenicity Island-1 and -2 encoded type III secretion systems. In Yashwant Kumar (ed.), *Salmonella* – A diversified Superbug, InTech, Rijeka (Croatia), p. 562.

Withanage, G. S., Wigley, P., Kaiser, P., Mastroeni, P., Brooks, H., Powers, C., Beal, R., Barrow, P., Maskell, D. and McConnell, I. (2005), Cytokine and chemokine responses associated with clearance of a primary *Salmonella enterica* serovar Typhimurium infection in the chicken and in protective immunity to re challenge. *Infect. Immun.*, 73:5173–82.

Yan, L., Wang, X., Guo, Y., Pei, X. and Yu, D. (2011), Comparison study of enrichment-PCR and traditional method for detection of *Salmonella* in poultry. *Wei Sheng Yan Jiu*, 40(3):348–51.

Yazdani, A., Asasi, K. and Nazifi, S. (2015), Evaluation of acute-phase proteins and inflammatory mediators changes in native chickens experimentally infected with *Salmonella* Typhimurium. *Comp. Clin. Pathol.*, 24(4):733–9.

Chapter 3

Safety management on the poultry farm

Jungsoo Joo, University of Maryland, USA; Aishwarya Pradeep Rao, University of Maryland and University of Arizona, USA; and Debabrata Biswas, University of Maryland, USA

1 Introduction

Flock health and infectious disease control, as well as meeting the demand of a growing and competitive market while adhering to high standards for the safety of products, are important for poultry farming, a key component of the US and global economies. The challenge to develop and implement management policies that most effectively focus on all the above and ensure a safe and secure egg and meat supply is even more complex than before due to the increasing concentration and intensification of poultry farming to meet burgeoning consumer demand. Furthermore, foodborne microbial pathogens transmitted via animals or animal products can result in disease outbreaks and human illnesses, and are primary food safety concerns (Goodwin and Shiptsova, 2002). According to a report published by the Centers for Disease Control and Prevention (CDC), poultry are the primary reservoir for two major foodborne bacterial pathogens, *Campylobacter* and *Salmonella* (Hanning et al., 2010; Salaheen et al., 2014, 2015, CDC, 2015). Continuously assuring consumers that their poultry products are safe will thereby remain a persistent issue for the poultry industry for the foreseeable future.

Globalization and technological advancements have increased export opportunities for the US poultry, but, via import of animals, the risk of introducing foreign animal diseases and zoonotic pathogens into the food supply chain could be economically devastating to the poultry industry. New technological advances in web-based communications are more important than ever for rapid identification of food safety issues and limiting prolonged

http://dx.doi.org/10.19103/AS.2016.0010.04

economic impact. To protect animal industries and consumers from poultry- and poultry-product-borne illnesses, farm safety management practices, including sanitary and phytosanitary (SPS) standards, are now included in most trade agreements. Phytosanitary requirements can be trade distorting and protectionist, accentuating, however, the need for harmonizing both standards and enforcement within regions and among trading partners around the world.

Most food-processing sectors have adopted essential control strategies comprising Hazard Analysis and Critical Control Points (HACCPs) and implementation of approved Standard Operating Procedures (SOPs), among other practice guidelines, in order to minimize risk potential during slaughter and processing. However, not all production sectors, specifically medium and small farms, are mandated to implement quality assurance protocols to address product quality and food safety issues, such as worker training, sanitation measures and proper farm management to reduce the colonization and infection of poultry with zoonotic pathogens in live flocks. Numerous outbreaks and foodborne illnesses linked to the poultry industry have accentuated the need for the integration of traceability and quality assurance systems into the supply chain. Due to the increase in globally exported and imported foods, a more stringent system is required to align a wide variety of regulatory systems in order to render them compliant to all inspection standards, even when producers are not required by regulations to do so. Even so, the potential for chaos remains if consistency is not maintained in food safety labelling, even as the globalization of food trade makes for a broader range of choices.

2 Safety practice in hatching and transport of chicks, sources and breeders

Producing and maintaining a healthy flock requires good husbandry practices, including the selection of healthy chicks. Production of high-quality, healthy birds from top-grade eggs depends upon careful attention to hatch, including environmental management, a practice that starts from the selection and purchase of eggs and ends with the delivery of chicks to their final destination (Kim and Kim, 2010). Major factors that are important for hatching eggs, which produce high-quality chicks are (i) breed/strain(s) of parent flock; (ii) quality of feed, water and housing of animals; (iii) vaccination, age and health performance of flock; (iv) size, weight, nutrient content and shell parameters of the eggs; (v) collection, disinfection, storage and transportation of eggs to the hatching facility; and (vi) proper management overseeing the hatching facility and hatch. The potential for contamination with microorganisms of eggs used for hatching starts as soon as an egg has been laid. Best safety practices are critical to maintaining chick health and preventing exposure to potential pathogens.

2.1 Breed/strain of parent flock

The genetic makeup and age of the parent stock/layer is important in selecting eggs for hatching. In addition, good layer rearing practices are also vital. In some cases, colonization and infection of parent stock with zoonotic and poultry pathogens can result in vertical pathogen transfer to the offspring. In addition to genetic factors, a range of maternal and early life environmental factors also influence the development, physiology and behaviour of the progeny (Lickliter, 2005).

2.2 Quality of feed, water and housing of flock

The parent stock flock must be housed in well-maintained, clean and properly ventilated houses with correct indoor temperature and light cycles and reliable litter removal systems. Contamination level and/or pathogenic microbial load in the nest can also play an important role in pathogen transmission to eggs and chicks. During or immediately following laying, pathogens can permeate the shell and enter into the egg. Feed and water are the most frequent sources of microorganisms (Dixon et al., 2016; Kim and Kim, 2010). Water and feed quality and their supplying systems must be carefully and continuously monitored. Ventilation is the most efficient method to reduce the amount of microorganisms in the air, which may also contaminate the hatching egg. Correct control and maintenance of airflow can significantly reduce levels of flies, dust and other pollutants in the hatching facility.

2.3 Vaccination, age and health performance of parent flock

Prenatal and neonatal periods have been found to be extremely important in shaping adult bird phenotypes (Rutherford et al., 2012). For chickens, both the pre-lay and egg-laying stages can influence post-hatch development, individual chick health performance, and morbidity and mortality. Lickliter (2005) and Henriksen et al. (2011) report that certain factors in the prenatal environment can have lasting effects on how the adult offspring respond to their housing environment. Maintaining uniformity and correct body weight is essential in feeding of the parent stock. Combined feed formulation and feeding management can ensure optimal nutrition and body weights in the parent flocks. It has already been shown that nutrition is the deciding variable facilitating productivity and profitability of the parent flock, as well as in the production of top-grade eggs, and, of importance, flock resistance to various infectious diseases (Henriksen et al., 2011).

Good safety management and biosecurity measures can prevent many diseases in parent flocks, but vaccination provides the best opportunity for a parent flock with potential exposure to infectious microorganisms to mount robust immunological defences and be actively protected against the spread of disease. Common diseases, including Marek's disease, Newcastle disease (ND), infectious bronchitis, avian encephalomyelitis and their causative agents, should be routinely monitored for and vaccination strategies should be implemented as the situation requires. Vaccination can prevent diseases when biosecurity measures alone do not provide the necessary protection.

2.4 Size, weight, nutritional content and shell of egg

Egg structure and its nutrient content/composition vary depending upon the breed/strain and individual layers. Egg structure including size, shell thickness and pore size, shape, shell membrane and albumin-to-yolk ratio are important factors in predicting hatch and health performance of the chick.

2.5 Collection, disinfection, storage and transportation of eggs to the hatching facility

The process of hatching egg collection from nests frequently introduces contamination with poultry disease-causing as well as zoonotic pathogens. Use of clean and disinfected containers and proper safety measures taken by the egg handler with regard to outer garments and correct glove use or hand-washing practices are also key to controlling microbial contamination from fomites. Disinfection of eggs with appropriate cleaning solutions and use of clean and disinfected vehicles for storage and transport of hatching eggs from breeder farm to hatchery also must be carefully monitored. It is essential that the egg processing building or area is designed, placed and constructed of materials that allow for proper egg sanitation procedures as well as easy and effective routine sanitation of the facility itself.

2.6 Managing the hatching facility and hatch

The hatchery environment and its hygienic conditions are another potential sources of contamination involving a variety of microorganisms that can cause diseases in a poultry farm and lead to the colonization of zoonotic pathogens resulting in infection by foodborne pathogens (Sheldon and Brake, 1991; Scott and Swetnam, 1993). In addition, a contaminated hatchery can cause significant economic losses for the poultry industry by increasing the mortality rate of newly hatched chicks (Funk and Irwin, 1955; Harry and Gordon, 1966). Therefore, an HACCP system specific to the hatching facility is essential to maintaining hatchery hygiene and hygiene of its equipment and facilities (Jordan et al., 2001). The hatchery's construction should be as vermin-proof as possible for protection against rodents, wild birds and insects that can be destructive as well as disease vectors. Sometimes, use of insecticides is necessary. Proper cleaning of vegetation, debris and unused equipment with disinfectants is important. Water source and its coliform load must be frequently monitored. Restricted Entry signs must be placed at critical points and entrance/egress must be controlled by a gatekeeper or locked door. Foot baths are essential and should contain a strong disinfectant. Positive airflow minimizes the risk of transfer of contamination from the hatcher back to the setters and the egg storage room. Use of air vents, positive pressure fans, exhaust fans and maintenance of negative air pressure in the dirty part of a hatching facility can improve hygienic conditions and control flies and other airborne pests. Producing the highest quality chicks requires regular monitoring of the microbial status of the hatchery, including the floor, waterers, incubators, batteries, and all other instruments and equipment. Various methods such as conventional culture or PCR techniques can be used to track bacterial and fungal levels.

3 Farm safety issues and recommendations for managers

The long-term economic, social and environmental sustainability of poultry farming requires proper on-farm safety management practices for health of the live flocks as well as the production of safe poultry products for consumers. A safe, nutritious and sustainable poultry product can be produced by following recommended farm management practices.

3.1 Worker/personnel safety management

Safety on the farm is best implemented through rules and regulations developed for all processes as well as by proper training of appropriate personnel for specific functions on the farm. It is of extreme importance that all personnel employed by the farm, regardless of specific occupations or tasks, be trained in safe, humane and approved procedures for handling of birds. They must also be adept at proper disposal of waste and hazardous materials that might otherwise cause disease in the birds or other personnel and in order to avoid any cross-contamination. One of the most effective ways to ensure a successful hygienic workflow is to ensure that all employees practice exemplary personal hygiene in addition to effective sanitation procedures on the farm, combined with the practice of meticulous record-keeping of all products used as well as any breach of protocols that could lead to safety problems. All employees at the farm need to be well-versed in all HACCP plans, SOPs and Good Manufacturing Practices (GMPs) that are in place at the processing facility.

Supervisors of all employees must take charge in ensuring that all safety protocols are adhered to. They must hold all employees to the highest standards of personal hygiene, and enforce absence from the processing facility if they show signs or symptoms of illness. All personnel involved in processes that may directly or indirectly affect product quality must practice sound personal hygiene such as effective hand-washing and garment maintenance regimes. Hygienic hand practices ranging from regular hand-washing, efficient cleaning products and monitored glove usage are of increased importance with poultry as it is considered to be a high-risk food. Proper work attire includes the use of hairnets, gloves and coats in order to minimize physical and biological hazards (MPPU Guide).

3.2 Farm environmental safety management

Maintaining an environment conducive to good safety practice is important not only in the hatching facility but also throughout the farm, processing areas, and slaughtering and other harvesting areas. Environmental safety is a key component in personnel and animal safety. Quarantine rooms are essential to prevent cross-contamination, allowing personnel who have been in contact with potential hazardous agents to be isolated for defined time windows prior to resuming interaction with animals. These rooms may also be used for birds and animals that are or may be sick or diseased in order to keep infections from spreading to the rest of the flock. Waste and manure management systems must be carefully maintained and serviced with detailed records of all procedures. Air systems and filters need to be constantly monitored and frequently changed to prevent cross-contamination between facilities. The macroenvironment of the birds and other animals must be regularly monitored and potential hazards such as loose wiring, manure, blood or feathers must be promptly eliminated. All temperature-controlled areas should be monitored for fluctuations in temperature and humidity outside predefined set points. All facilities must have appropriate signs and symbols indicating usage and contents, as well as requirements for personnel and environmental parameters.

3.3 Wild animal and bird infiltration management

Animals and birds on a poultry farm must be treated in the most humane way possible. All personnel must be trained to respect and handle the birds in a way that minimizes stress and harm. Also, keeping the birds healthy is the first step towards preventing any diseases that may result in loss of birds. To prevent infections by pathogenic bacterial pathogens such as *Salmonella*, *Campylobacter* and other microorganisms, only authorized personnel in proper clothing and protection ought to be allowed access to birds. Young birds are particularly prone to infections and so must be kept separate from adult birds. Supervisors must also ensure that feed and water access for the birds is regulated and monitored in order to prevent overfeeding and contamination. Since most of these pathogens are normal poultry microbiota, it is impossible to completely eliminate them, but strategies must be in place to reduce cross-contamination risks. All poultry farmers should monitor their facilities for potential routes of pathogen exposure, as well as ensure personnel are not exposed to sick birds during culling or processing. Vaccination and regular veterinary care along with record-keeping of all incidences and treatments must be supervised and mandated by authorized personnel. It is also crucial to employ specialized personnel trained to identify and prevent infiltration of pests in every facility on the farm, as well as deploy corrective measures to minimize damage to the facility, personnel and, more importantly, the flock.

3.4 Pest control

Pest control means regulation of pest infiltrations down to acceptable levels, as total pest eradication is unrealistic. Pests can cause stress to birds, spread disease, steal feed, damage structures and equipment, and annoy neighbours. Pests of primary concern are flies, beetles, lice and mites (Darre and Rock, 1995). The common pest and their involvement in transmitting diseases/pathogens are listed in Table 1. While pests cannot be completely eliminated, it is possible – and necessary – to maintain minimal numbers through manure management, sanitation and pesticide treatment.

Flies are the most common pests found in poultry farms, as they associate with manure. Flies can transmit microorganisms such as bacteria, viruses, and fungi, and spread infectious disease among animals. In addition, flies are a nuisance both on the farm and in neighbourhoods, and if unchecked, can cause litigation issues (Strother, 2008). Therefore, fly control is not only important for biosecurity but also for maintaining a good relationship with the surrounding community. Fly populations can be controlled by means of dry manure management, utilization of natural fly predators and chemical intervention. Beetles and their larvae carry many poultry pathogens and parasites, particularly Marek's disease (Strother, 2008). Beetle control is difficult but can be achieved by insecticide treatment of all wooden components of poultry housing.

The most common ectoparasites of poultry are mites and lice (Darre and Rock, 1995). Ectoparasites limit bird growth and egg production, damage plumage and cause skin irritations resulting in undue stress levels in the bird. Large numbers of bloodsucking mites, in addition to causing stress, can significantly reduce weight and seminal fluid volumes (Kaufman et al., 2000). Mites most often populate surrounding woodwork, clods of dirt, manure or even litter during the day and feed on animals at night. Pesticides are the most effective for eliminating mites and lice. Different types of pesticides such as dusts, wet table powder, liquid sprays and resin strips are available depending on the types of infestations that must be controlled (Mccrea et al., 2005).

Wild birds are of particular concern, as they can spread several infectious diseases including mycoplasma pneumonia, salmonellosis, avian influenza (AI) and exotic ND. Of importance, AI and ND are virus mediated and thus considered the most dangerous infectious poultry diseases in the world. These viruses can live on manure, egg containers, farming materials, processing equipment, and people and can infiltrate not just a single farm or community but also cross local and international barriers and spread throughout the global market (APHIS, 2014). Rodents are also highly mobile vectors for bordetellosis, leptospirosis, erysipelas, salmonellosis, fowl pox and fowl cholera (Loven

Table 1 Common poultry pests and their role in transmitting pathogens to flocks

Pest			
Type	Spices	Pathogens	References
Flies	• *Musca domestica* (House fly) • *Hydrotaea* spp. (Black dump fly) • *Fannia canicularis* (Little house fly) • *Calliphoridae nigribarbis* (Blow fly) • *Sphaeroceridae* (Small dung fly)	• Bacteria (*Salmonella* spp., *Campylobacter* spp., *E. coli* etc.) • Viruses (New castle disease virus, avian influenza A viruses etc.) • Rickettsia, protozoa, fungi, parasites	• Förster et al., 2007 • Sawabe et al., 2011 • Bahrndorff et al., 2013
Beetles	• *Dermestes maculatus* (Hide beetle) • *Alphitobius diaperinus* (Lesser mealworm)	• Bacteria (*Salmonella* spp., *E. coli* etc.) • Virus (Marek's disease virus, fowl pox virus etc.) • Parasite (tapeworms)	• Kaufma et al., 2000 • Zheng et al., 2012
Mites	• *Dermanyssus gallinae* (Red Mite)	• Bacteria (*Salmonella* spp., *E. coli, Staphylococcus* spp., *Streptococcus* spp. etc.) • Viruses (Newcastle disease virus, fowl poxvirus, Marek's disease virus etc.)	• Sparagano, et al., 2014 • Huong, et al., 2014
Lice	• *Menacanthus stramineus* (Chewing lice) • *Menacanthus cornutus* (Chewing lice)	• Itself is pathogen (anemia and heavy multi-focal lesion)	• Audi and Asmau, 2014
Rodents	• *Mus musculus* (House Mouse) • *Rattus norvegicus* (Brown rat) • *Rattus rattus* (Black rat)	• Bacteria (*Salmonella* spp., *Bordetella avium, Leptospira* spp., *Erysipelas* spp. etc.) • Virus (fowl pox virus, fowl cholera virus)	• Loven and Williams, 2010
Wild birds	• Numerous local birds and migratory birds	• Bacteria (*Mycoplasma* spp. etc.) • Virus (New castle disease virus, avian influenza A viruses etc.) • Parasite (mites and lice)	• Darre and Rock, 1995

and Williams, 2010). To protect poultry from wild bird and rodent-mediated diseases, a four-step control process is required, which includes sanitation, wild-bird- and rodent-proof construction, external animal population reduction, and constant re-evaluation (The Poultry Site, 2015).

3.5 Housing, cleaning practices and disease control

Suitable housing is important for poultry health, welfare and optimal production. Poultry may be raised in non-caged (broiler, meat) and caged (layer, egg) production systems. Regardless of the system, poultry management must ensure proper ventilation, lighting, temperature, adequate space and clean litter. Cleaning and disinfecting the poultry house between flocks is of primary importance for the prevention of infection as well as maintaining a good relationship with the surrounding community. Commercial (large-scale) broiler and layer production systems have different housing needs from small-scale farm housing. For small-scale poultry production, portable houses are the best; however, this is not practical for large-scale commercial poultry farming, which requires special attention to planning and design in order to maintain proper facility sanitation (Fanatico, 2007), which is essential for keeping poultry birds healthy and productive.

Broiler houses are large-span structures bedded with litter (straw, wood chips or wood shavings, etc.) on the floor whereas layer farms use cages organized into two to four tiers, with several birds raised in each cage and 5–10 layers occupying each colony (Stewart-Brown, 2015). Broiler litter floors are less likely to attract flies because they remain dry; however, this system does attract beetles. Conversely, in layer farms, manure accumulates with the litter under the cages and is more inviting to flies (Kaufman et al., 2000). In either situation, cleaning and disinfection of housing facilities are critical to poultry health and welfare, as well as biosecurity. Proper cleaning involves removing litter and dirt from a surface and cleaning with water and detergent first, followed by disinfection with water and a disinfectant. The performance of these procedures in this precise order is the most economical use of labour and materials. Once removed, litter should be dried and stored or disposed of as far as possible from the poultry house (Bolder, 2002).

Optimal stocking density is essential for the poultry health and welfare. Too many birds raised in the same house is stressful with regard to animal interactions and results in increased temperature and humidity, which increase the likelihood of flock morbidity. Optimal space parameters for meat strain (broilers) and egg strain (layers) birds are provided in Tables 17 and 18 in Stewart-Brown (2015), which must be taken into account when designing poultry housing. Light cycle management is another critical factor for bird activation; proper lengths of light periods are necessary for brain stimulation of seasonal reproduction, while dark periods are required for melatonin production and rest (Fanatico, 2007). A good ventilation system will maintain air quality and litter condition. Optimal O_2, CO_2, CO and NH_3 concentrations, as well as humidity, must be carefully controlled as increased levels of NH_3, CO_2 and humidity can trigger ascites disease, foot pad and eye burns, breast blisters and other skin irritations, and decreased weight in birds. (Cobb, 2013).

3.6 Management of feed quality and safety

Nutrition requirements for poultry depend upon strain or breed as well as their life cycle stages. Provision of complete and balanced nutrition improves poultry health and product quality. The USDA specifies select nutrient requirements for various poultry and farming systems (Sutton et al., 2003). Texas A&M University (College Station, TX) supplies more detailed nutrient requirements (Lee, 2003). Nutrient imbalances or deficiencies reduce growth and increase levels of morbidity and mortality (Lee, 2003).

Animal feed itself can be a source of various microorganisms, in addition to those naturally introduced by soil, rain or insects, yet most have little effect on poultry health (Maciorowski et al., 2007) Harmful microorganisms can proliferate in feed, however, due to a contaminated farm environment, improper manure management and pests. Therefore, feed should be dried rapidly and stored for short time periods, and supplemented with zinc bacitracin, bacteriophages and bacteriocin- or propionic acid-producing bacteria, as well as mineral acids, short-chain fatty acids, isopropanol, aldehydes and trisodium phosphate to inhibit microbial pathogen colonization (Maciorowski et al., 2007).

3.7 Water safety and risk management

Water is the most important nutrient for animals, and birds should have unlimited access to clean and high-quality water for poultry healthy and product quality. An *ad libitum* (unlimited) water supply is essential, and the USDA has specific recommendations for the number and placement of drinkers (CFSPH, 2013). Water quality determination consists of several factors including total number of bacteria, pH and mineral elements (Ca, Na, Fe, etc.). Water quality testing should be performed on a periodic basis no less than once per year. Whenever bacterial contamination is detected or prior to the arrival of new birds, water sanitation and water line cleaning procedures are a must for protection against microbial colonization and biofilm build-up (Cobb, 2013). Sanitizer efficacy can be altered by water pH, mineral content and bacterial load; therefore, it is critical to choose the correct sanitizer for the types of water contaminants present (Fairchild, 2009).

3.8 Regular maintenance and record-keeping

All employees on a poultry farm must be well versed and compliant with all safety regulations, which themselves should be constantly reviewed and updated by farm supervisors; moreover, farm safety management is a separate field that requires trained and dedicated personnel. Record-keeping protocols are key, ensuring detailed accounts are maintained and up-to-date information is always available to supervisors, inspectors and other employees at all times. Maintenance of records includes information on sources of birds, and dates of veterinary health checks and treatments administered, as well as tests and test results. In addition, all equipment must undergo regular maintenance checks, inspections and servicing in order to ensure smooth functionality and minimize safety incidents or other mishaps.

4 Farm biosecurity management

Biosecurity management is one of the most important and integral processes in poultry farming. Farming environments must be carefully planned and designed as to allow for effective biosecurity protocols prior to initiating any poultry production enterprise. Biosecurity failures can lead to the spread of pests and disease, and even depopulation of entire flocks. If infectious diseases spread to other farms, the producer is at risk for loss of credibility for their poultry products, causing decreases in product consumption and potentially disastrous income losses.

Biosecurity involves protection from disease through utilization of best practices for safety and sanitation in the farming environment (APHIS, 2014). Poultry biosecurity management such as prevention of infection diseases, injury and illness monitoring, and disposal of biohazardous treatments and waste (vaccines, medications and disinfectants, etc.) are essential for smooth operation of profitable poultry farming. Regular monitoring is the most important and basic procedure for biosecurity management, and early and accurate diagnosis of disease can rescue an entire flock as well as prevent the spread of disease to other farms. Mobley and Kahan (2007) provide an excellent primer of common symptoms and possible causes for common diseases. Employees should be well-acquainted with the symptoms of ill-health which must be immediately reported to facility health management and or the veterinarian, and clear, easy-to-follow procedures should be in place for prompt reporting of disease outbreaks to responsible parties such as the Agricultural Extension office, local and state agricultural departments, and the USDA, as well as local animal health diagnostic laboratories (APHIS, 2014).

4.1 Health monitoring and disease prevention

Poultry infectious disease control

Viruses, bacteria, fungi and parasites cause poultry infectious diseases, and in addition to carriage by vectors such as pests, rodents, and wild birds, trucks, equipment, and people can harbour and transmit these microorganisms as well. Regular monitoring and sanitation of all potential disease entry routes are necessary strategies for prevention of infectious disease and following adverse events. Symptoms of common infectious diseases in poultry are listed in Tables 1, 2 and 3 in Butcher et al. (2015).

Viral infection

The most serious viral infectious diseases are AI and ND virus (NDV). AI type A viruses that cause AI are classified depending upon intensity of the pathogen, either low pathogenic (LPAI) or high pathogenic (HPAI) viruses (Lee et al., 2016). AI viruses carried by migratory birds can infect domestic poultry and have resulted in the death or depopulation of around 50 million birds in the United States since first being identified (Lee et al., 2016). Asymptomatic infected migratory birds can nonetheless rapidly spread disease through flocks, as well as carrying illnesses between facilities. This zoonotic virus can also infect humans; the largest recorded AI outbreak, in 2004, was caused by H5N1, an HPAI, and took the lives of 37 people (WHO, 2005); as such, AI is a serious threat to poultry farms and public health.

ND is caused by paramyxoviridae, an avulavirus, and classified based on virulence: lentogenic (low virulence), and meso- and velogenic (both highly virulent) (USDA, 2016). ND is transmitted between birds via faeces and feathers, although transmission routes can also include egg containers, equipment, vehicles and people. Infected poultry exhibit nervous symptoms such as gasping, coughing, drooping and dragging of wings and legs, head and neck twisting, and circling and swelling of the eyes and neck. While ND transmission can be inhibited via pest control, wild bird control, disinfectants, and proper hygiene and sanitation, there is no treatment currently available for the disease (USDA, 2016).

As treatment options are limited, effective biosecurity programmes must be in place in order to restrict the routes for transmission or spread of these infections, including contact with wild or infected birds, contaminated equipment, vehicles and people (Capua and Marangon, 2009). To reduce wild bird disease transmission, it is important to promptly remove standing water, prevent livestock contact with untreated surface water, reduce or eliminate potential wild bird food resources, remove external perching and nesting venues, and install mock predators such as raptor statues or scarecrows (USDA, 2015). It is also helpful to refrain from keeping or coming into contact with pet birds, not to visit other poultry farms unless necessary, restrict access by persons who have been out of country within the previous seven days and post clear biosecurity signage at farm entry points dictating biosafety protocols including proper attire (Martin, 2011).

Bacterial infection

The most pathogenic bacteria in poultry is *Salmonella* spp. This causes pullorum disease and fowl typhoid in chickens and turkeys. Pullorum disease is most prevalent among poults (under 3 weeks of age) whereas adult birds are more resistant to infection by *Salmonella* spp. Symptoms of infection include lethargy, poor grooming, white diarrhoea and laboured breathing. Fowl typhoid usually occurs in young adult animals (past 12 weeks of age). Symptoms include reduced feed consumption, increased water consumption, and green or yellow diarrhoea. *Salmonella* is highly transmissible and can be found on birds, eggs, containers, equipment, vehicles and people (Berry and Whitenack, 1991)

Parasitic infection

Parasite infections cause blackhead disease and coccidiosis. Blackhead disease is transmitted via contaminated water, feed and eggs. Infected birds are lethargic and develop yellow diarrhoea. It results in high mortality among young turkey poults, although less so in chickens. Chemotherapeutics are effective against blackhead disease and can be given in starter and grower turkey feeds (MDS, 2013). Genus *Eimeria* causes coccidiosis transmitted primarily through faeces which contain sporulated *Eimeria* oocysts. A healthy immune system is the best defence against coccidiosis, and vaccines against the disease as well as medication with ionophore antibiotics are beneficial to the promotion of sustainable immunity (MDS, 2013).

4.2 Litter and carcass management

Proper disposal of litter and carcasses through composting, rendering, in-ground pits, incineration or burial is a key component of biosafety in poultry farming practice (The Virginia Farmstead Assessment System, 2009). During an outbreak, complete depopulation and carcass destruction, followed by total facility cleaning and disinfection, are recommended for halting the spread of the disease. Bird carcasses can be disposed of via on-site burial, incineration, rendering and landfilling, although incineration is considered the most effective (Air Burner, 2015; Flory et al., 2006; USDA, 2015). Used litter also must be disposed of in the same manner as bird carcasses, along with any objects or equipment that cannot be properly cleaned and disinfected.

4.3 Training and emergency preparedness

Emergencies such as mortalities, manure leakage, hazardous materials spills and natural disasters all require specific and well-outlined action plans. Farm producers should develop emergency action plans for several emergency situations. All procedures must be explained and demonstrated to employees and posted in a common location, with frequent retesting and refreshment courses. SOPs for detailed emergency action plans are essential (FAD PReP, 2013).

5 Regulatory acts, enforcement and monitoring

All processes and personnel on a poultry farm must be regularly monitored by the respective authorities. Records of all incidences and corrective measures as well as routine maintenance tasks must be available to all inspecting agencies. A fully compliant poultry farm requires adherence to three tiers of regulations overseeing all protocols and personnel. These three tiers comprise SPS measures, SOPs and SSOPs (Sanitary SOPs), and the HACCP systems.

Though the USDA, FDA and State Departments of Agriculture are primarily involved in enforcements, certain farms are exempt from regular inspections and other regulatory systems due to their size and annual gross incomes. On an eligible poultry farm, both FSIS (Food Safety and Inspection Service) and APHIS (Animal and Plant Health Inspection Services), divisions of the USDA, regulate and oversee the enforcement and approve of SOPs and HACCP plans.

5.1 SPS measures

The main aim of having effective SPS protocols in place in a dynamic environment such as a poultry farm is to ensure food safety as informed and enacted through current scientific techniques. The effective SPS measures involve numerous regulatory interactions that entail authorities looking into the proper import and export of production animals and breeding stock, as well as quality of chickens and other birds with respect to consumer preferences and health benefits. According to the 2014 SPS report, policies placed under the SPS umbrella were explicitly designed to preserve animal and bird health, rather than affect competition with local producers, by preventing the spread of plant and animal diseases. Whereas on the national level, these regulations are applied at international ports of entry, at the farm level, these measures act to prevent the spread of disease between farms as well as among animals within a facility. Sick birds are immediately quarantined and other birds in the flock are given preventative antibiotics as necessitated by duration of exposure. Regular tests are carried out to ensure the absence of toxins and other harmful substances that may alter product quality or otherwise adversely affect the consumer. All tests are performed with certified reagents and chemicals and duplicated for accuracy, accompanied by detailed record-keeping.

With regard to individual bird health, regular veterinary evaluations are performed along with quality and quantity of feed monitoring. Also, the dosages, types and lengths of treatment of antibiotics are regulated in order to prevent any harmful effects being passed on to meat consumers. Generally, only generally recognized as safe (GRAS) materials are used with strict adherence to record-keeping in order to maintain transparency.

5.2 SOPs and SSOPs

Poultry production is a dynamic industry that involves many processes from raw material to finished product, each of which must be regulated and monitored in order to minimize stress and improper handling. Both SOP and SSOP extend into all aspects, including bird housing maintenance and repair, feed and water apparatus maintenance and cleaning with certified chemicals, and macroenvironment monitoring for the presence of any harmful or toxic chemicals. Further, SOPs also apply to feeding schedules and regular removal of manure, feathers and blood to facilitate maintenance of a disease-free environment, and to inter-farm and intra-farm interactions and personnel for prevention of cross-contamination from sources within and from other facilities. Visual inspections as well as microbiological testing procedures are established and updated as per requirements by most of the processing facilities. Of paramount importance is record-keeping of all of the above activities in order to provide proper documentation upon inspection.

In processing level, SOPs for clothing and hygiene extend to personnel with both direct and indirect contact with animals, including those employed for cleaning and maintenance. All facilities, including egg collection centres, must maintain SOPs for regular sampling and testing to uphold standards and keep proper records of all procedures carried out and tests performed. Both SOP and SSOP for all procedures and routines are formulated by specially trained personnel and extend to employee training and continued education. Feed and water quantities and timings are built into SOPs as directed by the veterinarian. Any process or facility that is temperature controlled requires its own set of SOPs for monitoring of all parameters.

To extend the processes, SOPs are also needed to carry out before, during and after harvest. For farms, most SOPs are directed towards pre-harvest procedures. These must be carefully planned and executed, and constantly reviewed and updated as indicated by relevant on-farm processing and mobile poultry processing unit (MPPU) guides.

5.3 The HACCP system

HACCP is globally acknowledged as the best system for ensuring product quality and safety in addition to its utilization as a tool to control hazards that can arise in food production and safety (Khandke and Mayes, 1998). The HACCP plan consists of a systematic approach comprising the identification, evaluation and control of all hazard vulnerabilities including biological, chemical and physical hazards at all stages of the food production line (Codex Alimentarius, 1997). It has been aptly described as a dependable and dynamic control system able to implement both preventative and intervention control strategies (Jouve, 1998). Further, HACCP plans have also been applied at various stages of production and processing to control the risk and spread of pathogens, particularly during slaughtering and processing.

Conducting a hazard analysis for possible risks in a system is the first step in executing a comprehensive HACCP plan. A 'hazard' is defined as 'the unacceptable contamination, unacceptable growth and/or unacceptable survival by microorganisms of concern to safety or spoilage and/or the unacceptable production or persistence in foods of products of microbial metabolism (e.g., toxins, enzymes, biogenic amines)' by the ICMSF. In 1997, the USDA defined a food safety hazard as a physical, chemical or biological property that renders a given food or that in which it is contained unfit for human consumption. The HACCP approach enables a producer to assess, address and control virtually all potential hazards associated with a particular product, thereby significantly reducing the risks associated therein (Goodrich et al., 2005). Food production industries inherently involve many processes, each with its own set of hazards; for example, raw materials that may have an association with pathogens, possible sites of contamination in the processing line and any breakdown in the chain of bird and product handling could lead to the growth and persistence of pathogenic and spoilage microorganisms.

Potential hazards are classified into three categories, namely, physical, chemical and biological. Physical hazards include equipment and facility materials such as wood, metal, glass, stones and large pests. Interaction with such hazards can result in accidental contamination and poor handling practices may be directly responsible for causing illnesses following human consumption (Harris, 1999). Washing and visual examination of foods are two ways in which consumers can protect themselves from such hazards, but at the industry level, more advanced methods such as metal detectors and low energy X-ray machines are employed. Training of food handlers and restrictions on jewellery and clothing are also useful control points.

Chemical hazards include chemicals that have been added to or are present in the meat. These chemicals, such as nitrates and preservatives, are considered as hazards only if found above permissible limits. Other chemicals such as pesticides used in the birds' surroundings and antibiotics administered to the birds must also be carefully managed. Chemical hazards are divided into two categories: naturally occurring toxins and added synthetic toxins. Naturally occurring toxins of concern are aflatoxins, whereas added and synthetic chemicals pertain to additives, as well as paint residues and machine lubricants. Since regulation of potentially toxic animal metabolites is virtually impossible, primary focus is given to controlling raw materials, processing equipment and water sources for the prevention of contamination.

Biological hazards comprise all pathogenic bacteria that have been implicated in food spoilage and food poisoning such as *Salmonella*, *Escherichia coli* and *Clostridium* species among bacteria, and fungi such as *Aspergillus*, *Penicillium* and *Rhizopus*, among others. *Salmonella* continues to be the leading cause of foodborne illness in the poultry industry, theoretically as a result of improper handling of raw materials along with improper hygiene and improper conditions for cooking, cooling and storage. As such, all food handlers must be well informed and trained on the routes and implications of these hazards and how to prevent them (Forsythe and Hayes, 1998).

The poultry industry comprises many individual processes that each must be carefully regulated to minimize contamination risk. Analysing the critical points in a process and implementing corrective procedures are two crucial principles of any HACCP plan. For instance, in preparing for slaughter, personnel must ensure that birds are free of external faecal matter and given limited or no feed and water in the 24 hours prior to euthanasia. During the killing and evisceration procedures, chickens must be hosed down to remove any remaining residue of faecal matter and dirt to minimize pathogen transfer between birds or from birds to personnel. Equipment is to be cleaned between each batch to reduce contamination risk and facilitate traceability. All procedures such as scalding and cooling that involve water and other products require specific temperature ranges and water quality that must be monitored regularly. During weighing, packing and labelling, it is imperative to organize facts and instructions so that accurate information is printed and lot numbers match the batch of whole chickens, giblets and other separately sold parts. Moreover, poultry farms and processing units are required to maintain detailed records of all incidents including corrective measures taken and personnel involved in order to ensure complete traceability in the event a recall is issued. It is the responsibility of the poultry farmer and supervisors to regularly review and update any testing methods or procedures as per regulation requirements, and maintain robust calibration and validation records.

6 Conclusions

In conclusion, we recommend scientific knowledge as a vital component in farm safety management for minimizing the transmission of poultry pathogens, reducing the risk of poultry-borne human enteric infections, optimizing the quality of poultry products, and maximizing meat and egg economic profit. Through well-defined and implemented safety management practices, farmers and administrators can improve the growth, health, and productive and reproductive performance of their flocks, with improvements to housing, feeding, breeding and disease control protocols, to meet rising consumer demand for white meat. The prevalence and level of pathogenic organisms including poultry pathogens and poultry-borne human enteric pathogens of poultry vary widely, and colonization and infection risks differ by time of year, region, animal safety management, and production and processing practices. Therefore, ensuring the safety of poultry is a responsibility shared by all industry stakeholders, including breeders, farmers, producers, processors, retailers, and federal and state agencies. It has been concluded that food safety cannot be ensured by spot action at the pre-harvest level on the farm, post-harvest level in the processing plant, or end-product testing. The best solution involves an unbroken chain of safety management at all stages from egg production through processing and packaging. Modern technology and web-based communications are informing consumer consciousness at unprecedented levels, resulting in the need for ever stronger assurances of food safety. To meet current consumer demands, the poultry industry in the United States and across the globe must continue to be the leaders in farm safety management improvements as the consumption of poultry worldwide continues to grow.

7 Where to look for further information

Poultry Diseases, 6th revised edition is published by Pattison Mark, et al., has several information about diseases of poultry. This book can be helpful for poultry farmer understanding to diseases of poultry.
USDA website (https://www.aphis.usda.gov/aphis/home/) provides useful resources about poultry health, welfare and farming.

8 Acknowledgements

This project was supported by D. Biswas's start up grant provided by the Department of Animal and Avian Sciences, Center for Food Safety and Security Systems at the University of Maryland College park, MD.

9 References

Air Burner. 2015. Avian influenza bird carcass disposal, Technical memorandum, Air Burner, Inc. Palm City, FL. http://www.airburners.com/DATA-FILES_Print/AB-Bird_Flu_%20Carcass_Disposal-v0306d.pdf

APHIS. 2014. Biosecurity Guide for poultry and bird owners. USDA, Program Aid No. 1885, pp. 1–34.

APHIS. 2015. Prevent Avian Influenza at Your Farm: Improve Your Biosecurity with Simple Wildlife Management Practices, USDA. https://www.aphis.usda.gov/publications/animal_health/2015/fs_hpai_wildlife_mgmt_practices.pdf

Audi, A. H., and A. M. Asmau. 2014. Prevalence of bird Louse, MenacanthusCornutus (Pthiraptera: Amblycera) in Four Selected Poultry Farms in Kano State, Nigeria. *Bayero J. Pure. App. Sci.* 7(1), 142–6.

Bahrndorff, S., L. Rangstrup-Christensen, S. Nordentoft and B. Hald. 2013. Foodborne disease prevention and broiler chickens with reduced *Campylobacter*infection. *Emerg. Infect. Dis.* 19, 425–30. 10.3201/eid1903.111593.

Berry, J. G. and D. Whitenack. 1991. Bacterial diseases of poultry excluding respiratory disease. Oklahoma Cooperative Extension. https://efotg.sc.egov.usda.gov/references/Public/OK/F-9109_OSU_Factsheet_Bacterial_Poultry_Diseases.pdf.

Bolder, N. M., and L. Leoux. 2002. General protocol for cleaning and disinfecting poultry house. *World Poult.* 18(12), 26–9.

Butcher, G. D., J. P. Jacob and F. B. Mather. 2015. Common poultry disease, Publication #PS47, IFAS Extension, University of Florida. http://edis.ifas.ufl.edu/ps044.

Capua, I., and S. Marangon. 2009. Control of avian influenza in poultry. *Emerg. Infect. Diseases* 12(9), 1319–24.

CDC-2015. Foodborne Illness Source Attribution Estimates for Salmonella, Escherichia coli O157 (E. coli O157), Listeria monocytogenes (Lm), and Campylobacter using Outbreak Surveillance Data. https://www.cdc.gov/foodsafety/pdfs/ifsac-project-report-508c.pdf.

CFSPH. 2013. Poultry Industry Manual, FAD PReP, NAHEMS, USDA. http://www.cfsph.iastate.edu/pdf/fad-prep-nahems-poultry-industry-manual.

Cobb. 2013. Broiler management guide. Cobb-vantress. http://www.cobb-vantress.com/

Codex Alimentarius Commission – Joint FAO/WHO Food Standards Program. 1997. Hazard analysis and critical control point (HACCP) system and guidelines for its application.

Darre, M. J., and J. S. Rock. 1995. *Pest Management on Poultry Farms*. Cooperative Extension System, University of Connecticut, Storrs, CT.

Dixon, L. M, N. H. C. Sparks and K. M. D. Rutherford. 2016. Early experiences matter: a review of the effects of prenatal environment on offspring characteristics in poultry. *Poult. Sci.* 95, 489–99.

FAD PReP. 2013. *Standard Operating Procedures (Sop): 15*. Cleaning and Disinfection, USDA.

Fairchild, B. D. 2009. *Considerations in Poultry Drinker Line Management*. Cooperative Extension Service, The University of Georgia, Athens, GA.

Fanatico, A. 2007. Poultry House Management for Alternative Production, NCAT. https://attra.ncat.org/attra-pub/summaries/summary.php?pub=229.

Flory, G. A., R. W. Peer and E. S. Bendfeldt. 2006. *Evaluation of Poultry Carcass Disposal Methods Used During an Avian Influenza Outbreak in Virginia in 2002*. Virginia Department of Environmental Quality, Richmond, VA

Förster. M, S. Klimpel, H. Mehlhorn, K. Sievert, S. Messler and K. Pfeffer. 2007. Pilot studies on synantropic flies (e.g. *Musca, Sarcophaga, Calliphora, Fania, Lucilia, Stomoxys*) as vectors of pathogenic microorganisms. *Parasitol. Res.* 101, 243–6.

Forsythe, S. J., and P. R. Hayes. 1998. Chapter 3: Food Spoilage. In S. J. Forsythe and P. R. Hayes (eds), *Food Hygiene, Microbiology and HACCP*, pp. 86–149. Aspen Publishers Inc., Gaithersburg, Maryland.

Funk, E. M., and R. M. Irwin, ed. 1955. Prevention and control of disease in the hatchery. In *Hatchery Operation and Management*, pp. 305–320. John Wiley and Sons Inc., New York, NY. https://babel.hathitrust.org/cgi/pt?id=coo.31924002980203;view=1up;seq=8.

Goodrich, R. M., K. R. Schneider and R. H. Schmidt. 2005. HACCP: An overview. Online document of a series for Food Safety and Toxicology, University of Florida. http://edis.ifas.ufl.edu/FS122.

Goodwin, H. L., and R. Shiptsova. 2002. Changes in market equilibria resulting from food safety regulation in the meat and poultry industries. *Int. Food Agribus. Man.* 5(1), 61–74.

Hanning, I., D. Biswas, P. Herrera, M. Roesler and S. C. Ricke. 2010. Prevalence and characterization of *Campylobacter jejuni* isolated from pasture flock poultry. *J. Food Sci.* 75, M496–502.

Harris, L. J. 1999. HACCP 101 Part II- Principle 1. Hazard Analysis. *Perishables Handling Quarterly Issue* 98, 5–7

Harry, E. G., and R. F. Gordon. 1966. Egg and hatchery hygiene. *Veterinarian* 4, 5–15.

Henriksen, R., S. Rettenbacher and T. G. G. Groothuis. 2011. Prenatal stress in birds: Pathways, effects, function and perspectives. *Neurosci. Biobehav. R.* 35, 1484–501.

Huong, C. T. T., T. Murano, Y. Uno, T. and Usui, T. 2014. Yamaguchi. Molecular detection of avian pathogens in poultry red mite (*Dermanyssus gallinae*) collected in chicken Farms. *J. Vet. Med. Sci.* 76, 1583–7.

Jordan, F., M. Pattison, D. Alexander and T. Faragher. 2001. Hatchery. In *Poultry Disease*. 5th edition, pp. 35–42. W. B. Saunders, London, UK.

Jouve, J. L. 1998. Principles of food safety legislation. *Food Control*, 9, 75–81.

Kaufman, P. E., D. A. Rutz, and C. W. Pitts. 2000. Pest Management Recommendations for Poultry. A Cornell and Penn State Cooperative Extension publication. http://extension.psu.edu/publications/xf0274/view.

Khandke, S. S., and T. Mayes. 1998. HACCP implementation: A practical guide to the implementation of the HACCP plan. *Food Control*, 19, 103–9.

Kim, J. H., and K. S. Kim. 2010. Hatchery hygiene evaluation by microbiological examination of hatchery samples. *Poult. Sci.* 89, 1389–98.

Lee, A. 2003. Nutrition and feeding of show poultry. Texas Agrilife Extension, Texas A&M System. College Station, TX. http://oaktrust.library.tamu.edu/handle/1969.1/87765.

Lee, D., J. Bahl, M. K. Torchetti, M. L. Killian, H. S. Ip, T. J. Deliberto and D. E. Swayne. 2016. Highly pathogenic avian influenza viruses and generation of novel reassortants, United States, 2014–2015, *Emer. Infect. Dis.* 22(7), 1283–5

Lickliter, R. 2005. Prenatal sensory ecology and experience: implications for perceptual and behavioral development in precocial birds. *Adv. Stud. Behav.* 35, 235–74.

Loven, J., and R. Williams. 2010. Controlling rodent in commercial poultry facilities, Purdue University. https://extension.entm.purdue.edu/publications/ADM-3.pdf.

Maciorowski, K. G., P. Herrera, F. T. Jones, S. D. Pillai and S. C. Ricke. 2007. Effects on poultry and livestock of feed contamination with bacteria and fungi. *Anim. Feed Sci. Technol.* 133, 109–36.

Mccrea, B., J. S. Jeffrey, R. A. Ernst and A. C. Gerry. 2005. Common lice and mites of poultry: identification and treatment. Division of Agriculture and Natural Resources, University of California, ISBN-13: 978-1-60107-331-0, Publication No. 8162.

MDS. 2013. Important poultry diseases, MDS animal health research, version 5th.

Martin, J. 2011. Biosecurity Guidelines Regarding On-Farm Visitors. http://agsci.psu.edu/aec/research-extension/aec-webinars-and-events/aec-webinars/ag-101/preparing-to-visit-a-farm/Guidelines%20For%20Visitors%20-%20Martin.pdf.

Mobley, R., and T. Kahan. 2007. Practical Management of Health Issues in a Poultry Production System: Symptoms, Sources, and Prevention of Common Diseases, Florida A&M University. https://www.famu.edu/cesta/main/assets/File/coop_extension/herds/Practical_Management_of_Health_Issues_in_a_Poultry_Production_System.pdf.

Rutherford, K. M. D., R. D. Donald, G. Arnott, J. A. Rooke, L. M. Dixon, J. J. M. Mehers, J. Turnbull and A. B. Lawrence. 2012. Farm animal welfare: assessing risks attributable to the prenatal environment. *Anim. Welf.* 21, 129–429.

Salaheen, S., N. Chowdhury, I. Hanning and D. Biswas. 2015. Zoonotic bacterial Pathogens and mixed crop-livestock farming. *Poultry Sci.* 94(6), 1398–410.

Salaheen, S., C. Nguyen, D. Hewes and D. Biswas. 2014. Cheap extraction of antibacterial compounds of berry pomace and their mode of action against the pathogen *Campylobacter jejuni*. *Food Control*. 46, 174–81.

Sawabe, K., K. Hoshino, H. Isawa, T. Sasaki, K. Kim, T. Hayashi, Y. Tsuda, H. Kurahashi and M. Kobayashi. 2011. Blow flies were one of the possible candidates for transmission of highly pathogenic H5N1 avian influenza virus during the 2004 outbreaks in Japan. *Influ. Res. Treat.* Article ID: 652652. http://dx.doi.org/10.1155/2011/652652.

Scott, T. A., and C. Swetnam. 1993. Screening sanitizing agents and methods of application for hatching eggs. 2. Effectiveness against microorganisms on the egg shell. *J. Appl. Poult. Res.* 2, 7–11.

Sheldon, B. W., and J. Brake. 1991. Hydrogen peroxide as an alternative hatching egg disinfectant. *Poult. Sci.* 70, 1092–8.

Sparagano, O. A. E., D. R. George, D. W. J. Harrington and A. Giangaspero. 2014. Significance and control of the poultry red mite, *Dermanyssus gallinae*. *Annu. Rev. Entomol.*, 59, 447–66.

Stewart-Brown. B. 2015. Management of Laying Chickens. The Merck Veterinary Manual. http://www.merckvetmanual.com/mvm/poultry/nutrition_and_management_poultry/management_of_laying_chickens.html.

Strother, G. R. 2008. Poultry pest management, Alabama A&M and Auburn University. https://store.aces.edu/ItemDetail.aspx?ProductID=13589.

Sutton, A., and C. H. Lander. 2003. Feed and animal management for poultry. Nutrient management technical note No. 4, NRCS, USDA.

The Poultry Site. 2015. Controlling Rodents on the Poultry Farm. http://www.thepoultrysite.com/articles/3527/controlling-rodents-on-the-poultry-farm/

The Virginia Farmstead Assessment System. 2009. Poultry litter management and carcass disposal. Virginia State University, Fact sheet No. 10, Publication No. 442-910. https://pubs.ext.vt.edu/442/442-910/442-910.html.

USDA. 2015. Avian influenza response: mass depopulation and carcass disposal. USDA.

USDA. 2016. Newcastle disease standard operating procedures: 1. Overview of etiology and ecology, FAD PReP, USDA, SOP001, version 2.0.

WHO. 2005. Avian influenza: assessing the pandemic threat. http://apps.who.int/iris/bitstream/10665/68985/1/WHO_CDS_2005.29.pdf .

Zheng, L., T. L. Crippen, C. L. Sheffield, T. L. Poole, Z. Yu and J. K. Tomberlin. 2012. Evaluation of *Salmonella* movement through the gut of the lesser mealworm, *Alphitobius diaperinus* (Coleoptera: Tenebrionidae). *Vector-Borne Zoonot.* 12, 287–92.

Chapter 4

The emergence of antibiotic resistance in poultry farms

Issmat I. Kassem, Yosra A. Helmy, Isaac P. Kashoma and Gireesh Rajashekara, The Ohio State University, USA

1 Introduction: current poultry farming and the use of antibiotics

The increasing human population and cognate demands on wholesome and nutritious foods have exerted an indelible impact on modern agricultural practices. In addition, producer and consumer awareness has shaped a modern agricultural stewardship that accounts for affordability without jeopardizing public health, resource conservation, animal health and humane farming ethos. Therefore, now more than ever, there is a need to optimize agricultural production to achieve sustainability and food security and safety. Indeed, this also applies to poultry production, which is continuously evolving in breadth and practice. For example, broiler farming has witnessed a continuous improvement, which is evidenced by producing heavier birds in shorter growth cycles. Specifically, a 1.13 kg chicken was produced in 112 days in 1925 in comparison to 70 days in 1950. As of 2010, production of a 2.27 kg chicken is possible in less than 50 days (Krishnasamy et al., 2015). The expansion in the poultry industry and consumer demand for animal proteins were also evident in a recent Census of Agriculture report that showed that there were 233 770 poultry farms in the United States in 2012. Furthermore, approximately 96.4 billion eggs, 8.69 billion broilers and 233 million turkeys were produced in 2015, and the combined value of production was estimated at $48 billion (United States Department of Agriculture, USDA, 2015).

Raising competitive commercial poultry requires maintaining a healthy flock and generating a safe product for consumption. However, the production of healthy and

http://dx.doi.org/10.19103/AS.2016.0010.05

safe poultry is challenged by a plethora of bacterial pathogens that pose a risk to the consumers and/or adversely impact the health of birds. Although varying measures to control these pathogens are implemented during production, it is well known that current control methods display varying efficiency against disparate pathogens and can be expensive or difficult to implement in modern large-scale poultry production (Carvalho et al., 2012; Sahin et al., 2015; Kassem et al., 2016a). Furthermore, it has been suggested that controlling pathogens, which cause disease in humans, early in poultry production (e.g. on-farm) is ideal for significantly reducing the risk to consumers of the post-harvest products. However, the poultry production environment is complex and dynamic and can include numerous reservoirs that can spread disease within and between flocks or transmit food-borne pathogens to consumable products. The proliferation of these pathogens can (1) jeopardize standard control efforts such as biosecurity and hygiene, (2) adversely affect the health of the birds or the safety of cognate products and also (3) exert a significant financial and resource burden on production due to decontamination, maintenance, and product recall. Therefore, there is a need to devise, improve and use efficient and straightforward approaches to control and mitigate the proliferation of pathogens in poultry production.

The use of antimicrobials in food animals has been a conventional practice for controlling diseases and enhancing productivity (CDC, 2014; Johnsen et al., 2009; Smith and Fratamico, 2010; Van Boeckel et al., 2015). This practice has been proven popular among producers, leading to favourable impacts on the food-animal industry. Indeed, global consumption of antimicrobials in food animal production was estimated at 63 151 (±1560) tons in 2010 and is projected to rise by 67% (105 596 ± 3605 tons) by 2030 (Van Boeckel et al., 2015). Two-thirds (66%) of the global increase in antimicrobial consumption is due to the growing number of animals raised for food production. The remaining third (34%) is associated with a shift in farming practices, with a larger proportion of animals projected to be raised in intensive farming systems by 2030. In 2010, the five countries with the largest shares of global antimicrobial consumption in food animal production were China (23%), the United States (13%), Brazil (9%), India (3%) and Germany (3%). By 2030, this ranking is projected to be as follows: China (30%), the United States (10%), Brazil (8%), India (4%) and Mexico (2%). Furthermore, to meet demands on animal protein, countries such as Brazil, Russia, India, China and South Africa (BRICS) have shifted towards highly cost-effective and vertically integrated intensive livestock production systems. Interestingly, antimicrobial consumption for animals in the BRICS countries is expected to grow by 99% by 2030, whereas their human populations are only expected to grow by 13% over the same period (World Bank, 2016).

Despite the aforementioned advantages, antimicrobial use in animal farming has been receiving increasing scrutiny. The latter is due to a strong link between antibiotic use in food production and the rise of antibiotic-resistant pathogens that affect public health (Marshall and Levy, 2011; Sahin et al., 2015). This is not surprising, because microorganisms, like all organisms, have evolved disparate mechanisms to cope with various challenges and enhance their survival under unfavourable conditions. Successful microorganisms need to compete for resources and survive different stresses such as antimicrobial drugs. It is known that many microorganisms (1) produce compounds (e.g. antibiotics) to inhibit competing species or (2) develop resistance to inhibitory compounds secreted by other species (Wright, 2007). Therefore, antibiotic resistance existed in nature long before the discovery of the first antibiotics and their subsequent development for use in human medicine and agriculture (Perry et al., 2016). Importantly, antibiotic resistance is not exclusive to bacterial pathogens and may occur in commensals and even in beneficial microorganisms (Wright, 2007). The collective of genetic determinants that drives antibiotic resistance (i.e. antibiotic resistance genetic mechanisms) in all bacterial species is defined as the resistome (D'Costa et al., 2006). This includes cryptic resistance genes and precursor genes that might evolve into effective resistance genes (Wright, 2007). Therefore, the resistome can be regarded as a reservoir of genes that can give rise to antibiotic-resistant bacteria, including important pathogens. Based on these observations, it was predicted early on after the discovery of the first antibiotic that resistance to medically and agriculturally important antibiotics will ultimately develop. However, the over-dependence on antibiotics to achieve targets beyond the immediate treatment of serious infectious diseases might have hastened the emergence of resistant pathogens to a point that threatens the efficacy of antibiotic therapy (Wright, 2007). For example, the use of antibiotics as growth promoters in animal production has been considered as a major contributor to the pool of antibiotic-resistant pathogens that might pose a risk to public health (Marshall and Levy, 2011). Consequently, a popular stance directed at reducing or eliminating the use of antibiotics in farming practices based on the assumption that antibiotic withdrawal will result in 'diluting' resistance in the microbial population, has gained traction (Wright, 2007). However, antimicrobial-resistant strains may develop compensatory mutations that might allow for the persistence of these strains even in the absence of the antimicrobial selective pressure (Andersson, 2003 and 2006). This, along with the aforementioned 'natural' occurrence of antibiotic resistance, suggests that the resistome might still contribute to the rise or persistence of antibiotic-resistant pathogens even in the absence of the use of antibiotics (Andersson, 2003 and 2006). Whereas the latter does not diminish the coveted decrease in cognate risks to public health that is associated with limiting antibiotic use in animal farming, it is crucial to also understand the nuances and mechanisms that govern the emergence of antibiotic resistance in different production practices (e.g. organic/antibiotic-free vs conventional farming). The problem of antibiotic resistance is complex and its mitigation is associated with significant ramifications on the economy, food safety and security, and public health. However, the emergence of antibiotic resistance threatens the global fight against infectious diseases. Therefore, the mitigation of antibiotic resistance requires urgent actions from all involved in modern agriculture, including the scientific community, governmental legislators, producers and consumers.

2 The different applications of antibiotics in poultry farming

Antibiotics can be broadly defined as compounds that are conventionally produced by or derived from microorganisms and that have the capacity to inhibit the growth of or destroy other microorganisms. The modern era of antibiotics started with the accidental discovery of penicillin by Sir Alexander Fleming in 1928 (Fleming, 1929). Antibiotics were first prescribed to treat serious human infections in the 1940s and have positively impacted modern medicine, saving millions of lives to date. Similarly, antibiotic use in feed for food animal production first started in the 1940s (Stokstad et al., 1949). Since then antibiotics have been conventionally used in food animals to treat clinical diseases (therapy), to prevent and control common disease events (prophylaxis and metaphylaxis) and to enhance animal growth (growth promoter). Because of these advantages, antibiotic use has been adopted in the production of food animals in the United States, throughout the developed world, and in lesser-developed countries (Krishnasamy et al., 2015). The different applications of antibiotics in animal production are described in more details below.

2.1 Antibiotic therapy in animal farming

It is the administration of antibiotics to animals or groups of animals in order to treat illness/infections. The main objectives of employing antibiotic therapy in farm animals are to (1) alleviate the symptoms and the suffering of the affected animal(s), (2) limit the transmission of the disease to otherwise healthy animals in the herd or flock and (3) destroy the aetiological agent. As disease decreases farm animals' performance and can result in mortality, therapeutic applications are also essential components to increase profitable and humane livestock production. Therapeutic uses of antibiotics must be under the advice and discretion of a practising animal physician/veterinarian and require a valid prescription. In addition, in the United States, antibiotics must be approved by the US Food and Drug Administration (FDA), which monitors antibiotic use in farm animals to ensure not only the effectiveness of treatment but also safety of the animals and associated consumable products.

2.2 Growth promotion

It refers to the administration of antibiotics, usually as an animal feed additive, over a period of time to improve the animals' physiological performance, mainly weight gain and feed efficiency. Roughly, in the United States, growth-promoting antibiotics are administered at concentrations below 200 grams per ton of feed to healthy animals over a period of more than 14 days (Graham et al., 2007). While the actual net benefit of using antibiotics in feed for farm animals is difficult to measure, this practice is likely spurred by economic gain and the need to cope with increasing consumer demands on animal proteins. The mechanisms that facilitate antibiotic-associated growth promotion are still not completely clear (Butaye et al., 2003; Graham et al., 2007). However, it is thought that growth-promoting effects of antibiotics are mediated through alterations in the normal intestinal microbial population of the animal, which result in more efficient digestion of animal feed and metabolic uptake of nutrients (Dhama et al., 2014). In addition, sub-therapeutic doses of antibiotics used for growth promotion in farm animals might suppress some pathogenic gut bacteria, and it was hypothesized that antibiotics lead to decreased illness allowing weight to accumulate faster (Izat et al., 1990).

In comparison with other antibiotic applications in farming, growth promotion is the most controversial, likely because (1) it is regarded primarily as a method for increasing economic gain in shorter growth cycles (Graham et al., 2007), (2) the expansion of animal farming operations requires unprecedented amounts of antibiotics to be used in growth promotion (approximately 3.1 million pounds to 25 million pounds are used as growth promoters annually in the United States which corresponded to 13–70% of all antibiotic use) (Graham et al., 2007), (3) antibiotics used for growth promotion are not as tightly regulated as the other antibiotic applications in farming and (4) it is thought that sub-therapeutic applications over a relatively prolonged period of time will exert selective pressure that allows the emergence and persistence of antibiotic-resistant pathogens that can affect animals and humans (Graham et al., 2007). Consequently, governmental authorities in Europe and the United States have adopted measures to ban or phase out the use of growth-promoting antibiotics in food animals (see below) (Casewell et al., 2003; FDA, 2013). This was judicially done as a precaution in order to preserve the effectiveness of antibiotics that are used to treat human infections.

2.3 Metaphylaxis and prophylaxis

These procedures are often recommended by veterinarians for animals that are generally determined to be at an unacceptable high risk of developing a bacterial disease. For example, an animal may receive antibiotics as a precaution after surgery or trauma (prophylaxis). Also, antibiotics may be administered to whole herd(s) or flock(s) that are at risk of exposure to an outbreak of infectious disease or those that have been under extremely unfavourable conditions (metaphylaxis) (Nickell and White, 2010). Typically, metaphylaxis and prophylaxis involve administering drugs at therapeutic levels for short periods of time (Scott and Fedorka-Cray, 2002). Currently, several antibiotics are available for metaphylactic use in animals, including Micotil® (tilmicosin), Nuflor® (florphenicol), Tetradure® (oxytetracycline 300 mg/ml), Excede® (ceftiofur 200 mg/ml) and Draxxin® (tulathromycin).

Judicious poultry production approaches the use of antibiotics very seriously, adhering to all regulations and cognizant of animal and consumer welfare and public health. Antibiotics have been commonly used to increase growth and feed efficiency and decrease mortality in broiler poultry production as early as the 1950s (Stokstad et al., 1949; Elliott et al., 1959). In 1951, the FDA approved the use of antibiotics as animal additives without veterinary prescription in the United States (Castanon, 2007; Jones and Ricke, 2003). Notably, approximately 50% increase in the average weight of broilers and 35% decrease in growth time and feed intake that are considered necessary to generate a pound of broiler meat occurred after the introduction of antibiotics (Boyd, 2001). Therefore, the use of antibiotics has been associated with increasing efficiency of production, which in turn contributes to the affordability of chicken products for consumers (Smith, 2002). Consequently, many groups of antimicrobials/antibiotics (arsenicals, polypeptides, glycolipids, tetracyclines, elfamycins, macrolides, lincosamides, ionophores/polyethers, penicillins, quinoxalines, streptogramins and sulfonamides) have been at one point registered for use as growth promoters in poultry and other food animals in Australia, the European Union (EU) and Canada (Sarmah et al., 2006). In the United States, arsenicals (arsanilic acid, roxarsone, cabarsone), polypeptides (bacitracin), glycolipids (bambermycins), tetracyclines (chlortetracycline), macrolides (oleandomycin, tylosin) and penicillins (penicillin) have been used as growth promoters in poultry (Sarmah et al., 2006).

Chicks and poults are extremely fragile in the first few weeks of production and antibiotics are often injected into eggs or chicks to improve early viability and prevent possible infections (Van Looveren et al., 2001; Sneeringer et al., 2015). Antibiotics are also used to prevent and treat disease in growing birds (Table 1). For example, tetracyclines and fluoroquinolones, among others, are used to treat chronic respiratory disease and salmonellosis in birds, while penicillins in drinking water have been used against necrotic enteritis (The Poultry Site, 2016). In the United States, arsanilic acid, bacitracin, erythromycin, lincomycin and tylosin have been used to prevent diseases in poultry (Sarmah et al., 2006). Some of the antibiotics such as tetracyclines, erythromycin and penicillins are used in the treatment of infections in humans (Table 1) (Landers et al., 2012). Therefore, minimum

Table 1 Classes of antimicrobials that are used in poultry production and their importance to human medicine as classified by the World Health Organization (WHO) (Aarestrup et al., 2008; WHO, 2001)

Classes of antibiotics	Application in poultry/food animal production	Importance for humans
Aminoglycosides	Disease treatment and prevention	Critical
Penicillins	Disease treatment and prevention and growth promotion	Critical
Third generation Cephalosporins	Disease treatment and prevention	Critical
Glycopeptides	Growth promotion	Critical
Macrolides	Disease treatment and prevention and growth promotion	Critical
Quinolones	Disease treatment and prevention	Critical
Streptogramins	Disease treatment and prevention and growth promotion	Critical
Spectinomycin	Disease prevention	Highly important
Sulfonamides	Disease treatment and growth promotion	Highly important
Tetracyclines	Disease treatment and prevention and growth promotion	Highly important
Polypeptides	Disease treatment and prevention and growth promotion	Important
Lincosamides	Disease treatment and prevention	Important
Bambermycin	Disease prevention and growth promotion	Not known
Ionophores	Disease prevention and growth promotion	Not known

withdrawal periods between the last use of the antibiotic and slaughter are implemented, and the meat is also screened for antibiotic residues. The concentration of antibiotics in the meat, if any, should fall within the levels accepted by the FDA standards; otherwise the meat will be rejected (Sneeringer et al., 2015).

Taken together, the aforementioned observations suggest that antibiotics have been an integral tool in conventional poultry production. Aside from public health repercussions, it is clear that devising approaches to maintain the efficacy of antibiotics is in the best interest of both the industry and consumers. Therefore, there is a pressing need for more evidence-based strategies on how to best use antibiotics in poultry production and on how to simultaneously not jeopardize food safety and security, public health and efficient production.

3 Human health risks associated with antibiotic use in poultry farming

One of the major advances in human medicine has been the discovery of antibiotics and the control of infectious diseases. Therefore, the emergence of antibiotic-resistant pathogens poses a grievous threat to human welfare across the globe. The World Health Organization (WHO) recognizes antibiotic resistance as one of the top health challenges facing the twenty-first century (Marshall and Levy, 2011; WHO, 2015). The overuse of antibiotics in intensively produced farm animals is now believed to play a major role in this global problem (Landers et al., 2012; WHO, 2015), impacting the pools of antibiotic-resistant bacteria that might affect humans (Elliot, 2015). Studies have shown a close association between the prevalence of livestock-associated antibiotic-resistant bacteria in animals and in humans (Vieira et al., 2011). A similar relation was noted between the levels of antibiotics use in animals at a population level and the prevalence of antibiotic-resistant bacteria in animals (Aarestrup, 2005) and in humans (Schwarz et al., 2001). Furthermore, Chantziaras et al. (2014) showed a strong correlation between consumption levels of eight classes of antimicrobials and the prevalence of antimicrobial-resistant commensal *Escherichia coli* in pigs, poultry and cattle.

Notably, the WHO highlighted that several antibiotic classes, designated as 'critically or highly important' to human health, have been also used in animal farming (Table 1). For example, in 2009, macrolides, penicillins and tetracyclines, which are categorized as critically important in human medicine, were the top grossing antibiotics in livestock-related sales (CDDEP, 2015). The WHO and other organizations also singled out some agents such as quinolones, third- and fourth-generation cephalosporins, and macrolide as top priorities for risk assessment (Elliott, 2015). This is not surprising, because the overuse of antibiotics in farm animals might lead to several related public health ramifications, including:

1 Antibiotic-resistant food-borne infections: The increase in antibiotic-resistant food-borne bacteria such as *Campylobacter*, *Salmonella* and *E. coli* will complicate the treatment of food-borne infections and diseases acquired from exposure to food animals and the farm environment.
2 The emergence or re-emergence of new or previously controlled antibiotic-resistant strains of bacteria. For example, the emergence of new strains of methicillin-resistant *Staphylococcus aureus* (also known as livestock-associated MRSA, LA-MRSA) that can be transmitted to the human population via contact with animals and the farm environment (Cuny et al., 2015).
3 The off-farm spread of resistance genes and antibiotics: The selective pressure of antibiotics in farm operations might enrich the pools of antibiotic-resistant genetic markers in farm animals and increase the potential transmission of these markers off-farm (via run-off and farm waste disposal). It has been highlighted previously that intensive animal farming operations can be a dominant cause for the proliferation of antibiotic resistance genes in an off-farm environment (Storteboom et al., 2010). Subsequently, these genes can be horizontally transferred to other non-food-animal associated bacteria, including pathogens, in the environment (Chee-Sanford et al., 2001; Storteboom et al., 2010). Furthermore, an estimated 75% of antimicrobial agents administered to confined poultry may be excreted back into the environment (Addison, 1984). Antimicrobial residues were detected in 67% of water samples that were collected from surface water and groundwater resources located in the proximity of intensive poultry operations (Campagnolo et al., 2002). Therefore, antibiotic run-off from poultry operations might select for antibiotic-resistant bacteria in off-farm/poultry processing niches.

The rise in antibiotic-resistant bacteria will increase the risk of exposure for the public. This can potentially occur via:

1 Direct contact with infected animals: Handling and direct contact with farm animals, including poultry, might increase the risk of exposure to antibiotic-resistant bacteria that might be excreted from the animals. For example, a study showed that resistant bacteria could be potentially transmitted between farm animals of the same species or of different species, and that farm workers were colonized for several weeks by *E. coli* that was likely acquired from a bull (Marshall et al., 1990). Studies in the Netherlands showed that the same genetic patterns of antibiotic resistance occurred in *E. coli* isolates that were collected from turkeys and broiler chickens and their farm- and slaughter workers (Van den Bogaard et al., 2001, 2002).
2 Consumption of food contaminated with antibiotic-resistant bacteria: Contamination of meat generally results from faecal material getting onto the poultry carcass during processing (slaughter and evisceration). This is important, because chickens colonized with food-borne pathogens such as *Campylobacter* are 30 times more likely to result in contaminated meat than are uninfected birds (EFSA, 2010). Furthermore, contaminated meat can also cross-contaminate other foods in domestic or restaurant/catering kitchens during meal preparations.
3 Exposure via environmental contamination: Land application of food animal manure as a fertilizer and off-farm run-off can be potential sources for the dissemination of antibiotics and antibiotic-resistant bacteria to the environment, resulting in the contamination of soils, and surface- and groundwater (Berglund, 2015; Pruden et al., 2013). For example, tetracycline resistance genes have been found in groundwater samples collected downstream from the slurry lagoon of a pig farm (Chee-Sanford et al., 2001). Interestingly, tetracycline-, erythromycin- and quinupristin/dalfopristin-resistant *Enterococcus* spp. have been isolated from the surfaces and internal air of cars driving behind poultry transport trucks (Rule et al., 2008). These antimicrobials are used to treat human infections and have been also approved for use in the production of broiler chickens (Rule et al., 2008). Furthermore, in the proximity of turkey farms that used the growth promoter avoparcin, 14% of people were found to carry *Enterococcus* species that were resistant to vancomycin, a clinically important drug that is chemically similar to avoparcin (Stobberingh et al., 1999).

Taken together, the aforementioned observations highlight that the use of antibiotics in farming is associated with complex public health ramifications that extend beyond the immediate farm environment and the food chain. This necessitates adopting serious proactive actions to control any potential undesirable impacts associated with the on-farm use of antibiotics. Perhaps, this also highlights the need for extension programmes that educate stakeholders and the public to the risks and different routes of exposure to antibiotics and antibiotic-resistant pathogens associated with poultry production.

4 Examples of antibiotic-resistant food-borne pathogens associated with poultry

There are many poultry-associated bacterial pathogens that can cause serious disease in humans and in the flocks. In this section, the focus will be on the emergence of antibiotic-resistant *Campylobacter* and *Salmonella*, two bacterial genera that include known food-borne human pathogens. Human infections with these pathogens have been often linked in high percentages to the consumption of poultry products.

4.1 *Campylobacter*

The genus *Campylobacter* consists of 26 species of which *C. jejuni* and *C. coli* are of major importance to public health (CDC, 2015). *Campylobacter* results in food-borne disease worldwide, and the consumption of undercooked poultry and other foods that are cross-contaminated with raw poultry during preparation is considered a major source of *Campylobacter* infections in humans (Sahin et al., 2015; Kassem et al., 2016a; Coker et al., 2002). The most common manifestation of *Campylobacter* infections in humans is gastroenteritis, which is characterized by symptoms that range from watery, non-bloody and non-inflammatory diarrhoea to severe inflammatory diarrhoea with abdominal cramps, pain and fever (Man, 2011). *Campylobacter*-associated post-infection complications include the Guillain-Barré syndrome, the Miller Fisher syndrome and reactive arthritis (Nachamkin et al., 2000). Although most *Campylobacter* infections are self-limiting, severe or prolonged infections may require antimicrobial therapy under certain scenarios. In these cases, erythromycin and fluoroquinolones such as ciprofloxacin are usually considered the drugs of choice (Allos, 2001; Smith and Fratamico, 2010). Other antimicrobials such as gentamicin, tetracycline, clindamycin and ampicillin may be used as alternative drugs for the treatment of systemic *Campylobacter* infections (Blaser, 2000). Therefore, the resistance of *Campylobacter* to fluoroquinolones and macrolides is a major concern for public health.

Campylobacter colonization of poultry is mostly asymptomatic and does not require antibiotic treatment. However, *Campylobacter* is highly prevalent in poultry, and antibiotics administered in poultry farming operations can select for antibiotic-resistant *Campylobacter* (Price et al., 2005; Schmidt, 2005). This is further evidenced in numerous studies from different countries that report the isolation of antibiotic-resistant *Campylobacter* from poultry products (Sahin et al., 2015). For example, a study conducted in Germany found that 42% and 71% of chicken-associated *C. jejuni* and *C. coli* were resistant to ciprofloxacin, respectively (Luber et al., 2003). In Poland, 47.9% of the *Campylobacter* strains isolated from poultry were resistant to ciprofloxacin in the mid-1990s (1994–6) in comparison to 90.2% approximately a decade later (2005–8) (Wozniak, 2011). Similarly, in the same time interval, erythromycin-resistant *Campylobacter* associated with poultry increased from 49.3 to 88.9% (Wozniak and Wieliczko, 2011). An increase in the incidences of antibiotic-resistant *Campylobacter* was also observed in the United States. Specifically, a comparison between 2001 and 2010 showed that the detection rate of ciprofloxacin- and erythromycin-resistant *C. jejuni* on chicken carcasses at slaughter increased from 20.3 to 23.1% and from 0 to 10%, respectively (FDA, 2012). Furthermore, 49% and 36% of organically and conventionally raised retail chicken carcasses were contaminated with erythromycin-resistant *Campylobacter*, respectively (Cui et al., 2005). A study on turkeys showed that erythromycin-resistant *C. coli* and *C. jejuni* occurred in 95.6% and 0% of conventionally raised birds, respectively (Lee et al., 2005). More recently, it was reported that turkey-associated *Campylobacter* isolates were resistant to tetracycline (95%), azithromycin (43%), ciprofloxacin (42%) and erythromycin (18%) (Kashoma et al., 2014). In China, 99.1% of *C. jejuni* and 100% of *C. coli* isolated from broiler chicken were resistant to ciprofloxacin, while 16% and 94.1% were resistant to erythromycin, 53.8% and 95.4% were resistant to gentamicin and 100% and 98.7% were resistant to tetracycline, respectively (Ma et al., 2014).

Taken together, the frequent occurrence of *Campylobacter* that are resistant to medically important antibiotics on poultry products is notable. This should motivate efforts to (1) better assess the contribution of the poultry-associated antibiotic-resistant *Campylobacter* to recalcitrant human infections and (2) devise approaches and monitoring programmes to mitigate the increase in antibiotic-resistant isolates in poultry production.

4.2 *Salmonella*

The genus *Salmonella* includes species and serovars that can cause illnesses in humans and animals. Human salmonellosis is a major worldwide disease with annual estimates of over 27 million cases and more than 200 000 deaths due to typhoid fever (Crump et al., 2004) and 93.8 million cases of gastroenteritis and 155 000 deaths due to non-typhoidal *Salmonella* (Majowicz et al., 2010). *Salmonella*-induced gastroenteritis is associated with diarrhoea, abdominal pain, fever and vomiting. Severe complications such as endocarditis, bacteraemia, meningitis and pneumonia may occur in the young, the elderly and the immunocompromised patients (Arshad et al., 2008; Hohmann, 2001). *S. enterica* is considered one of the most prevalent causes of gastroenteritis worldwide (Majowicz et al., 2010). Patients with non-complicated *Salmonella* infections can be treated with supportive therapy without resorting to the use of antimicrobial agents. However, antimicrobials such as ampicillin, chloramphenicol, trimethoprim-sulfamethoxazole and ciprofloxacin can be recommended for high-risk patients (Hohmann, 2001; McDermott, 2006). There is strong link between human salmonellosis and the consumption of contaminated poultry products, including eggs and broiler meat (CDC, 2011; Currie et al., 2005; Tribe et al., 2002; Altekruse et al., 2006; Gallegos-Robles et al., 2008; Hellberg and Chu, 2016; Mohle-Boetani et al., 2009; Nunes et al., 2003).

Antibiotic-resistant *Salmonella* occur in contaminated poultry and might be transmitted to humans. One striking example is a study that was conducted in Turkey. The authors reported that *Salmonella* serotypes that were isolated from chicken carcasses exhibited resistance to penicillin (100% of isolates), vancomycin (92.6%), erythromycin (89.7%), ampicillin (85.2%), tetracycline (67.6%), streptomycin (61.7%), neomycin (55.8%), gentamicin (14.7%), chloramphenicol (10.2%), cefotaxime (2.9%) and others (Yildirim et al., 2011). In a study that targeted *Salmonella* isolated from processed poultry, it was reported that 79.8% and 53.4% of the isolates were resistant to at least one antimicrobial and three or more antimicrobials, respectively. The isolates were resistant to tetracycline (73.4% of isolates), ampicillin (52.9%), amoxicillin–clavulanic acid (52%), ceftiofur (51.7%), streptomycin (35.2%) and sulfisoxazole (21.8%) (Parveen et al., 2007). Furthermore, the proportion of ceftriaxone-resistant S. Heidelberg from humans increased to over 20% in 2009 (FDA, 2009), which mirrored similar increases in ceftriaxone-resistant isolates that were retrieved from chickens and turkeys and from retail chicken and ground turkey. This is very important, because S. Heidelberg can cause more severe illness in comparison with other serotypes, and it is the most common serotype retrieved from humans in the United States. However, the FDA has recently reported that ceftriaxone-resistant *Salmonella* declined to 18% in retail chicken meats in 2014 and was as low as 5% during the first half of 2015 (FDA, 2015), while these strains have declined by 80% in ground turkey since 2011. The FDA also reported that ciprofloxacin and azithromycin resistance was not detected in *Salmonella* isolated from poultry. However, it is essential to note that in the previous studies, the prevalence of *Salmonella* had also decreased to 6% in ground turkey and 9% in retail chicken by 2014. Therefore, the reduction in antimicrobial-resistant *Salmonella* might be attributed to the successful *Salmonella* control programmes that are adopted by the poultry industry. It would be very interesting and useful to investigate the fitness of the antibiotic-resistant *Salmonella* under these programmes in order to better understand the impact on antibiotic resistance.

The aforementioned observations strongly suggest that there is a need for (1) continuous surveillance for the emergence of antibiotic resistance in *Salmonella* in poultry and (2) a deep understanding of the characteristics of these strains and their survival and persistence under current poultry farming and processing practices.

5 Regulations, recommendations and future trends

Antimicrobial resistance has been detected in at least 70% of the pathogens associated with domestically acquired food-borne illnesses that occur annually in the United States (Cosby et al., 2015). Treatment of infections caused by antimicrobial-resistant pathogens is very expensive (Cosgrove, 2006). For example, it has been estimated that drug-resistant non-typhoidal *Salmonella* exerts $365 million in medical costs annually in the United States alone (CDC, 2013). Projecting these statistics on a global scale will strongly highlight food-producing practices such as poultry production as hot spots for mitigating the emergence of antibiotic resistance. However, this must be motivated by two major concerns that (1) there is a peremptory need to conserve the efficacy of current antibiotics used in human medicine and (2) food animal/poultry production, for economic and food security purpose, should not be detrimentally or needlessly affected by attempts to restrict the use of antibiotics. Therefore, careful planning and science-driven and timely actions will be a cornerstone for these efforts.

Governmental agencies in the United States and Europe have already initiated rules and guidelines that are meant to regulate the use of antibiotics in food animal production. In the United States, the FDA has considered imposing limits on growth-promoting antibiotics since the 1970s. In 2005, the FDA banned enrofloxacin (a fluoroquinolone that was administered for therapeutic uses, not growth promotion) in food animals on the grounds that its use contributed to fluoroquinolone resistance in human pathogens (FDA, 2005). Recently, the FDA passed voluntary guidance (Guidance for Industry, GFI, #209 and 213) that target pharmaceutical companies that sell veterinary antibiotics, recommending that they voluntarily increase veterinary oversight of antibiotics (by requiring prescriptions) and change the labels on their drugs so that their use for 'production', or growth promotion, is no longer allowed. Companies have 3 years (until 2018) to comply with the guidance (FDA, 2013). The FDA also prohibit the extra-label (off-label) use of certain antibiotics in food-producing animals, which includes using the drug at unapproved dosage levels, as a growth promoter or for disease prevention, and using drugs meant for one species on another. The use of chloramphenicol for any reason is prohibited (FDA, 2012).

In 2006, the EU banned all antibiotic growth promoters. Since the ban of avilamycin, erythromycin, vancomycin and virginiamycin as antibiotic growth promoters in Denmark, antibiotic resistance levels in humans have decreased, suggesting that the ban has had the desired effect. Based on the council regulation established in 1990, EEC No. 2377/90, the commission regulation (EU) 37/2010 outlines maximum levels of antibiotics in food products of animal origin. This commission regulation also includes a list of several antimicrobials, including chloramphenicol, dapsone dimetridazole, metronidazole, nitrofurans and ronidazole, that are banned from use in food products because safe levels have not been determined (European Commission 2010). In November 2011, the EU put forward a five-year plan to fight antimicrobial resistance. The plan was based on 12 key actions that included enhancing regulations on medicines used in veterinary practices and on medicated feed (Action # 2), promoting the judicious use of antibiotics in veterinary medicine (Action # 3) and bolstering surveillance systems on the use of antimicrobials in animal medicine (Action # 10) (European Commission 2011). The World Health Organization Regional Office for Europe also released a strategic action plan on antibiotic resistance in 2011 (WHO, 2011). The objectives include the prevention and control of the development and spread of antibiotic resistance in the veterinary and agricultural sectors. The following recommendations to conserve antibiotic effectiveness in humans and animals have been proposed by the World Organisation for Animal Health (OIE) (OIE, 2015):

1 Track rates of veterinary antibiotic use, resistance, and residues through a nationwide surveillance and monitoring system
2 Change incentives to discourage unnecessary antibiotic use in animals
3 Educate farmers, veterinarians, and consumers on the dangers of antibiotic resistance
4 Phase out the sub-therapeutic use of antibiotics in animals.

While the impact of the aforementioned actions is predicted to be favourable in reducing antibiotic resistance in pathogens, it is also argued that environmental reservoirs of antibiotic-resistant pathogens might not be immediately affected by antimicrobial withdrawal (Johnsen et al., 2009; Smith and Fratamico, 2010). Mechanisms/mutations that lead to antibiotic resistance and have little or no fitness costs are likely to persist in the absence of antibiotic pressure (Melnyk et al., 2015). A meta-analysis showed the latter to be true for several drug classes and species of bacteria that did not exhibit an apparent fitness cost (Melnyk et al., 2015). For example, a fluoroquinolone-resistant *Campylobacter* rapidly outcompeted susceptible strains and persisted in chickens in the absence of antibiotic pressure (Luo et al., 2005). Interestingly, the withdrawal of fluoroquinolone has had a limited effect on the prevalence of fluoroquinolone-resistant *Campylobacter* in poultry in the United States (Luangtongkum et al., 2009). However, an erythromycin-resistant *Campylobacter* exhibited defective fitness and was outcompeted by a susceptible strain in the chicken host, which confirms that fitness will depend on the type of resistance mechanism or mutation (Luangtongkum et al., 2012).

It is important to also note that antibiotic-resistant pathogens have been frequently isolated on certified organic farms that by regulation must not use antimicrobials (Kassem et al., 2016b; Luangtongkum et al., 2006). Therefore, it is of high merit to highlight in these instances the unexplored diversity and ubiquity of antibiotic resistance genetic markers (the resistome) in a variety of sources such as water, manure and soil, and the natural role of antibiotic resistance in enhancing the competitiveness of bacteria (Noyes et al., 2016). Taken together, the aforementioned observations have raised many questions, mainly (1) how would eliminating antimicrobials affect poultry production? (2) will antibiotic resistance be sufficiently reduced and when would this reduction become apparent after withdrawal of antibiotics? and (3) what are the sources and mechanisms that drive emergence of antibacterial resistance in the absence of antibiotic use in poultry production? Answering these questions will allow the judicial management of antibiotics in poultry production and clarify the factors that will effectively mitigate antibiotic resistance.

Growth promotion and controlling infectious diseases in poultry are not restricted to the use of antibiotics. In recent years, several farming approaches and applications that are antibiotic-independent have more or less emerged. For example, efficient biosecurity and hygiene protocols; the use of prebiotics and probiotic bacteria for competitive exclusion; bacteriophages; and additives such as organic acids, genetic resistance, vaccination, and organic- and antibiotic-free farming (Sahin et al., 2015). These and others can be used to replace or reduce the dependence on antibiotics. The merits of reducing antibiotic resistance are readily obvious. Human mortality and morbidity due to infectious diseases were significantly higher in the pre-antibiotic era. The discovery of antibiotics has been one of the singular breakthroughs in human medicine. Consequently, serious efforts must be directed to conserve the efficacy of these compounds and abuse might lead to undesirable consequences and the loss of the bliss of an antibiotic paradise.

6 Where to look for further information

More information on antibiotic resistance and its impact on public health can be found at the WHO website (http://www.who.int/antimicrobial-resistance/en/), which includes important information on the WHO's Global Action Plan (GAP) on antibiotic resistance and its implementation in addition to numerous resources and publications (e.g. World Antibiotic Awareness Week), and reports on global surveillance initiatives. The WHO also lists 'other organizations and partners' (http://www.who.int/antimicrobial-resistance/Other_Organizations_and_Partners/en/) that are involved in antimicrobial resistance surveillance, advocacy and education. These include the Alliance for the Prudent Use of Antibiotics (APUA), Alliance Working for Antibiotic Resistance Education (AWARE), CDC, Asian Network for Surveillance of Resistant Pathogens (ANSORP), European Antimicrobial Resistance Surveillance System (EARSS) and the United States' National Antimicrobial Resistance Monitoring System (NARMS). The Food and Agriculture Organization of the United Nations (FAO) has recently published its action plan on antibiotic resistance, which aimed to support the food and agriculture sectors in implementing GAP (http://www.fao.org/3/a-i5996e.pdf). The FAO's plan addressed 'Four Focus Areas' in combating antibiotic resistance, specifically awareness, practices, governances and evidence (building local capacity for surveillance and monitoring of antimicrobial use and resistance in food and agriculture).

In the United States, the NARMS was established in 1996 in collaboration between CDC, the FDA, the USDA and public health departments at the state and local level. The CDC's website dedicated to NARMS (https://www.cdc.gov/narms/) contains important resources, reports and publications on antibiotic resistance, including links that focus extensively on the use of antibiotics in food animals (http://www.cdc.gov/narms/animals.html). Similarly, the United Kingdom maintains a website that details its efforts to 'slow the growth' of antibiotic resistance (https://www.gov.uk/government/collections/antimicrobial-resistance-amr-information-and-resources), including resources for healthcare and livestock professionals.

The primary academic literature is rich with uncountable authoritative publications that tackle the history, evolution, mechanisms and regulation of antibiotic resistance among other topics. Many of these publications were included in the reference list of this chapter. The resources listed above are a small sample that were selected to suite disparate readers. The focus was on material and initiatives that highlight the complexity, scale and burden of antibiotic resistance and the cognate efforts to fight this global threat to public health.

7 References

Aarestrup, F. M. 2005. Veterinary drug usage and antimicrobial resistance in bacteria of animal origin. *Basic Clin. Pharmacol. Toxicol.*, 96, 271–81.

Aarestrup, F. M., Wegener, H. C. and Collignon, P. 2008. Resistance in bacteria of the food chain: epidemiology and control strategies. *Expert. Rev. Anti Infect. Ther.*, 6, 733–50.

Addison, J. B. 1984. Antibiotics in sediments and run-off waters from feedlots. *Residue Rev.*, 92, 1–28.

Allos, B. M. 2001. *Campylobacter jejuni* infections: update on emerging issues and trends. *Clin. Infect Dis.*, 32, 1201–6.

Altekruse, S. F., Bauer, N., Chanlongbutra, A., Desagun, R., Naugle, A., Schlosser, W., Umholtz, R. and White, P. 2006. *Salmonella* enteritidis in broiler chickens, United States, 2000–2005. *Emerg. Infect Dis.*, 12, 1848–52.

Andersson, D. I. 2003. Persistence of antibiotic resistant bacteria. *Curr. Opin. Microbiol.*, 6, 452–6.

Andersson, D. I. 2006. The biological cost of mutational antibiotic resistance: any practical conclusions? *Curr. Opin. Microbiol.*, 9, 461–5.

Arshad, M. M., Wilkins, M. J., Downes, F. P., Rahbar, M. H., Erskine, R. J., Boulton, M. L., Younus, M. and Saeed, A. M. 2008. Epidemiologic attributes of invasive non-typhoidal *Salmonella* infections in Michigan, 1995–2001. *Int. J. Infect Dis.*, 12, 176–82.

Berglund, B. 2015. Environmental dissemination of antibiotic resistance genes and correlation to anthropogenic contamination with antibiotics. *Infect. Ecol. Epidemiol.*, 5, 28564.

Blaser, M. J. 2000. *Campylobacter jejuni* and related species. In Mandell, G. L., Bennett, J. E. and Dolin, R. (eds), *Mandell, Douglas, and Bennett's Principles and Practice of Infectious Diseases*, vol. 2, 5th ed., pp. 2276–85. Churchill Livingstone, Philadelphia, PA.

Boyd, W. 2001. Making meat: science, technology, and American poultry production. *Technol. Cult.*, 42, 631–64.

Butaye, P., Devriese, L. A. and Haesebrouck, F. 2003. Antimicrobial growth promoters used in animal feed: effects of less well known antibiotics on gram-positive bacteria. *Clin. Microbiol. Rev.*, 16, 175–88.

Campagnolo, E. R., Johnson, K. R., Karpati, A., Rubin, C. S., Kolpin, D. W., Meyer, M. T., Esteban, J. E., Currier, R. W., Smith, K., Thu, K. M. and McGeehin, M. 2002. Antimicrobial residues in animal waste and water resources proximal to large-scale swine and poultry feeding operations. *Sci. Total Environ.*, 299, 89–95.

Carvalho, C. M., Kropinski, A. M., Ferreira, E. C., Azeredo, J. and Santos, S. B. 2012. Phages as therapeutic tools to control major foodborne pathogens: *Campylobacter* and *Salmonella*. *Intech Open Access Publisher*, pp. 179–214. http://cdn.intechopen.com/pdfs/32285.pdf (Accessed September 2016).

Casewell, M., Friis, C., Marco, E., Mcmullin, P. and Phillips, I. 2003. The European ban on growth-promoting antibiotics and emerging consequences for human and animal health. *J. Antimicrob. Chemother.*, 52, 159–61.

Castanon, J. I. 2007. History of the use of antibiotic as growth promoters in European poultry feeds. *Poult. Sci.*, 86, 2466–71.

CDC (Centers for Disease Control and Prevention). 2011. Outbreak of *Salmonella* serotype enteritidis infections associated with raw almonds – United States and Canada, 2003–2004. http://www.cdc.gov/mmwr/preview/mmwrhtml/mm5322a8.html (Accessed August 2016).

CDC. 2013. Antibiotic resistance threats in the United States. http://www.cdc.gov/drugresistance/pdf/ar-threats-2013-508.pdf (Accessed August 2016)

CDC. 2014. Antibiotic resistance from the farm to the table. http://www.cdc.gov/foodsafety/from-farm-to-table.html (Accessed August 2016).

CDC. 2015. Foodborne Disease Active Surveillance Network (FoodNet). http://www.cdc.gov/foodnet/surveillance.html (Accessed August 2016).

CDDEP (Center for Disease Dynamics, Economics, and Policy). 2015. Antibiotic use and resistance in food animals, current policy and recommendations. http://www.cddep.org/sites/default/files/india_abx_report.pdf (Accessed August 2016).

Chantziaras, I., Boyen, F., Callens, B. and Dewulf, J. 2014. Correlation between veterinary antimicrobial use and antimicrobial resistance in food-producing animals: a report on seven countries. *J. Antimicrob. Chemother.*, 69, 827–34.

Chee-Sanford, J. C., Aminov, R. I., Krapac, I. J., Garrigues-Jeanjean, N. and Mackie, R. I. 2001. Occurrence and diversity of tetracycline resistance genes in lagoons and groundwater underlying two swine production facilities. *Appl. Environ. Microbiol.*, 67, 1494–502.

Coker, A. O., Isokpehi, R. D., Thomas, B. N., Amisu, K. O. and Obi, C. L. 2002. Human campylobacteriosis in developing countries. *Emerg. Infect. Dis.*, 8, 237–44.

Cosby, D. E., Cox, N. A., Harrison, M. A., Wilson, J. L., Buhr, R. J. and Fedorka-Cray, P. J. 2015. *Salmonella* and antimicrobial resistance in broilers: A review. *J. Appl. Poult. Res.*, 24, 408–26.

Cosgrove, S. E. 2006. The relationship between antimicrobial resistance and patient outcomes: mortality, length of hospital stay, and health care costs. *Clin. Infect. Dis.*, 42 (Suppl. 2), S82–9.

Crump, J. A., Luby, S. P. and Mintz, E. D. 2004. The global burden of typhoid fever. *Bull. World Health Organ.*, 82, 346–53.

Cui, S., Ge, B., Zheng, J. and Meng, J. 2005. Prevalence and antimicrobial resistance of *Campylobacter* spp. and *Salmonella* serovars in organic chickens from Maryland retail stores. *Appl. Environ. Microbiol.*, 71, 4108–11.

Cuny, C., Wieler, L. H. and Witte, W. 2015. Livestock-Associated MRSA: The Impact on Humans. *Antibiotics (Basel)*, 4, 521–43.

Currie, A., Macdougall, L., Aramini, J., Gaulin, C., Ahmed, R. and Isaacs, S. 2005. Frozen chicken nuggets and strips and eggs are leading risk factors for *Salmonella* Heidelberg infections in Canada. *Epidemiol. Infect.*, 133, 809–16.

D'costa, V. M., McGrann, K. M., Hughes, D. W. and Wright, G. D. 2006. Sampling the antibiotic resistome. *Science*, 311, 374–7.

Dhama, K., Tiwari, R., Khan, R. U., Chakraborty, S., Gopi, M., Karthik, K., Saminathan, M., Desingu, P. A. and Sunkara, L. T. 2014. Growth promoters and novel feed additives improving poultry production and health, bioactive principles and beneficial applications: The trends and advances – a review. *Int. J. Pharmacol.*, 10, 129–59.

EFSA (European Food Safety Authority). 2010. Analysis of the baseline survey on the prevalence of *Campylobacter* in broiler batches and of *Campylobacter* and *Salmonella* on broiler carcasses in the EU, 2008. *EFSA J.*, 8(8), 1522. http://www.efsa.europa.eu/en/efsajournal/doc/1522.pdf (Accessed July 2016).

Elliott, K. 2015. Antibiotics on the Farm: Agriculture's Role in Drug Resistance. Policy Paper 059. Center for Global Development, Washington, DC.

Elliott, S. D. and Barnes, E. M. 1959. Changes in serological type and antibiotic resistance of Lancefield group D streptococci in chickens receiving dietary chlortetracycline. *J. Gen. Microbiol.*, 20, 426–33.

European Commision. 2010. Commission Regulation (EU) No 37/2010. Official Journal of the European Union, no. 2377. http://ec.europa.eu/health/files/eudralex/vol-5/reg_2010_37/reg_2010_37_en.pdf (Accesed August 2016).

European Commission. 2011. Communication from the Commission to the European Parliament and the Council: Action Plan against the Rising Threats from Antimicrobial Resistance. http://

ec.europa.eu/dgs/health_food-safety/docs/communication_amr_2011_748_en.pdf (Accessed August 2016).

FDA (Food and Drug Administration). 2005. Final decision of the FDA commissioner. Withdrawal of approval of the new animal drug application for enrofloxacin in poultry. Rockville (MD): Center for Veterinary Medicine, 27 July 2005. Docket no. 2000N-1571. http://www.fda.gov/AnimalVeterinary/SafetyHealth/RecallsWithdrawals/ucm042004.htm (Accessed August 2016).

FDA. 2009. NARMS (National Antimicrobial Resistance Monitoring System) at Work. http://www.fda.gov/AnimalVeterinary/SafetyHealth/AntimicrobialResistance/NationalAntimicrobialResistanceMonitoringSystem/ucm431899.htm (Accessed August 2016).

FDA. 2012. National Antimicrobial esistance Monitoring System-Enteric Bacteria (NARMS): 2010 Executive Report. Rockville, MD: U.S. Department of Health and Human Services, Food and Drug Administration.

FDA. 2013. Phasing out certain antibiotic use in farm animals. http://www.fda.gov/ForConsumers/ConsumerUpdates/ucm378100.htm (Accessed 4 August 2015).

FDA. 2015. 2014–2015 Retail Meat Interim Report. http://www.fda.gov/downloads/AnimalVeterinary/SafetyHealth/AntimicrobialResistance/NationalAntimicrobialResistanceMonitoringSystem/UCM498134.pdf (Accessed August 2016).

Fleming, A. 1929. On the antibacterial action of cultures of a penicillium, with special reference to their use in the isolation of B. influenzae. *Bull. World Health Organ.*, 79, 780–90.

Gallegos-Robles, M. A., Morales-Loredo, A., Alvarez-Ojeda, G., Vega, P. A., Chew, M. Y., Velarde, S. and Fratamico, P. 2008. Identification of *Salmonella* serotypes isolated from cantaloupe and chile pepper production systems in Mexico by PCR-restriction fragment length polymorphism. *J. Food Prot.*, 71, 2217–22.

Graham, J. P., Boland, J. J. and Silbergeld, E. 2007. Growth promoting antibiotics in food animal production: an economic analysis. *Public Health Rep.*, 122, 79–87.

Hellberg, R. S. and Chu, E. 2016. Effects of climate change on the persistence and dispersal of foodborne bacterial pathogens in the outdoor environment: A review. *Crit. Rev. Microbiol.*, 42, 548–72.

Hohmann, E. L. 2001. Nontyphoidal salmonellosis. *Clin. Infect. Dis.*, 32, 263–9.

Izat, A. L., Colberg, M., Reiber, M. A., Adams, M. H., Skinner, J. T., Cabel, M. C., Stilborn, H. L. and Waldroup, P. W. 1990. Effects of different antibiotics on performance, processing characteristics, and parts yield of broiler chickens. *Poult. Sci.*, 69, 1787–91.

Johnsen, P. J., Townsend, J. P., Bohn, T., Simonsen, G. S., Sundsfjord, A. and Nielsen, K. M. 2009. Factors affecting the reversal of antimicrobial-drug resistance. *Lancet. Infect Dis.*, 9, 357–64.

Jones, F. T. and Ricke, S. C. 2003. Observations on the history of the development of antimicrobials and their use in poultry feeds. *Poult. Sci.*, 82, 613–7.

Kashoma, I. P., Kumar, A., Sanad, Y. M., Gebreyes, W., Kazwala, R. R., Garabed, R. and Rajashekara, G. 2014. Phenotypic and genotypic diversity of thermophilic *Campylobacter* spp. in commercial turkey flocks: a longitudinal study. *Foodborne Pathog. Dis.*, 11, 850–60.

Kassem, I. I., Kehinde, O., Helmy, Y. A., Pina-Mimbela, R., Kumar, A., Chandrashekhar, K. and Rajashekara, G. 2016a. *Campylobacter* in Poultry: The Conundrums of Highly Adaptable and Ubiquitous Foodborne Pathogens. In Soon, J. M., Manning, L. and Wallace, C. A. (eds), *Foodborne Diseases Case Studies of Outbreaks in the Agri-Food Industries*, pp. 79–112. CRC Press, Florida.

Kassem, I. I., Kehinde, O., Kumar, A. and Rajashekara, G. 2016b. Antimicrobial-resistant *Campylobacter* in organically and conventionally raised layer chickens. *Foodborne Pathog. Dis.* doi:10.1089/fpd.2016.2161.

Krishnasamy, V., Otte, J. and Silbergeld, E. 2015. Antimicrobial use in Chinese swine and broiler poultry production. *Antimicrob. Resist. Infect. Control.*, 4, 17.

Landers, T. F., Cohen, B., Wittum, T. E. and Larson, E. L. 2012. A review of antibiotic use in food animals: perspective, policy, and potential. *Public Health Rep.*, 127, 4–22.

Lee, B. C., Reimers, N., Barnes, H. J., D'Lima, C., Carver, D. and Kathariou, S. 2005. Strain persistence and fluctuation of multiple-antibiotic resistant *Campylobacter coli* colonizing turkeys over successive production cycles. *Foodborne Pathog. Dis.*, 2, 103–10.

Luangtongkum, T., Jeon, B., Han, J., Plummer, P., Logue, C. M. and Zhang, Q. 2009. Antibiotic resistance in *Campylobacter*: emergence, transmission and persistence. *Future Microbiol.*, 4, 189–200.

Luangtongkum, T., Morishita, T. Y., Ison, A. J., Huang, S., Mcdermott, P. F. and Zhang, Q. 2006. Effect of conventional and organic production practices on the prevalence and antimicrobial resistance of *Campylobacter* spp. in poultry. *Appl. Environ. Microbiol.*, 72, 3600–7.

Luangtongkum, T., Shen, Z., Seng, V. W., Sahin, O., Jeon, B., Liu, P. and Zhang, Q. 2012. Impaired fitness and transmission of macrolide-resistant *Campylobacter jejuni* in its natural host. *Antimicrob. Agents Chemother.*, 56, 1300–8.

Luber, P., Wagner, J., Hahn, H. and Bartelt, E. 2003. Antimicrobial resistance in *Campylobacter jejuni* and *Campylobacter coli* strains isolated in 1991 and 2001–2002 from poultry and humans in Berlin, Germany. *Antimicrob. Agents Chemother.*, 47, 3825–30.

Luo, N., Pereira, S., Sahin, O., Lin, J., Huang, S., Michel, L. and Zhang, Q. 2005. Enhanced in vivo fitness of fluoroquinolone-resistant *Campylobacter jejuni* in the absence of antibiotic selection pressure. *Proc. Natl. Acad. Sci. U.S.A.*, 102, 541–6.

Ma, L., Wang, Y., Shen, J., Zhang, Q. and Wu, C. 2014. Tracking *Campylobacter* contamination along a broiler chicken production chain from the farm level to retail in China. *Int. J. Food Microbiol.*, 181, 77–84.

Majowicz, S. E., Musto, J., Scallan, E., Angulo, F. J., Kirk, M., O'Brien, S. J., Jones, T. F., Fazil, A., Hoekstra, R. M. and International Collaboration on Enteric Disease Burden of Illness, S. 2010. The global burden of nontyphoidal *Salmonella* gastroenteritis. *Clin. Infect. Dis.*, 50, 882–9.

Man, S. M. 2011. The clinical importance of emerging *Campylobacter* species. *Nat. Rev. Gastroenterol. Hepatol.*, 8, 669–85.

Marshall, B., Petrowski, D. and Levy, S. B. 1990. Inter- and intraspecies spread of Escherichia *coli* in a farm environment in the absence of antibiotic usage. *Proc. Natl. Acad. Sci. U.S.A.*, 87, 6609–13.

Marshall, B. M. and Levy, S. B. 2011. Food animals and antimicrobials: impacts on human health. *Clin. Microbiol. Rev.*, 24, 718–33.

Mcdermott, P. F. 2006. Antimicrobial Resistance in notyphoidal Salmonellae. In Aaretrup, F. M. (ed.), *Antimicrobial Resistance in Bacteria of Animal Origin*, pp. 293–314. ASM Press, Washington, DC.

Melnyk, A. H., Wong, A. and Kassen, R. 2015. The fitness costs of antibiotic resistance mutations. *Evol. Appl.*, 8, 273–83.

Mohle-Boetani, J. C., Farrar, J., Bradley, P., Barak, J. D., Miller, M., Mandrell, R., Mead, P., Keene, W. E., Cummings, K., Abbott, S., Werner, S. B. and Investigation, T. 2009. *Salmonella* infections associated with mung bean sprouts: epidemiological and environmental investigations. *Epidemiol. Infect.*, 137, 357–66.

Nachamkin, I., Allos, B. M. and Ho, T. W. 2000. *Campylobacter jejuni* infection and the association with Guillain-Barre´ syndrome. In Nachamkin, I. and Blaser, M. J. (eds), *Campylobacter*, 2nd ed., pp. 155–75. American Society for Microbiology, Washington, DC.

Nickell, J. S. and White, B. J. 2010. Metaphylactic antimicrobial therapy for bovine respiratory disease in stocker and feedlot cattle. *Vet. Clin. North Am. Food Anim. Pract.*, 26, 285–301.

Noyes, N. R., Yang, X., Linke, L. M., Magnuson, R. J., Cook, S. R., Zaheer, R., Yang, H., Woerner, D. R., Geornaras, I., Mcart, J. A., Gow, S. P., Ruiz, J., Jones, K. L., Boucher, C. A., Mcallister, T. A., Belk, K. E. and Morley, P. S. 2016. Characterization of the resistome in manure, soil and wastewater from dairy and beef production systems. *Sci. Rep.*, 6, 24645.

Nunes, I. A., Helmuth, R., Schroeter, A., Mead, G. C., Santos, M. A., Solari, C. A., Silva, O. R. and Ferreira, A. J. 2003. Phage typing of *Salmonella* enteritidis from different sources in Brazil. *J. Food Prot.*, 66, 324–7.

OIE (World Organisation for Animal Health). 2015. Combating Antimicrobial Resistance and Promoting the Prudent Use of Antimicrobial Agents in Animals. http://www.oie.int/fileadmin/Home/eng/Our_scientific_expertise/docs/pdf/AMR/A_RESO_AMR_2015.pdf (Accessed August 2016).

Parveen, S., Taabodi, M., Schwarz, J. G., Oscar, T. P., Harter-Dennis, J. and White, D. G. 2007. Prevalence and antimicrobial resistance of *Salmonella* recovered from processed poultry. *J. Food Prot.*, 70, 2466–72.

Perry, J., Waglechner, N. and Wright, G. 2016. The Prehistory of Antibiotic Resistance. *Cold Spring Harb. Perspect Med.*, 6, pii: a025197.

Price, L. B., Johnson, E., Vailes, R. and Silbergeld, E. 2005. Fluoroquinolone-resistant *Campylobacter* isolates from conventional and antibiotic-free chicken products. *Environ. Health Perspect.*, 113, 557–60.

Pruden, A., Larsson, D. G., Amezquita, A., Collignon, P., Brandt, K. K., Graham, D. W., Lazorchak, J. M., Suzuki, S., Silley, P., Snape, J. R., Topp, E., Zhang, T. and Zhu, Y. G. 2013. Management options for reducing the release of antibiotics and antibiotic resistance genes to the environment. *Environ. Health Perspect.*, 121, 878–85.

Rule, A. M., Evans, S. L. and Silbergeld, E. K. 2008. Food animal transport: a potential source of community exposures to health hazards from industrial farming (CAFOs). *J. Infect. Public Health*, 1, 33–9.

Sahin, O., Kassem, I., Shen, Z., Lin, J., Rajashekara, G. and Zhang, Q. 2015. *Campylobacter* in Poultry: Ecology and Potential Interventions. *Avian Dis.*, 59(2), 185–200.

Sarmah, A. K., Meyer, M. T. and Boxall, A. B. 2006. A global perspective on the use, sales, exposure pathways, occurrence, fate and effects of veterinary antibiotics (VAs) in the environment. *Chemosphere*, 65, 725–59.

Schmidt, C. W. 2005. Poultry's Persistence Problem: Drug-Resistant *Campylobacter* in Chicken. *Environ. Health Perspect.*, 113(5), A325.

Schwarz, S., Kehrenberg, C. and Walsh, T. R. 2001. Use of antimicrobial agents in veterinary medicine and food animal production. *Int. J. Antimicrob. Agents*, 17, 431–7.

Scott, A. M. and Fedorka-Cray, P. J. 2002. Antimicrobial Use and Resistance in Animals. *Clin. Infect. Dis.*, 34 (Suppl. 3), S93–S106. doi: 10.1086/340246.

Smith, J. A. 2002. The future of poultry production in the USA without antibiotics. *Poult. Int.*, 41, 68–9.

Smith, J. L. and Fratamico, P. M. 2010. Fluoroquinolone resistance in *Campylobacter*. *J. Food Prot.*, 73, 1141–52.

Sneeringer, S., Macdonald, J., Key, N., McBride, W. and Mathews, K. 2015. Economics of antibiotic use in US livestock production. USDA Economic Research Service Report, 200. http://ers.usda.gov/media/1950577/err200.pdf (Accessed August 2016).

Stobberingh, E., Van den Bogaard, A., London, N., Driessen, C., Top, J. and Willems, R. 1999. Enterococci with glycopeptide resistance in turkeys, turkey farmers, turkey slaughterers, and (sub)urban residents in the south of The Netherlands: evidence for transmission of vancomycin resistance from animals to humans? *Antimicrob. Agents Chemother.*, 43, 2215–21.

Stokstad, E. L., Jukes, T. H., Pierce, J., Page Jr., A. C. and Franklin, A. L. 1949. The multiple nature of the animal protein factor. *J. Biol. Chem.*, 180, 647–54.

Storteboom, H., Arabi, M., Davis, J. G., Crimi, B. and Pruden, A. 2010. Tracking antibiotic resistance genes in the South Platte River basin using molecular signatures of urban, agricultural, and pristine sources. *Environ. Sci. Technol.*, 44, 7397–404.

The Poultry Site 2016. http://www.thepoultrysite.com/ (Accessed August 2016).

Tribe, I. G., Cowell, D., Cameron, P. and Cameron, S. 2002. An outbreak of *Salmonella* typhimurium phage type 135 infection linked to the consumption of raw shell eggs in an aged care facility. *Commun. Dis. Intell. Q. Rep.*, 26, 38–9.

USDA, United States Department of Agriculture, 2015. USDA Poultry Production Data Factsheet, http://www.usda.gov/documents/nass-poultry-stats-factsheet.pdf (Accessed 4 August 2016).

Van Boeckel, T. P., Brower, C., Gilbert, M., Grenfell, B. T., Levin, S. A., Robinson, T. P., Teillant, A. and Laxminarayan, R. 2015. Global trends in antimicrobial use in food animals. *Proc. Natl. Acad. Sci. U.S.A.*, 112, 5649–54.

Van den Bogaard, A. E., London, N., Driessen, C. and Stobberingh, E. E. 2001. Antibiotic resistance of faecal *Escherichia coli* in poultry, poultry farmers and poultry slaughterers. *J. Antimicrob. Chemother.*, 47, 763–71.

Van den Bogaard, A. E., Willems, R., London, N., Top, J. and Stobberingh, E. E. 2002. Antibiotic resistance of faecal enterococci in poultry, poultry farmers and poultry slaughterers. *J. Antimicrob. Chemother.*, 49, 497–505.

Van Looveren, M., Chasseur-Libotte, M. L., Godard, C., Lammens, C., Wijdooghe, M., Peeters, L. and Goossens, H. 2001. Antimicrobial susceptibility of nontyphoidal *Salmonella* isolated from humans in Belgium. *Acta Clin. Belg.*, 56, 180–6.

Vieira, A. R., Collignon, P., Aarestrup, F. M., Mcewen, S. A., Hendriksen, R. S., Hald, T. and Wegener, H. C. 2011. Association between antimicrobial resistance in *Escherichia coli* isolates from food animals and blood stream isolates from humans in Europe: an ecological study. *Foodborne Pathog. Dis.*, 8, 1295–301.

WHO (World Health Organization). 2001. Antibiotic resistance: synthesis of recommendations by expert policy groups. Alliance for the Prudent Use of Antibiotics. http://whqlibdoc.who.int/hq/2001/WHO_CDS_CSR_DRS_2001.10.pdf (Accessed August 2016).

WHO. 2011. European strategic action plan on antibiotic resistance. http://www.euro.who.int/__data/assets/pdf_file/0008/147734/wd14E_AntibioticResistance_111380.pdf (Accessed August 2016).

WHO. 2015. WHO multi-country survey reveals widespread public misunderstanding about antibiotic resistance http://www.who.int/mediacentre/news/releases/2015/antibiotic-resistance/en/(Accessed August 2016).

Wozniak, A. 2011. Fluoroquinolones resistance of *Campylobacter jejuni* and *Campylobacter coli* isolated from poultry in 1994–1996 and 2005–2008 in Poland. *Bull. Vet. Ins. Pulawy*, 50, 15–20.

Wozniak, A. and Wieliczko, A. 2011. Tetracycline, erythromycin, and gentamicin resistance of *Campylobacter jejuni* and *Campylobacter coli* isolated from poultry in Poland. *Bull. Vet. Ins. Pulawy.*, 55, 1.

World Bank. 2016. data.Worldbank.org (Accessed July 2016).

Wright, G. D. 2007. The antibiotic resistome: the nexus of chemical and genetic diversity. *Nat. Rev. Microbiol.*, 5, 175–86.

Yildirim, Y., Gonulalan, Z., Pamuk, S. and Ertas, N. 2011. Incidence and antibiotic resistance of *Salmonella* spp. on raw chicken carcasses. *Food Res. Int.*, 44, 725–8.

Chapter 5

Alternatives to antibiotics in preventing zoonoses and other pathogens in poultry: prebiotics and related compounds

S. C. Ricke, University of Arkansas, USA, A. V. S. Perumalla, Kerry, USA; and Navam S. Hettiarachchy, University of Arkansas, USA

1 Introduction

While bird performance is economically important to the poultry industry, food-borne pathogens that have been identified with poultry, such as *Salmonella* and *Campylobacter*, continue to be a major issue of concern. Although more is being understood about their abilities to colonize the gastrointestinal (GI) tracts of chickens and the subsequent corresponding interaction with the host, questions still remain (Park et al., 2008; Dunkley et al., 2009; Horrocks et al., 2009; Finstad et al., 2012; Howard et al., 2012; Foley et al., 2011, 2013). For example, there are increasing concerns about the risk of developing cross-resistance and multiple antibiotic resistances in pathogenic bacteria of the host animal. In the last few decades, several antibiotics and chemotherapeutic agents have been used in prophylactic doses in poultry feed to improve the poultry performance and thereby achieve economic benefits (Jones and Ricke, 2003). As a result, there is potential for outbreaks of particular enteric diseases originating from food animal sources such as poultry that would greatly compromise public health treatment efforts

http://dx.doi.org/10.19103/AS.2016.0010.06

due to acquired antibiotic resistance (Jones and Ricke, 2003; Chen et al., 2015). These enteric pathogens could lower the productivity as well as increase bird mortality and associated contamination of poultry products for human consumption. These concerns have challenged the meat animal industry to seek a wide range of feed amendment alternatives to not only maintain live animal production proficiency but also retain quality attributes of the resulting food product (Jones and Ricke, 2003; Hajati and Rezaei, 2010; Alloui et al., 2013; Ricke, 2015).

Among the attributes of health and quality that are of concern to the public as well as the processor is microbial contamination that cannot only be a cause of diminished shelf-life, but, depending on the nature of the contaminant, also a food safety risk. While several postharvest strategies are currently implemented to decontaminate and thus minimize spoilage costs and economic losses caused by food-borne diseases/recalls (Ricke et al., 2005; Sofos et al., 2013), preharvest food safety options remain limited. Preharvest food safety control measures, including vaccination, competitive exclusion cultures and the use of natural antimicrobial feed additives such as plant extract compounds, bacteriophage, bacteriocins and their combinations, just to name a few, have all either been proposed, experimentally tested or in some cases currently implemented as means to limit either the colonization and/or carriage of food-borne pathogens such as *Salmonella* associated with poultry (Fuller, 1989; Joerger, 2003; Patterson and Burkholder, 2003; Rastall, 2004; Ricke and Pillai, 1999; Revolledo et al., 2006; Vandeplas et al., 2010; Hume, 2011; Callaway and Ricke, 2012; Ricke et al., 2012a; O'Bryan et al., 2015; Ricke, 2015; Rivera Calo et al., 2015). A number of competitive exclusion cultures, also known as probiotics, have been developed over the years for poultry, some of which consist of a complex consortia of microorganisms, while others rely only on a single organism such as *Bacillus subtilis* (Stavric, 1992; Nisbet et al., 1994; 1996a,b; Ricke and Pillai, 1999; Nisbet, 2002; Siragusa and Ricke, 2012; Ricke and Saengkerdsub, 2015).

Prebiotics were originally defined by Gibson and Roberfroid (1995) as 'a non-digestible food ingredient that beneficially affects the host by selectively stimulating the growth and/or activity of one or a limited number of GI microflora', but that definition has been further refined as more has become understood about the interaction between the gut microbiome and dietary components (Gibson and Roberfroid, 1995; Wang, 2009; Bird et al., 2010; Hutkins et al., 2016). More simply put, prebiotics can generally be considered as a class of compounds that specifically provide substrates (energy) for the beneficial GI microbiome (Hajati and Rezaei, 2010). Prebiotics are non-digestible ingredients; passing through the small intestine to the lower gut where they undergo hydrolysis and subsequent fermentation by colonic and/or cecal bacteria to produce short-chain fatty acids (SCFAs), namely acetate, butyrate and propionate that, in turn, can be absorbed by the host (Demigné et al., 1986; Józefiak et al., 2004; Wang, 2009; Bird et al., 2010).

Prebiotics used in poultry nutrition for preharvest safety include non-digestible carbohydrates such as fructooligosaccharide (FOS) products (oligofructose, inulin-type fructans found in many vegetables), *trans*-galactooligosaccharides (TOS), glucooligosaccharides, glycooligosaccharides, lactulose, lactitol, maltooligosaccharides, soy-oligosaccharides, xylo-oligosaccharides, stachyose, raffinose and yeast cell walls (mannan-oligosachharides (MOSs)) and other fermentable carbohydrates which are not or only minimally digested in the small intestine (Monsan and Paul, 1995; Flickinger et al., 2003; Collins and Gibson, 1999; Ricke, 2015; Roto et al., 2015). In general, the most common oligosaccharides are inulin and its hydrolysates and oligofructans, present in chicory, onion, garlic, asparagus, artichoke, leek, bananas, tomatoes and many other plants (Flickinger et al., 2003; Ricke, 2015). Such prebiotic oligosaccharides can be produced by extraction from plant sources, microbial synthesis or enzymatic hydrolysis of polysaccharides (Crittenden and Playne, 1996; Ricke, 2015). In recent years, use of prebiotics has become a promising approach and offers a potential alternative to growth-promoting antibiotics especially in the monogastric small intestine (Yang et al., 2009; Hume, 2011; Callaway and Ricke, 2012; Alloui et al., 2013; Ricke, 2015; Roto et al., 2015).

2 Beneficial effects of prebiotics: general mechanisms of action

Beneficial effects of prebiotics include altering the GI microbiome, stimulating the immune system, preventing colon cancer, reducing growth and host invasion by such bacteria as *Salmonella* and pathogenic *Escherichia coli* (Cummings and Macfarlane, 2002; Bengmark, 2012; Roto et al., 2015). In general, lactobacilli and bifidobacteria are the two common microbial groups thought to be specifically stimulated by prebiotic supplements (Kaplan and Hutkins, 2000; Ricke, 2015).

Prebiotics can act by physically binding pathogens, increasing the osmotic potential in the lumen of the intestine and/or through metabolites that are produced by the intestinal microbiome which use the prebiotic compounds as specific substrates for their own metabolism (Hajati and Rezaei, 2010). Carbohydrates and proteins indigestible to the host undergo microbiota-dependent hydrolysis and degradation in the lower intestine to produce SCFA and branched chain fatty acids, respectively, that generates Adenosine Triphosphate (ATP) via fermentation pathways (Schell et al., 2002; Ten Bruggencate et al., 2003; Apajalahti, 2005; Rossi et al., 2005; Ricke, 2015; Roto et al., 2015). Even when not utilized for energy production, some prebiotics such as MOSs exert effects to the host by binding to the mannose receptors and inhibit pathogen colonization of the gut (Spring et al., 2000; Patterson and Burkholder, 2003). Overall, there is evidence that incorporating prebiotics as feed supplements facilitates antagonistic effects on enteric pathogens by competitive exclusion, promotion of enzyme reactions (Hajati and Rezaei, 2010), immune stimulation (Monsan and Paul, 1995; Patterson and Burkholder, 2003) and an increase in resistance to pathogen colonization (Bengmark, 2001; Patterson and Burkholder, 2003; Ricke, 2015). Factors influencing the effect of prebiotics include type of diet (amount of non-digestible carbohydrates), type and level of supplements used, animal parameters (species, age, sex and stage of production) (Yang et al., 2009) and hygiene status of the farm (Verdonk et al., 2005; Yang et al., 2009). Specific actions of particular oligosaccharides displaying prebiotic properties are discussed in more detail in the following sections.

3 Non-digestible carbohydrates as prebiotics

These types of carbohydrates are non-digestible in the small intestine; however, they are hydrolysed by colonic bacteria of the large intestine and cecal bacteria depending on the host (Józefiak et al., 2004; Ricke, 2015). The main categories of non-digestible oligosaccharides (NDOs) commonly used in prebiotic supplements include carbohydrates whose monosaccharide units are fructose, galactose, glucose and/or xylose (Bird, 1999).

Indigestible carbohydrates can act as prebiotics when they undergo saccharolytic fermentation in the hindgut and ceca to produce SCFA (Bird, 1999; Józefiak et al., 2004; Apajalahti, 2005; Bird et al., 2010; Hutkins et al., 2016). These fatty acids exhibit beneficial effects including lowering the pH of the environment, improving mineral absorption, inhibiting acid-sensitive pathogens and enhancing the proliferation of indigenous bacteria such as bifidobacteria (Patterson and Burkholder, 2003; Apajalahti, 2005; Bengmark, 2012; Roto et al., 2015).

Non-digestible carbohydrates include NDOs, non-starch polysaccharides and resistant starch (Bird, 1999; Bird et al., 2010; Hutkins et al., 2016). In the beginning, NDOs were used as low-calorie, low cariogenic, sucrose substitutes and bulking agents in the food industry (Crittenden and Playne, 1996; Ricke, 2015). Commercially available prebiotics include FOS, lactulose, isomaltooligosaccharides (IMO), galactooligosaccharides (GOS), TOS, inulin and oligofructose (Collins and Gibson, 1999). Except for FOS, these compounds do not necessarily meet the traditional criteria for prebiotics originally stated by Gibson and Roberfroid (1995) because no specific mechanism of action of colonic fermentation of these complex carbohydrates was known. However, as the gut microbiota and their various metabolic interactions have become better understood, the definition of prebiotics has evolved and now also includes a much wider range of compounds such as NDOs (Bird, 1999; Patterson and Burkholder, 2003; Roberfroid, 2000, 2007; Bird et al., 2010; Bengmark, 2012; Ricke, 2015; Hutkins et al., 2016).

There has been an increasing trend in poultry nutrition research to formulate feeds rich in NDOs that can act as prebiotic supplements (Patterson and Burkholder, 2003; Hajati and Rezaei, 2010; Alloui et al., 2013). Since these relatively resistant oligosaccharides escape the small intestine and can reach the ceca largely intact, they could potentially support the growth of beneficial bacteria including bifidobacteria and lactobacilli as well as other fermentative bacteria and thus exhibit prebiotic properties, as described in the previous section (Bird, 1999; Gomes and Malcata, 1999; Józefiak et al., 2004; Ricke et al., 2013; Lindberg, 2014). As non-traditional cereal grains, cereal coproducts and protein crops have become a greater part of the primary nutritional sources for energy and protein in poultry, increases in the intake of fibre and non-starch carbohydrates may have detrimental impacts on bird performance (Knudsen, 2014); however, with additional research on further processing and extraction approaches, these feedstuffs may offer opportunities to recover components with prebiotic properties. For example, Herfel et al. (2013) demonstrated that stabilized rice bran improved weanling pig performance by exhibiting prebiotic characteristics and Kumar et al. (2012) reported a reduction of *S.* Typhimurium faecal shedding and an increase in colonization of indigenous lactobacilli in mice fed with rice bran. The opportunity to isolate such components and screen them *in vitro* under conditions that simulate the GI tract as well as more complete analyses of gut contents (metabolomics, proteomics and others) from poultry fed diets with these dietary ingredients should help to better refine the prebiotic characteristics and develop more targeted benefits (Bird, 1999; Rastall and Maitin, 2002; Donalson et al., 2007, 2008a; Dunkley et al., 2007a,b; Bengmark, 2012; Park et al., 2013; Ricke et al., 2013; Lindberg, 2014).

4 Fructooligosaccharides

FOSs are one of the more frequently used prebiotics (Flickinger et al., 2003; Ricke, 2015). Common sources include onions, Jerusalem artichokes, bamboo shoots, chicory roots and bananas (Flickinger et al., 2003; Ricke, 2015). FOSs are non-digestible and non-absorbable oligosaccharides consisting of short-chain polymers of 1 to 2 linked fructose units (short-chain FOS, SCFO) that can be produced commercially either by hydrolysis of inulin or by enzymatic synthesis from sucrose or lactose (Patterson and Burkholder, 2003; Ricke, 2015). FOSs are resistant to enzymatic degradation and absorption due to the presence of β-linkages, and hence pass through the upper GI tract to reach the colon ceca, where they undergo fermentation (Gibson and Roberfroid, 1995; Gibson, 1999; Flickinger et al., 2003; Xu et al., 2003; Józefiak et al., 2004). The end products of FOS fermentation include SCFAs, which have the ability to inhibit *Salmonella*, *E. coli* and *Clostridium perfringens* (Cummings et al., 2001; Cummings and Macfarlane, 2002; Flickinger et al., 2003; Patterson and Burkholder, 2003; Ricke 2003a). In addition, FOS can act as a fermentable substrate and thus select beneficial intestinal bacteria such as bifidobacteria and lactobacilli (Gibson and Wang, 1994; Ricke, 2015), which, in turn, reduce the pathogens by competitive exclusion (Gibson and Roberfroid, 1995; Flickinger et al., 2003; Xu et al., 2003; Ricke, 2015).

The use of FOS in poultry has been reviewed extensively elsewhere (Ricke, 2015) and will be discussed only briefly in the current review. FOSs can be selectively fermented by most strains of bifidobacteria and support proliferation of bifidobacteria in poultry (Kaplan and Hutkins, 2000; Rossi et al., 2005; Fukuda et al., 2011; Ricke, 2015). Bifidobacteria are anaerobic, gram-positive bacteria found in the GI tract of various warm-blooded animals. The presence of bifidobacteria has been demonstrated to suppress potential pathogens by producing antimicrobials or by lowering the pH through the rapid production of organic acids such as acetate (Fukuda et al., 2011).

Numerous studies have demonstrated the ability of FOS supplementation to reduce the colonization of *Salmonella* in the GI tract (Bailey et al., 1991; Fukata et al., 1999; Donalson et al., 2008b; Ricke, 2015). Dietary amendments to limit *Salmonella* colonization and infection in commercial birds has proven to be particularly critical for laying hens undergoing induced moults (Ricke 2003b; Ricke et al., 2013). Historically, the moulting of commercial laying hens was initiated by complete removal of feed for a number of days to shut down the reproductive tract and cease egg production as a prelude to starting an additional egg laying cycle (Ricke, 2003b). However, emptying of the intestinal tract creates a microenvironment favourable for *S.* Enteritidis colonization and subsequent systemic invasion into the reproductive tract (Durant et al., 1999; Ricke, 2003b; Ricke et al., 2013). Supplying low-energy high-fibre diets such as alfalfa during the moult period proved to be effective in limiting *Salmonella* colonization and infection (Ricke et al., 2013). Although Donalson et al. (2008b) reported that laying hens fed with alfalfa-based diets supplemented with FOS reduced the colonization of *S.* Enteritidis in ovaries and livers when compared to the hens subjected to feed withdrawal, some inconsistencies among independent trials were observed. Clearly, more needs to be understood about the impact of FOS on not only the gut microbiota, but also the physiological status of the gut architecture, along with parameters such as passage rate, and how this impacts the residence time of dietary FOS (Ricke, 2015). For example, Hanning et al. (2012) reported increased villi length and intestinal crypt depth in pasture flock-raised Naked Neck birds fed with FOS, compared to control birds by the end of an eight-week growth cycle.

In a few studies, FOS in their various forms have been used to limit establishment of certain pathogens in poultry. Short-chain FOS and MOS supplementation (4 g/kg) of a dextrose-isolated soy protein diet decreased cecal *C. perfringens* population levels in some of the trials reported by Biggs et al. (2007). When feeding male broilers, Xu et al. (2003) observed that adding 4.0 g/kg FOS enhanced the growth of *Bifidobacterium* and

Lactobacillus, but inhibited *E. coli* populations in small intestinal and cecal contents. Yusrizal and Chen (2003) reported that supplementation of broiler diets with fructans resulted in an increase in lactobacilli counts in the gut and decrease in *Campylobacter* and *Salmonella* in some of the reported trials.

Over time, the mechanisms by which FOS select for beneficial bacteria and create an adverse gut environment for *Salmonella* have become better understood, but the same is not true for other pathogens. In part, this is the case because fewer studies have been done that looked at bacteria other than *Salmonella*. It is important that the effect of FOS on these pathogens also be explored, because the interactions of the indigenous microbiome and these pathogens may differ from that experienced by *Salmonella*. For example, Indikova et al. (2015) have suggested that *Campylobacter* may, to some extent, actually be protected by the microbial consortia that is associated within mixed microbial populations. If these relationships consistently exist in the GI tract, it will be critical to identify the individuals among these potentially supportive microbial communities and use this information to design prebiotics that more effectively target either members antagonistic to *Campylobacter* or that limit those members that are most mutualistically supportive of *Campylobacter*.

5 Yeast-derived components and mannan derivatives as pro- and prebiotics

Yeasts are eukaryotic and unicellular fungi that have been used for both preventive and therapeutic effects against diarrhoea and other GI disturbances (Auclair, 2000; Roto et al., 2015). *Saccharomyces cerevisiae* cells have been extensively examined and commercialized as a potential probiotic source because of their long history in the food and beverage industry that made them readily available (Roto et al., 2015). Live yeast cells gained early popularity as an animal feed additive with various health and animal performance benefits associated with administration either alone or in combination with other probiotic bacteria (Bradley et al., 1994; Auclair, 2000; Fonty and Chaucheyras-Durand, 2006; Roto et al., 2015). Most of the work has been focused on live yeasts as probiotics for ruminants (Roto et al., 2015). Several benefits to rumen function have been noted, including positive impacts on fibre-degrading bacteria, stabilization of rumen pH and enhancement of lactate-utilizing ruminal bacteria (Jouany, 2001; Fonty and Chaucheyras-Durand, 2006).

The presence of lactate-utilizing ruminal bacteria can be critical for ruminants under certain dietary conditions. Lactate accumulation in the rumen occurs after sudden dietary shifts to highly fermentable carbohydrates that can be readily fermented by lactate producers such as *Streptococcus bovis* and other lactic acid bacteria (Herrera et al., 2009). If left unchecked, lactic acid accumulation can become a severe condition in ruminants leading to lactic acidosis, acidification of the rumen and in some cases death of the animal (Herrera et al., 2009). When the lactate-utilizing ruminal bacterium *Selenomonas ruminantium* (Ricke et al., 1996) was co-cultured with a commercial *S. cerevisiae* filter-sterilized filtrate, Nisbet and Martin (1991) observed increases on growth and uptake of radiolabelled lactate by the selenomonad. More recently, Ding et al. (2014) demonstrated that when live *S. cerevisiae* were included in the diets of beef steers, the percentages of lactate-producing bacteria were reduced, while the percentage of *Selenomonas ruminantium* was increased.

In addition to serving as probiotics, yeast cell components can be used as prebiotics. The principal components, glucans and mannans, present in the yeast cell wall along with the fermentation and metabolite products have been extensively examined for prebiotic properties (Baurhoo et al., 2009; Roto et al., 2015). The *S. cerevisiae* cell wall consists of an outer layer composed mostly of glycosylated mannoproteins (90% carbohydrate) and an inner layer (50–60% of the cell wall dry weight) composed of β 1,3-glucan and chitin (Klis et al., 2002). Yeast mannans that are closely associated with cell wall proteins are one of the primary soluble polysaccharides that can be extracted from the cell wall (Lee and Ballou, 1965; Stewart et al., 1968). Although mannan structures are yeast species specific, they are generally characterized as macromolecular polymers in which the mannose backbone is linked via α-(1-6) bonds with side chain-bearing short chains (average of 2–3 mannose units containing α-(1-2) and α-(1-3) linkages) attached to the backbone via α-(1-2) linkages (Jones and Ballou, 1968; Raschke et al., 1973).

MOSs as an isolated component of the yeast outer cell wall were introduced as a feed additive in the early 1990s (Hooge, 2004). The exact mechanism of inhibition of pathogenic bacteria is still uncertain, but several general modes of actions have been proposed and can be summarized as follows: (a) MOS may bind the bacteria with type-1 fimbriae, thereby inhibiting them from binding to carbohydrate moieties of the intestinal lining (Hooge, 2004). Steer et al. (2000) has pointed out that this may be a key mechanism, as most pathogens possess carbohydrate-specific binding proteins for attachment to host cells, and any compound that interferes with that cannot only potentially prevent colonization, but also dislodge already attached cells. (b) Agglutination or clumping of pathogenic cells with type-1 fimbriae and subsequent expulsion from the incubation solution, thus reducing subclinical or lethal infection (Spring et al., 2000; Steer et al., 2000). (c) Immune system modulation (Tizard et al., 1989; Hajati and Rezaei, 2010; Alloui et al., 2013). (d) Modulation of intestinal brush border and mucin enzyme expression (Hooge, 2004; Hajati and Rezaei, 2010). (e) Mannan-derivative induction of phenotypic and functional maturation in mouse dendritic cells (Sheng et al., 2006). The immune modulatory effects of MOS have been attributed to several aspects, as summarized by Tizard et al. (1989), including (1) binding to mannose-binding proteins, (2) attenuating presentation of antigens (from pathogens) to activate macrophages and (3) induction of interleukin-l release.

As a prebiotic, MOS has been examined for both bird health and performance as well as resistance to colonization by food-borne pathogens. Based on an overall statistical meta-analysis across broiler pen trials conducted globally from 1993 to 2003, Hooge (2004) concluded that dietary supplementation of MOS decreased mortalities, while achieving similar body weights and feed conversion ratios (FCRs) when compared to birds receiving antibiotic-supplemented diets. This retention of bird performance appears to hold true when MOS has been combined with immuno-based treatments such as vaccines or introduction of infective microorganisms. For example, Yang et al. (2008b) reported improvements in FCR and/or body weight gains in birds fed with MOS and challenged with pathogenic *E. coli*, compared to the negative control. Likewise, when Nollet et al. (2007) fed MOS to birds vaccinated for *Eimeria* coccidia, followed by a coccidian challenge, they found that MOS improved the FCR of the vaccinated and *Eimeria*-challenged broilers from 15 to 42 days of age, compared to the birds receiving the same coccidia/vaccination combination, but no MOS while remaining the same as the FCR for non-vaccinated/non-challenged birds.

Oyofo et al. (1989a,b) were able to demonstrate that D-mannose could block *in vitro* attachment on chicken small intestinal epithelial cells and reduce cecal colonization of

S. Typhimurium in challenged birds. However, MOSs impact on cecal microbiota and the cecal microenvironment has been somewhat inconsistent. Spring et al. (2000) did see some reductions in *S.* Typhimurium and *S.* Dublin-challenged broiler chicks fed with MOS, but no effect on cecal concentrations of lactobacilli, enterococci, anaerobic bacteria, lactate, SCFA or cecal pH. Populations of lactobacilli and coliforms were reduced in the ileum of birds fed with 2 g/kg of MOS, but less impact was observed in the ceca (Yang et al., 2008a). Environmental management conditions may be a factor as well. For example, when birds were raised in the presence of new litter and fed with 2 g/kg of MOS, Yang et al. (2007) did not detect any differences in total anaerobes, *C. perfringens* or lactic acid bacterial populations in the ileum, duodenum or ceca. Using a *C. perfringens* challenge model, Thanissery et al. (2010) observed only a slight reduction of intestinal *C. perfringens* in broiler chicks fed with a commercial yeast extract on days 1 and 7 post-challenge.

6 Galactooligosaccharide and isomaltooligosaccharide

GOSs are generated from lactose by a glycosyl transfer of one or more D-galactosyl units onto a chain of galactose units (Crittenden and Playne, 1996; Mahoney, 1998; Boon et al., 2000; Tzortzis et al., 2005; Macfarlane et al., 2008). The composition of GOS varies depending on the degree of polymerization ranging from two to eight monomeric units (Macfarlane et al., 2008). The impact of GOS on the host may depend on a number of factors, but they are believed to preferentially stimulate the growth of organisms such as bifidobacteria and lactobacilli by providing a fermentable substrate and thus generating an inhibitory GI microenvironment for pathogens (Gomes and Malcata, 1999; Rastall and Maitin, 2002; Tzortzis et al., 2005; Macfarlane et al., 2008; Saminathan et al., 2011; Alloui et al., 2013). Other inhibitory mechanisms may be in play as well. Shoaf et al. (2006) demonstrated that GOS could reduce adherence of enteropathogenic *E. coli* to epithelial cell tissue cultures, leading them to suggest that GOS could act as a structural mimic of the pathogen binding sites on the GI epithelial surface. However, such binding may be both host- and pathogen strain-specific as Naughton et al. (2001) did not observe decreases in *S.* Typhimurium or non-pathogenic *E. coli* association with pig intestinal cells when a commercial GOS product was included. Intestinal architecture could also be influenced, as Hanning et al. (2012) reported that at eight weeks pasture, flock-raised Naked Neck broilers fed with GOS exhibited the deepest intestinal crypt depth of the intestinal villi, when compared to birds fed either with FOS, plum fibres or with no feed additive.

IMOs are glucosyl saccharides linked by α 1-6 glucosidic bonds and occur in the form of compounds such as isomaltose, panose and isomaltotriose (Kuriki et al., 1993). They possess a degree of polymerization anywhere from 2 to 6 and can be derived from a variety of sources, including starch, maltose, dextran and sucrose (Crittenden and Playne, 1996; Aslan and Tanriseven, 2007). IMOs have been proposed as a potential prebiotic for poultry use (Chung and Day, 2004). Based on this premise, Chung and Day (2004) screened a branched α-IMO produced by *Leuconostoc mesenteroides* and demonstrated that unidentified poultry cecal isolates as well as *Bifidobacterium longum* could use the IMO but *E. coli* and *S.* Typhimurium could not. When cecal isolates and *S.* Typhimurium were grown in mixed culture with IMO, the *Salmonella* populations were reduced (Chung and Day, 2004). In a follow-up *in vivo* study, broilers challenged with *S.* Typhimurium and fed either 1, 2 or 4% wt/wt IMO exhibited a 2-log CFU/ml reduction of cecal *Salmonella* only at the 1% IMO level versus the control (no IMO), while an increase in cecal bifidobacteria populations was similar for all levels of IMO (Thitaram et al., 2005). However, Zhang et al. (2003) reported that while IMO inclusion in broiler diets improved performance during the first three weeks, no detectable changes occurred in the levels of *Lactobacillus*, *E. coli* and total aerobes from the crop and the ceca.

7 Guar gum as a potential prebiotic source

Guar gum is a highly viscous soluble fibre produced from the seed of the Indian cluster bean (*Cyamopsis tetragonoloba*) and has been used widely in the food industry as an emulsion stabilizer and thickener (Tuohy et al., 2001; Slavin and Greenberg, 2003). Guar gum is a polysaccharide, consisting of a 1-4 linked β-D-mannopyranose backbone with a branched 1-6-α-D-galactopyranose. Guar meal is a by-product of guar gum manufacture and exhibits two deleterious properties, namely, it acts as a trypsin inhibitor and contains residual gum (approximately 18–20%) (Couch et al., 1966; Nagpal et al., 1971; Lee et al., 2003). Since guar gum is a highly branched polysaccharide, it can be categorized in the 'well-fermentable' fibre category which includes pectin, acacia, polydextrose, inulin and oligosaccharides, which when fermented, result in large quantities of SCFAs, including acetic, butyric and propionic acids in varying proportions (Bengmark, 1998; Bird, 1999; Józefiak et al., 2004; Ricke et al., 2013; Lindberg, 2014; Hutkins et al., 2016).

Feeding guar gum and guar meal to broilers has reportedly caused growth depression (Vohra and Kratzer, 1964a,b; Verma and MacNab, 1982). Over the years, this was explained by the presence of gum residue as the former is responsible for delayed gastric emptying and increased intestinal transit time because of its naturally high viscosity (Anderson and Warnick, 1964; Vohra and Kratzer, 1964a,b; Blackburn and Johnson, 1981). Guar gums also elicit some beneficial effects in some hosts such as decreased plasma cholesterol (Moriceau et al., 2000; Yamamoto et al., 2000), decreased postprandial serum glucose (Fairchild et al., 1996; Ou et al., 2001) and suppression of colon cancer (Heitman et al., 1992).

Guar meal is a processed by-product of guar beans rich in crude protein and consists of two fractions: a high protein germ fraction and a low protein psyllium husk (Van Etten et al., 1961; Couch et al., 1967; Lee et al., 2003). The gum content of typical guar meals is anywhere from 18 to 20% and contains the trypsin inhibitor (Couch et al., 1966; Nagpal et al., 1971). When considered for meeting amino acid requirements of poultry, analysis of guar meal reveals that it is rich in arginine, but low in sulphur amino acids (Verma and MacNab, 1984). Leucine and isoleucine are low when compared to soya bean and groundnut protein sources (Verma and MacNab, 1984). By and large, the germ fraction is considered the more beneficial fraction compared to the hull fraction for inclusion in broiler diets (Lee et al. 2003). Feeding high concentrations (10–15%) of guar meal caused diarrhoea, depressed growth and increased mortality in broilers, and reduced egg production in layers (Patel and McGinnis, 1985), but was an effective moult induction diet for initiating a second egg production cycle in laying hens (Zimmermann et al., 1987). The upper limit for feeding guar germ fraction to broiler chickens can be increased somewhat with the addition of an industrial source of β-mannanase in the feed (Lee et al., 2005).

Partially hydrolyzed guar gum (PHGG), a mixture of galactomannans, low in molecular weight and viscosity, is obtained by selectively cleaving the backbone chain of the guar gum using endo-β-D-mannanase and can be added to feed formulations without retaining the deleterious effects of the gum residue (Slavin and Greenberg, 2003; Yoon et al., 2008). PHGG has demonstrated therapeutic effects including reducing incidence or duration of diarrhoea (Hohmann et al., 1994; Slavin and Greenberg, 2003; Alam et al., 2000, 2005) and improving the GI microbiome by selection of beneficial microorganisms such as bifidobacteria and lactobacilli spp. in human subjects (Okubo et al., 1994; Tuohy et al., 2001). Feeding PHHG (0.025%) to pullets for six weeks proved to be beneficial as it decreased *S.* Enteritidis colonization in internal organs and intestine, reduced incidence on the egg surface and increased the cecal bifidobacteria and lactobacilli spp. (Ishihara et al., 2000). Moreover, feeding Sprague–Dawley rats with PHHG enhanced IgA productivity in the spleen and mesenteric lymph node lymphocytes, when compared to cellulose-fed rats (Yamada et al., 1999).

8 Synbiotics: combining pre- and probiotics for enhanced nutritional supplements

As there is evidence that probiotics can have similar effects on bird performance, health and prevent colonization by food-borne pathogens as prebiotics, it makes sense to combine selected pre- and probiotics to achieve an even more effective feed additive. Since many prebiotics serve as substrates for certain GI organisms, their effects might be enhanced by combining them with selective probiotic cultures that would employ synergism in their action for better protective effects against food-borne pathogens. These combinations have been referred to as synbiotics and are defined as nutritional supplements which contain a mixture of prebiotics and probiotics that act synergistically and deliver beneficial effects to the host by improving the survival and establishment of live microbial dietary supplements in the GI tract (Gibson and Roberfroid, 1995; Schrezenmeir and de Vrese, 2001; Rastall and Maitin, 2002; Patterson and Burkholder, 2003; Rastall, 2004; Bengmark, 2012). However, achieving synergistic effects in the host requires developing optimal combinations of prebiotic and probiotic (Rastall et al., 2005).

While conceptionally the idea of a symbiotic appears straightforward, mechanistically it can be more complicated. This is because inclusion of a prebiotic involves not only manipulation of the indigenous microbiome and host responses such as the immune system, but the type of prebiotic that is provided as well as the particular bacteria serving as the target probiotic (Bomba et al., 2002; Bengmark, 2012). For example, *in vitro* mixed culture studies conducted by Fooks and Gibson (2002) demonstrated that oligosaccharides supported growth of *Lactobacillus plantarum* and *Bifidobacterium bifidum* and inhibited *E. coli*, *Campylobacter jejuni* and *S.* Enteritidis, but the extent of reduction was dependent on the type of oligosaccharide. Herrera et al. (2012) not only noted a similar carbohydrate specificity in *in vitro* co-cultures of *Streptococcus bovis* and *S.* Typhimurium, but also observed that allowing time for the *Streptococcus bovis* to grow and ferment the carbohydrate prior to inoculating *S.* Typhimurium was critical to maximizing reduction of the pathogen. Based on their *in vitro* screening study with poultry *Lactobacillus* isolates, Saminathan et al. (2011) concluded that not only were there differences in isolate responses to different prebiotic substrates, but there were strain differences in the ability to grow on a particular prebiotic.

A novel way to optimize prebiotic/probiotic combinations is to simply employ the prebiotic to selectively isolate GI tract bacteria that then become probiotic candidates possessing the desired properties of interest in the corresponding host animal. A classic example of this approach is using lactose as the carbon source to isolate a defined cecal population that would limit *Salmonella* colonization in poultry flocks. Lactose as a poultry feed candidate to target GI tract bacteria is attractive because its utilization by the bird is generally low at all times and it cannot be used by *S.* Typhimurium (Atkinson et al., 1957; Gutnick et al., 1969; Moran, Jr., 1985; Waldroup et al., 1992; Kermanshahi and Rostami, 2006). Lactose either directly fed, added to water or as a part of a feed supplement such as whey, was initially shown to inhibit *Salmonella* colonization in broilers and believed to be fermented by cecal bacteria to generate organic acids inhibitory to *Salmonella* (Bilgili and Moran, Jr., 1990; Corrier et al., 1990a,b; DeLoach et al., 1990; Kermanshahi and Rostami, 2006). The relationship between the cecal microbiome and lactose was further substantiated when dietary lactose was combined with anaerobic cecal bacterial inocula and introduced to broiler chickens and turkey poults (Hinton, Jr. et al., 1991; Ziprin et al., 1990; Corrier et al., 1991a,b). The cecal inocula had been generated by incubating a mixed cecal culture for 24 h in a Viande–Levure (VL) broth with lactose and no glucose (Corrier et al., 1991b). In order to stabilize the cecal inocula, Nisbet et al. (1993a,b) took this a step further and used continuous flow (CF) cultures containing lactose VL broth in place of glucose to select for a cecal lactose-utilizing consortium that stabilized after reaching steady state following several growth vessel turnovers. When they inoculated chickens, significant *S.* Typhimurium reduction was achieved with the CF culture combined with lactose, when compared to control birds, and these birds also yielded higher levels of cecal propionic acid (Nisbet et al., 1993b). As further refinements were introduced to avoid dietary dependence on relatively expensive lactose, it eventually became clear that developing and maintaining a metabolically stable cecal microbial consortium that could consistently generate propionic acid in the ceca of inoculated birds was a critical element for achieving potential commercial success (Corrier et al., 1995; Nisbet et al., 1996a,b).

Other symbiotic combinations have also proven to be beneficial to the poultry host and/or inhibitory to pathogen colonization and some have since been commercialized. These studies are too numerous to list and discuss here. However, overall outcomes tend to vary with the type of application and composition of the symbiotic. This is not necessarily surprising, given the extra layers of complexity associated not just with the different chemical compositions of individual prebiotics but also considering the biological variability associated with the probiotic live cultures, and finally the immense and still poorly understood GI tract microbiome. Host factors such as the immune response to a shifting GI tract microbiome as the host animal matures are critical as well. The interaction between the prebiotic and the probiotic introduces additional mechanistic complexities related to stability of the prebiotic in the digestive tract and its relative fermentability by the GI microbiome. This interaction has potential commercial implications. One of the benefits of including prebiotics as part of the feed amendment is the potential to protect probiotic bacteria from hostile host environments such as the acidic gastric compartment as they travel through the upper intestinal tract to the target areas of the lower GI tract (Rastall and Maitin, 2002). Obviously, the degree of protection will potentially influence the consistency of the symbiotic.

Finally, the response of individual pathogen strains to external environmental stresses and interventions employed in the food processing industry are not always the same and strain variation must be taken into account (Lianou and Koutsoumanis, 2013). This no doubt holds true for *Salmonella* responses to the presence of prebiotics and synbiotics as well. For example, *Salmonella* serovar and strain differences in susceptibility to antagonistic metabolic products such as fermentation acids generated by probiotic cultures do occur and therefore should probably be taken into account (Ricke, 2003a; Dunkley et al., 2009; Foley et al., 2013; Ricke, 2014). As interest in prebiotics and synbiotics grows, it will be important to screen the candidates against not only multiple *Salmonella* serovars but several representative strains for each individual serovar as well.

9 Summary and conclusions

Prebiotics are a class of feed ingredients that can act as substrates in the host to produce various health benefits. Obviously, as more has become known about the interactions and activities occurring in the host microbiome, this definition has evolved considerably and will continue to do so. Regardless of the definition, the goal of prebiotic supplementation remains, namely improving the growth performance, health status and decreasing the food-borne pathogen load of the host animal. As antibiotics are removed from the diets of food animals, achieving these goals with non-antibiotic alternatives will become commercially imperative. Likewise with the increasing interest in non-traditional animal production such as organic and natural animal husbandry, such alternatives will be particularly attractive to the consumer of these products (Ricke et al., 2012b). Optimizing combinations of prebiotics and probiotic cultures into synbiotics offers an opportunity to enhance the growth of beneficial bacteria such as bifidobacteria and lactobacilli spp. in a more controlled fashion that will ultimately produce more predictable synergistic effects in the host. However, such combinations, while conceptually quite attractive, are fraught with complexities that require a better understanding not only of the GI microbiome, but of the host animal responses as well. Likewise, further research is needed to gain a better understanding of the mechanism(s) of action against food-borne pathogens such as *Salmonella* in the GI microbiome, and the pathogen's capabilities to express resistance, and variability among strains within a pathogen species.

Other factors need to be optimized as well to achieve more effectiveness for these types of products and to make them more commercially attractive. Certainly, further chemical refinement of prebiotic compounds to be more mechanistically selective and consistently predictable towards targeted GI tract microorganisms would be helpful. For example, Maillard reaction products may offer some interesting multifunctional properties including antimicrobial activities directed against food-borne pathogens (Ricke, 2015). Chalova et al. (2012) used transcriptomic microarrays to demonstrate that *S.* Typhimurium growing on lysine–glucose Maillard compounds expressed genes related to an energy starvation mode. This is an outcome that offers intriguing potential, and as genomic information further expands for both pathogens and GI bacteria the ability to refine such compounds to favour the beneficial bacteria while simultaneously being antagonistic to the pathogen population could be explored and commercially developed. Likewise, designing prebiotics that are more resistant to degradation until they reach their respective target GI tract compartment would probably improve efficacy. These chemical modifications would in part support an increasing interest in more novel sources of NDO-based prebiotics.

Probably one of the areas of most interest is retaining efficacy of the biologically complex product during all phases of its production and eventual delivery to the target group of food animals. Using advanced technologies such as biofunctionalized carriers or nanoparticles (Taylor et al., 2004; Luo et al., 2005) that could serve as multifunctional/multipurpose entities and thus be used to retain diverse activities ranging from prebiotic properties to multiple hurdle antimicrobial activities (Ricke and Hanning, 2013) that exclusively target food-borne pathogens is attractive. Multifunctional carriers offer a conceptually attractive 'next generation' approach to further optimize the efficacy and precision for prebiotics. However, cost-effectiveness for generating such particles on a large scale, as well as regulatory issues, will have to be considered before these can be practical for food animal production systems. Such particles would probably also need to be environmentally friendly, which means they would need to be biodegradable with a relatively short half-life. Regardless of the potential future research directions, prospects appear to be promising for prebiotics playing a major commercial role in poultry and animal production systems.

10 Where to look for further information

Callaway, T. R. and Ricke, S. C. (eds) (2012), *Direct Fed Microbials/Prebiotics for Animals: Science and Mechanisms of Action*. Springer Science, New York, NY.

Gibson G. R. and Roberfroid, M. B. (1995), Dietary modulation of the human colonic microflora: Introducing the concept of prebiotics. *J. Nutr. 125*, 1401–12.

Hutkins, R. W., Krumbeck, J. A., Bindels, L. B., Cani, P. D., Fahey, Jr., G. C., Goh, Y. J., Hamaker, B., Martens, E. C., Mills, D. A., Rastal, R. A., Vaughan, E. and Sanders, M. E. (2016), Prebiotics: Why definitions matter. *Curr. Opin. Biotech. 37*, 1–7.

Patterson, J. A. and Burkholder, K. (2003), Application of prebiotics and probiotics in poultry production. *Poult. Sci. 82*, 627–31.

Rastall, R. (2004), Bacteria in the gut: Friends and foes and how to alter the balance. *J. Nutr. 134*, 2022S–6S.

Rastall, R. A. and Maitin, V. (2002), Prebiotics and synbiotics: Towards the next generation. *Curr. Opin. Biotechnol. 13*, 490–6.

Rastall, R. A., Gibson, G. R., Gill, H. S., Guarner, F., Klaenhammer, T. R., Pot, B., Reid, G., Rowland, I. R. and Sanders, M. E. (2005), Modulation of the microbial ecology of the human colon by probiotics and synbiotics to enhance human health: An overview of enabling science and potential applications. *FEMS Microbiol. Ecology 52*, 145–52.

Ricke, S. C. (2015), Potential of fructooligosaccharide prebiotics in alternative and nonconventional poultry production systems. *Poult. Sci. 94*, 1411–18.

Roberfroid, M. B. (2007), Prebiotics: The concept revisited. *J. Nutr. 137*, 830S–7S.

Roto, S. M., Rubinelli, P. M. and Ricke, S. C. (2015), An introduction to the avian gut microbiota and the effects of yeast-based prebiotic compounds as potential feed additives. *Front. Vet. Sci. 2 (Article 28)*, 1–18. doi: 10.3389/fvets.2015.00028.

Wang, Y. (2009), Prebiotics: Present and future science and technology. *Food Res. Int. 42*, 8–12.

11 References

Alam, N. H., Meier, R., Schneider, H., Sarker, S. A., Bardhan, P. K., Mahalananbis, D., Fuchs, G. J. and Gyr, N. (2000), Partialy hydrolyzed guar gum-supplemented oral rehydration solution in the treatment of acute diarrhea in children. *J. Ped. Gastroenterol. Nutr. 31*, 503–7.

Alam, N. H., Meier, R., Sarker, S. A., Bardhan, P. K., Schneider, H. and Gyr, H. (2005), Partially hydrolysed guar gum supplemented comminuted chicken diet in persistent diarrhoea: A randomised controlled trial. *Arch. Dis. Child. 90*, 195–9.

Alloui, M. N., Szczurek, W. and Świątkiewicz, S. (2013), The usefulness of prebiotics and probiotics in modern poultry nutrition: A review. *Ann. Anim. Sci. 13*, 17–32.

Anderson, J. O. and Warnick, R. E. (1964), Value of enzyme supplements in rations containing certain legume seed meals or gums. *Poult. Sci. 43*, 1091–7.

Apajalahti, J. (2005), Comparative gut microflora, metabolic challenges, and potential opportunities. *J. Appl. Poult. Res. 14*, 444–53.

Aslan, Y. and Tanriseven, A. (2007), Immobilization of *Penicillium lilacinum* dextranase to produce isomaltooligosaccharides from dextran. *Biochem. Eng. J. 34*, 8–12.

Atkinson, R. L., Kratzer, F. H. and Stewart, G. F. (1957), Lactose in animal and human feeding. *J. Dairy Sci. 50*, 1114–32.

Auclair, E. (2000), Yeast as an example of the mode of action of probiotics in monogastric and ruminant species. III Conference show feed manufacturing in the Mediterranean region improving food safety. From feed to food. *Reus, Espana. 54*, 45–53.

Bailey, J. S., Blankenship, L. C. and Cox, N. A. (1991), Effect of fructooligosaccharide on *Salmonella* colonization of the chicken intestine. *Poult. Sci. 70*, 2433–8.

Baurhoo, B., Goldflus, F. and Zhao, X. (2009), Purified cell wall of *Saccharomyces cerevisiae* increases protection against intestinal pathogens in broiler chickens. *Int. J. Poult. Sci. 8*, 133–7.

Bengmark, S. (1998), Immunonutrition: Role of biosurfactants, fiber, and probiotic bacteria. *Nutrition 14*, 585–94.

Bengmark S. (2001), Pre-, pro- and synbiotics. *Curr. Opin. Clin. Nutr. Metab. Care. 4*, 571–9.

Bengmark, S. (2012), Integrative medicine and medicine-The role of pre-, pro- and synbiotics. *Clin. Translational Med. 1:6*, 1–13 (http://www.clintransmed.com/content/1/1/6).

Biggs, P., Parsons, C. M. and Fahey, Jr., G. C. (2007), The effects of several oligosaccharides on growth performance, nutrient digestibilities, and cecal microbial populations in young chicks. *Poult. Sci. 86*, 2327–36.

Bilgili, S. F. and Moran, Jr., E. T. (1990), Influence of whey and probiotic-supplemented withdrawal feed on the retention of *Salmonella* intubated into market age broilers. *Poult. Sci. 69*, 1670–4.

Bird, A. R. (1999), Prebiotics: A role for dietary fibre and resistant starch? *Asia Pacific J. Clin. Nutr. 8 (Suppl.)*, S32–6.

Bird, A. R., Conlon, M. A., Christopher, C. T. and Topping, D. L. (2010), Resistant starch, large bowel fermentation and a broader perspective of prebiotics and probiotics. *Benef. Microbes 1*, 423–31.

Blackburn, N. A. and Johnson, J. T. (1981), The effect of guar gum on the viscosity of the gastrointestinal contents and on glucose uptake from the perfused jejunum in the rat. *Br. J. Nutr. 46*, 239–46.

Bomba, A., Nemcová, R., Gancarčiková, S., Herich, R., Guba, P. and Mudroňová, D. (2002), Improvement of the probiotic effect of micro-organisms by their combination with maltodextrins, fructooligosaccharides and polyunsaturated fatty acids. *Br. J. Nutr. 88 (Suppl. 1)*, S95–9.

Boon, M. A., van't Riet, K. and Janssen, A. E. M. (2000), Enzymatic synthesis of oligosaccharides: Product removal during a kinetically controlled reaction. *Biotechnol. Bioeng. 70*, 411–20.

Bradley, G. L., Savage, T. F. and Timm, K. I. (1994), The effects of supplementing diets with *Saccharomyces cerevisiae* var. *boulardii* on male poult performance and ileal morphology. *Poult. Sci. 73*, 1766–70.

Callaway, T. R. and Ricke, S. C. (eds) (2012), *Direct Fed Microbials/Prebiotics for Animals: Science and Mechanisms of Action*. Springer Science, New York, NY.

Chalova, V. I., Hernández-Hernández, O., Muthaiyan, A., Sirsat, S. A., Natesan, S., Sanz, M. L., Javier Moreno, F., O'Bryan, C. A., Crandall, P. G. and Ricke, S. C. (2012), Growth and transcriptional response of *Salmonella* Typhimurium LT2 to glucose-lysine-based Maillard reaction products generated under low water activity conditions. *Food Res. Int. 45*, 1044–53.

Chen, C.-Y., Yan, X. and Jackson, C. (eds) (2015), *Antimicrobial Resistance and Food Safety: Methods and Techniques*. Academic Press/Elsevier Science, Amsterdam, The Netherlands.

Chung, C.-H. and Day, D. F. (2004), Efficacy of *Leuconostoc mesenteroides* (ATCC 13146) isomaltooligosaccharides as a poultry prebiotic. *Poult. Sci. 83*, 1302–6.

Collins, M. D. and Gibson, G. R. (1999), Probiotics, prebiotics, and synbiotics: Approaches for modulating the microbial ecology of the gut. *Am. J. Clin. Nutr. 69 (Suppl)*, 1052S–7S.

Corrier, D. E., Hinton, Jr., A., Ziprin, R. L., Beier, R. C. and DeLoach, J. (1990a), Effect of dietary lactose on cecal pH, bacteriostatic volatile fatty acids, and Salmonella typhimurium colonization of broiler chicks. *Avian Dis. 34*, 668–76.

Corrier, D. E., Hinton, Jr., A, Ziprin, R. L. and DeLoach, J. (1990b), Effect of dietary lactose on Salmonella colonization of market-age broiler chickens. *Avian Dis. 34*, 668–76.

Corrier, D. E., Hargis, B., Hinton, Jr., A., Lindsey, D., Caldwell, D., Manning, J. and DeLoach, J. (1991a), Effect of anaerobic cecal microflora and dietary lactose on colonization resistance of layer chicks to invasive Salmonella enteritidis. *Avian Dis. 35*, 337–43.

Corrier, D. E., Hinton, Jr., A., Kubena, L. F., Ziprin, R. L. and DeLoach, J. (1991b), Decreased *Salmonella* colonization in turkey poults inoculated with anaerobic cecal microflora and provided dietary lactose. *Poult. Sci.* 70, 1345–50.

Corrier, D. E., Nisbet, D. J., Scanlan, C. M., Hollister, A. G. and DeLoach, J. R. (1995), Control of *Salmonella typhimurium* colonization in broiler chicks with a continuous-flow characterized mixed culture of cecal bacteria. *Poult. Sci. 74*, 916–24.

Couch, J. R., Creger, C. R. and Bakshi, Y. K. (1966), Trypsin inhibitor in guar meal. *Proc. Soc. Exp. Biol. Med. 123*, 263–5.

Couch, J. R., Bakshi, Y. K., Ferguson, T. M., Smith, E. B. and Creger, C. R. (1967), The effect of processing on the nutritional value of guar meal for broiler chicks. *Br. Poult. Sci. 8*, 243–50.

Crittenden, R. G. and Playne, M. J. (1996), Production, properties, and applications of food-grade oligosaccharides. *Trends Food Sci. Tech. 7*, 353–61.

Cummings, J. H. and Macfarlane, G. T. (2002), Gastrointestinal effects of prebiotics. *Br. J. Nutr. 87*, S145–51.

Cummings, J. H., Macfarlane, G. T. and Englyst, H. N. (2001), Prebiotic digestion and fermentation. *Am. J. Clin. Nutr. 73*, S415–20.

DeLoach, J. R., Oyofo, B. A., Corrier, D. E., Kubena, L. F., Ziprin, R. L. and Norman, J. O. (1990), Reduction of Salmonella typhimurium concentration in broiler chickens by milk or whey. *Avian Dis. 34*, 389–92.

Demigné, C., Yacoub, C. and Rémésy, C. (1986), Effects of absorption of large amounts of volatile fatty acids on rat liver metabolism. *J. Nutr. 116*, 77–86.

Ding, G., Chang, Y., Zhao, L., Zhou, Z., Ren, L. and Meng, Q. (2014), Effect of *Saccharomyces cerevisiae* on alfalfa nutrient degradation characterstics and rumen microbial populations of steers fed diets with different concentrate – to – forage ratios. *J. Anim. Sci. Biotechnol. 5*, 24 (http://www.jasbsci.com/content/5/1/24).

Donalson, L. M., Kim, W. K., Chalova, V. I., Herrera, P., Woodward, C. L., McReynolds, J. L., Kubena, L. F., Nisbet, D. J. and Ricke, S. C. (2007), *In vitro* anaerobic incubation of *Salmonella enterica* serotype Typhimurum and laying hen cecal bacteria in poultry feed substrates and a fructooligosacharide prebiotic. *Anaerobe 13*, 208–14.

Donalson, L. M., Kim, W. K., Chalova, V. I., Herrera, P., McReynolds, J. L., Gotcheva, V. G., Vidanović, D., Woodward, C. L., Kubena, L. F., Nisbet, D. J. and Ricke, S. C. (2008a). In vitro fermentation response of laying hen cecal bacteria to combinations of fructooligosaccharide (FOS) prebiotic with alfalfa or a layer ration. *Poult. Sci. 87*, 1263–75.

Donalson, L. M., McReynolds, J. L., Kim, W. K., Chalova, V. I., Woodward, C. L., Kubena, L. F., Nisbet, D. J. and Ricke, S. C. (2008b), The influence of a fructooligosacharide prebiotic combined with alfalfa molt diets on the gastrointestinal tract fermentation, *Salmonella* Enteritidis infection and intestinal shedding in laying hens. *Poult. Sci. 87*, 1253–62.

Dunkley, K. D., Dunkley, C. S., Njongmeta, N. L., Callaway, T. R., Hume, M. E., Kubena, L. F., Nisbet, D. J. and Ricke, S. C. (2007a), Comparison of in vitro fermentation and molecular microbial profiles of high-fiber feed substrates (HFFS) incubated with chicken cecal inocula. *Poult. Sci. 86*, 801–10.

Dunkley, K. D., McReynolds, J. L., Hume, M. E., Dunkley, C. S., Callaway, T. R., Kubena, L. F., Nisbet, D. J. and Ricke, S. C. (2007b), Molting in *Salmonella* Enteritidis-challenged laying hens fed alfalfa crumbles.II. Fermentation and microbial ecology response. *Poult. Sci. 86*, 2101–9.

Dunkley, K. D., Callaway, T. R., Chalova, V. I., McReynolds, J. L., Hume, M. E., Dunkley, C. S., Kubena, L. F., Nisbet, D. J. and Ricke, S. C. (2009), Foodborne *Salmonella* ecology in the avian gastrointestinal tract. *Anaerobe 15*, 26–35.

Durant, J. A., Corrier, D. E., Byrd, J. A., Stanker, L. H. and Ricke, S. C. (1999), Feed deprivation affects crop environment and modulates *Salmonella enteritidis* colonization and invasion of leghorn hens. *Appl. Environ. Microbiol. 65*, 1919–23.

Fairchild, R. M., Ellis, P. R., Byrne, A. J., Luzio, S. D. and Mir, M. A. (1996), A new breakfast cereal containing guar gum reduces postprandial plasma glucose and insulin concentrations in normal-weight human subjects. *Br. J. Nutr. 76*, 63–73.

Finstad, S., O'Bryan, C. A., Marcy, J. A., Crandall, P. G. and Ricke, S. C. (2012), *Salmonella* and broiler production in the United States: Relationship to foodborne salmonellosis. *Food Res. Int. 45*, 789–94.

Flickinger, E. A., Van Loo, J. and Fahey, Jr., G. C. (2003), Nutritional responses to the presence of inulin and oligofructose in the diets of domesticated animals: A review. *Crit. Revs. Food Sci. Nutr. 43*, 19–60.

Foley, S. L., Johnson, T. J., Ricke, S. C., Nayak, R. and Danzeisen, J. (2013), *Salmonella* pathogenicity and host adaptation in chicken-associated serovars. *Microbiol. Mol. Biol. Revs. 77*, 582–607.

Foley, S., Nayak, R., Hanning, I. B., Johnson, T. L., Han, J. and Ricke, S. C. (2011), Population dynamics of *Salmonella enterica* serotypes in commercial egg and poultry production. *Appl. Environ. Microbiol. 77*, 4273–9.

Fonty, G. and Chaucheyras-Durand, F. (2006), Effects and modes of action of live yeasts in the rumen. *Biologia, Bratislava 61/6*, 741–50.

Fooks, L. J. and Gibson, G. R. (2002), In vitro investigations of the effect of probiotics and prebiotics on selected human intestinal pathogens. *FEMS Microbiol. Ecol. 39*, 67–75.

Fukata, T., Sasai, K., Miyamoto, T. and Baba, E. (1999), Inhibitory effects of competitive exclusion and fructooligosaccharide, singly and in combination, on *Salmonella* colonization of chicks. *J. Food Prot. 62*, 229–33.

Fukuda, S., Toh, H., Hase, K., Oshima, K., Nakanishi, Y., Yoshimura, K., Tobe, T., Clarke, J. M., Topping, D. L., Suzuki, T., Taylor, T. D., Itoh, K, Kikuchi, J., Morita, H., Hattori, M. and Ohno, H. (2011), Bifidobacteria can protect from enteropathogenic infection through production of acetate. *Nature 469*, 543–7.

Fuller, R. (1989), Probiotics in man and animals. *J. Appl. Bacteriol. 66*, 365–78.

Gibson, G. R. and Wang, X. (1994), Inhibitory effects of bifidobacteria on other colonic bacteria. *J. Appl. Bacteriol. 77*, 412–20.

Gibson G. R. and Roberfroid, M. B. (1995), Dietary modulation of the human colonic microflora: Introducing the concept of prebiotics. *J. Nutr. 125*, 1401–12.

Gibson, G. R. (1999), Dietary modulation of the human gut microflora using the prebiotics oligofructose and inulin. *J. Nutr. 129*, 1438–41.

Gomes, A. M. P. and Malcata, F. X. (1999), *Bifidobacterium* spp. and *Lactobacillus acidophilus*: Biological, biochemical, technological and therapeutical properties relevant for use as probiotics. *Trends Food Sci. Technol. 10*, 139–57.

Gutnick, D., Calvo, J. M., Klopowski, T. and Ames, B. N. (1969), Compounds which serve as the sole source of carbon or nitrogen for *Salmonella typhimurium* LT-2. *J. Bacteriol. 100*, 215–19.

Hajati, H. and Rezaei, M. (2010), The application of prebiotics in poultry production. *Int. J. Poult. Sci. 9*, 298–304.

Hanning, I., Clement, A., Owens, C., Park, S. H., Pendleton, S., Scott, E. E., Almeida, G., Gonzalez Gil, F. and Ricke, S. C. (2012), Assessment of production performance in two breeds of broilers fed prebiotics as feed additives. *Poult. Sci. 91*, 3295–9.

Heitman, D. W., Hardman, W. E. and Cameron, L. L. (1992), Dietary supplementation with pectin and guar gum on 1,2-dimethylhydrazine-induced colon carcinogenesis in rats. *Carcinogenesis 13,* 815–18.

Herfel, T., Jacobi, S., Lin, X., van Heugten, E., Fellner, V. and Odle, J. (2013), Stabilized rice bran improves weaning pig performance via a prebiotic mechanism. *J. Anim. Sci. 91,* 907–13.

Herrera, P., Kwon, Y. M., Maciorowski, K. G. and Ricke, S. C. (2009), Ecology and pathogenicity of gastrointestinal *Streptococcus bovis. Anaerobe 15,* 44–54.

Herrera, P., O'Bryan, C. A., Crandall, P. G. and Ricke, S. C. (2012), Growth response of *Salmonella enterica* Typhimurium in co-culture with ruminal bacterium *Streptococcus bovis* is influenced by time of inoculation and carbohydrate substrate. *Food Res. Int. 45,* 1054–7.

Hinton, Jr., A., Corrier, D. E., Ziprin, R. L., Spates, J. O. and DeLoach, J. R. (1991), Comparison of the efficacy of cultures of cecal anaerobes as inocula to reduce *Salmonella typhimurium* colonization in chicks with or without dietary lactose. *Poult. Sci. 70,* 67–73.

Hohmann, H.-H., Kemen, M., Fuessenenich, C., Senkal, M. and Zumtobel, V. (1994), Reduction in diarrhea incidence by soluble fiber in patients receiving total or supplemental enternal nutrition. *J. Parenteral Enternal Nutr. 18,* 486–90.

Hooge, D. M. (2004), Meta-analysis of broiler chicken pen trials evaluating dietary mannan oligosaccharide, 1993-2003. *Int. J. Poult. Sci. 3,* 163–74.

Horrocks, S. M., Anderson, R. C., Nisbet, D. J. and Ricke, S. C. (2009), Incidence and ecology of *Campylobacter* in animals. *Anaerobe 15,* 18–25.

Howard, Z. R., O'Bryan, C. A., Crandall, P. G. and Ricke, S. C. (2012), *Salmonella* Enteritidis in shell eggs: Current issues and prospects for control. *Food Res. Int. 45,* 755–64.

Hume, M. E. (2011), Historic perspective: Prebiotics, probiotics, and other alternatives to antibiotics. *Poult. Sci. 90,* 2663–9.

Hutkins, R. W., Krumbeck, J. A., Bindels, L. B., Cani, P. D., Fahey, Jr., G. C., Goh, Y. J., Hamaker, B., Martens, E. C., Mills, D. A., Rastal, R. A., Vaughan, E. and Sanders, M. E. (2016), Prebiotics: Why definitions matter. *Curr. Opin. Biotech. 37,* 1–7.

Indikova, I., Humphrey, T. J. and Hilbert, F. (2015), Survival with a helping hand: *Campylobacter* and microbiota. *Front. Microbiol. 6 (Article 1266),* 1–6. doi: 10.3389/fmicb.2015.01266.

Ishihara, N., Chu, D.-C., Akachi, S. and Juneja, L. R. (2000), Preventive effect of partially hydrolyzed guar gum on infection of *Salmonella enteritidis* in young and laying hens. *Poult. Sci. 79,* 689–97.

Joerger, R. D. (2003), Alternatives to antibiotics: Bacteriocins, antimicrobial peptides and bacteriophages. *Poult. Sci. 82,* 640–7.

Jones, F. T. and Ricke, S. C. (2003), Observations on the history of the development of antimicrobials and their use in poultry feeds. *Poult. Sci. 82,* 613–17.

Jones, G. H. and Ballou, C. E. (1968), Isolation of an α-mannosidase which hydrolyzes yeast mannan – Structure of the backbone of yeast mannan. *J. Biol. Chem. 243,* 2442–6.

Jouany, J. P. (2001), A new look at yeast cultures as probiotics for ruminants. *Feed Mix 9,* 17–19.

Józefiak, D., Rutkowski, A. and Martin, S. A. (2004), Carbohydrate fermentation in the ceca: A review. *Anim. Feed Sci. Technol. 113,* 1–15.

Kaplan, H. and Hutkins, R. W. (2000), Fermentation of fructooligosaccharides by lactic acid bacteria and bifidobacteria. *Appl. Environ. Microbiol. 66,* 2682–4.

Kermanshahi, H. and Rostami, H. (2006), Influence of supplemental dried whey on broiler performance and cecal flora. *Int. J. Poult. Sci. 5,* 538–43.

Klis, F. M., Mol, P., Hellingwerf, K. and Brul, S. (2002), Dynamics of cell wall structure in *Saccharomyces cerevisiae. FEMS Microbiol. Revs. 26,* 239–56.

Knudsen, K. E. B. (2014), Fiber and nonstarch polysaccharide content and variation in common crops used in broiler diets. *Poult. Sci. 93,* 2380–93.

Kumar, A., Henderson, A., Forster, G. M., Goodyear, A. W., Weir, T. L., Leach, J. E., Dow, S. W. and Ryan, E. P. (2012), Dietary rice bran promotes resistance to *Salmonella enterica* serovar Typhimurium colonization in mice. *BMC Microbiol. 12,* 71 (http://www.biomedcentral.com/1471–2180/12/71).

Kuriki, T., Yanase, M., Takata, H., Takesada, Y., Imanaka, T. and Okada, S. (1993), A new way of producing isomalto-oligosaccharide syrup by using the transglycosylation reaction of neopullulanase. *Appl. Environ. Microbiol. 59*, 953–9.

Lee, J. T., Bailey, C. A. and Cartwright, A. L. (2003), Guar meal germ and hull fractions differently affect growth performance and intestinal viscosity of broiler chickens. *Poult. Sci. 82*, 1589–95.

Lee, J. T., Connor-Appleton, S., Bailey, C. A. and Cartwright, A. L. (2005), Effects of guar meal by-product with and without β-mannanase hemicell on broiler performance. *Poult. Sci. 84*, 1261–7.

Lee, Y.-C. and Ballou, C. E. (1965), Preparation of mannobiose, mannotriose, and a new mannotetraose from *Saccharomyces cerevisiae* mannan. *Biochemistry 4*, 257–64.

Lianou, A. and Koutsoumanis, K. P. (2013), Strain variability of the behavior of foodborne bacterial pathogens: A review. *Int. J. Food Microbiol. 167*, 310–21.

Lindberg, J. E. (2014), Fiber effects in nutrition and gut health in pigs. *J. Anim. Sci. Biotechnol. 5*, 15 (http://www.jasbsci.com/content/5/1/15).

Luo, P. G., Tzeng, T.-R. J., Qu, L, Lin, Y., Caldwell, E., Latour, R. A., Stutzenberger, F. and Sun, Y. P. (2005), Quantitative analysis of bacterial aggregation mediated by bioactive nanoparticles. *J. Biomed. Nanotech. 2*, 1–5.

Macfarlane, G. T., Steed, H. and Macfarlane, S. (2008), Bacterial metabolism and health-related effects of galacto-oligosaccharides and other prebiotics. *J. Appl. Microbiol. 104*, 305–44.

Mahoney, R. R. (1998), Galactosyl-oligosaccharide formation during lactose hydrolysis: A review. *Food Chem. 63*, 147–54.

Monsan, P. and Paul, F. (1995), Enzymatic synthesis of oligosaccharides. *FEMS Microbiol. Revs 16*, 187–92.

Moran, Jr., E. T. (1985), Digestion and absorption of carbohydrates in fowl and events through perinatal development. *J. Nutr. 115*, 665–74.

Moriceau, S., Besson, C., Levrat, M.-A., Moundras, C., Rémésy, C., Morand, C. and Demigné, C. (2000), Cholesterol-lowering effects of guar gum: Changes in bile acid pools and intestinal reabsorption. *Lipids 35*, 437–44.

Nagpal, M. L. Agrawal, O. P. and Bhatia, I. S. (1971), Chemical and biological examination of guar-meal (*Cyamopsis tetragonoloba* L.). *Ind. J. Anim. Sci. 41*, 283–93.

Naughton, P. J., Mikkelsen, L. L. and Jensen, B. B. (2001), Effects of nondigestible oligosaccharides on *Salmonella enterica* serovar Typhimurium and nonpathogenic *Escherichia coli* in the pig small intestine in vitro. *Appl. Environ. Microbiol. 67*, 3391–5.

Nisbet, D. J. and Martin, S. A. (1991), Effect of a *Saccharomyces cerevisiae* culture on lactate utilization by the ruminal bacterium *Selenomonas ruminantium. J. Anim. Sci. 69*, 4628–33.

Nisbet, D. J., Corrier, D. E. and DeLoach, J. R. (1993a), Effect of mixed cecal microflora maintained in continuous culture and of dietary lactose on Salmonella typhimurium colonization in broiler chicks. *Avian Dis. 37*, 528–35.

Nisbet, D. J., Corrier, D. E., Scanlan, C. M., Hollister, A. G., Beier, R. C. and DeLoach, J. R. (1993b), Effect of a defined continuous-flow derived bacterial culture and dietary lactose on Salmonella typhimurium colonization in broiler chickens. *Avian Dis. 37*, 1017–25.

Nisbet, D. J., Ricke, S. C., Scanlan, C. M., Corrier, D. E., Hollister, A. G. and DeLoach, J. R. (1994), Inoculation of broiler chicks with a continuous-flow derived bacterial culture facilitates early cecal bacterial colonization and increases resistance to *Salmonella typhimurium. J. Food Prot. 57*, 12–15.

Nisbet, D. J., Corrier, D. E., Ricke, S. C., Hume, M. E., Byrd, J. A. and DeLoach, J. R. (1996a), Maintenance of the biological efficacy in chicks of a cecal competitive-exclusion culture against *Salmonella* by continuous-flow fermentation. *J. Food Prot. 59*, 1279–83.

Nisbet, D. J., Corrier, D. E., Ricke, S. C., Hume, M. E., Byrd, J. A. and DeLoach, J. R. (1996b), Cecal propionic acid as a biological indicator of the early establishment of a microbial ecosystem inhibitory to *Salmonella* in chicks. *Anaerobe, 2*, 345–50.

Nisbet, D. (2002), Defined competitive exclusion cultures in the prevention of enteropathogen colonisation in poultry and swine. *Antonie van Leeuwenhoek. 81*, 481–6.

Nollet, L., Huyghebaert, G. and Spring, P. (2007), Effect of dietary mannan oligosaccharide (Bio-Mos) on live performance of broiler chickens given an anticoccidial vaccine (Paracox) followed by a mild coccidial challenge. *J. Appl. Poult. Res. 16*, 397–403.

O'Bryan, C. A., Pendleton, S. J., Crandall, P. G. and Ricke, S. C. (2015), Potential of plant essential oils and their components in animal agriculture – *in vitro* studies on antibacterial mode of action. *Front. Vet. Sci. 2 (Article 35)*, 1–8. doi: 10.3389/fvets.2015.00035.

Okubo, T., Ishihara, N., Takahashi, H., Fujisawa, T., Kim, M., Yamamoto, T. and Mitsuoka, T. (1994), Effects of partially hydrolyzed guar gum on human intestinal microflora and its metabolism. *Biosci. Biotech. Biochem. 58*, 1364–9.

Ou, S., Kwok, K., Li, Y. and Fu, L. (2001), In vitro study of possible role of dietary fiber in lowering postprandial serum glucose. *J. Agric. Food Chem. 49*, 1026–9.

Oyofo, B. A., DeLoach, J. R., Corrier, D. E., Norman, J. O., Ziprin, R. L. and Mollenhauer, H. H. (1989a), Prevention of *Salmonella typhimurium* colonization of broilers with D-mannose. *Poult. Sci. 68*, 1357–60.

Oyofo, B. A., Droleskey, R. E., Norman, J. O., Mollenhauer, H. H., Ziprin, R. L., Corrier, D. E. and DeLoach, J. R. (1989b), Inhibition by mannose of *in vitro* colonization of chicken small intestine by *Salmonella typhimurium*. *Poult. Sci. 68*, 1351–6.

Park, S. H., Hanning, I., Perrota, A., Bench, B. J., Alm, E. and Ricke, S. C. (2013), Modifying the gastrointestinal ecology in alternatively raised poultry and the potential for molecular and metabolomic assessment. *Poult. Sci. 92*, 546–61.

Park S. Y., Woodward, C. L., Kubena., L. F., Nisbet., D. J., Birkhold, S. G. and Ricke, S. C. (2008), Environmental dissemination of foodborne *Salmonella* in preharvest poultry production: Reservoirs, critical factors, and research strategies. *Crit. Revs. Environ. Sci. Technol. 38*, 73–111.

Patel, M. B. and McGinnis, J. (1985), The effect of autoclaving and enzyme supplementation of guar meal on the performance of chicks and laying hens. *Poult. Sci. 64*, 1148–56.

Patterson, J. A. and Burkholder, K. (2003), Application of prebiotics and probiotics in poultry production. *Poult. Sci. 82*, 627–31.

Raschke, W. C., Kern, K. A., Antalis, C. and Ballo, C. E. (1973), Genetic control of yeast mannan structure. Isolation and characterization of mannan mutants. *J. Biol. Chem. 248*, 4660–6.

Rastall, R. (2004), Bacteria in the gut: Friends and foes and how to alter the balance. *J. Nutr. 134*, 2022S–6S.

Rastall, R. A. and Maitin, V. (2002), Prebiotics and synbiotics: Towards the next generation. *Curr. Opin. Biotechnol. 13*, 490–6.

Rastall, R. A., Gibson, G. R., Gill, H. S., Guarner, F., Klaenhammer, T. R., Pot, B., Reid, G., Rowland, I. R. and Sanders, M. E. (2005), Modulation of the microbial ecology of the human colon by probiotics and synbiotics to enhance human health: An overview of enabling science and potential applications. *FEMS Microbiol. Ecology 52*, 145–52.

Revolledo, L., Ferreira, A. J. P. and Mead, G. C. (2006), Prospects in *Salmonella* control: Competitive exclusion, probiotics, and enhancement of avian intestinal immunity. *J. Appl. Poult. Res. 15*, 341–51.

Ricke, S. C., Martin, S. A. and Nisbet, D. J. (1996), Ecology, metabolism, and genetics of ruminal selenomonads. *Crit. Revs. Microbiol. 22*, 27–65.

Ricke, S. C. and Pillai, S. D. (1999), Conventional and molecular methods for understanding probiotic bacteria functionality in gastrointestinal tracts. *Crit. Revs. Microbiol. 25*, 19–38.

Ricke, S. C. (2003a), Perspectives on the use of organic acids and short chain fatty acids as antimicrobials. *Poult. Sci. 82*, 632–9.

Ricke, S. C. (2003b), The gastrointestinal tract ecology of *Salmonella* Enteritidis colonization in molting hens. *Poult. Sci. 82*, 1003–7.

Ricke, S. C., Kundinger, M. M., Miller, D. R. and Keeton, J. T. (2005), Alternatives to antibiotics: Chemical and physical antimicrobial interventions and foodborne pathogen response. *Poult. Sci. 84*, 667–75.

Ricke, S. C., Hererra, P. and Biswas, D. (2012a), Chapter 23. *Bacteriophages for potential food safety applications in organic meat production*. In S. C. Ricke, E. J. Van Loo, M. G. Johnson and C. A. O' Bryan (eds), *Organic Meat Production and Processing*. Wiley Scientific/IFT, New York, NY, pp. 407–24.

Ricke, S. C., Van Loo, E. J., Johnson M. G. and O'Bryan, C. A. (eds) (2012b), *Organic Meat Production and Processing*. Wiley Scientific/IFT, New York, NY.

Ricke, S. C., Dunkley, C. S. and Durant, J. A. (2013), A review on development of novel strategies for controlling *Salmonella* Enteritidis colonization in laying hens: Fiber – based molt diets. *Poult. Sci. 92*, 502–25.

Ricke, S. C. and Hanning, I. B. (2013), Chapter 9. *Food safety applications of nanoparticles*. In R. Asmatulu (ed.), *Nanotechnology Safety*. Elsevier, Amsterdam, The Netherlands, pp. 115–25.

Ricke, S. C. (2014), Application of molecular approaches for understanding foodborne *Salmonella* establishment in poultry production. *Adv. Biol. Vol.* Article ID 813275, p. 25. doi:10.1155/2014/813275.

Ricke, S. C. (2015), Potential of fructooligosaccharide prebiotics in alternative and nonconventional poultry production systems. *Poult. Sci. 94*, 1411–18.

Ricke, S. C. and Saengkerdsub, S. (2015), Chapter 19. *Bacillus probiotics and biologicals for improving animal and human health: Current applications and future prospects*. In V. Ravishankar Rai and A. Jamuna Bai (eds), *Beneficial Microbes in Fermented and Functional Foods*. CRC Press/Taylor & Francis Group, Boca Raton, FL, pp. 341–60.

Rivera Calo, J., Crandall, P. G., O'Bryan, C. A. and Ricke, S. C. (2015), Essential oils as antimicrobials in food systems – A review. *Food Control. 54*, 111–19.

Roberfroid, M. B. (2007), Prebiotics: The concept revisited. *J. Nutr. 137*, 830S–7S.

Roberfroid, M. B. (2000), Prebiotics and probiotics: Are they functional foods? *Am J Clin Nutr. 71*, 1682–7.

Rossi, M., Corradini, C., Amaretti, A., Nicolini, M., Pompei, A., Azanoni, S. and Matteuzzi, D. (2005), Fermentation of fructooligosaccharides and inulin by bifidobacteria: A comparative study of pure and fecal cultures. *Appl. Environ. Microbiol. 71*, 6150–8.

Roto, S. M., Rubinelli, P. M. and Ricke, S. C. (2015), An introduction to the avian gut microbiota and the effects of yeast-based prebiotic compounds as potential feed additives. *Front. Vet. Sci. 2 (Article 28)*, 1–18. doi: 10.3389/fvets.2015.00028.

Saminathan, M., Sieo, C. C., Kalavathy, R., Abdullah, N. and Ho, Y. W. (2011), Effect of prebiotic oligosaccharides on growth of *Lactobacillus* strains used as a probiotic for chickens. *African J. Microbiol. Res. 5(1)*, 57–64.

Schell, M. A., Karmirantzou, M., Snel, B., Vilanova, D., Berger, B., Pessi, G., Zwahlen, M.-C., Desiere, F., Bork, P., Delley, M., Pridmore, R. D. and Arigoni, F. (2002), The genome sequence of *Bifidobacterium longum* reflects its adaptation to the human gastrointestinal tract. *PNAS 99*, 14422–7.

Schrezenmeir, J. and de Vrese, M. (2001), Probiotics, prebiotics, and synbiotics – approaching a definition. *Am. J. Clin. Nutr. 73*, 361S–4S.

Sheng, K.-C., Pouniotis, D. S., Wright, M. D., Tang, C. K., Lazoura, E., Pietersz, G. A. and Apostolopoulos, V. (2006), Mannan derivatives induce phenotypic and functional maturation of mouse dendritic cells. *Immunology 118*, 372–83.

Shoaf, K., Mulvey, G. L., Armstrong, G. D. and Hutkins, R. W. (2006), Prebiotic galactooligosaccharides reduce adherence of enteropathogenic *Escherichia coli* to tissue culture cells. *Infect. Immun. 74*, 6920–8.

Siragusa, G. R. and Ricke, S. C. (2012), Chapter 20. *Probiotics as pathogen control agents for organic meat production*. In S. C. Ricke, E. J. Van Loo, M. G. Johnson and C. A. O' Bryan (eds), *Organic Meat Production and Processing*. Wiley Scientific/IFT, New York, NY, pp. 331–49.

Slavin, J. L. and Greenberg, N. A. (2003), Partially hydrolyzed guar gum: Clinical nutrition uses. *Nutrition 19*, 549–52.
Sofos, J. N., Flick, G., Nychas, G.-J., O'Bryan, C. A., Ricke, S. C. and Crandall, P. G. (2013), Chapter 6. *Meat, Poultry, and Seafood*. In M. P. Doyle and R. L. Buchanan (eds), *Food Microbiology-Fundamentals and Frontiers, 4th Edition*. American Society for Microbiology, Washington, D.C., pp. 111–67.
Spring, P., Wenk, C., Dawson, K. A. and Newman, K. E. (2000), The effects of dietary mannanoligosaccharides on cecal parameters and the concentrations of enteric bacteria in the ceca of *Salmonella*-challenged broiler chicks. *Poult. Sci. 79*, 205–11.
Stavric S. (1992), Defined cultures and prospects. *Int. J. Food Microbiol. 15*, 245–63.
Steer, T., Carpenter, H., Tuohy, K. and Gibson, G. R. (2000), Perspectives on the role of the human gut microbiota and its modulation by pro and prebiotics. *Nutr. Res. Rev. 13*, 229–54.
Stewart, T. S., Mendershausen, P. B. and Ballou, C. E. (1968), Preparation of a mannopentaose, mannohexaose, and mannoheptaose from *Saccharomyces cerevisiae* mannan. *Biochemistry 7*, 1843–54.
Taylor, S., Qu, L., Kitaygorodskiy, A., Teske, J., Latour, R. A. and Sun, Y. P. (2004), Synthesis and characterization of peptide-functionalized polymeric nanoparticles. *Biomacromolecules 5*, 245–8.
Ten Bruggencate, S. J. M., Bovee-Oudenhoven, I. M. J., Lettink-Wissink, M. L. G. and Van der Meer, R. (2003), Dietary fructo-oligosaccharides dose-dependently increase translocation of *Salmonella* in rats. *J. Nutr. 133*, 2313–18.
Thanissery, R., McReynolds, J. L., Conner, D. E., Macklin, K. S., Curtis, P. A. and Fasina, Y. O. (2010), Evaluation of the efficacy of NuPro yeast extract in reducing intestinal *Clostridium perfringens* levels in broiler chickens. *Poult. Sci., 89*, 2380–8.
Thitaram, S. N., Chung, C.-H., Day, D. F., Hinton Jr., A., Bailey, J. S. and Siragusa, G. R., (2005), Isomaltooligosaccharide increases cecal bifidobacteria population in young broiler chickens. *Poult. Sci. 84*, 998–1003.
Tizard, I. R., Carpenter, R. H., McAnalley, B. H. and Kemp, M. C. (1989), The biological activities of mannans and related complex carbohydrates. *Mol. Biother. 1*, 290–6.
Tuohy, K. M., Kolida, S., Lustenberger, A. M. and Gibson, G. R. (2001), The prebiotic effects of biscuits containing partially hydrolysed guar gum and fructo-oligosaccharides – A human volunteer study. *Br. J. Nutr. 86*, 341–8.
Tzortzis, G., Goulas, A. K. and Gibson, G. R. (2005), Synthesis of prebiotic galactooligosaccharides using whole cells of a novel strain, *Bifidobacterium bifidum* NCIMB 41171. *Appl. Micobiol. Biotech. 68*, 412–16.
Vandeplas, S., Dauphin, R. D., Beckers, Y., Thonart, P. and Théwis, A. (2010), *Salmonella* in chicken: Current and developing strategies to reduce contamination at farm level. *J. Food Prot. 73*, 774–85.
Van Etten, C. H., Miller, R. W., Wolff, I. A. and Jones, Q. (1961), Amino acid composition of twenty-seven selected seed meals. *Agric. Food Chem. 9*, 79–82.
Verdonk, J. M. A. J., Shim, S. B., Van Leeuwen, P. and Verstegen, M. W. A. (2005), Application of inulin-type fructans in animal feed and pet food. *Br. J. Nutr. 93*, S125–38.
Verma, S. V. S. and MacNab, J. M. (1982), Guar meal in diets for broiler chickens. *Br. Poult. Sci. 23*, 95–105.
Verma, S. V. S. and MacNab, J. M. (1984), Chemical, biochemical and microbiological examination of guar meal. *Ind. J. Poult. Sci. 19*, 165–70.
Vohra, P. and Kratzer, F. H. (1964a), The use of guar meal in chicken rations. *Poult. Sci. 43*, 502–3.
Vohra, P. and Kratzer, F. H. (1964b), Growth inbitory effect of certain polysaccharides for chickens. *Poult. Sci. 43*, 1164–70.
Waldroup, A. L., Yamaguchi, W., Skinner, J. T. and Waldroup, P. W. (1992), Effects of dietary lactose on incidence and levels of salomonellae on carcasses of broiler chickens grown to market age. *Poult. Sci. 71*, 288–95.

Wang, Y. (2009), Prebiotics: Present and future science and technology. *Food Res. Int. 42*, 8–12.

Xu, Z. R., Hu, C. H., Xia, M. S., Zhan, X. A. and Wang, M. Q. (2003), Effects of dietary fructooligosaccharide on digestive enzyme activities, intestinal microflora and morphology of male broilers. *Poult. Sci. 82*, 1030–6.

Yamada, K., Tokunaga, Y., Ikeda, A., Ohkura, K., Mamiya, S., Kaku, S., Sugano, M. and Tachibana, H. (1999), Dietary effect of guar gum and its partially hydrolyzed product on the lipid metabolism and immune function of Sprague-Dawley rats. *Biosci. Biotech. Biochem. 63*, 2163–7.

Yamamoto, Y., Sogawa, I., Nishina, A., Saeki, S., Ichikawa, N. and Iibata, S. (2000), Improved hypolipidemic effects of xanthan gum-galactomannan. *Biosci. Biotech. Biochem. 64*, 2165–71.

Yang, Y., Iji, P. A., Kocher, A., Mikkelsen, L. L. and Choct, M. (2007), Effects of mannanoligosaccharide on growth performance, the development of gut microflora, and gut function of broiler chickens raised on new litter. *J. Appl. Poult. Res. 16*, 280–8.

Yang, Y., Iji, P. A., Kocher, A., Thomsen, E., Mikkelsen, L. L. and Choct, M. (2008a), Effects of mannanoligosaccharide in broiler chicken diets on growth performance, energy utilisation, nutrient digestibility and intestinal microflora. *Br. Poult. Sci. 49*, 186–94.

Yang, Y., Iji, P. A., Kocher, A., Mikkelsen, L. L. and Choct, M. (2008b), Effects of mannanoligosaccharide and fructooligosaccharide on the response of broilers to pathogenic *Escherichia coli* challenge. *Br. Poult. Sci. 49*, 550–9.

Yang, Y., Iji, P. A. and Choct, M. (2009), Dietary modulation of gut microflora in broiler chickens: A review of the role of six kinds of alternatives to in-feed antibiotics. *World's Poult. Sci. J. 65*, 97–114.

Yoon, S.-J., Chu, D.-C. and Juneja, L. R. (2008), Chemical and physical properties, safety and application of partially hydrolized guar gum as dietary fiber. *J. Clin. Biochem. Nutr. 42*, 1–7.

Yusrizal, X. N. and Chen, T. C. (2003), Effect of adding chicory fructans in feed on fecal and intestinal microflora and excreta volatile ammonia. *Int. J. Poult. Sci. 2*, 188–94.

Zhang, W. F., Li, D. F., Lu, W Q. and Yi, G. F. (2003), Effects of isomalto-oligosaccharides on broiler performance and intestinal microflora. *Poult. Sci. 82*, 657–63.

Zimmermann, N. G., Andrews, D. K. and McGinnis, J. (1987), Comparison of several induced molting methods on subsequent performance of single comb white leghorn hens. *Poult. Sci. 66*, 408–17.

Ziprin, R. L., Corrier, D. E., Hinton, Jr., A., Beier, R. C., Spates, G. E. and DeLoach, J. R. (1990), Intracloacal Salmonella typhimurium infection of broiler chickens: Reduction of colonization with anaerobic organisms and dietary lactose. *Avian Dis.* 34, 749–53.

Chapter 6

Safety management and pathogen monitoring in poultry slaughterhouse operations: the case of the United States

Manpreet Singh and Estefanía Novoa Rama, Purdue University, USA

1 Introduction: food safety concerns in poultry processing

Foodborne illnesses are considered to be a major source of public health concerns worldwide. Although these are preventable, the costs associated with reducing the incidence of foodborne disease outbreaks, by continuously developing policies, regulations and implementing more effective technologies in the food industry, are quite high. With the consumers' increasing demand for natural, nutritious, wholesome and readily available food products, the industry faces challenges to successfully prevent foodborne illnesses in a cost-effective manner. The Centers for Disease Control and Prevention (CDC) estimates

http://dx.doi.org/10.19103/AS.2016.0010.08

that each year, there are approximately 48 million cases of foodborne illnesses causing about 128 000 hospitalizations and 3000 deaths only in the United States (CDC, 2015). Roughly 9.4 million of those cases are caused by known pathogens and in 2013, the CDC reported 818 foodborne disease outbreaks resulting in 13 360 illnesses and 16 deaths. Approximately 35% of these outbreaks were caused by pathogens such as *Salmonella* spp., *Escherichia coli* (STEC), *Clostridium perfringens* and *Campylobacter* spp. When organized by food category, the highest outbreak-associated illnesses were related to chicken meat (23%), with *Salmonella* in chicken being the leading pathogen–food category pair resulting in outbreak-associated illnesses.

Almost all pathogens associated with raw poultry have reservoirs in healthy animals from which they spread to humans, often times not causing illness in the infected animal. During processing, the likelihood of contamination of poultry carcasses by harmful infectious agents is quite high. There are a number of factors such as initial concentration of bacteria, presence of faecal matter or rises in ambient temperatures and humidity, which are difficult to control and contribute significantly to the ease of spread and growth of bacteria on the birds along the process. Consequently, the development of new control strategies to minimize contamination in the food production chain and prevent disease transmission becomes a challenge.

While most of the efforts to reduce the incidence of foodborne illness are being made by the processing industry, the risk of disease outbreak associated with the final consumer should also be evaluated. Interestingly, findings from a recent survey conducted in the United States showed that most consumers are not completely aware of safe handling and proper preparation procedures of raw meat. Improper thawing of meat and washing or rinsing raw poultry before cooking are examples of common practices that can lead to cross-contamination and dissemination of pathogens to other foods or kitchen surfaces (Kosa et al., 2015). These incorrect practices may lead to increasing numbers of disease outbreaks that take place in restaurants and private homes. In 2010, the CDC reported that 48% of outbreaks, with a known single setting, were caused by food consumed in a restaurant or deli, and 21% were due to food consumed in a private home. However, the pressure and responsibility of risk prevention continues to lie on the food manufacturer (Gould et al., 2013).

With the increased global demand for poultry meat, and development of highly intensive systems, controlling poultry-borne pathogens has become a worldwide public health-related concern. Continuous technological innovations and development of rigorous and more effective regulations through integrated efforts of the poultry industry and regulation agencies perpetuate the improvement of growth, harvest and production of poultry meat. Despite these advances, the public health concerns associated with microbiological hazards still prevail at all steps of the processing chain. This chapter focuses on the principal foodborne pathogens relevant to poultry safety and the management of effective measures for surveillance and control in the United States.

2 Common foodborne pathogens associated with poultry

Because the slaughtering process involves transport and handling of live animals, there is already a significant number of native bacteria and yeast that are carried into the slaughterhouse facilities on the feathers, feet and intestines of the birds. Most of the general microbiota present in poultry are mainly responsible for the spoilage of the product and are not necessarily pathogenic. However, a considerable number of bacteria associated with poultry are of food safety concern. The main pathogens responsible for causing foodborne illnesses related to the consumption of poultry products are *Salmonella* spp., *Campylobacter jejuni*, *Listeria monocytogenes*, *Clostridium perfringens* and *Staphylococcus aureus* (Waldroup, 1996).

2.1 *Salmonella* spp.

The *Salmonella* genus comprises a large number of illness-inducing serotypes with shared characteristics. As a member of the Enterobacteriaceae family, *Salmonella* are Gram-negative, mesophilic, facultative, rod-shaped bacilli adapted to grow in a wide variety of environments. *Salmonella* is known to have more than 2500 serotypes, with less than 100 of those accounting for human disease infections (CDC, 2013). According to the nature of the disease caused in humans, *Salmonella* serotypes can be classified as 'typhoidal' or 'non-typhoidal'. Typhoidal *Salmonella* is represented by the species *S. typhi*, which causes typhoid fever, and *S. paratyphi*, which is responsible for causing paratyphoid fever in humans. These species have reservoirs in humans, and they spread via the faecal-oral route; however, they are rarely found in industrialized countries.

Non-typhoidal *Salmonella*, on the other hand, causes gastroenteritis in humans, and it is represented by many serotypes of the species *S. enterica*. The most common serotypes responsible for non-typhoidal salmonellosis are Enteritidis, Typhimurium, Newport, Javiana and Heidelberg (CDC, 2014), and they can all be isolated from the intestinal tract of both humans and animals, with poultry being a frequent reservoir (Bryan et al., 1995). Gastroenteric salmonellosis can be described as the inflammation of the lower intestinal tract that may result in mild to severe diarrhoeal illness. The infectious dose ranges from 10^4 to 10^6 cells and symptoms may include diarrhoea, vomiting, abdominal cramps and fever, lasting several days.

Salmonella is ubiquitous and can easily survive in a wide variety of environments; however, its main reservoir is the gastrointestinal (GI) tract of animals, and it is mainly transmitted to humans through consumption of foods or water that has been contaminated with animal faeces (Bäumler et al., 2000). In the poultry industry *Salmonella* is commonly found on final raw products, mainly due to contamination of flocks with faecal matter during the different stages of production. While efforts to control *Salmonella* are mainly directed at reducing the risk of contamination in slaughterhouses, a pre-harvest control approach is necessary to reduce the persistence of the pathogen in the feed, water and litter (Russell, 2002).

2.2 *Campylobacter jejuni*

Campylobacter are Gram-negative spiral rods known to cause disease in both animals and humans. They are mesophilic and grow favourably under microaerophilic conditions, allowing them to quickly multiply in the GI tract of humans once ingested through consumption of contaminated foods. The principal species responsible for inducing disease in humans are *C. jejuni*, the most prevalent species, followed by *C. coli* and *C. lari*. According to the World Health Organization, *Campylobacter* are the leading cause of bacterial diarrhoeal disease worldwide and can cause infection at very small doses of less than 500 organisms (WHO, 2012). Gastroenteritis caused by *Campylobacter* is quite similar to salmonellosis, with diarrhoea, cramping, abdominal pain and fever being the main symptoms lasting about a week.

Campylobacter jejuni is most commonly found in the GI tract of birds, and it can easily spread between flocks through common water sources or contact with infected faeces. During bird slaughter, *C. jejuni* can contaminate the surface of poultry carcasses when released from the cloaca of birds in processes such as defeathering (Berrang et al., 2001). Furthermore, *C. jejuni* can become attached to the chicken skin and can persist in the carcass even after washing operations are conducted (Jang et al., 2007).

2.3 *Listeria monocytogenes*

Listeria monocytogenes is considered an opportunistic pathogen, causing disease mainly in immunocompromised individuals such as pregnant women, newborns, the elderly and HIV-positive patients. It is a Gram-positive, rod-shaped and environmental pathogen, and therefore can be introduced in the processing plant through soil, water, dust, air and plant personnel. *Listeria monocytogenes* is psychrotrophic; it can easily persist in cool and humid areas inside the processing facilities and cause post-thermal processing contamination on fully cooked, ready-to-eat (RTE) refrigerated meat and poultry products. The majority of listeriosis cases are invasive infections, spreading beyond the GI tract. Listeriosis can develop into life-threatening illnesses such as septicaemia and meningitis on immunocompromised patients. However, immunocompetent patients may only experience gastroenteritis, fever or even show no symptoms (Riedo et al., 1994).

As mentioned earlier, RTE meat and poultry products can become contaminated with *L. monocytogenes* after the products have been thermally processed, posing a major risk for the consumer. In an effort to reduce the risk and incidence of listeriosis, the United States Department of Agriculture (USDA) implemented a 'zero tolerance' rule of *L. monocytogenes* on RTE meat and poultry products, which is still maintained to this day in the meat and poultry industry.

2.4 *Clostridium perfringens*

With an estimated 1 million cases of foodborne illnesses caused by *Clostridium perfringens* every year, this Gram-positive, anaerobic and spore-forming bacillus is amongst the highest prevalent foodborne pathogens in the United States (Grass et al., 2013). *Clostridium perfringens* can be commonly found in the intestines of humans and animals, and it is also widely distributed in the environment. Because of its very nature, this pathogen can be frequently isolated from raw meat and poultry. In contrast to other common foodborne pathogens previously discussed, *C. perfringens* causes disease in humans through ingestion of large numbers of viable vegetative cells that are able to produce enough enterotoxin in the intestines to induce illness. Symptoms of this toxicoinfection include diarrhoea and abdominal cramps that begin only after 6 hours of exposure and last for less than 24 hours. The greatest risk associated with *C. perfringens* relates to its spore-forming ability that allows the organism to persist in the final product even after the vegetative cells are killed. These spores can resist high temperatures and survive harsh environmental conditions, suggesting its ability to survive commercial cooking temperatures. These spores can germinate and multiply in cooked food products that have been improperly cooled or temperature abused; in poultry meat establishments, this scenario can easily occur when poultry products are cooked in large batches, not allowing for thorough heating and quick cooling. Because of the organism's short generation time in its optimal temperature growth rate (43–47°C), rapid cooling of thermally processed poultry products is crucial to prevent germination and outgrowth of the pathogen.

2.5 *Staphylococcus aureus*

Staphylococcus aureus is a Gram-positive, aerobic, coccus-shaped bacteria commonly found on the skin and in mucosal areas of up to 25% of healthy people and animals (CDC, 2013). Only certain strains of *S. aureus* are considered pathogenic, and these strains can induce illness in humans by producing enterotoxins during their regular growth cycle. *Staphylococcus aureus* causes gastroenteritis and this disease is caused by the consumption of foods contaminated with these toxins. For *S. aureus* to produce enough enterotoxin to induce illness, a large number of viable bacterial cells (more than 10^6 CFU/g) need to be present in the food.

In poultry, *S. aureus* is present as part of its natural microbiota; however, it should be noted that the strain responsible for causing illness is usually of human origin. Therefore, the operators that handle food products are the main source of contamination with *S. aureus* in poultry slaughterhouses. The implementation of effective hygiene standards and sanitation standard operating procedures (SOPs) is thus crucial in reducing disease incidence.

3 Effects of processing operations on carcass contamination: from live transportation to carcass washing

Many factors should be considered when discussing carcass contamination during poultry processing. The steps during poultry processing are designed to periodically reduce the incidence of spoilage and pathogenic bacteria while maintaining the nutritional and sensory quality of the product. Pathogen monitoring and control may be achieved in the poultry industry through combined efforts between regulatory agencies and processors. The implementation of effective hazard analysis and critical control point (HACCP) systems and competent performance standards establish a preventive approach to achieve food safety at the processing plants. However, as previously discussed, microbiological food safety should also be controlled at the farm level, implementing a holistic farm-to-table approach to minimize pathogen prevalence.

3.1 Live transportation and handling

Transportation and handling of live birds can be considered the first step of poultry slaughter. The initial bacterial load carried by the birds upon arrival is determinant for the design of processing steps intended to reduce microbial levels to a minimum, while controlling the risk of cross-contamination. Since the arrival of live birds to slaughterhouses comprises the main entry for potential pathogenic bacteria, it is important to follow appropriate handling and transport conditions to reduce this incidence. Physical separation between flocks, keeping the birds clean and dry during transport and unloading and minimizing stress associated with feed withdrawal and confinement during transport are examples of practices that need to be controlled before slaughter to dramatically reduce the incidence of pathogens during poultry processing. In recent years, transportation of birds to slaughter facilities in small plastic coops has been replaced by large transport-container systems that are highly automated. The level of automation and large aperture of these modern containers allows for facilitated loading and unloading operations that, overall, cause less stress and harm to the birds (Bolder, 1998, 2007).

During transportation, cross-contamination between birds is likely to occur; containers rapidly become contaminated with pathogens of faecal origin that can be easily transferred from one flock to the next and eventually impact the incidence of these pathogens in subsequent processing steps (Rigby et al., 1980; Berrang et al., 2003; Slader et al., 2002; Rasschaert et al., 2007). While appropriate cleaning and sanitation of transport containers can significantly reduce this likelihood (Slader et al., 2002), pathogenic bacteria of concern such as *Salmonella* can persist in transport containers even after washing and sanitizing (Rigby et al., 1980). The use of dirty, contaminated crates or containers leads to external contamination of the birds and can significantly reduce the effectiveness of the pathogen control systems implemented in the slaughterhouses. If containers are not properly washed and sanitized between loads, contamination of flocks that were initially negative for certain pathogenic bacteria can also occur, resulting in increased pathogenic load on the birds upon arrival at the slaughter facilities (Berrang et al., 2003).

Other factors contributing to cross-contamination of live birds during transport include environmental conditions, close crowding and feed withdrawal time. Feed withdrawal is a common practice in poultry production. Feed is withdrawn from the birds 8–12 hours before slaughter in an effort to empty out the GI contents of the birds during transportation and holding periods to prevent chances of cross-contamination during evisceration. While being a recommended practice (NCC, 1992), studies have shown that feed withdrawal practices may actually result in increased incidence of pathogens, such as *Salmonella* and *Campylobacter* in the crop and ceca of young chickens, possibly due to increased stress (Corrier et al., 1999; Ramirez et al., 1997; Mulder, 1995). However, effective interventions such as adding lactic acid to drinking water during feed withdrawal have shown positive results in improving these practices and reducing crop contamination with *Salmonella* and *Campylobacter* (Byrd et al., 2001).

Once the birds arrive at the slaughterhouse, it is important that the steps of unloading and shackling are done in a manner that reduces stress levels. Birds are usually transported through the plant via conveyors, where opportunities for contamination by aerosols or condensation may increase (Bolder, 2007). The level of mechanization of the initial steps of poultry slaughter varies greatly between low- and high-capacity processing plants; however, physical separation between 'clean' and 'dirty' employees and positive airflow towards the outside of the facility to avoid condensation formation are conditions that should be controlled in all plants to effectively minimize the risk of contamination.

3.2 Stunning and bleeding

Before slaughter, the birds are stunned, to render them unconscious and insensitive to pain. While gas-stunning tunnels are becoming popular in the poultry industry due to animal welfare concerns in respect to shackling of birds before electrical stunning, stunning with electrical currents continues to be a widely used method in poultry slaughter. After electrical stunning, the birds remain unconscious for a duration of 60–90 seconds; during this time they should be unable to stand when removed from the shackle. Immediately after the birds are stunned, a cut in the neck is performed, manually or automatically, which allows them to bleed out. Neck cutting and bleeding ensures that the birds have stopped breathing before reaching the scalder (9 CFR 381.65 (b)) and are, therefore, killed humanely.

3.3 Scalding

The purpose of scalding is to facilitate feather removal. Applying high heat to the carcasses causes the feather follicles to dilate by denaturing the proteins responsible for attachment and therefore facilitates the following step of defeathering. Scalding is performed by immersing the chicken carcasses in tanks filled with hot water at 50–60°C for up to 3 minutes. Different combinations of time and temperature are used which account for hard or soft scalding methods (62–64°C for 45 seconds and 53.35°C for 120 seconds, respectively). For fresh poultry production, soft scalding is usually preferred because it leaves the outer skin layer intact, an essential quality attribute for consumer acceptance. Scalding can significantly reduce dirt, faecal material and other debris that can be carried on the feathers and skin of the birds. Thus, the process serves a dual purpose by contributing to a significant reduction in the incidence of *Salmonella* and *Campylobacter* on the birds when temperatures of 55–60°C are used (Yang et al., 2001).

All birds pass through common water tanks during scalding; therefore, scalding water becomes a significant contact point between birds that can result in cross-contamination and pathogen spread. With highly prevalent pathogens like *Salmonella* and *Campylobacter* surviving when organic matter is present, scalding water with high levels of faecal contamination can be a significant problem (Berrang et al., 2000). To avoid this, most slaughterhouses, by recommendation of the National Chicken Council (NCC), use countercurrent systems in the scalders that allow water to move through the system against incoming carcass flow (NCC, 1992). By doing so, water is continuously fed from clean-to-dirty ends of the scalder, thereby allowing carcasses to be washed by cleaner water as they move through the tanks. Other factors such as high flow rates, continuous agitation, use of multiple-tank scalders and water pH can also affect pathogen growth (Cason et al., 2000, 2001). However, scalding may have an insignificant effect on the overall microbial quality of poultry meat (Bolder, 2007; FAO/WHO, 2009).

3.4 Defeathering

Immediately after scalding, the birds are defeathered in a picker. The picker consists of sets of automatic rubber picking fingers that mechanically remove the feathers as the carcasses pass through them. While defeathering can result in lower microbial counts due to a high prevalence of bacteria in bird feathers, it can also be an important source of cross-contamination. Cross-contamination between carcasses can easily occur when contaminated rubber picking fingers come in contact with 'clean' carcasses. Apart from being difficult to clean and sanitize, rubber fingers can spread faecal material and other contaminants to the processing environment or other carcasses, due to continuous agitation and rubbing of the carcasses (Allen et al., 2003). Additionally, the ability of pathogenic bacteria to get entrapped in feather follicles once the feathers are removed should be considered, since it may result in high persistence of bacteria and inefficacy of interventions (NACMCF, 1997). Good recommended practices to reduce the incidence of cross-contamination in this step consist of appropriate sanitation and maintenance of the equipment, as well as prevention of feather build-up through continuous rinsing of equipment and carcasses (NCC, 1992). Post defeathering, washing of carcasses with antimicrobial solutions can also be performed to reduce pathogen prevalence.

3.5 Evisceration

The evisceration process relies on a series of interconnected equipment that performs the necessary procedures to remove the viscera from the body cavity of the carcasses. This process poses a risk of cross-contamination as the GI tract of birds harbours large numbers of both pathogenic and spoilage bacteria. Proper maintenance and operation of the equipment is crucial in this step to effectively remove the viscera while avoiding undesirable cuts or damage to the GI system, which can lead to potential leakage of its content and contamination of equipment and other carcasses. Another factor to be considered during evisceration is the uniformity in bird size of the flock, since equipment cannot easily be adjusted through the automated process. Feed withdrawal times also play a role in proper evisceration with gut strength and integrity rapidly decreasing when birds are excessively feed-deprived before slaughter (Bolder, 2007).

In order to comply with the 'Zero Fecal' standard established by the USDA-FSIS, appropriate safety management practices must be followed during evisceration, including online and offline reprocessing of visibly contaminated carcasses; therefore, continuous monitoring of the carcasses as they pass through the evisceration line is necessary. A number of interventions can be implemented to reduce the incidence of cross-contamination during this step. When handling carcasses during evisceration, plant personnel must maintain high hygiene standards. Continuous rinsing and sanitizing of the equipment should also be performed to minimize bacterial attachment and remove all faecal material (NCC, 1992; Bolder, 2007). Furthermore, the use of organic acids, chlorine and other antimicrobial solutions for whole-carcass rinsing after evisceration, as well as for offline reprocessing, might aid in the reduction of bacterial load and control cross-contamination (FAO/WHO, 2009; NCC, 1992; Waldroup et al., 1992; Fletcher et al., 1997).

3.6 Crop removal

The process of crop removal has gained interest regarding vis-a-vis cross-contamination among processors in recent years. Although cecal and intestinal damage has been considered a major source of cross-contamination in poultry slaughter, crop removal also poses significant risk. A study conducted by Byrd et al. (2002) showed that upon evisceration, approximately half of the carcasses can be contaminated with crop contents as well as contents from the upper GI tract. Multiple studies show that the incidence of *Campylobacter* and *Salmonella* contamination of crop contents could exceed that of the cecal and intestinal contents (Byrd et al., 1998; Hargis et al., 1995). Additionally, the same study by Hargis et al. (1995) showed that during processing, the crops were more prone to be ruptured than the ceca by up to 86-fold. These studies further highlight the significance of proper crop removal as a control procedure to reduce both cross-contamination and pathogen load on the broilers at the time of transport.

3.7 Washing

Carcass washing after evisceration is an important step to assure that no visibly contaminated carcasses enter the chiller. Washing reduces faecal material and overall organic load present on both inside and outside the chicken carcasses. A wide variety of washing procedures are performed during poultry processing as a multiple hurdle approach to control the risk of pathogen contamination of carcasses. As stated above, rinsing the carcasses and equipment with chlorinated water during defeathering, evisceration and crop removal operations is widely recommended to reduce cross-contamination and pathogen incidence on carcasses (NCC, 1992). Prior to chilling, several methods of carcass washes can be followed, including inside/outside washing, multiple washers and whole-carcass rinses. A dose of 20–50 ppm of free available chlorine in the water is recommended to significantly reduce the incidence of coliforms and pathogenic bacteria such as *Salmonella* and *E. coli* (Loretz et al., 2010; NCC, 1992; Waldroup et al., 1992).

4 Effects of processing operations on carcass contamination: chilling and post-chill processing

4.1 Chilling

Temperature is a decisive factor affecting bacterial growth. Varying temperatures to which poultry and poultry products are subjected to directly influence the types of bacteria that will prevail in the final food product. All bacteria, pathogenic or not, have optimum growth temperature ranges in which they can grow and multiply. While most bacteria associated with spoilage of poultry products are psychrophilic or psychrotrophic, growing at temperatures that range from −15 to 43°C, pathogenic bacteria are usually mesophilic, showing optimum growth at temperatures that range from 25 to 43°C. With the exception of *L. monocytogenes*, pathogen proliferation is significantly delayed at refrigeration temperatures. In order to meet the chilling requirements specified in 9 CFR 381.66 (b), the USDA-FSIS issued a series of guidelines for the modernization of poultry slaughter inspection (USDA-FSIS, 2014). According to these guidelines, minimizing pathogen incidence on poultry products can be achieved by quickly dropping the temperature of the product to 4°C, slowing down the growth and multiplication rates of mesophilic bacteria, while keeping the internal temperature from rising above 10°C during subsequent processing steps. During poultry processing this is achieved by immersing the carcasses in tanks containing ice water which is usually chlorinated (25–50 ppm) in a process known as 'immersion chilling', or by exposing the carcasses to forced cold-air currents in large rooms, which is known as 'air chilling'.

4.1.1 Immersion chilling

Immersion chilling is the most commonly used chilling method for poultry in the United States. This process consists of immersing carcasses in large cold water tanks chlorinated up to 50 ppm to reduce the spread and cross-contamination of carcasses by pathogenic bacteria. While regulatory standards allow for a maximum of 50 ppm of free available chlorine in the water, levels of 25 ppm have been shown to be effective against *Salmonella* (Lillard, 1980). Immersion chilling is usually carried out in countercurrent water flow chillers, where the carcasses are transported through a one- or two-bath system in colder and cleaner water. By immersing the washed and eviscerated poultry carcasses in cold water, both inside and outside surfaces of the carcass are rapidly cooled, slowing down the growth and multiplication of mesophilic bacteria that may be present. Carcass decontamination in processing plants is commonly performed through immersion chilling by adding antimicrobial solutions in the chilling water. Studies have shown that this method of application may be more effective than applying chemicals by sprays or through air chilling as contact time is reduced (Sakhare et al., 1999; Bolder et al., 2006). However, immersion chilling can be a significant source of cross-contamination during poultry processing; as carcasses are immersed in the water, the potential for washing bacteria off the surface of one carcass that have become attached to the surface of another carcass is significant. Studies conducted by Waldroup (1992, 1993) showed increased incidence of both *Salmonella* spp. and *Campylobacter* spp. on broiler carcasses from pre- to post-chill operations by 20 and 4%, respectively. Overall, this results in the need to combine chilling operations with the use of antimicrobials, such as chlorine, to effectively reduce contamination (Lillard, 1980); however, it should be noted that even when chlorinated, chilling water may be an important source of cross-contamination (Allen et al., 2000). In recent years, studies showing the effectiveness of peracetic acid (peroxyacetic acid; PAA) in reducing *Salmonella* and *Campylobacter* from poultry (Bauermeister et al., 2008) have led to a wider use of PAA in the industry as compared to chlorine. Cetylpyridinium chloride (CPC) has also gained interest as a non-chlorine-based alternative in the poultry industry. In a study conducted by Riedel et al. (2009), the use of CPC showed greater than 4.2 log CFU/ml reductions of *C. jejuni*, thus proving to be the most effective of 11 chemical interventions tested.

4.1.2 Air chilling

Air chilling is commonly used in poultry processing in Europe. The process involves cold-air blast systems through which carcasses flow, holding carcasses in a large cold room where they are exposed to temperatures ranging from 2 to 4°C. For evaporative air chilling, the carcasses are sprayed with water for an increased cooling effect as water evaporates from the surface. A disadvantage of regular air chilling and evaporative air chilling is the potential for cross-contamination through air currents and water droplets during spraying (Mead et al., 2000). Furthermore, during air chilling there is a concern with lower duration of cold exposure as compared to immersion chilling, which may result in failure to reach optimum internal temperatures (<4.4°C) rapidly (Bolder, 2007).

Comparisons between air and immersion chilling show mixed results. Extensive research has been conducted (Sanchez et al., 2002; Huezo et al., 2007), which will be discussed later in this chapter. Overall, the potential risk for cross-contamination during both immersion and air chill operations is significant (Mead et al., 2000; Allen et al., 2000). Therefore, maintaining strict sanitary conditions in both systems and implementing a multiple hurdle approach with the use of antimicrobial solutions is essential for minimizing the risk of contamination and effective pathogen control in the plant.

4.2 Post-chilling processing

Processing operations after chilling include whole-carcass packaging, portioning and deboning operations, and partial or full cooking for the production of comminuted and RTE poultry products. As the development of further-processed poultry products evolves, new risks appear, jeopardizing the wholesomeness of the products as they become available for the consumer.

The persistence of pathogenic bacteria as poultry carcasses enter post-chilling operations is of significant concern. Bacterial retention on the carcass accounts for an increased risk of pathogen persistence (Waldroup, 1996). When pathogens become attached and embedded in the poultry skin, they become more difficult to kill (Tamblyn et al., 1997). This occurs as carcasses are washed and immersed in water during initial processing steps. The water films formed on the surface of the carcasses increase the opportunity for attachment of bacteria to poultry skin (Thomas et al., 1981; Lillard, 1985; Conner et al., 1994). During cutting, deboning and grinding operations, the risk for cross-contamination is enhanced, and pathogens can easily spread to equipment, materials and other poultry parts (Kim et al., 1996; Wu et al., 2014). It is crucial that these steps are performed without delay to avoid proliferation of bacteria; furthermore, operators must follow strict sanitary practices, such as cleaning and sanitizing of contact surfaces as required, to minimize the risk of cross-contamination.

With an increased demand of wholesome, nutritious and readily available food products by consumers, the poultry industry faces new challenges to assure the production of safe products with extended shelf life. As the consumption of refrigerated RTE products expands, so does the risk of illnesses attributed to *L. monocytogenes*. Contamination of RTE products with *L. monocytogenes* usually happens during post processing operations such as slicing, peeling or packaging (Beresford et al., 2001). Since *L. monocytogenes* can survive at refrigeration temperatures and RTE cooked poultry products are usually consumed without further cooking, processors search for effective post-package decontamination procedures and novel ingredient formulations that can significantly reduce the threat (Zhu et al., 2005). As with other pathogens of concern, reducing the incidence of *L. monocytogenes* in RTE meats and poultry products to meet regulatory requirements does not just depend on the use of pathogen reduction techniques after processing, but also on the implementation of well-written procedures that effectively minimize the impact of the pathogen along the processing line.

5 Regulatory standards for food safety management during poultry slaughter

With the growing worldwide concern for food safety issues and increasing demands for poultry meat amongst consumers comes the expanding need for regulatory policies applicable to all meat and poultry establishments designed to ensure the production of safe food products and reduce the risk of foodborne illness associated with the consumption of these products. In 1996, the USDA Food Safety and Inspection Service (USDA-FSIS) developed a set of regulations titled 'Pathogen Reduction, Hazard Analysis and Critical Control Point (PR/HACCP) Systems' for the poultry-processing industry in the United States. The goal of the FSIS when providing these regulatory policies was to ensure that all meat and poultry establishments take appropriate and feasible measures at each step of the processing line by developing technologies and new procedures for the prevention of hazards. This set of new regulations for federally regulated processors consisted of the following requirements:

- Development and implementation of written sanitation SOPs.
- Regular microbial testing (generic *Escherichia coli* testing) to verify the adequacy of the establishments' process controls for the prevention and removal of faecal contamination and associated bacteria.

- Establishment of pathogen reduction performance standards for *Salmonella* that all slaughterhouses and further processing plants must meet.
- Development and implementation of an HACCP plan.

The FSIS's efforts to reduce accidental contamination of poultry carcasses during poultry slaughter and processing has led them to develop a series of other regulations, guidelines and performance standards for poultry processors. In 1998, the FSIS issued the 'Poultry Post-mortem Inspection and Re-inspection-Enforcing the Zero Tolerance for Visible Fecal Material' directive. This directive stated that all carcasses entering the chilling tank must be free of visible faecal material, considered an indicator for pathogen contamination, thus reducing the likelihood of cross-contamination of other carcasses. The zero tolerance for faecal material standard, along with the PR/HACCP final rule, which included a *Salmonella* performance standard, was a major incentive for the poultry industry that resulted in the development and implementation of new technologies and procedures to reduce microbial contamination.

Since the *Salmonella* Verification Program was established, in 1996, the USDA-FSIS has published a number of Federal Register notices in order to set more rigorous performance standards for the control of *Salmonella* and *Campylobacter* in poultry. In order to meet public health goals, the FSIS has developed a system that allows for continuous evaluation of the *Salmonella* performance standards. This system works by placing poultry plants in a sampling schedule based on their past performance regarding pathogen control. Specifically, the FSIS schedules approximately 75 *Salmonella* verification sample sets each month and allocates sampling resources by prioritizing the testing of plants categorized by product class, expected number of *Salmonella* isolates and process control, in the following order:

I Category 3: Plants that have highly variable process controls for *Salmonella* reduction. These are plants with positive *Salmonella* results exceeding the USDA-FSIS performance standards in the most recently completed sample set.

II Category 2: Plants that have variable process control. In these plants, the positive *Salmonella* results are higher than 50% of the performance standard in the most recently completed sample set, showing intermediate control for *Salmonella*. In this case, product class is taken into account by prioritizing testing at processing plants.

III Category 1: Plants with consistent process control for *Salmonella* reduction. These plants have positive *Salmonella* results at 50% or less of the performance standard in the two most recently completed sample sets.

In most recent years, new performance standards have been issued by the USDA-FSIS for the control of *Salmonella* and *Campylobacter* in poultry slaughter plants. In 2011, the FSIS published a Federal Register titled 'New Performance Standards for *Salmonella* and *Campylobacter* in Young Chicken and Turkey Slaughter Establishment: Response to Comments and Announcement of Implementation Schedule', which introduced a series of changes with respect to the requirements PR/HACCP final rule; these changes included, among others:

- Moving all poultry plants to the highest scheduling priority due to the initiation of *Campylobacter* testing for young chicken (broilers) and turkey products.
- New *Salmonella* standards accepting 5 positive samples in a 51-sample set for broilers and 4 positive samples in a 56-sample set for turkeys.

- Setting the sample set criteria for tracking *Campylobacter* to 1 ml sample portions, and setting the standards to accepting 8 positive samples per 51-sample set for broilers, and 3 positive samples per 56-sample set for turkeys.

In 2014, the FSIS issued the Modernization of Poultry Slaughter Inspection Final Rule that establishes a New Poultry Inspection System (NPIS) for broiler and turkey processing plants. This rule was issued to 'facilitate pathogen reduction in poultry products, improve the effectiveness of poultry slaughter inspection, make better use of the agency's resources, and remove unnecessary regulatory obstacles to innovation' (USDA-FSIS, 2014). Key elements of the NPIS included:

- Shifting agency resources to conduct more offline inspection activities that have been proven to be more effective in ensuring food safety, greater compliance with sanitation and HACCP regulations, carcasses with lower levels of visible faecal material and lower levels of *Salmonella*.
- Allowing the establishment personnel to have direct control of the sorting process within their facilities in order to have more flexibility to implement and assess new technologies that will benefit their performance.
- Developing, implementing and maintaining written procedures to ensure that carcasses contaminated with visible faecal material do not enter the chiller, as well as procedures to prevent contamination of carcasses and parts by enteric pathogens and faecal material throughout the processing plant. Having these procedures incorporated into their HACCP plans and monitored daily thus replaced the regulation that requires poultry-processing plants to test for generic *E. coli*, allowing plants to use more relevant indicators for process control.
- Authorizing broiler processing plants to operate at a maximum line speed of 140 birds per minute (bpm), while ensuring process control.

The continued efforts of USDA-FSIS to reduce foodborne illnesses and outbreaks to meet the goals for Healthy People 2020 have resulted in a new set of *Salmonella* and *Campylobacter* performance standards for raw chicken and not-RTE (NRTE) comminuted poultry products (USDA-FSIS, 2015). These new requirements and standards were set to achieve a reduction in human illness from *Salmonella* by about 25% and a 33% reduction for *Campylobacter*. The Federal Register notice also introduced a new moving window approach by using the results of routine sampling throughout the year, to determine the effectiveness of the plant's processes to address pathogen contamination on poultry carcasses and other derived products.

The USDA-FSIS has further implemented regulations to help reduce the incidence of *L. monocytogenes* on RTE poultry products. This pathogen rapidly emerged as a source of concern in processed meat and poultry products during the 1980s. With the intention of reducing the number of listeriosis outbreaks, the Federal Meat Inspection Act and the Poultry Products Inspection Act established a 'zero tolerance' policy for *L. monocytogenes* in RTE, which significantly decreased the incidence of illness caused by the bacteria. While the incidence of listeriosis has decreased by about 42% from 1996 to 2012 (CDC, 2014), the CDC estimates that approximately 1860 listeriosis cases occur per year, with more than 13% fatalities in the United States (Scallan et al., 2011). In 2003, the USDA-FSIS published a final rule for the control of *L. monocytogenes* in RTE meat and poultry products (9 CFR, 430) which stated that all 'establishments producing post-lethality exposed RTE products' were obliged to develop and maintain written procedures included in their HACCP plan, SSOPs and other prerequisite programs to effectively control *L. monocytogenes*. Furthermore, FSIS proposed detailed requirements for environmental testing of *Listeria* spp. to verify the effectiveness of the control measures implemented by the establishments to reduce the incidence of *L. monocytogenes* in RTE meat and poultry products. Poultry processors must work in collaboration with regulatory agencies to develop effective programmes that assure the production of safe and wholesome poultry products. Combining the processors' efforts of implementing well-developed HACCP plans, sanitation standard operating procedures (SSOPs) and good manufacturing practices (GMPs) with appropriate verification inspections conducted by the USDA-FSIS will guarantee that regulatory requirements are met.

6 GMP and the HACCP system

6.1 GMPs-HACCP prerequisites

Effective surveillance, control and quantification of food safety risks occurring at each step of processing require an understanding of the mechanisms by which contamination occurs. Once fully understood, we can start developing measures to control pathogen contamination in the processing plant. The key point for the regulation of poultry slaughter and processing is that all processing plants are required to conduct hazard analysis as part of their HACCP system. The hazard analysis is required to include 'food safety hazards that can occur before, during, and after entry into the establishment' (9 CFR 417.2 (a)). Pathogens of public health impact, such as *Salmonella* and *Campylobacter*, are common hazards that establishments producing raw poultry products should control and prevent in the processing environment through the integrated HACCP approach and GMPs or other prerequisite programmes.

The HACCP system is designed for application at every stage of the processing chain, from slaughter and processing to storage and distribution. For the production of safe and wholesome foods, it is necessary that each one of these segments is controlled by basic environmental and production conditions and satisfactory practices. Traditionally known as GMPs, the importance of these conditions and practices in serving as the HACCP foundation was stressed in a document issued in 1997 by the National Advisory Committee on Microbiological Criteria for Foods (NACMCF), where they were referred to as 'prerequisite programs'. Prerequisite programmes are, therefore, essential for the establishment of successful HACCP plans. The role of these programmes in food safety systems is to provide support and supplement the development, implementation and maintenance of well-established HACCP plans. Understanding how these programmes interconnect, work together and overall impact the food safety system is crucial for identifying weaknesses and making appropriate decisions that will directly influence process control and pathogen reduction practices.

It is important to understand how prerequisite programmes support the HACCP system and how management of both strategies differs. As opposed to HACCP, prerequisite programmes encompass food safety concerns of the entire establishment, as well as product quality and process control, and are not usually meant to manage specific process hazards. Examples of prerequisite programmes include GMPs, cleaning and sanitation

(SSOPs), pest control, processing procedures and product traceability. These programmes provide written documentation describing adequate practices that support the decision-making process during hazard analysis. They must be effectively designed and consistently implemented to provide reliable environmental and operational procedures that ultimately prevent food safety hazards from being likely to occur on an ongoing basis. When prerequisite programmes are not well implemented or effectively targeted to prevent potential hazards, they fail to support hazard analysis. Consequently, a negative impact on the establishment's HACCP arises, leading to an increase in the number of critical control points (CCPs), which are directly regulated by the HACCP plan. Therefore, the importance of reliable and adequate prerequisite programmes must be stressed as they alleviate process control and result in less complex and easier-to-manage HACCP plans.

In order to prevent contamination or adulteration of the food products and maintain process control, prerequisite programmes must meet certain requirements regardless of what they address. In general, these programmes should be based on written procedures with specified critical operational parameters; they should be continuously validated by supporting documentation, subjected to record-keeping that demonstrates adequate implementation as well as effectiveness in hazard prevention and, most importantly, provide corrective actions, including reassessment, needed in response to an 'unforeseen hazard' (9 CFR 417.3 (b)).

Written documentation such as records, SSOPs and other data are essential for the decision-making process in hazard analysis. Inspectors, management personnel and other responsible employees may review these results and records not only to validate and verify facility-wide hazard control, but also to identify trends or potential weaknesses that may result in non-compliance.

6.1.1 Establishment grounds and facilities

Establishment grounds and facilities must be maintained to avoid unsanitary conditions, product adulteration or any other provisions that may compromise the overall hygiene of the premises. Buildings must be constructed with durable materials that facilitate cleaning, sanitation and maintenance procedures. Establishments must also prevent pest harbourage and breeding within its facilities and in close proximity to the processing plant. Pest control is usually achieved by the implementation of a pest management programme.

All rooms and structures of the facilities should be constructed in a manner that allows for easy inspection with sufficient space to facilitate processing, handling and storage of food products under sanitary conditions. As a preventive measure to avoid cross-contamination and/or product adulteration, construction design should allow for separation of rooms in which edible products are processed from others where materials, chemicals and non-edible products are stored or handled. Moreover, facilities should allow for smooth product flow, allergen management and employee traffic. To prevent microbial dissemination across the processing plant, products should always flow from high microbiologically charged areas to those with lower microbial loads. Adequate ventilation that prevents unwanted condensation, as well as sufficient lighting in all areas of the processing plant, is also an essential measure provided to maintain sanitary conditions and prevent product adulteration (9 CFR 416.2).

Special attention must be paid to walls, ceilings and floors within the facilities. As mentioned earlier, these structures must be built of durable materials but should also be easy to clean and sanitize, be impervious to moisture and prevent the accumulation or entry of microorganisms and/or pests. Windows and doors, while facilitating ventilation, may serve as entry points or pest harbourage. These should remain closed during normal plant operations, and they should be designed to prevent accumulation of dust, dirt and debris while facilitating cleaning and sanitation. Condensation and standing water are major harbourage for bacteria and can be prevented by avoiding overhead piping, false ceilings and constructing sloped and well-drained floors (9 CFR 416.2).

Water in poultry-processing plants is another major issue of concern. Bacteria and other contaminants can easily spread in this medium throughout the plant, and can eventually result in the adulteration of the product. Plumbing systems installed in the facilities should allow for a potable water supply that reaches all areas where needed. From processing product to cleaning and sanitation, water is used in a wide variety of daily operations in the processing plant. Therefore, it is crucial that plumbing systems are designed to prevent cross-connection between pipes carrying wastewater or sewage and those systems that carry water used in the slaughter process. Adequate drainage must also be provided to prevent the accumulation of standing water, which can harbour bacteria. All water that comes in contact with the food product should be constantly monitored for microbial, physical and chemical contamination through an adequate water programme, usually implemented in the processing plant as a prerequisite (9 CFR 416.2).

6.1.2 Equipment and utensils

All equipment and utensils used in the processing plant, including those where raw ingredients are handled, should be designed in such a way that facilitates thorough cleaning and sanitation operations and it should be constructed of materials that do not contribute to product adulteration or contamination during processing, handling or storage. If storage of utensils is needed, it should be done in clean and dry receptacles separated from areas where edible products are processed, stored or handled to prevent adulteration. These receptacles should be clearly identified for their use and should not be used to store any edible product. All pieces of equipment must be constructed, located and operated in a way that does not interfere with inspection or maintenance procedures. To ensure adequate processing and minimize chemical or physical hazards due to equipment wear, all pieces of equipment and utensils should be under periodic preventive maintenance. All maintenance operations, preventive and corrective, should be specified in a written maintenance plan, including past records, preventive maintenance schedules and specific procedures (9 CFR 416.3).

6.1.3 Employee hygiene

Employee hygiene is a key aspect of ensuring wholesome poultry production and avoiding contamination in the processing plant. All employees must have appropriate training in food safety-related issues and personal hygiene. Common hygienic practices such as hand washing and appropriate garment cleaning must be followed by all employees directly or indirectly coming in contact with poultry in order to minimize adulteration risks. Furthermore, disease control practices should be followed to prevent the creation of unsanitary conditions due to employees carrying infectious disease or other sources of microbial contamination (9 CFR 416.5). To ensure that these practices are followed by all, new and existing employees, the processing plant must assign responsibilities and implement a formal food safety training programme consisting of up-to-date hygienic practices, training schedules and other documents that verify received training.

6.1.4 Cleaning and sanitation

Cleaning and sanitation practices are managed differently in each processing plant; however, all establishments must develop, implement and maintain written SSOPs that address them. Cleaning and sanitation are essential practices for microbial control since pathogens can easily multiply and spread through surfaces that have been in contact with poultry and poultry products or areas with excessive water. The Code of Federal Regulations (CFR) states that all surfaces of facilities, equipment and utensils that may or may not come in direct contact with food 'must be cleaned and sanitized as frequently as necessary to prevent the creation of insanitary conditions and the adulteration of the product' (9 CFR, Part 416). The frequency, responsible personnel, materials and chemical agents used to carry out these operations must be detailed in the SSOP. Furthermore, SSOPs should be routinely evaluated for their effectiveness and be subject to daily record-keeping in order to facilitate verification procedures.

6.1.5 Pest control

Infestation by pests includes insects, rodents and birds. Through integrated pest management (IPM) programmes, establishments can effectively reduce the number of incidences related to public health pests (Ehler, 2006). Pests are considered populations of disease vectors that can compromise the overall safety of the food products; therefore, the development and implementation of a good pest control plan is essential (9 CFR, Part 416). The IPM approach uses procedures that focus on prevention, reduction and elimination of favourable infestation conditions. While this approach includes the use of chemical pesticides, the main focus of the programme is directed to the establishment of routine inspections, appropriate identification procedures, setting threshold levels based on potential health impacts, implementing effective control measures and, most importantly, constant evaluation and validation of the IPM plan (Ehler, 2006). Inspection and verification activities should be conducted by both plant personnel responsible for pest management and trained personnel from outsider sources. In both cases, all procedures, frequency of analysis, incidences and corrective measures applied must be appropriately written and recorded in SSOPs and record forms that are integrated in the HACCP plan.

6.2 The HACCP system

Hazard analysis is considered to be the foundation of an establishment's food safety system. It is a science-based process control system that emphasizes a preventive approach, rather than reactionary, for pathogen control in the poultry industry. Through a thorough hazard analysis, the HACCP system monitors the risks associated with poultry production and serves as an effective safety management tool for processors. As opposed to previously discussed prerequisite programmes and GMPs, the HACCP plan focuses on each product being processed in the plant and emphasizes control measures of attributes affecting food safety. Furthermore, it establishes food safety thresholds, known as CCPs, designed to eliminate or reduce the hazards to an acceptable level throughout the entire slaughter process.

The importance of well-developed prerequisite programmes in supporting the implementation of a successful HACCP system has already been stressed. If establishments determine that hazards are not reasonably likely to occur due to control through prerequisite programmes or SSOP, they are required to have records associated with these programmes that support their decision-making process and demonstrate that the programmes are being effectively implemented on an ongoing basis. The first step in designing an appropriate HACCP plan is to develop a flow diagram of all the processing steps for all products being fabricated in the plant. A simple diagram will facilitate the identification and assessment of potential hazards associated with each step. After performing a thorough hazard analysis, the HACCP team will identify which steps are more likely to occur and pose a risk for severe illness, known as CCPs and establish limits and preventive measures to eliminate the risk and prevent deviations. Validation and verification of the HACCP system must be conducted periodically. Responsible personnel must ensure that the system is being followed and updated at every step of the processing line on a daily basis. The maintenance of well-written SSOPs, GMPs and record forms that assess if CCPs are being monitored and appropriate corrective measures are being taken when critical limits are not met is essential to carry out a successful verification process. Overall, the validity of the HACCP plan should be routinely assessed through integrated efforts of the processors and USDA-FSIS trained personnel. This approach will allow for the development of more effective regulations and practices for the constant improvement of the food safety system in the poultry industry.

7 Antimicrobial treatments used in poultry slaughter: chemical and biological treatments

Generally recognized as safe (GRAS) compounds for pathogen reduction in meat and poultry-processing plants include chemicals such as organic acids, trisodium phosphate (TSP) and chlorinated compounds that can be applied by spraying or immersion during processing. Biological treatments such as the use of bacteriophages and its bacteriocins are also used in the poultry industry solely or in combination with GRAS compounds as effective interventions.

7.1 Chemical treatments

7.1.1 Chlorine-based treatments

The efficacy of a wide variety of chemical treatments has been extensively studied throughout the years. Among the antimicrobials used currently in the poultry industry, chlorine-based interventions continue to be the most commonly used. Studies have shown that the use of chlorine treatments of 20–50 ppm, by water immersion or spraying, result in significant reductions of aerobic bacteria, coliforms and *E. coli* (Loretz et al., 2010). Reductions up to 2.6 log CFU/ml and 1.1 log CFU/ml were obtained on the same study when artificially inoculated carcass with *Campylobacter* spp. and *Salmonella*, respectively, were sprayed with 55 ppm chlorinated water.

While there are many beneficial aspects of using chlorine treatments, such as relatively low cost and rapid mode of action, it should be noted that there is a significant loss of effectiveness against pathogens at pH values above 7.0–7.5 due to the breakdown of hypochlorous acid to form less effective hypochlorite ions. Other factors that can reduce the efficacy of chlorine treatments include high levels of organic matter, initial bacterial load, high temperatures and presence of trace minerals in the water.

Chlorine dioxide, on the other hand, has certain advantages over hypochlorous acid as it is considered to be more stable and effective against *Salmonella* (Villareal et al., 1990) and its efficacy is not reduced by organic matter load (Vandekinderen et al., 2009). A study conducted by Lillard (1980) showed significant reduction of *Salmonella* counts on broiler carcasses that were chilled by immersion in water treated with chlorine dioxide at concentrations of 3–5 ppm. In more recent studies, bacterial reductions of 0.2–0.4 log CFU/ml were observed on poultry carcasses sprayed with 500–1200 ppm chlorine dioxide. However, a combination of chlorine dioxide treatments with chlorine resulted in higher reductions (Stopforth et al., 2007).

7.1.2 Acidified sodium chlorite

Acidified sodium chlorite (ASC) is a strong oxidizer obtained when mixing sodium chloride with an acid, usually citric or phosphoric acid. The use of ASC in the United States is permitted at concentrations between 500 and 1200 ppm (USDA, 2016). Sexton et al. (2007) evaluated the efficacy of ASC treatment and observed a reduction of *Salmonella* incidence from 90 to 10% when broiler carcasses were immersed in a 900 ppm solution of ASC during a 20-second period after chilling. Significant reductions of coliforms and Enterobacteriaceae, by 0.9–2.0 and 1.5 log units, respectively, were also obtained by Loretz et al. (2010) on poultry carcasses and parts treated with 500 ppm ASC.

7.1.3 Trisodium phosphate

TSP is a highly alkaline antimicrobial used in the poultry-processing industry at concentrations of 8–12%. Because of its characteristic surfactant properties, it is very effective in reducing attachment of bacteria to the skin of poultry carcasses, thus facilitating the washing of bacterial cells from the carcass surface. Early studies showed significant reductions of up to 2 log CFU/cm^2 of *S.* Typhimurium on chicken carcasses when treated with 10% TSP (Lillard, 1994). Jassas (2008) conducted a study that illustrated the efficacy of this antimicrobial against *E. coli* and aerobic total counts on chicken carcasses. The study consisted of immersing artificially inoculated chicken portions in TSP of variable concentrations of 8–12%, followed by dipping the chicken meat in tap water for 20 seconds. Results showed reductions of up to 1.6 and 1.7 log CFU/cm^2 for *E. coli* and total aerobic counts, respectively, when 12% TSP was used. Antimicrobial activity of TSP against other pathogens, including *Campylobacter*, *Listeria* and *Salmonella* species, has also been demonstrated (Loretz et al., 2010).

7.1.4 Organic acids

The antimicrobial activity of organic acids such as lactic, acetic and citric has been extensively studied throughout the years (Izat et al., 1989; Ismail et al., 2001). The low cost of organic acids and their efficacy against pathogens such as *Salmonella* (Smulders et al., 1986, 1998) has made them a desirable choice for pathogen control in poultry-processing plants. Mulder et al. (1987) showed reductions of *S.* Typhimurium by 1–2 log units on artificially inoculated chicken carcasses when immersed in a solution of 1% lactic acid for 10 minutes. Lactic acid has also been shown to be effective against *S.* Enteritidis,

E. coli O157:H7 and *L. monocytogenes* in a study conducted by Anang et al. (2007) causing reductions of 1.71, 2.59 and 1.97 log CFU/g, respectively, in artificially inoculated chicken breasts. Studies on the efficacy of acetic acid in reducing total aerobic counts, coliforms and Enterobacteriaceae have showed no significant results (Fabrizio et al., 2002; Sakhare et al., 1999). However, the use of a mixture of the peroxy compound with acetic acid with hydrogen peroxide known as PAA has been proven to be more successful for pathogen control (Purnell et al., 2014). Bauermeister et al. (2008) conducted a study that showed significant reductions of *Salmonella* and *Campylobacter* on poultry carcasses immersed in chilled water treated with PAA; concentrations as low as 0.0025% and 0.02% of PAA showed effectiveness against *Salmonella* and *Campylobacter*, respectively. The effectiveness of citric acid against *S.* Typhimurium on chicken breasts and *S.* Enteritidis on whole chicken has also been documented (Over et al., 2009; del Rio et al., 2007).

7.2 Biological treatments

With an increased concern of antimicrobial-resistant pathogenic bacteria, demands for the development and implementation of natural antimicrobial interventions have increased for the poultry industry. Examples of such interventions include the use of organic acids and essential oils against pathogenic bacteria. However, in an effort to move away from chemical interventions, the development of novel biological interventions, such as the use of bacteriophages and bacteriocins is gaining interest.

Bacteriophages are viruses that are capable of infecting bacteria and are widely distributed in the environment. Bacteriophages can penetrate the cell membrane and successfully replicate within the bacterium resulting in the death of the infected bacteria. Many studies have been conducted in recent years that illustrate the efficacy of bacteriophages against pathogens of concern to the poultry industry. Goode et al. (2003) reported significant reductions in pathogens *S.* Enteritidis and *C. jejuni* of artificially inoculated chicken skin. The efficacy of lytic bacteriophages in this study was found to rise as the multiplicity of infection was increased. Similar results were reported by Atterbury et al. (2003) with *C. jejuni* reductions of 1 log. More recently, a study on the antimicrobial effect of anti-*Salmonella* phage combined with several organic acids demonstrated *Salmonella* reductions of up to 2.4 log units on chicken skin and 2.6 log units on skinless chicken breasts (Sukumaran et al., 2015). The antimicrobial potential of bacteriocins isolated from lactic acid bacteria against *C. jejuni* has also been studied. In a study conducted by Stern et al. (2006) *C. jejuni* colonization reductions of one million-fold were observed in young chickens treated with bacteriocin-amended feed for 3 days.

8 Antimicrobial treatments used in poultry slaughter: physical treatments

Physical treatments include the use of water washes, with or without chemical solutions, commonly distributed through poultry slaughter as independent processing steps. As discussed earlier in this chapter, water washes can dramatically reduce the bacterial load present both on the surface and inside cavities of the chicken carcass. However, it can also be a significant source of cross-contamination. A multiple hurdle approach is generally used by poultry processors to minimize and/or eliminate the risks of cross-contamination.

Temperature of the water used also plays an important role in pathogen control using physical treatments. Since most pathogenic bacteria are mesophilic, the use of both low- and high-temperature interventions may reduce risk of microbial hazards during slaughter. While hot water and steam can effectively kill pathogenic and other bacteria responsible for spoilage, maintaining the product at low temperatures (<4.4°C) through the processing line until it reaches the consumer will be crucial to minimize the incidence of these microbes. Another physical treatment of relevant importance is the use of radiation. The basis of this method relies on using safe doses that can significantly reduce pathogen contamination without raising the temperature of the product, thus maintaining its sensory quality attributes.

8.1 Immersion chilling

Immersion chilling after washing of carcass is an effective physical treatment against foodborne pathogens. By rapidly dropping the temperature of poultry carcasses, the incidence of certain pathogenic bacteria is minimized; however, most importantly, it facilitates temperature control during subsequent processing steps including storage. While it is a complicated task to maintain the carcasses at a temperature below 4°C, it is still expected that carcasses do not surpass temperatures of 10°C during processing. This translates to a need of minimizing processing times, which can sometimes be a hurdle for the poultry industry. Adequate chilling mechanisms are therefore crucial to preventing proliferation of bacteria and reducing the risk.

The effectiveness of immersion chilling as an intervention for pathogen reduction in combination with chemical treatments has been widely studied (Ismail et al., 2001; del Rio et al., 2007; Loretz et al., 2010). Chemical solutions used as treatments for pathogen reduction are usually applied in the poultry industry by adding them to water tanks where the carcasses are immersed or by directly spraying them on the carcass surface. When comparing these two methods of application, there are many factors that need consideration and may directly affect the efficacy of the treatments, such as initial bacterial load, exposure time to the antimicrobials or temperature (Loretz et al., 2010). Many studies have shown an increased effectiveness of immersion treatments when compared to spraying (Sakhare et al., 1999; Bolder, 2007), possibly due to full coverage of the product. However, there have been studies that show no differences between treatments (Sinhamahapatra et al., 2004).

8.2 Air chilling

Air chilling consists of exposing carcasses to cold-air currents to rapidly drop their temperature to control the growth of bacteria. As mentioned in previous segments, air chilling can result in incomplete or slower chilling of carcasses (Bolder, 2007) and has the potential to cause excessive desiccation of the carcass surface resulting in an undesirable loss of quality. Mixed results have been observed when comparing air chilling to immersion chilling. The main advantage of air chilling resides in a lower potential for cross-contamination due to less physical contact between carcasses (Allen et al., 2000). However, Sanchez et al. (2002) observed that air-chilled carcasses had higher aerobic and coliform counts than those chilled by immersion in cold water, suggesting that cross-contamination can still occur when carcasses are sprayed with water. The same study showed that air chilling, on the other hand, significantly reduced the incidence of

Salmonella to lower levels than those obtained using immersion chilling. Overall, this intervention has significant antimicrobial activity, showing reductions of approximately 1 log for coliforms, *E. coli* and *Salmonella* on poultry carcasses (Huezo et al., 2007) and reductions of 1.4 log CFU/ml for *C. jejuni* on breast skin (James et al., 2007).

8.3 Hot water and steam

Exposure of mesophilic and psychrophilic bacteria to high temperatures for short periods is a powerful strategy, widely used in the food processing industry, for killing pathogenic bacteria. Many studies have shown the effectiveness of using hot water and steam to reduce the incidence of pathogenic and spoilage bacteria associated with poultry (Avens et al., 1999; Nutsch et al., 1997). However, finding an adequate time–temperature combination that can effectively kill bacteria without compromising the overall quality of the product can be a challenge (Smulders et al., 1998). High reductions of aerobic bacteria, coliforms, Enterobacteriaceae and *E. coli* on poultry carcasses were found by Loretz et al. (2010) when carcasses were treated with hot water at 90°C for 0.2–0.4 min. Avens et al. (2002) observed reductions of more than 3.3 log units in aerobic bacteria from poultry carcasses when treated with steam at 98°C for a 3-minute period. However, deterioration in chicken skin appearance was observed when exposed to steam at 90°C for 12 seconds, while longer exposure resulted in significant damage (Avens et al., 2002).

8.4 Electrolyzed water

Electrolyzed water is characterized by having a low pH (2.3–2.7) and a high redox potential, making it an effective agent with a high potential for killing bacteria. Used mainly as an immersion solution, electrolyzed water is considered an emerging technology in the poultry industry for pathogen control. Its effectiveness against *C. jejuni*, with reductions of up to 3 log units, in artificially contaminated poultry carcasses has been thoroughly studied (Fabrizio et al., 2002; Park et al., 2002). Significant reductions for aerobic bacteria, coliforms, *E. coli* and *S.* Typhimurium on poultry carcasses were also observed when combining spraying electrolyzed water with further acidic electrolyzed water immersion (Fabrizio et al., 2002).

8.5 Ozone processing

Ozone use has GRAS status in the United States since 1984, when it was approved by the USDA for its use in recycling water from poultry chill tanks (Kim et al., 1999). It is highly reactive and penetrable and its use does not result in the formation of toxic compounds. Yousef et al. (2001) studied the efficacy of gaseous and liquid ozone in reducing *S.* Enteritidis from externally contaminated eggshells, showing reductions by 2.2–2.7 log units when eggs were treated with ozone for 1–20 minutes without pressure. Fabrizio et al. (2002) reported insignificant reductions of *E. coli*, coliforms, aerobic bacteria and *S.* Typhimurium on poultry carcasses treated with ozonated water.

8.6 Irradiation

Food irradiation is an established process for pathogen reduction and shelf-life extension that consists of exposing foods to high levels of ionizing radiant energy, usually gamma rays. The use of irradiation in poultry was first approved by the USDA in 1990, at dosage levels of 1.5–3 KGy, which can effectively reduce or eliminate common enteric pathogens such as *C. jejuni*, *S. aureus*, *L. monocytogenes*, *Salmonella* spp. and *E. coli* O157:H7 (Thayer, 1995). These low-dose treatments of ionizing radiation have been shown to effectively reduce *Salmonella* on chilled poultry by 3 log units (Farkas, 1998). The main advantage of food irradiation is its efficacy to reduce the number of foodborne pathogens without significantly raising the temperature of the food product, thus preventing off-flavours and odours that result in a loss of quality. The main source of ionizing radiation is gamma radiation generated by radioisotopes (cobalt-60 and caesium-137). The process of exposing poultry to gamma rays is usually done in USDA-approved irradiation facilities, where the packaged products are exposed to low-dose irradiation while being carefully controlled by responsible plant personnel and USDA inspectors.

Another method of inactivating microorganisms is through electron beam irradiation, consisting of linear accelerators used as the source for ionizing radiation. The efficacy of this method is comparable to the application of gamma rays (Tahergorabi et al., 2012) and has also been proven to be effective against *Campylobacter*, generic *E. coli* and coliforms. A study conducted by Lewis et al. (2002) on boneless and skinless chicken breasts treated with electron beam irradiation at doses of 1.0 and 1.8 KGy showed reductions of up to 5 logs for *Campylobacter* and *Salmonella* and up to 3 logs for other bacteria. Additionally, sensory evaluation showed that no significant loss of quality was observed until 14 days after the treatment. More recently, Sarjeant et al. (2005) showed reductions of *S.* Typhimurium on artificially inoculated fresh chicken breasts by 4 log units, using variable doses, from 1 to 3 KGy, of electron beam irradiation.

While being a highly effective intervention against foodborne pathogens, food irradiation still faces a great challenge in consumer acceptance; thus its use in the poultry industry is still not a common practice (Thayer, 1995). As stated earlier, irradiation doses approved by use from the USDA and FDA do not pose a risk for consumers (9 CFR, Part 424). Appropriately irradiated poultry products maintain their nutritional value and wholesomeness without compromising the overall quality of the product; therefore, continued efforts need to be directed towards educating consumers on the practice and its applications in the food industry.

9 Assessing the efficacy of pathogen reduction systems: microbial testing

In a need to validate and verify the efficacy of pathogen reduction systems implemented in the poultry-processing plant through integrated efforts of a well-established HACCP plan, written SSOPs, GMPs and antimicrobial interventions, adequate and regular microbiological testing must be conducted. An appropriate microbiological testing method along with a statistically valid sampling plan should allow for assessment of microbiological hazards in the food products and the processing operations, and validation of the strategies applied for pathogen control.

In correspondence to product assessment, proper sampling techniques and procedures are necessary to guarantee accuracy in the results of microbial testing. It should also be considered that the sample collection method will directly influence the ability for bacterial detection on poultry carcasses. When developing a proper sampling plan, sample size and location are primary aspects to be considered. Since microbial testing of the entire lot is not plausible, samples that are representative of the whole lot should be obtained and analysed. Sample handling and collection procedures should also be established and carried out by trained personnel to avoid inaccuracy of results. It is also important that collected samples are sent for testing without delay and under proper transportation conditions. In general, the procurement of reliable quantitative data to evaluate process control will highly depend on the collection, handling, storage and transportation of samples. In order to successfully use this data, all these procedures must be carefully monitored, controlled and avoid high-temperature fluctuations, sample leakage and other unwanted occurrences that may lead to unreliable results.

In poultry slaughter, whole-carcass rinses for broilers, swab samples for turkeys and defined volume samples for ground and portioned product are often used. For microbiological testing, both quantitative and qualitative analyses are performed. Under the most recent final rule for poultry inspection, all establishments are required to incorporate their microbiological sampling plans into their HACCP plans, SSOPs or other prerequisite programmes (USDA-FSIS, 2014). In addition to maintaining written procedures to prevent contamination of carcasses as part of their HACCP plans, poultry processors are also responsible for developing and implementing a microbiological sampling plan, which requires inclusion of carcass sampling at pre chill and post chill (79 CFR 49565). As far as indicator organisms are concerned, FSIS recommends testing for generic *E. coli*, aerobic plate counts, Enterobacteriaceae and total coliforms. *Listeria monocytogenes* performance standards are also used to validate the procedures used by the processor in handling cooked products after processing, thus ensuring a safe product.

Overall, these performance standards are under constant revision; therefore, poultry processors must implement an ongoing microbiological testing programme that will allow for compliance with regulatory requirements and guarantee the production of wholesome poultry and poultry products. To fit the demands of the consumers and regulatory requirements, the processor must design a microbiological testing plan that meets these objectives and that can be easily integrated into the normal operations of the plant. This will result in the development of a preventive approach towards safety management operations.

10 Conclusions

Pathogen monitoring and control in poultry slaughterhouses is essential to meet the ultimate goal of producing safe and wholesome products. The implementation of a multiple hurdle approach comprising an effective HACCP plan system, prerequisites, antimicrobial interventions and an appropriate microbiological testing plan facilitates poultry processors tasks of ensuring food safety throughout the entire process while complying with regulatory guidelines. Effectiveness of pathogen control systems during processing depends on the incoming load of pathogenic bacteria on live birds. Therefore, adopting a pre-harvest pathogen control approach is crucial and should be a part of the overall food safety programme. Ultimately, the production of poultry and poultry products that are safe for human consumption will rely on understanding the mechanisms of growth, infection and persistence of pathogens in poultry. Continuous research, technology advances and education on safe practices are critical for minimizing foodborne illnesses.

11 Where to look for further information

A good introduction to the subject is available in the book titled 'Poultry meat processing' (CRC Press) edited by Casey M. Owens, Christine Z. Alvarado and Alan R. Sams, which provides a comprehensive review of poultry processing from the industry perspective as well as a reference for researchers in the areas of sensory and microbial quality of poultry meat. More specifically, an overview of antimicrobial interventions for pathogen monitoring in poultry processing can be found in Loretz et al. (2010).

The USDA is a key source of information on poultry meat processing and quality in the United States (http://www.usda.gov). The USDA-FSIS constitutes the public health agency providing guidelines and regulations for industry professionals, academic researchers and consumers (http://www.fsis.usda.gov), while the USDA-ARS serves as an important source for research trends and popular topics regarding poultry science (http://www.ars.usda.gov).

Information on foodborne infections and up to date disease outbreak surveillance reports can be found in the CDC archives (http://www.cdc.gov), as well as introductory information on foodborne pathogens and illness prevention guidelines for consumers.

12 References

Allen, V. M., Tinker, D. B., Wathes, C. M. and Hinton, M. H. (2003). Dispersal of micro-organisms in commercial defeathering systems. *British Poultry Science*, *44*(1), 53–9.

Allen, V. M., Corry, J. E., Burton, C. H., Whyte, R. T. and Mead, G. C. (2000). Hygiene aspects of modern poultry chilling. *International Journal of Food Microbiology*, *58*(1), 39–48.

Anang, D. M., Rusul, G., Bakar, J. and Ling, F. H. (2007). Effects of lactic acid and lauricidin on the survival of *Listeria monocytogenes*, *Salmonella* Enteritidis and *Escherichia coli* O157: H7 in chicken breast stored at 4°C. *Food Control*, *18*(8), 961–9.

Atterbury, R. J., Connerton, P. L., Dodd, C. E., Rees, C. E. and Connerton, I. F. (2003). Application of host-specific bacteriophages to the surface of chicken skin leads to a reduction in recovery of *Campylobacter jejuni*. *Applied and Environmental Microbiology*, *69*(10), 6302–6.

Avens, J. S., Albright, S. N., Morton, A. S., Prewitt, B. E., Kendall, P. A. and Sofos, J. N. (2002). Destruction of microorganisms on chicken carcasses by steam and boiling water immersion. *Food Control*, *13*(6), 445–50.

Avens, J. S. and Morton, A. S. (1999). Virtual eradication of psychrophilic and mesophilic bacteria on turkey carcass tails by steam and boiling water immersion. *The Journal of Applied Poultry Research*, *8*(2), 205–13.

Bauermeister, L. J., Bowers, J. W. J., Townsend, J. C. and McKee, S. R. (2008). The microbial and quality properties of poultry carcasses treated with peracetic acid as an antimicrobial treatment. *Poultry Science*, *87*(11), 2390–8.

Bäumler, A. J., Hargis, B. M. and Tsolis, R. M. (2000). Tracing the origins of *Salmonella* outbreaks. *Science*, *287*(5450), 50–2.

Beresford, M. R., Andrew, P. W. and Shama, G. (2001). *Listeria monocytogenes* adheres to many materials found in food-processing environments. *Journal of Applied Microbiology*, *90*(6), 1000–5.

Berrang, M. E., Northcutt, J. K., Fletcher, D. L. and Cox, N. A. (2003). Role of dump cage fecal contamination in the transfer of *Campylobacter* to carcasses of previously negative broilers. *The Journal of Applied Poultry Research*, *12*(2), 190–5.

Berrang, M. E., Buhr, R. J., Cason, J. A. and Dickens, J. A. (2001). Broiler carcass contamination with *Campylobacter* from feces during defeathering. *Journal of Food Protection*, *64*(12), 2063–6.

Berrang, M. E., Dickens, J. A. and Musgrove, M. T. (2000). Effects of hot water application after defeathering on the levels of *Campylobacter*, coliform bacteria, and *Escherichia coli* on broiler carcasses. *Poultry Science*, *79*(11), 1689–93.

Bolder, N. M. (2007). Microbial challenges of poultry meat production. *World's Poultry Science Journal*, *63*(3), 401–11.

Bolder, N. M., Lipman, L. J. and Putirulan, F. F. (2006). Chemical carcass decontamination to control *Salmonella* and *Campylobacter* in poultry meat. In *Proceeding European Poultry Congress*, Verona.

Bolder, N. M. (1998). The microbiology of the slaughter and processing of poultry. *The Microbiology of Meat and Poultry*. Blackie Academic and Professional, London, UK, 158–73.

Bolder, N. M. (1997). Decontamination of meat and poultry carcasses. *Trends in Food Science and Technology*, *8*(7), 221–7.

Bryan, F. L. and Doyle, M. P. (1995). Health risks and consequences of *Salmonella* and *Campylobacter jejuni* in raw poultry. *Journal of Food Protection*, *58*(3), 326–44.

Byrd, J. A., Hargis, B. M., Corrier, D. E., Brewer, R. L., Caldwell, D. J., Bailey, R. H., McReynolds, J. L., Herron, K. L. and Stanker, L. H. (2002). Fluorescent marker for the detection of crop and upper gastrointestinal leakage in poultry processing plants. *Poultry Science*, *81*(1), 70–4.

Byrd, J. A., Hargis, B. M., Caldwell, D. J., Bailey, R. H., Herron, K. L., McReynolds, J. L., Brewer, R. L., Anderson, R. C., Bischoff, K. M., Callaway, T. R. and Kubena, L. F. (2001). Effect of lactic acid administration in the drinking water during preslaughter feed withdrawal on *Salmonella* and *Campylobacter* contamination of broilers. *Poultry Science*, *80*(3), 278–83.

Byrd, J. A., Corrier, D. E., Hume, M. E., Bailey, R. H., Stanker, L. H. and Hargis, B. M. (1998). Incidence of *Campylobacter* in crops of preharvest market-age broiler chickens. *Poultry Science*, *77*(9), 1303–5.

Cason, J. A., Buhr, R. J. and Hinton, A. (2001). Unheated water in the first tank of a three-tank broiler scalder. *Poultry Science*, *80*(11), 1643–6.

Cason, J. A., Hinton, A. and Ingram, K. D. (2000). Coliform, *Escherichia coli*, and *Salmonellae* concentrations in a multiple-tank, counterflow poultry scalder. *Journal of Food Protection*, *63*(9), 1184–8.

Centers for Disease Control and Prevention. (2015). Surveillance for foodborne disease outbreaks-United States, 2013 Annual report. http://www.cdc.gov/foodsafety/pdfs/foodborne-disease-outbreaks-annual-report-2013-508c.pdf (accessed 16 April 2016).

Centers for Disease Control and Prevention. (2014). Foodborne Diseases Active Surveillance Network (FoodNet): FoodNet Surveillance Report for 2014 (Final Report). http://www.cdc.gov/foodnet/pdfs/2014-foodnet-surveillance-report.pdf (accessed 16 April 2016).

Centers for Disease Control and Prevention. (2013). Surveillance for foodborne disease outbreaks-United States, 2009–2010. MMWR. Morbidity and Mortality Weekly Report, *62*(3), 41.

Conner, D. E. and Bilgili, S. F. (1994). Skin attachment model for improved laboratory evaluation of potential carcass disinfectants for their efficacy against *Salmonella* attached to broiler skin. *Journal of Food Protection*, *57*(8), 684–8.

Corrier, D. E., Byrd, J. A., Hargis, B. M., Hume, M. E., Bailey, R. H. and Stanker, L. H. (1999). Presence of *Salmonella* in the crop and ceca of broiler chickens before and after preslaughter feed withdrawal. *Poultry Science*, *78*(1), 45–9.

del Río, E., Muriente, R., Prieto, M., Alonso-Calleja, C. and Capita, R. (2007). Effectiveness of trisodium phosphate, acidified sodium chlorite, citric acid, and peroxyacids against pathogenic bacteria on poultry during refrigerated storage. *Journal of Food Protection*, *70*(9), 2063–71.

Ehler, L. E. (2006). Integrated pest management (IPM): definition, historical development and implementation, and the other IPM. *Pest Management Science*, *62*(9), 787–9.

Fabrizio, K. A., Sharma, R. R., Demirci, A. and Cutter, C. N. (2002). Comparison of electrolyzed oxidizing water with various antimicrobial interventions to reduce *Salmonella* species on poultry. *Poultry Science*, *81*(10), 1598–605.

Farkas, J. (1998). Irradiation as a method for decontaminating food: a review. *International Journal of Food Microbiology*, *44*(3), 189–204.

Fletcher, D. L., Craig, E. W. and Arnold, J. W. (1997). An evaluation of on-line 'reprocessing' on visual contamination and microbiological quality of broilers. *The Journal of Applied Poultry Research*, *6*(4), 436–42.

Goode, D., Allen, V. M. and Barrow, P. A. (2003). Reduction of experimental *Salmonella* and *Campylobacter* contamination of chicken skin by application of lytic bacteriophages. *Applied and Environmental Microbiology*, *69*(8), 5032–6.

Gould, L. H., Walsh, K. A., Vieira, A. R., Herman, K., Williams, I. T., Hall, A. J. and Cole, D. (2013). Surveillance for foodborne disease outbreaks – United States, 1998–2008. *MMWR Surveill Summ*, *62*(2), 1–34.

Grass, J. E., Gould, L. H. and Mahon, B. E. (2013). Epidemiology of foodborne disease outbreaks caused by *Clostridium perfringens*, United States, 1998–2010. *Foodborne Pathogens and Disease*, *10*(2), 131–6.

Hargis, B. M., Caldwell, D. J., Brewer, R. L., Corrier, D. E. and DeLoach, J. R. (1995). Evaluation of the chicken crop as a source of *Salmonella* contamination for broiler carcasses. *Poultry Science*, *74*(9), 1548–52.

Huezo, R., Northcutt, J. K., Smith, D. P., Fletcher, D. L. and Ingram, K. D. (2007). Effect of dry air or immersion chilling on recovery of bacteria from broiler carcasses. *Journal of Food Protection*, *70*(8), 1829–34.

Ismail, S. A. S., Deak, T., El-Rahman, H. A., Yassien, M. A. M. and Beuchat, L. R. (2001). Effectiveness of immersion treatments with acids, trisodium phosphate, and herb decoctions in reducing populations of *Yarrowia lipolytica* and naturally occurring aerobic microorganisms on raw chicken. *International Journal of Food Microbiology*, *64*(1), 13–19.

Izat, A. L., Colberg, M., Adams, M. H., Reiber, M. A. and Waldroup, P. W. (1989). Production and processing studies to reduce the incidence of *Salmonellae* on commercial broilers. *Journal of Food Protection*, *52*(9), 670–3.

James, C., James, S. J., Hannay, N., Purnell, G., Barbedo-Pinto, C., Yaman, H., Araujo, M., Gonzalez, M. L., Calvo, J., Howell, M. and Corry, J. E. L. (2007). Decontamination of poultry carcasses using steam or hot water in combination with rapid cooling, chilling or freezing of carcass surfaces. *International Journal of Food Microbiology*, *114*(2), 195–203.

Jang, K., Kim, M., Ha, S., Kim, K., Lee, K., Chung, D., Kim, C. and Kim, K. (2007). Morphology and adhesion of *Campylobacter jejuni* to chicken skin under varying conditions. *Journal of Microbiology and Biotechnology*, *17*(2), 202.

Jasass, F. B. (2008). Effectiveness of trisodium phosphate, lactic acid, and acetic acid in reduction of *E. coli* and microbial load on chicken surfaces. *African Journal of Microbiology Research*, *2*(3), 50–5.

Kim, J. G., Yousef, A. E. and Dave, S. (1999). Application of ozone for enhancing the microbiological safety and quality of foods: a review. *Journal of Food Protection*, *62*(9), 1071–87.

Kim, J. W. and Slavik, M. F. (1996). Cetylpyridinium chloride (CPC) treatment on poultry skin to reduce attached *Salmonella*. *Journal of Food Protection*, *59*(3), 322–6.

Kosa, K. M., Cates, S. C., Bradley, S., Chambers, I. V. and Godwin, S. (2015). Consumer-reported handling of raw poultry products at home: results from a national survey. *Journal of Food Protection*, *78*(1), 180–6.

Lewis, S. J., Velasquez, A. and Cuppett, S. L. (2002). Effect of electron beam irradiation on poultry meat safety and quality. *Poultry Science*, *81*(6), 896–903.

Lillard, H. S. (1994). Effect of trisodium phosphate on *Salmonellae* attached to chicken skin. *Journal of Food Protection*, *57*(6), 465–9.

Lillard, H. S. (1985). Bacterial cell characteristics and conditions influencing their adhesion to poultry skin. *Journal of Food Protection*, *48*(9), 803–7.

Lillard, H. S. (1980). Effect on broiler carcasses and water of treating chiller water with chlorine or chlorine dioxide. *Poultry Science*, *59*(8), 1761–6.

Line, J. E., Bailey, J. S., Cox, N. A. and Stern, N. J. (1997). Yeast treatment to reduce *Salmonella* and *Campylobacter* populations associated with broiler chickens subjected to transport stress. *Poultry Science*, *76*(9), 1227–31.

Loretz, M., Stephan, R. and Zweifel, C. (2010). Antimicrobial activity of decontamination treatments for poultry carcasses: a literature survey. *Food Control*, *21*(6), 791–804.

Mead, G. C., Allen, V. M., Burton, C. H. and Corry, J. E. L. (2000). Microbial cross-contamination during air chilling of poultry. *British Poultry Science*, *41*(2), 158–62.

Mulder, R. W. A. W. (1995). Impact of transport and related stresses on the incidence and extent of human pathogens in pigmeat and poultry. *Journal of Food Safety*, *15*(3), 239–46.

Mulder, R. W. A. W., Van der Hulst, M. C. and Bolder, N. M. (1987). Research note: *Salmonella* decontamination of broiler carcasses with lactic acid, L-cysteine, and hydrogen peroxide. *Poultry Science*, *66*(9), 1555–7.

National Advisory Committee on Microbiological Criteria for Foods. (1997). Generic HACCP application in broiler slaughter and processing. *Journal of Food Protection*, *60*(5), 579–604.

National Chicken Council (1992). Good manufacturing practices for fresh broiler products. http://www.usapeec.org/p_documents/newsandinfo_160404101434.pdf (accessed 20 April 2016).

Nutsch, A. L., Phebus, R. K., Riemann, M. J., Schafer, D. E., Boyer, J. E., Wilson, R. C. and Kastner, C. L. (1997). Evaluation of a steam pasteurization process in a commercial beef processing facility. *Journal of Food Protection*, *60*(5), 485–92.

Over, K. F., Hettiarachchy, N., Johnson, M. G. and Davis, B. (2009). Effect of organic acids and plant extracts on *Escherichia coli* O157: H7, *Listeria monocytogenes*, and *Salmonella* Typhimurium in broth culture model and chicken meat systems. *Journal of Food Science*, *74*(9), M515–21.

Park, H., Hung, Y. C. and Brackett, R. E. (2002). Antimicrobial effect of electrolyzed water for inactivating *Campylobacter jejuni* during poultry washing. *International Journal of Food Microbiology*, *72*(1), 77–83.

Purnell, G., James, C., James, S. J., Howell, M. and Corry, J. E. (2014). Comparison of acidified sodium chlorite, chlorine dioxide, peroxyacetic acid and tri-sodium phosphate spray washes for decontamination of chicken carcasses. *Food and Bioprocess Technology*, *7*(7), 2093–101.

Ramirez, G. A., Sarlin, L. L., Caldwell, D. J., Yezak, C. R., Hume, M. E., Corrier, D. E. and Hargis, B. M. (1997). Effect of feed withdrawal on the incidence of *Salmonella* in the crops and ceca of market age broiler chickens. *Poultry Science*, *76*(4), 654–6.

Rasschaert, G., Houf, K. and De Zutter, L. (2007). Impact of the slaughter line contamination on the presence of *Salmonella* on broiler carcasses. *Journal of Applied Microbiology*, *103*(2), 333–41.

Riedel, C. T., Brøndsted, L., Rosenquist, H., Haxgart, S. N. and Christensen, B. B. (2009). Chemical decontamination of *Campylobacter jejuni* on chicken skin and meat. *Journal of Food Protection*, *72*(6), 1173–80.

Riedo, F. X., Pinner, R. W., de Lourdes Tosca, M., Cartter, M. L., Graves, L. M., Reeves, M. W., Weaver, R. E., Plikaytis, B. D. and Broome, C. (1994). A point-source foodborne listeriosis outbreak: documented incubation period and possible mild illness. *Journal of Infectious Diseases*, *170*(3), 693–6.

Rigby, C. E., Pettit, J. R., Baker, M. F., Bentley, A. H., Salomons, M. O. and Lior, H. (1980). Flock infection and transport as sources of *Salmonellae* in broiler chickens and carcasses. *Canadian Journal of Comparative Medicine*, *44*(3), 328.

Russell, S. (October 2002). Intervention strategies for reducing *Salmonella* prevalence. *Watt Poultry U.S.A.*, (2002): 28–44.

Sakhare, P. Z., Sachindra, N. M., Yashoda, K. P. and Rao, D. N. (1999). Efficacy of intermittent decontamination treatments during processing in reducing the microbial load on broiler chicken carcass. *Food Control*, *10*(3), 189–94.

Sanchez, M. X., Fluckey, W. M., Brashears, M. M. and McKee, S. R. (2002). Microbial profile and antibiotic susceptibility of *Campylobacter* spp. and *Salmonella* spp. in broilers processed in air-chilled and immersion-chilled environments. *Journal of Food Protection*, *65*(6), 948–56.

Sarjeant, K. C., Williams, S. K. and Hinton, A. (2005). The effect of electron beam irradiation on the survival of *Salmonella enterica* serovar Typhimurium and psychrotrophic bacteria on raw chicken breasts stored at four degrees celsius for fourteen days. *Poultry Science*, *84*(6), 955–8.

Scallan, E., Hoekstra, R. M., Angulo, F. J., Tauxe, R. V., Widdowson, M. A., Roy, S. L., Jones, J. L. and Griffin, P. M. (2011). Foodborne illness acquired in the United States – major pathogens. *Emerging Infectious Diseases*, *17*(1).

Sexton, M., Raven, G., Holds, G., Pointon, A., Kiermeier, A. and Sumner, J. (2007). Effect of acidified sodium chlorite treatment on chicken carcases processed in South Australia. *International Journal of Food Microbiology*, *115*(2), 252–5.

Sinhamahapatra, M., Biswas, S., Das, A. K. and Bhattacharyya, D. (2004). Comparative study of different surface decontaminants on chicken quality. *British Poultry Science*, *45*(5), 624–30.

Slader, J., Domingue, G., Jørgensen, F., McAlpine, K., Owen, R. J., Bolton, F. J. and Humphrey, T. J. (2002). Impact of transport crate reuse and of catching and processing on *Campylobacter* and *Salmonella* contamination of broiler chickens. *Applied and Environmental Microbiology*, *68*(2), 713–19.

Smulders, F. J. M. and Greer, G. G. (1998). Integrating microbial decontamination with organic acids in HACCP programmes for muscle foods: prospects and controversies. *International Journal of Food Microbiology*, *44*(3), 149–69.

Smulders, F. J. M., Barendsen, P., Van Logtestijn, J. G., Mossel, D. A. A. and Van Der Marel, G. M. (1986). Review: Lactic acid: Considerations in favour of its acceptance as a meat decontamininant. *International Journal of Food Science & Technology*, *21*(4), 419–36.

Stern, N. J., Svetoch, E. A., Eruslanov, B. V., Perelygin, V. V., Mitsevich, E. V., Mitsevich, I. P., Pokhilenko, V. D., Levchuk, V. P., Svetoch O. E. and Seal, B. S. (2006). Isolation of a *Lactobacillus salivarius* strain and purification of its bacteriocin, which is inhibitory to *Campylobacter jejuni* in the chicken gastrointestinal system. *Antimicrobial Agents and Chemotherapy*, *50*(9), 3111–16.

Stopforth, J. D., O'connor, R., Lopes, M., Kottapalli, B., Hill, W. E. and Samadpour, M. (2007). Validation of individual and multiple-sequential interventions for reduction of microbial populations during processing of poultry carcasses and parts. *Journal of Food Protection*, *70*(6), 1393–401.

Sukumaran, A. T., Nannapaneni, R., Kiess, A. and Sharma, C. S. (2015). Reduction of *Salmonella* on chicken meat and chicken skin by combined or sequential application of lytic bacteriophage with chemical antimicrobials. *International Journal of Food Microbiology*, *207*, 8–15.

Tahergorabi, R., Matak, K. E. and Jaczynski, J. (2012). Application of electron beam to inactivate *Salmonella* in food: Recent developments. *Food Research International*, *45*(2), 685–94.

Tamblyn, K. C. and Conner, D. E. (1997). Bactericidal activity of organic acids in combination with transdermal compounds against *Salmonella* Typhimurium attached to broiler skin. *Food Microbiology*, *14*(5), 477–84.

Thayer, D. W. (1995). Use of irradiation to kill enteric pathogens on meat and poultry. *Journal of Food Safety*, *15*(2), 181–92.

Thomas, C. J. and McMeekin, T. A. (1981). Attachment of *Salmonella* spp. to chicken muscle surfaces. *Applied and Environmental Microbiology*, *42*(1), 130–4.

United States Department of Agriculture. (2016). Safe and suitable ingredients used in the production of meat, poultry, and egg products. Directive 7120.1. Food Safety and Inspection Service, Washington, DC.

United States Department of Agriculture. (2015). Changes to the *Salmonella* and *Campylobacter* verification testing program: proposed performance standards for *Salmonella* and *Campylobacter* in not-ready-to-eat comminuted chicken and turkey products and raw chicken parts and related agency verification procedures and other changes to agency sampling; Notice. Food Safety and Inspections Service. *Federal Register*, *80*, 3940–50.

United States Department of Agriculture. (2014). Modernization of poultry slaughter inspection; Final rule. Food Safety and Inspections Service. *Federal Register*, *79*, 49566–637.

United States Department of Agriculture. (2013). Food ingredients and sources of radiation listed and approved for use in the production of meat and poultry products: final rule. Food Safety and Inspections Service. *Federal Register*, *78*, 14636–40.

United States Department of Agriculture. (2011). New performance standards for *Salmonella* and *Campylobacter* in young chicken and turkey slaughter establishments: Response to comments and announcement of implementation schedule, notice. Food Safety and Inspections Service. *Federal Register*, *76*, 15282–90.

United States Department of Agriculture. (2003). Control of *Listeria monocytogenes* in post-lethality exposed ready-to-eat products. Food Safety and Inspection Service. *Federal Register, 66*(34224), 652–3.

United States Department of Agriculture. (2001). Poultry products inspection regulations-Operations and procedures, generally. Food Safety and Inspection Service. *Federal Register, 66*(1771), 448–9.

United States Department of Agriculture. (2001). Poultry products inspection regulations-Temperatures and chilling and freezing procedures. Food Safety and Inspection Service. *Federal Register, 66*(1771), 449–50.

United States Department of Agriculture. (1998). Poultry post-mortem inspection and reinspection – nforcing zero tolerance for visible fecal material. Directive 6150.1. Food Safety and Inspection Service, Washington, DC.

United States Department of Agriculture. (1996). Pathogen reduction: Hazard Analysis and Critical Control Point (HACCP) systems; Final rule. Food Safety and Inspection Service. *Federal Register, 61*, 38806–944.

United States Department of Agriculture. (1996). Sanitation. Food Safety and Inspection Service. *Federal Register, 61*(38868), 617–21.

United States Department of Agriculture. (1996). Hazard analysis and critical control point (HACCP) systems. Food Safety and Inspection Service. *Federal Register, 61*(38868), 622–6.

Vandekinderen, I., Devlieghere, F., Van Camp, J., Kerkaert, B., Cucu, T., Ragaert, P., De Bruyne, J. and De Meulenaer, B. (2009). Effects of food composition on the inactivation of foodborne microorganisms by chlorine dioxide. *International Journal of Food Microbiology, 131*, 138–44.

Villarreal, M. E., Baker, R. C. and Regenstein, J. M. (1990). The incidence of *Salmonella* on poultry carcasses following the use of slow release chlorine dioxide (Alcide). *Journal of Food Protection, 53*(6), 465–501.

Waldroup, A. L. (1996). Contamination of raw poultry with pathogens. *World's Poultry Science Journal*, *52*(1), 7–26.

Waldroup, A. L., Rathgeber, B. M., Hierholzer, R. E., Smoot, L., Martin, L. M., Bilgili, S. F., Fletcher, D. L., Chen, T. C. and Wabeck, C. J. (1993). Effects of reprocessing on microbiological quality of commercial prechill broiler carcasses. *The Journal of Applied Poultry Research*, *2*(2), 111–16.

Waldroup, A. L., Rathgeber, B. M., Forsythe, R. H. and Smoot, L. (1992). Effects of six modifications on the incidence and levels of spoilage and pathogenic organisms on commercially processed postchill broilers. *The Journal of Applied Poultry Research*, *1*(2), 226–34.

World Health Organization (2012). The global view of campylobacteriosis: report of an expert consultation. http://apps.who.int/iris/bitstream/10665/80751/1/9789241564601_eng.pdf (accessed 5 April 2016).

World Health Organization, Food and Agricultural Organization & Series (MRA) (2009). *Salmonella* and *Campylobacter* in chicken meat. http://www.fao.org/3/a-i1133e.pdf (accessed 5 April 2016).

Wu, D., Alali, W. Q., Harrison, M. A. and Hofacre, C. L. (2014). Prevalence of *Salmonella* in neck skin and bone of chickens. *Journal of Food Protection*, *77*(7), 1193–7.

Yang, H., Li, Y. and Johnson, M. G. (2001). Survival and death of *Salmonella* Typhimurium and *Campylobacter jejuni* in processing water and on chicken skin during poultry scalding and chilling. *Journal of Food Protection*, *64*(6), 770–6.

Yousef, A. E. and Rodriguez-Romo, L. A. (2004). *U.S. Patent No. 6,800,315*. U.S. Patent and Trademark Office, Washington, DC.

Zhu, M., Du, M., Cordray, J. and Ahn, D. U. (2005). Control of *Listeria monocytogenes* contamination in ready-to-eat meat products. *Comprehensive Reviews in Food Science and Food Safety*, 4(2), 34–42.

Chapter 7

Inspection techniques for poultry slaughterhouse operations: the case of the European Union

Janne Lundén, University of Helsinki, Finland

1 Introduction

Poultry slaughtering includes complex and challenging operations from a food safety point of view. The food business operator (FBO) is responsible for identifying food safety risks and managing them (FSMA, 2011; EC, 2002). The FBO needs to correct any non-compliance occurring in the operations by his own initiative. The official veterinarians (OVs) inspect the food businesses and make sure that food safety legislation is followed. To do this, the OV needs to have sufficient knowledge for carrying out inspections and the corresponding tools for enforcement (Lundén et al., 2007; Buntain, 2004). A precondition for carrying out reliable and efficient inspections is that the OV has the right to enter all production areas and inspect documents related to food production (EC, 2004a). This right should be stipulated in the legislation.

Official control in poultry slaughterhouses is, in principle, similar to official control in other food establishments: the OV ensures that the operations are compliant with food safety legislation (EC, 2004a). However, the complex production processes, the presence of pathogens in live animals and the usually continuous presence of official control in poultry slaughterhouses are a challenging combination for official control.

http://dx.doi.org/10.19103/AS.2016.0010.10

Inspections should be carried out with such depth that the OV can get a comprehensive picture on how the FBO is managing the operations. Any non-compliance should be detected (Evira, 2015) and also signs of food frauds should be recognized (Obbink et al., 2014). To achieve these aims, the inspections need to be of high quality (Evira, 2015; Lundén et al., 2007).

This chapter does not include ante- and post-mortem inspections as they are described in detail in the legislation and performed or supervised by the official control. This chapter will focus on the food safety inspections in poultry slaughterhouses. Food safety inspections can be defined as the OV's verification of FBO's self-inspection and actions taken by the OV on the basis of control findings. This includes inspection on how the FBO manages maintenance of slaughterhouse facilities, slaughter hygiene, pest control and traceability, just to mention a few. An integral part of food safety inspections is not only to verify how the FBO is complying with food safety regulations but also any actions taken by the OV due to observed non-compliance. Food safety inspections should be carried out in an efficacious and consistent manner. This chapter describes different inspection techniques that official control can use in poultry slaughterhouses, common non-compliance observed in poultry slaughterhouses and actions taken by the OV due to non-compliance.

2 Goals and prerequisites of poultry slaughterhouse inspections

2.1 Verifying compliance and facilitating business

The aims of food safety inspections in poultry slaughterhouses are the same as in any food establishment. The aims can be divided into two main categories. Firstly, food safety inspections are carried out to make sure that poultry slaughterhouse operators are complying with food safety legislation. This includes safeguarding the consumer, which is the most important task for the official control (EC, 2002). In addition to ensuring food safety, the official control must also ensure that consumers are not misled with regard to e.g. the origin of animal species. The consumer has minimal possibilities to make sure that the food is safe and therefore have to rely on the FBO, possible third-party audits, and the independently operating official control. Secondly, although safeguarding consumer safety is the top priority, the OV also has an important role in facilitating food businesses' international trade. International trade requires certifications for meat and products thereof to make sure that the products are wholesome and subject to official control (Kontio, 2014). In addition, the facilitation of business can include, for instance, guidance to the FBO about legislative requirements, which is particularly important for smaller businesses (Buckley, 2015; Yapp and Fairman, 2006).

2.2 Prerequisites for high-quality inspections

The OV has a challenging task to make sure that food safety is maintained and also to take into account the possibility to carry on the business. The legislation is not always detailed and safe food can be produced in many different ways (EC, 2004c). The goal of the legislation to produce safe food can be reached in many different ways and the OV

Figure 1 Four cornerstones of high-quality official inspections and enforcement of food safety legislation in poultry slaughterhouses.

needs to be able to assess if the FBO's approach is satisfactory (Evira, 2015). Therefore, the OV needs a broad understanding and knowledge of both poultry meat safety and the legislative requirements. Further, the OV need to have skills in using control techniques and applying control actions (Nevas et al., 2013; Lundén et al., 2007; EC, 2004a; Buntain, 2004). The OV should also communicate the inspection results to the FBO in an understandable way to achieve a desired outcome at inspections (Buckley, 2015; Nevas et al., 2013; Lundén et al., 2007; EC, 2004a). Such knowledge and skills are prerequisites for high-quality inspections (Fig. 1).

3 The frequency and focus of food safety inspections

OVs are usually present in poultry slaughterhouses when slaughtering is ongoing; however, the presence can vary between countries as well as the size of the slaughterhouse (EC, 2004b). The frequency of food safety inspections is not easy to determine. Several aspects must be considered and official control organizations have developed systems to define the frequency. A mutual principle is that the frequency should be risk based (Evira, 2015; NFAS, 2013; EC, 2004a; Gagnon et al., 2000) so that increased risks increase the inspection frequency. In practice, the frequency of food safety inspections is based on factors related to the size of the operations, the nature of the operations and the history of compliance (Evira, 2015; EC, 2004a). The frequency of inspections

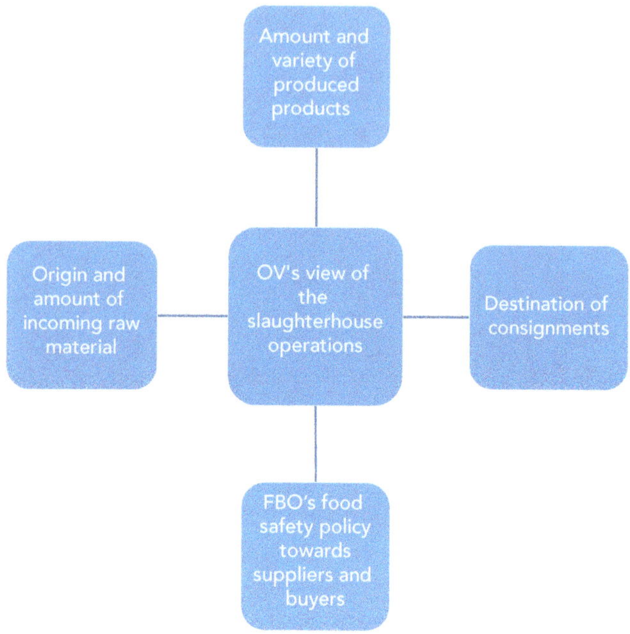

Figure 2 The official veterinarian (OV) should obtain a broad perspective of the slaughterhouse's operations.

is important, but the efficacy of the inspections cannot simply be measured by the number of performed inspections (Husu-Kallio, 2011). Correction of non-compliance as shown in a restaurant setting may not be improved by the increase of inspection frequency (Newbold et al., 2008). Even more important than the frequency is what the OV does during the inspection and how the OV reacts to possible non-compliance (Luukkanen and Lundén, 2015; Newbold et al., 2008).

Whatever the inspection frequency is, it should be such that the OV can obtain a transparent view of the operations. The OV should know what the slaughterhouse operations comprise in general (Fig. 2), but also focus on essential details (Evira, 2015). The focus of the inspections should be on issues causing possible meat safety risks, misleading of the consumer and on previously observed non-compliance (Evira, 2015; EC, 2004a). The implementation of Hazard Analysis and Critical Control Point (HACCP) is considered essential for managing the risks in slaughterhouses (Nørrung and Buncic, 2008) and is therefore a high-priority control subject. Habib et al. (2012) observed a strong connection between operational hygiene scores (concerning for example, adequate space, sanitation, condition of surfaces, slaughter and working hygiene), given by official control and the *Campylobacter* prevalence in broiler slaughterhouses. This relationship indicates that the focus of the inspections is relevant for food safety. Furthermore, for official control to have any real impact on food safety, non-compliance should be corrected by firm measures (Luukkanen and Lundén, 2015; Läikkö-Roto et al., 2015).

4 Inspection techniques and documentation of inspection results

The inspection of the facilities and operations can be seen as a snapshot of the situation at that specific time (Gagnon et al., 2000), but inspections usually also comprise elements that describe long-term activities of the food business. Such elements include the condition of surfaces, level of cleanliness, self-inspection documentation and the correction of earlier detected non-compliance. Inspections can be conducted using checklists or templates to aid the inspection process and to make sure that all relevant aspects are considered (Läikkö-Roto et al., 2015; FVO, 2014b; Gagnon et al., 2000). Checklists or templates can enhance the consistency of inspections as they direct the inspector to inspect the same issues and to report the inspection results in a similar manner (Läikkö-Roto et al., 2015). Templates may also increase the efficacy of the inspections by giving a certain mode of action and a structured way of carrying out the inspections and the subsequent reporting of inspections (Läikkö-Roto et al., 2015). Templates and given modes of action should, however, not lead to rigidity in the inspection process. The OV should be able to apply to surprising situations in the slaughterhouse and refocus the inspection if necessary (Evira, 2015).

Table 1 Inspection techniques in poultry slaughterhouses categorized by information source

Source of inspection data	Inspection technique	Example
Food business operator	Interview/dialog	Discussion with manager/other personnel
	Inspection of self-inspection documentation	Cold storage temperature documentation
	Checking the information on company internet pages	Product spectrum
Official veterinarian's own observations	Observation of facilities	Slaughter line
	Observation of operations	Slaughter hygiene
	Ability testing	Traceability test
	Inspection of self-inspection plan	Temperature control plan
	Inspection of self-inspection documentation	Automatic temperature control logs
	Measurement by official veterinarian	Measurement of cold storage temperature
	Cross-inspection using data from two or more sources	Temperature self-inspection data and official veterinarian's own temperature measurement data
	Sampling	Surface hygiene samples, water content samples

Source: Gagnon (2000), EC (882/2004) and FSIS (2015).

Slaughterhouse activities and meat trade in general involves large economic interests and therefore during routine inspections, the OV should also keep in mind the possibility of fraudulent activities. There is every possibility that fraudulent activities will take place in the meat business and slaughterhouses (Obbink et al., 2014; Premanandh, 2013) and such activities might have international dimensions (EC, 2015). Nevertheless, fraudulent acts always originate at the local level and therefore OVs inspecting slaughterhouses and other food premises are placed in key positions to prevent and detect food frauds (Premanandh, 2013). The routine inspections should be of such high quality that signs of fraudulent actions will be revealed at some point of time. This means that inspections need to be unexpected (EC, 2004a), thorough and in-depth so that common fraudulent activities like discrepancies in traceability, mislabelling or other intentional violations (EC, 2015; Obbink et al., 2014; Premanandh, 2013) could be detected.

Inspection techniques should be used in a versatile manner to verify compliance in a reliable way (Evira, 2015). Inspection techniques can be divided into two main categories depending on the source of the inspection data (Table 1). The OV can receive information from the FBO or by making his own observations (EC, 2004a). Even if the OV receives good support from a compliant and responsible FBO, the OV must make his own observations. The OV cannot rely solely on the information provided by the FBO. If this happens, it can be concluded that the official control is not credible in the eyes of the public or in the eyes of the FBO. Therefore, the knowledge of, and the correct use of, inspection techniques are very important.

The inspection findings should convince the OV that the FBO is complying with requirements. If the OV is not satisfied with the information he has received during the inspection and if he senses that something is amiss, then he should hold further inspections. Also, the compliance history of the FBO may influence the inspection (Läikkö-Roto et al., 2015; EC, 2004c; Gagnon et al., 2000), leading to possibly a more in-depth inspection of a non-compliant FBO. The following examples are given on how to use different inspection techniques.

4.1 Interviewing personnel

Information received from the FBO by talking to or interviewing personnel is valuable when gathering information about how the FBO handles the operations, recent updates in the operations and any hands-on problems influencing hygiene. Non-specific answers should be elaborated by the FBO and documented proof should be displayed whenever possible. Both closed and open-ended questions should be used. The outcome of the interview should give the OV an impression that the issue is thoroughly inspected and relevant information has emerged. If not, the OV should pursue the issue by further questions or other inspection techniques (Evira, 2015). Interviews, however, cannot be the only technique, because the OV must also collect information that is not provided by the FBO (EC, 2004a).

4.2 Review of self-inspection documentation

Inspection of self-inspection documentation generated by the FBO is carried out to make sure that the FBO is carrying out the procedures described in the self-inspection plan (EC, 2004a). The FBO documents the self-inspection procedures such as HACCP procedures by making entries of the monitored issue. The entries can be made manually or they can be generated automatically. This can include temperature documentation, documentation on received raw material, traceability documents and sampling results. Major differences can be found between the self-inspection results and the observations made by the food control authority as noted in temperature control documentation at the retail level despite the assessments made by automatic self-inspection systems (Lundén et al., 2014). The self-inspection results may not reflect the actual situation due to many reasons like errors in the measurement device or in the use of it. The errors may be intentional or unintentional. The OV should take into account possible falsification of the self-inspection documents (Gagnon et al., 2000) like entries marked for the next day or other falsification of the self-inspection outcome. Therefore, it is of utmost importance that the OV does not regard the self-inspection results as automatically correct, but also make his own measurements (Lundén et al., 2014). Such cross-inspection is important for official control to evaluate the reliability and functionality of self-inspection.

4.3 Review of internet presence

The exploring of the company's internet pages or social media can reveal useful knowledge about the company especially for OVs who are not familiar with the company. Internet pages can reveal issues related to products and marketing that have not been displayed at inspections. For example, the FBO may produce a product type that is not approved for the establishment, but marketed on the internet. It can therefore be beneficial to explore the internet pages occasionally.

4.4 Observation of facilities and operations

Observation of facilities and operations is important to gain a transparent picture of the premises and how food safety aspects are taken into consideration in the activities of the slaughterhouse. In particular, facility-related issues that may lead to the contamination of products such as inadequate size of the premises, poor possibility to work hygienically and poor condition of surfaces (Zoonosis Center, 2016; Petran et al., 2012; Buchholz et al., 2002) are observed during the on-site inspection. The principle should be that hygienic working is made as easy as possible. The space should be large enough and designed to enable hygienic working practices at the individual work posts (EC, 2004c). The routes of personnel, product and waste material should consider cross-contamination and surfaces should be in good condition. The most effective and reliable way to assess the management of cross-contamination is by following the activities in the premises for a sufficient amount of time. It can also be assessed by studying the layout and floor plan, but only by observing the operations reveal the actual practices on floor level. The OV should make sure that the inspection covers all facilities, processing machines and other equipment (Evira, 2015; EC, 2004a).

In the slaughterhouses, the observation of slaughter hygiene is of utmost importance because meat can be contaminated with pathogenic bacteria such as *Salmonella* and *Campylobacter* due to poor hygiene practices (Nørrung and Buncic, 2008; WHO, 2006). Poor hygiene in slaughterhouses can also lead to serious outbreaks (Pennington, 2009) and should therefore be a frequent target for official control. Slaughter hygiene is inspected by observing the slaughter process and working hygiene. When observing working hygiene it is important to follow the work practices in such a way that a transparent picture is attained. Hand hygiene practices should be evaluated because it has an important role to play in the matter of food safety (Todd et al., 2010). Also, the use of knife sterilizers should be followed until the true practices are displayed. Therefore, the inspection of operations require more than just a walk-through of the slaughter line; it requires time that has to be acknowledged already when planning the resources of the official control.

The inspection of cleanliness is best done after sanitation and before initiation of the operations. This inspection should be comprehensive and more than just a 'look'. The sanitation of processing machines can be challenging due to complex design. Any accumulation of dirt should be recognized (Wirtanen and Salo, 2014) and eliminated by the FBO. In particular, conveyor belts and cutting devises have been shown to be difficult to clean (Wirtanen and Salo, 2014). This means that the OV needs to inspect processing machines not just on the outside but also on the inside and evaluate surfaces in direct and indirect contact with the meat, in particular.

4.5 Ability testing

Ability testing is an in-depth inspection method where the FBO is requested to display how a certain food safety procedure is managed. Traceability testing is a typical capability test where the FBO demonstrates how the traceability is conducted in practice. The FBO should have considerable interest in ensuring that the traceability is working in practice due to safety and economic reasons (Pozo and Schroeder, 2016), but the OV cannot rely on assumptions. Instead he should inspect that the FBO can trace products in practice. This can be done by requesting the FBO to trace and display the list of consignees of a hand-picked lot.

4.6 Inspection of the written self-inspection plan

Inspection of the written self-inspection plan gives the OV an understanding how the FBO has planned to manage the risks involved in production. The self-inspection plan includes written instructions on sanitation, pest control, sampling, traceability and HACCP (EC, 2004c). The self-inspection plan should describe sufficiently possible risks in the production and how the risks are managed (EC, 2004c). Although the plan is important, the execution of the plan is even more important. A well-prepared plan does not mean that the activities are carried out adequately (Lundén et al., 2014). It is extremely important to inspect the actual activities described in the plan.

4.7 Cross-inspection

Acquiring data from two or more sources, that is, cross-inspection, can be a valuable technique in verifying the FBO's self-inspection results. Cross-inspection aims at acquiring data from another source besides the FBO. This is not always possible, but when possible and at situations when the OV is unsure about the reliability of the data generated by the FBO, cross-inspection should be used. The simplest way of cross-inspection involves data generated by the OV such as temperature measurements and sampling which are then compared with the data produced by the FBO (Lundén et al., 2014). Also data generated by other parties can be used when inspecting the disposal of non-edible by-products. Non-edible by-products have to be disposed properly to eliminate any health threats and contagious animal disease agents (EC, 2009). Proper disposal is expensive, however, and it may be tempting to dispose such material in an improper way. Cross-inspection can be used to make sure that the material is transported to the waste treatment plant by comparing the amount of slaughtered animals and the amount of delivered non-edible by-products to the waste treatment plant. (OV can assess the expected amount of non-edible by-products from the number of slaughtered animals.) The OV can further request information from the waste management plant on how much non-edible by-products has been delivered from the slaughterhouse and compare that to the consignment documents sent from the slaughterhouse to the waste management plant. These inspection procedures will take time, but will give a good idea about the correctness of the disposal of non-edible by-products.

4.8 Sampling and measurements

Sampling and measurements are useful in cases where visual observation of premises, activities and self-inspection need further assessment. Sampling by the OV can be used as verification of successful sanitation by taking surface samples for chemical or microbiological analysis indicating cleanliness (Wirtanen and Salo, 2014). Other samples taken by the OV can include microbiological surface samples of carcasses for assessment of the slaughter hygiene, samples to determine the amount of retained water in the carcasses and possible samples for determination of animal species in product preparations. A typical measurement that the OV should carry out is temperature measurements of the cold storage and the meat, as temperature can have a major impact on the growth of pathogens. Sampling and measurements done by the OV are independent from the slaughterhouse (EC, 2004a) and describes the situation at the time of sampling and measuring. Whether the OV is taking his own samples and carrying out measurements or not, he should occasionally inspect the sampling and measurement procedures done by the FBO to make sure that relevant and representative samples are taken and measurements made.

4.9 Documentation and disclosure of the inspection results

The documentation of the inspection results should be done so that the FBO understands what has been inspected, what the non-compliance issues are and what is expected of him (Yapp and Fairman, 2006). In the case of non-compliance, the OV should explain to the FBO why a non-compliance should be corrected, as this has been observed to influence the efficacy of the inspections (Läikkö-Roto et al., 2015). Several countries also publish the results of the inspections (Anon., 2013, 2015; DVFA, 2011) to enhance the transparency of official control (EC, 2004a). Some of the countries also publish poultry slaughterhouse inspection results on the internet (Anon., 2013; DVFA, 2011). The disclosure systems have been suggested to have beneficial effects on compliance (Waters et al., 2013; DVFA, 2011) and should therefore be considered as an additional tool for official control.

5 Observing and responding to non-compliance in poultry slaughterhouses

5.1 Observed non-compliance in poultry slaughterhouses

Non-compliance related to slaughter and working hygiene and the condition of surfaces seem to be common in poultry slaughterhouses regardless of country and control organization (FVO, 2004, 2011, 2014a,b; Pinillos and Jukes, 2008). Slaughter hygiene has been observed to be compromised by poor design of the slaughter line causing faecal contamination, poor possibilities for hand washing and poor working practices (FVO, 2011). Faecal contamination is of major concern in poultry slaughterhouses and appear to be difficult to correct as it is a recurring problem (Luukkanen and Lundén, 2016; FVO, 2011, 2014a,b). Other observed non-compliance issues have been inadequate space, temperature abuse of carcasses, neglecting corrective actions described in the HACCP plan and general dirtiness of the premises (Luukkanen and Lundén, 2016; FVO, 2004, 2011, 2014a,b).

5.2 Actions taken by official veterinarians (OVs) due to non-compliance

The OV needs to take appropriate actions upon noticing non-compliance at inspections (Buntain, 2004; EC, 2004a). The actions taken depend on the nature of the observed non-compliance and the history of compliance of the FBO. The actions taken should be gradual and in proportion to the non-compliance: the actions should be efficient enough to make the FBO to correct the violation but not oversized (Fig. 3) (Kettunen et al., 2015; Lepistö and Hänninen, 2011; EC, 2004a). The actions taken should be swift when food safety is clearly threatened and milder when food safety is not directly challenged. Actions available for OVs are presented in Table 2. Actions may differ between countries, and also the terminology of the measures used can vary.

The minimal action is to guide the FBO about the food safety requirements. Guiding may not be seen as a control action at all; however, guidance of FBOs should be appreciated

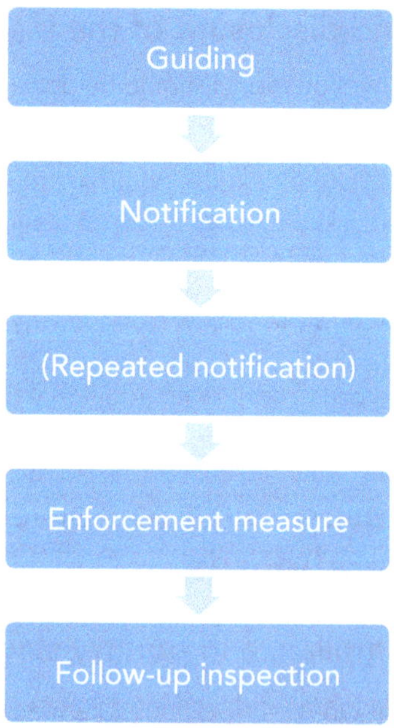

Figure 3 Stepwise and gradual use of control methods.

Table 2 Actions taken by official veterinarian due to non-compliance

Action	Example
Guidance	Guiding concerning food safety requirements
Oral notification	Minor non-compliance concerning sanitation
Written notification (or warning)	Minor to major non-compliance
Oral or written notification with a time limit for correction of non-compliance	Any non-compliance requiring a time limit
Order	Ordering to improve sanitation
Prohibition	Prohibition of using or placing a product on the market
Seizure	Seizure of a food lot (taking the lot into the possession of the authority)
Destruction of food stuffs	Destruction of a food lot unfit for human consumption
Suspension of operations or part of operations	Severe uncleanliness of premises, serious pest problem, unapproved operations
Economic sanctions	
• Threat of a penalty payment	To enhance the effect of an order to correct surfaces in poor condition
• Fine	Fining the FBO to increase compliance in the future
Withdrawal of approval	Major non-compliance that are not corrected by other measures
Notification to police	Serious and/or intentional non-compliance

Source: EC (882/2004), FSIS (2015), Kettunen et al. (2015) and Luukkanen and Lundén (2016).

highly due to several reasons. In particular, small and medium-sized food businesses may experience challenges in understanding the requirements of food safety legislation and therefore need guidance in the interpretation of the regulations (Buckley, 2015; Kotisalo et al., 2015; Nevas et al., 2013; Yapp and Fairman, 2006). Instead of immediately moving into enforcement measures, the first action taken should be guidance, unless stricter actions are warranted due to the nature of the non-compliance or previous compliance history (Kettunen et al., 2015; Buckley, 2015; Lindblad and Berking, 2013). An assistive approach with the first action being guidance is also appreciated by the FBOs (Buckley, 2015) and is considered efficient (Reske et al., 2007). In this way the FBOs are also given a possibility to correct the non-compliance without strict administrative procedures. Competent, high-quality guidance has also been shown to increase the safety of products by increasing the knowledge of the FBOs (Nevas et al., 2013).

Oral notification can be used while carrying out the inspection. Oral notification as the sole control action should be used only in very minimal non-compliance situations, because it has not been observed to be as effective as other control actions (Luukkanen and Lundén, 2016). A much more efficacy control action is a written notification (also called a 'warning' in some countries) combined with a time limit for correction of the non-compliance (Luukkanen and Lundén, 2016; Läikkö-Roto et al., 2015). In slaughterhouses where OVs use more oral notifications without time limits, non-compliance tend to be more common (Luukkanen and Lundén, 2016; FVO, 2011). Therefore, it is important not only to use oral notification to communicate the observed non-compliance to the FBO at the inspection level, but also to document the non-compliance and to use time limits.

Enforcement measures such as orders, prohibitions and economic sanctions should be used only when guiding and negotiation is not leading to correction of the non-compliance (Kettunen et al., 2015; Lindblad and Berking, 2013). The OV, however, must have the readiness and skills to use enforcement measures without delay when needed. The use of enforcement measures should also be gradual and proportional to the non-compliance (Kettunen et al., 2015; Lepistö and Hänninen, 2011). Enforcement measures have been used in non-compliance related to slaughter hygiene, working hygiene and poor condition of surfaces and are considered effective when used (Luukkanen and Lundén, 2016; Kettunen et al., 2015; Lindblad and Berking, 2013).

6 Summary

Official on-site control should use different inspection techniques to obtain a comprehensive and transparent picture of the operations and of the management of the food safety risks. The inspection techniques aim at collecting data from different sources. FBOs' interviews and inspection of the FBOs' self-inspection plan and documentation are used to assess the compliance of the slaughterhouse. Inspecting only data obtained from the FBO such as self-inspection documentation, however, may give an incorrect impression of the status of the slaughterhouse. Therefore, the OV must also make his own observations and measurements when applicable to get an objective picture of the premises and the operations. The inspection of slaughter and working hygiene, which often have been observed as insufficient in poultry slaughterhouses, should include observation of the operations for a sufficient time to get a true picture of the hygiene practices. Other inspection techniques include capability testing such as FBOs' ability to trace products and cross-inspection, which utilizes data from two or more sources. Also sampling can be used to verify the adequacy of sanitation, animal species and water content of the meat.

Research has shown that soft control actions such as guiding the FBO have beneficial effects on compliance. Especially small- and medium-sized enterprises need more guidance in interpreting sometimes complex and ambiguous food safety legislation. Such guidance can be given at on-site inspections. In cases of non-compliance, the OV can use as a first enforcement measure notifications combined with discussion about how and when the FBO intends to correct the non-compliance. Notifications should include a time limit for correction of the non-compliance because it has been shown to facilitate compliance. Follow-up inspections should be done or in some other way verify that the non-compliance has been eliminated. When notifications are not leading to correction of the non-compliance or the non-compliance is causing a serious health threat, the OV should consider using more efficient enforcement measures such as orders, prohibitions, economic sanctions or suspension of operations. Negligence, intentional non-compliance and fraudulent activities can also be grounds for notification to the police. The inspection results are documented and required corrections should be well justified to the FBO to increase the motivation to correct the non-compliance. The outcomes of the inspection reports are published in some countries, and this can enhance compliance.

7 Future trends in research

The actions of the OVs should lead to the correction of non-compliance. It appears, however, that official control should possess more efficacy when non-compliance activities are observed frequently in slaughter hygiene. The use of control actions should be further elucidated and new control actions developed.

Slaughterhouses are in a key position in the meat business and possibly involved intentionally or unintentionally in fraudulent activities. The official control, including the OV on site, needs to have efficient tools to recognize fraudulent activities. In particular, early signals of fraud related to slaughter business should be recognized for future research efforts.

The cooperation and often daily presence of the OV in the slaughterhouse is challenging. OVs need to cooperate with the FBO, but at the same time maintain a distance from the FBO to ensure objectivity at inspections. How the OV and the entire official control organization handles this should be investigated. Furthermore, official control in slaughterhouses can be organized in different ways with a different amount of control responsibility shifted to the FBO (Luukkanen et al., 2015; Berzins et al., 2014; Anon., 2014). Research has shown that balancing between these responsibilities may lead to challenges and mistrust (Anon., 2014) and further investigation is needed to recognize well-functioning models.

8 Where to look for further information

Scientific research concerning inspection techniques and their efficacy in slaughterhouses is unfortunately still quite sparse and numerous papers deal with restaurant inspection. Results from those papers can in some cases be applied to the slaughterhouse context. However, this should be done with consideration and understanding of official control issues both in slaughterhouses and retail. The principles of inspecting are, in general, the same regardless of the type of food business, but the official control in slaughterhouses have many distinctive features such as the inspection frequency and presence on-site. Inspection manuals, guidelines and checklists for carrying out inspections have been prepared by competent authorities in different countries and can be found on the internet. Some text books consider official control in slaughterhouses and can offer deeper insight into the role and complexity of the control in slaughterhouses (Ninios et al., 2014; Kaario and Lundén, 2010).

9 References

Anon. (2013), Oiva smileys are an indication of food safety, Brochure, *Finnish Food Safety Authority*, 08/2013, http://www.evira.fi/portal/en/about+evira/publications/?a=category&cid=25 (accessed 30 March 2016).

Anon. (2014), Summary: risks in the meat supply chain, *Dutch Safety Board*, http://www.onderzoeksraad.nl/uploads/phase-docs/560/38357a3466aerisico-s-vleesketen-en-web.pdf (accessed 29 March 2016).

Anon. (2015), Food hygiene rating schemes, *Food Standards Agency, UK*, http://www.food.gov.uk/business-industry/hygieneratings (accessed 30 March 2016).

Berzins, A., Lundén, J. and Korkeala, H. (2014), Organization of official control, in T. Ninios, J. Lundén, H. Korkeala and M. Fredriksson-Ahomaa (eds), *Meat Inspection and Control in the Slaughterhouse*, Hoboken: Wiley Blackwell, pp. 556–61.

Buchholz, U., Run, G., Kool, J. L., Fielding, J. and Mascola, L. (2002), A risk-based restaurant inspection system in Los Angeles County, *J. Food Prot.*, 65, 367–72.

Buckley, J. (2015), Food safety regulation and small processing: A case study of interactions between processors and inspectors, *Food Policy*, 51, 74–82.

Buntain, B. (2004), Emerging challenges in public health protection, food safety, and security: Veterinary needs in the USDA's food safety and inspection service, *J. Vet. Medic. Educ.*, 31, 334–40.

DVFA. (2011), Smileys keep food safety high in Denmark, *The Danish Veterinary and Food Administration*, http://www.findsmiley.dk/en-US/Forside.htm (accessed 30 March 2016).

EC. (2002), Regulation (EC) No 178/2002 of the European Parliament and of the Council laying down the general principles and requirements of food law, establishing the European Food Safety Authority and laying down procedures in matters of food safety, http://eur-lex.europa.eu/LexUriServ/LexUriServ.do?uri=OJ:L:2002:031:0001:0024:en:PDF (accessed 21 February 2016).

EC. (2004a), Regulation (EC) No 882/2004 of the European Parliament and of the Council on official controls performed to ensure the verification of compliance with feed and food law, animal health and animal welfare rules, http://eur-lex.europa.eu/LexUriServ/LexUriServ.do?uri=OJ:L:2004:165:0001:0141:EN:PDF (accessed 20 January 2016).

EC. (2004b), Regulation (EC) No 854/2004 of the European Parliament and of the Council laying down specific rules for the organisation of official controls on products of animal origin intended for human consumption, http://eur-lex.europa.eu/LexUriServ/LexUriServ.do?uri=OJ:L:2004:226:0083:0127:EN:PDF (accessed 22 March 2016).

EC. (2004c), Regulation (EC) No 852/2004 of the European Parliament and of the Council on the hygiene of foodstuffs, http://eur-lex.europa.eu/LexUriServ/LexUriServ.do?uri=OJ:L:2004:139:0001:0054:EN:PDF (accessed 19 January 2016).

EC. (2009), Regulation (EC) No 1069/2009 of the European parliament and of the council laying down health rules as regards animal by-products and derived products not intended for human consumption and repealing Regulation (EC) No 1774/2002 (Animal by-products Regulation), http://eur-lex.europa.eu/LexUriServ/LexUriServ.do?uri=OJ:L:2009:300:0001:0033:EN:PDF (accessed 29 March 2016).

EC. (2015), Food fraud network activity report 2015, *European Commission*, http://ec.europa.eu/food/safety/docs/official-controls_food-fraud_network-activity-report_2015.pdf (accessed 29 March 2016).

Evira. (2015), Instructions of the risk-based food safety control in food establishments. *Evira Instructions*, http://www.evira.fi/files/attachments/fi/evira/lomakkeet_ja_ohjeet/elintarvikkeet/laitokset/eviran_ohje_16044_1.pdf (accessed 21 March 2016).

FSIS. (2015), Inspection methods: Regulatory process overview 4/4/2015, *Food Safety and Inspection Service*, http://www.fsis.usda.gov/wps/portal/fsis/topics/inspection/workforce-training/regional-on-site-training/inspection-methods (accessed 23 March 2016).

FSMA. (2011), Food safety modernization act, https://www.gpo.gov/fdsys/pkg/PLAW-111publ353/pdf/PLAW-111publ353.pdf (accessed 30 January 2016).

FVO. (2004), Final report of an audit carried out in the United States of America from 7 to 11 June 2004 in order to evaluate the operation of control systems in the fresh poultry meat sector under the auspices of the agreement between the European Community and the United States of America on sanitary measures to protect public and animal health in trade in live animals and animal products, DG(SANCO) 7303-2004–MR, *European Commission, Directorate General for Health and Food Safety*, http://ec.europa.eu/food/fvo/audit_reports/details.cfm?rep_id=1309 (accessed 30 March 2016).

FVO. (2014a), Final report of an audit carried out in Germany from 31 March to 10 April 2014 in order to evaluate the food safety control system in place governing the production and placing on the market of poultry meat and products derived therefrom, DG(Sanco) 2014-7156–MR, *European Commission, Directorate General for Health and Food Safety*, http://ec.europa.eu/food/fvo/audit_reports/details.cfm?rep_id=3333 (accessed 30 March 2016).

FVO. (2014b), Final report of an audit carried out in Poland from 18 to 28 November 2014 in order to evaluate the food safety control system in place governing the production and placing on the market of poultry meat and products derived therefrom, DG(SANTE) 2014-7160–MR, *European Commission, Directorate General for Health and Food Safety*, http://ec.europa.eu/food/fvo/audit_reports/details.cfm?rep_id=3514 (accessed 30 March 2016).

FVO. (2011), Final report of an audit carried out in the Netherlands from 16 to 25 November 2011 in order to evaluate the food safety control system in place governing the production and placing on the market of poultry meat and products derived therefrom, DG(SANCO) 2011-6008–MR, *European Commission, Directorate General for Health and Food Safety*, http://ec.europa.eu/food/fvo/audit_reports/details.cfm?rep_id=2998 (accessed 30 March 2016).

Gagnon, B., McEachern, V. and Bray, S. (2000), The role of the Canadian government agency in assessing HACCP, *Food Control*, 11, 359–64.

Habib, I., Berkvens, D., De Zutter, L., Dierick, K., Van Huffel, X., Spreybroeck, N., Geeraerd, A. and Uyttendaele, M. (2012), *Campylobacter* contamination in broiler carcasses and correlation with slaughterhouses operational hygiene inspection, *Food Microbiol.*, 29, 105–12.

Husu-Kallio, J. (2011), Audits and guides for good practice, *Slov. Vet. Res*, 48, 77.

Kaario, N. and Lundén, J. (2010), Official control of slaughterhouses and processing plants, in I. G. Legarreta and Y. H. Hui (eds), *Handbook of Poultry Science and Technology*, New York: John Wiley & Sons, Inc., pp. 107–19.

Kettunen, K., Nevas, M. and Lundén, J. (2015), Effectiveness of enforcement measures in local food control in Finland, *Food Control*, 56, 41–6.

Kontio. (2014), International trade, in T. Ninios, J. Lundén, H. Korkeala and M. Fredriksson-Ahomaa (eds), *Meat Inspection and Control in the Slaughterhouse*, Hoboken: Wiley Blackwell, pp. 643–50.

Kotisalo, N., Luukkanen, J., Fredriksson-Ahomaa, M. and Lundén, J. (2015), Effects of centralizing meat inspection and food safety inspections in Finnish small-scale slaughterhouses, *Food Policy*, 55, 15–21.

Läikkö-Roto, T., Mäkelä, S., Lundén, J., Heikkilä, J. and Nevas, M. (2015), Consistency in inspection processes of food control officials and efficacy of official controls in restaurants in Finland, *Food Control*, 57, 341–50.

Lepistö, O. and Hänninen, M. L. (2011), Effects of legal aspects on the use of compulsory procedures in environmental health and food control, *J. Environ. Health Res.*, 11, 127–34.

Lindblad, M. and Berking, C. (2013), A meat control system achieving significant reduction of visible faecal and ingesta contamination of cattle, lamb and swine carcasses at Swedish slaughterhouses, *Food Control*, 30, 101–5.

Lundén, J., Björkroth, J. and Korkeala, H. (2007), Meat inspection education in Finnish veterinary curriculum, *J. Vet. Medic. Educ.*, 34, 205–11.

Lundén, J., Vanhanen, V., Kotilainen, K. and Hemminki, K. (2014), Retail food stores' internet-based own-check databank records and health officers' on-site inspection results for cleanliness and food holding temperatures reveal inconsistencies, *Food Control*, 35, 79–84.

Luukkanen, J. and Lundén, J. (2016), Compliance in slaughterhouses and control measures applied by official veterinarians, *Food Control*, 68, 133–8.

Luukkanen, J., Kotisalo, N., Fredriksson-Ahomaa, M. and Lundén, J. (2015), Distribution and importance of meat inspection tasks in Finnish high-capacity slaughterhouses, *Food Control*, 57, 246–51.

Nevas, M., Kalenius, S. and Lundén, J. (2013), Significance of official food control in food safety; food business operators' perceptions, *Food Control*, 31, 59–64.

Newbold, K. B., McKeary, M., Hart, R. and Hall, R. (2008), Restaurant inspection frequency and food safety compliance, *J. Environ. Health.*, 71, 56–61.

NFAS. (2013), Risk classification of establishments and determining of inspection time, *National Food Agency, Sweden,* 29 November 2013, p. 50, http://www.livsmedelsverket.se/globalassets/produktion-handel-kontroll/vagledningar-kontrollhandbocker/vagledning-riskklassning-livsmedelsanlaggningar-berakning-kontrolltid.pdf (accessed 22 March 2016).

Ninios, T., Lundén, J., Korkeala, H. and Fredriksson-Ahomaa, M. (2014), *Meat Inspection and Control in the Slaughterhouse,* Hoboken: Wiley Blackwell, UK, p. 698.

Nørrung, B. and Buncic, S. (2008), Microbial safety of meat in the European Union, *Meat Sci.,* 78, 14–24.

Obbink, N., Frissen, J. M. and Post, S. B. (2014), Food frauds, in T. Ninios, J. Lundén, H. Korkeala and M. Fredriksson-Ahomaa (eds), *Meat Inspection and Control in the Slaughterhouse,* Hoboken: Wiley Blackwell, UK, pp. 628–38.

Pennington, H. (2009), The public inquiry into the September 2005 outbreak of *E. coli* O157 in South Wales, http://www.reading.ac.uk/foodlaw/pdf/uk-09005-ecoli-report-summary.pdf (accessed 23 March 2016).

Petran, R. L., White, B. W. and Hedberg, C. W. (2012), Health department inspection criteria more likely to be associated with outbreak restaurants in Minnesota, *J. Food Prot.,* 75, 2007–15.

Pinillos, R. G. and Jukes, D. J. (2008), Hygiene assessment system (HAS) scores – an analysis of the available data from English slaughterhouses, *Food Control,* 19, 806–16.

Pozo, V. F. and Schroeder, T. C. (2016), Evaluating the costs of meat and poultry recalls to food firms using stock returns, *Food Policy,* 59, 66–77.

Premanandh, J. (2013), Horse meat scandal – a wake-up call for regulatory authorities, *Food Control,* 34, 568–9.

Reske, K., Jenkins, T., Fernandez, C., VanAmber, D. and Hedberg, C. (2007), Beneficial effects on implementing an announced restaurant inspection program, *J. Environ. Health,* 69, 27–34.

Todd, E. C. D., Michaels, B. S., Greig, J. D., Smith, D., Holah, J. and Bartleson, C. A. (2010), Outbreaks where food workers have been implicated in the spread of foodborne disease. Part 7. Barriers to reduce contamination of food by workers, *J. Food Prot.,* 73, 1552–65.

Waters, A. B., Van Derslice, J., Porucznik, C. A., Kim, J., De Legge, R. and Durrant, L. (2013), Impact of internet posting of restaurant inspection scores on critical violations, *J. Environ. Health,* 75, 8–12.

WHO. (2006), Five keys to safer food manual. *World Health Organization Department of Food Safety, Zoonoses and Foodborne Diseases,* http://www.who.int/foodsafety/publications/consumer/manual_keys.pdf (accessed 8 February 2016).

Wirtanen, G. and Salo, S. (2014), Use of meat inspection data, in T. Ninios, J. Lundén, H. Korkeala and M. Fredriksson-Ahomaa (eds), *Meat Inspection and Control in the Slaughterhouse,* Hoboken: Wiley Blackwell, UK, pp. 453–71.

Yapp, C. and Fairman, R. (2006), Factors affecting food safety compliance within small and medium-sized enterprises: Implications for regulatory and enforcement strategies, *Food Control,* 17, 42–51.

Zoonosis Center. (2016), Food poisonings, *Finnish Food Safety Authority Evira and National Institution for Welfare and Health,* http://www.zoonoosikeskus.fi/portal/fi/ruokamyrkytykset (accessed 8 March 2016).

Chapter 8

Ensuring safety in chilling and freezing of poultry meat

Alma Delia Alarcon-Rojo and Ana Luisa Renteria-Monterrubio, Universidad Autónoma de Chihuahua, Mexico

1 Introduction

Meat and meat products contribute valuable nutrients to human health, such as proteins, minerals, vitamins and micronutrients (e.g. potassium, phosphorus, vitamins B1 and B2 etc.). According to FAO (2014), poultry meat was the most traded (13.5 million tonnes) in the world with the highest production increment (1.6% from year 2013 to 2014). Poultry consumption has increased almost continuously since 1965 (44.4 pounds per capita to 100.5 pounds per capita) (The National Chicken Council, 2016).

The increment in poultry demand has forced producers to adjust their production techniques. However, such adjustments can no longer be directed towards a more intense caged production, as consumers are now more concerned about animal welfare, environmental responsibility and safety. Hence, in order to achieve sustainable production, the poultry industry needs to redesign its methods and technologies to fulfil its current demands.

As mentioned, safety is one of the major concerns of society. Currently, it is necessary not only to control traditional microorganisms but emerging pathogens as well. These pathogens can be of greater virulence and resistance (e.g. antibiotics and disinfectants) and of lower infectious doses (Andersen et al., 2006; Jarquin et al., 2015; Logue et al., 2003; Marotta et al., 2015; Mor-Mur and Yuste, 2010; Sofos, 2008; Wieczorek et al., 2013). Spoilage bacteria are not a direct cause of illnesses, but when they get associated with

http://dx.doi.org/10.19103/AS.2016.0010.11

pathogens, they might become more resistant (Culotti and Packman, 2014, 2015). Poultry meat is a common source of spoilage bacteria and food-borne pathogens and is regularly associated with food-borne illnesses (Mor-Mur and Yuste, 2010; Park, 2002) and outbreaks (Berrang et al., 2000).

Some of the consumer demands have translated into more modern processing plants, as it has been suggested that automation of the various stages of slaughter improves the microbial quality of poultry products. However, automation did not influence the type of hazards and the number of critical control points, but reduced indicator microorganisms approximately by 1 $\log_{10}$ CFU/g and pathogens (Tsola et al., 2008).

Transfer of bacteria can be at any point of the production system, including raw ingredients, food contact and non-food contact surfaces, environment and finished products (Allen et al., 2007; Berrang et al., 2000; Fox et al., 2015; Griekspoor et al., 2010). If preventive measures are taken at the processing plants, poultry products might not be subjected to illnesses even if the prevalence of microorganisms is high in chickens. (Rorvik et al., 2003). Until more recent times processing plants have not been able to reduce the incidence of pathogenic bacteria effectively (Hermans et al., 2011) and facilities are potential transfer routes of microorganisms (Fox et al., 2015). Hence, the farm-to-fork system is the most effective for controlling microorganisms (Hafez, 1999; Mead, 2004). Even though a single approach for risk assessment would be desirable, it would be difficult to adopt as every country and particularly every factory has its own practices. Hence, a unified risk assessment would be very complicated (Nauta et al., 2009).

Several methods have been developed to control the growth of microorganisms in poultry and poultry products with various results (Hermans et al., 2011). Interventions (Loretz et al., 2010) can be divided as physical (e.g. water-based, irradiation, ultrasound, UV light, modified atmosphere packaging, freezing) and chemical (e.g. organic acids, chlorine- and phosphate-based, natural preservatives). They can be used prior to slaughter (Hafez, 1999; Rubio-Garcia et al., 2015), during the process (Bashor et al., 2004) and after slaughter (Avens et al., 2002; Boyle et al., 1988; Chaves et al., 2011; del Rio et al., 2007a). Any programme employed must be cost-effective and easily applied at the processing plants. For example, Belgium leaders and experts in the poultry meat industry have selected spraying (lactic acid or electrolysed water), radiation and crust freezing as the most effective interventions and of these, electrolysed water seems to have the best cost–benefit ratio (Gellynck et al., 2008).

The objective of this chapter is to present a review of safety practices and potential risks associated with chilling and freezing of poultry meat. It covers relevant research on the improvement of poultry safety during chill processing, the post-slaughter handling risk and treatment, and the most important poultry pathogens and their control.

2 Chilling for poultry meat preservation

Scalding is needed invariably during poultry slaughter as the feathers must be loosened; however, this process could increase the number of bacteria and the speed of the muscle metabolism. Scalding must be as short and mild as possible to avoid adverse sensorial effects.

Historically, refrigeration has been a technology used for food preservation that limits the growth of food-borne and spoilage bacteria. Normally, chilling retards the metabolic pathways of microorganisms; however, the response to chilling depends greatly on the type of microorganism (Tholozan et al., 1998; Voidarou et al., 2007).

The development and application of chilling methods for end-product decontamination and preservation in the processing plants are still under review. Chilling (e.g. water immersion, evaporative cooling and air chilling) might be used by itself or as part of more complex processes (Allen et al., 2000a; Mead et al., 2000; Mercogliano et al., 2014).

Air chilling has resulted in a ten-fold reduction in numbers of *Pseudomonas*, total aerobic mesophiles and presumptive coliforms, and more importantly, this reduction is maintained through the storage time (Allen et al., 2000a, b; Gonzalez-Miret et al., 2006). Water sprays have proven ineffective against bacteria, even when enriched with chlorine (Allen et al., 2000a, b), but have shown a reduction against pathogens when peracetic acid (PAA) was added, though not as effective in comparison with water/PAA immersion baths (Smith et al., 2015). The effectiveness of chlorine might increase when used in an immersion bath (Trindade et al., 2012). It has been suggested that ozone can substitute chlorine as a sanitizing agent in a water bath, not only because of its antimicrobial characteristics (Trindade et al., 2012) but also because ozone produces lower levels of the biogenic amines, cadaverine and putrescine, and consequently renders a longer shelf-life (Mercogliano et al., 2014). Another possible substitute of chlorine is electrolysed water, as it has been demonstrated to reduce the number of *Campylobacter jejuni* up to 3 logs and also prevent cross-contamination (Park et al., 2002).

Consumers are demanding more environmentally friendly products, enhancing the efficiency of water use in poultry production by decreasing the amount of water during carcass washing is a method of reducing adverse environmental impacts. It has been demonstrated that reducing the level of water in the chilled-immersion bath from 6.7 to 3.3 L/kg has the same effect against bacteria (Northcutt et al., 2008a, b) and it has also been demonstrated that multiple washes do not reduce the bacterial number, unless the washes are acidified with sodium chlorite or trisodium phosphate (Bashor et al., 2004). Water can be saved when it is treated with ozone and UV light, which, apart from reducing the microbial counts (more than 99.99%), allows a 90% water recirculation that translates into 426000 L/day less water used in the slaughter process (Diaz et al., 2001).

Chilling, apart from being an unavoidable step in the production process, influences the taste, texture and overall appearance of poultry. Hence, any defect during refrigeration may produce low-quality products, such as tough meat in extremely rapid-chilled carcasses or PSE meat in slow-chilled processes (James et al., 2006).

3 Other post-slaughter interventions for the decontamination and preservation of poultry meat

Post-processing food handling is an important factor for reducing food-borne-linked infections and outbreaks. Interventions, either chemical or physical (Table 1), must (1) observe regional and/or international regulations, (2) decrease the number of microorganisms, (3) not be inactivated by organic matter (e.g. faeces, blood, fat etc.), and (4) not have a negative impact in the sensorial characteristics of the product.

As before, water-based interventions are very popular. Immersion baths at this point of the process are regularly conducted for short periods of time, in hot/boiling water (e.g. 70°C/40 s) with or without additives. Most of the results when using the combination water bath – chemical are favourable and the advantages of that combination are maintained during

Table 1 Interventions used to decrease the number of microorganisms in poultry carcasses and products

	Authors
Chemical	
Chloride-based	(Bashor et al., 2004; Park et al., 2002; Shin et al., 2011)
Phosphate-based	(Bashor et al., 2004; Purnell et al., 2004)
Organic acids	(Gellynck et al., 2008; Okolocha and Ellerbroek, 2005; Purnell et al., 2004)
Hydrogen peroxide	(Parikh et al., 2012; Vadhanasin et al., 2004)
Peracetic acid	(Purnell et al., 2004; Vadhanasin et al., 2004)
Ozone	(Vadhanasin et al., 2004)
Electrolysed water	(Gellynck et al., 2008; Loretz et al., 2010; Park et al., 2002)
Natural preservatives	(Ntzimani et al., 2010; Piskernik et al., 2011)
Physical	
Irradiation	(Gomes and da Silva, 2006; Loretz et al., 2010; Sarjeant et al., 2005)
High-power pulse light	(Paskeviciute et al., 2011)
Ultrasound	(Koolman et al., 2014)
Ultraviolet light	(Parikh et al., 2012)
Modified atmosphere packaging	(Shin et al., 2011)

the storage period. The number of bacteria, either pathogens *Campylobacter* (Purnell et al., 2004), *Salmonella* (Okolocha and Ellerbroek, 2005) or total aerobic counts, is reduced (del Rio et al., 2007b). The combination of immersion bath and chemicals may be synergistic and reduce the number and/or prevalence of pathogens (Vadhanasin et al., 2004); however, the effectiveness of chilling (air or immersion) without added chemicals has been demonstrated even against pathogens such as *Campylobacter* and *Salmonella* (Huezo et al., 2007).

Even though pasteurization in hot water (75–80°C for 20–30 s) and steam for less than 12 s might have an adverse effect on the sensorial characteristics of the carcass, they do reduce the numbers of *C. jejuni* and *Escherichia coli* (Corry et al., 2007; James et al., 2007; Loretz et al., 2010; Tompkins et al., 2008), they might be a good option for finished or processed products.

Freezing is a preservation method that regularly stops the metabolism of the microorganisms; however, susceptible organisms might be killed by this process. Freezing resistance may be modified with the addition of chemicals, as some bacteria are sublethally damaged (Archer, 2004). As freezing by itself does not kill all bacteria, its use as a single method should be assessed closely (Chaves et al., 2011; Hanel and Atanassova, 2007), and specifically for each bacteria, because normally the response depends on the microorganism (Hughes et al., 2009; Tholozan et al., 1998). For example, *Campylobacter* does not grow below 30°C, but it can survive freezing temperatures (Georgsson et al., 2006; Sampers et al., 2010; Zhao et al., 2003).

Not only does irradiation reduce the initial numbers of pathogens (as *Salmonella enterica* serovar Typhimurium) and psychrotrophic bacteria in poultry products, but its effect might also last between 28 and 90 days at chilling or frozen temperatures respectively. While maintaining the sensorial quality, caution must be taken to use the right dosage, as numbers might increase when it is not high enough (Gomes and da Silva, 2006; Sarjeant et al., 2005). Other pathogens regularly linked to poultry that are susceptible to irradiation are *Staphylococcus aureus*, *E. coli* and *S.* Typhimurium and as before the effect might remain for up to 28 days in refrigeration (Filet et al., 2000).

High-power pulsed light is an environmentally friendly technique that has proved effective against *S. enterica* serovar Typhimurium and *Listeria monocytogenes*. The technique does not change the sensorial characteristics of the meat nor the lipid peroxidation (Paskeviciute et al., 2011). These two pathogens may also be susceptible to modified atmosphere packaging combined with chlorine dioxide; however, the reduction did not reach more than 1 log CFU/g (Shin et al., 2011).

Every method by itself might be powerful enough to reduce or eliminate bacteria; however taking advantage of the hurdle effect could be more effective. For example, the use of UV light and hydrogen peroxide separately in recycled chill brine resulted in a less reduction of *L. monocytogenes* than when they were combined (Parikh et al., 2012). Ultrasonication and chemical additives in combination demonstrated a greater impact on the decontamination of poultry carcass. Apart from the antimicrobial effect, the technology has a positive cost-effective ratio (Koolman et al., 2014).

4 Review of important poultry pathogens and their response to control measures

Apart from the spoilage microorganisms, bacteria, such as *E. coli* O157:H7, *Salmonella*, *Campylobacter* and *L. monocytogenes* are pathogens that continuously affect the safety of raw and processed poultry meat and consequently are major contributors to human food-borne illnesses (Hermans et al., 2011; Mandrell and Wachtel, 1999; Mor-Mur and Yuste, 2010; Wieczorek and Osek, 2015). For example, *Salmonella* and *Campylobacter* caused an estimated 1.8 million food-borne infections, most of them associated with poultry and poultry products (Kosa et al., 2015). Apart from the health risks, costs linked to poultry food-borne *Campylobacter* reached $699 million (Bryan and Doyle, 1995).

Pathogens linked to poultry can survive chilling temperatures, even when they are not considered psychrophiles or psychrotrophics. The mechanisms of survival have not been fully understood, but several studies have tried to explain them. Ligowska et al. (2011) suggested that *C. jejuni* benefited in the presence of drip as chances of survival improved when cultured in meat juice rather than in laboratory medium, as indicated by the expression of the *luxS* gene. Genotypes (*C. jejuni*) isolated from broiler are the common isolates, hence the environment plays a very important role in their survival (Griekspoor et al., 2010). Apart from the environment, other potential mechanisms for cold tolerance are acquisition or biosynthesis of cryoprotectant molecules (e.g. polynucleotide phosphorylase (PNPase)), maintenance of translation initiation, presence of specific genes (e.g. *rponN*), and efficiency and modification of the lipid composition of the membrane (Haddad et al., 2009; Hughes et al., 2009; Hwang et al., 2011; Stintzi, 2003). The resistance responses are triggered under different conditions (Sagarzazu et al., 2010).

Pathogens are frequently isolated from chicken intestines (Marotta et al., 2015; Wieczorek et al., 2013; Wieczorek and Osek, 2015). Even when the flocks are free of pathogens, carcasses might become positive after slaughter (Wieczorek and Osek, 2015), and the contamination is normally associated with the slaughter process. Reduction in the number of bacteria might be achieved when intervention processes are combined at the farm level and in the slaughterhouse (Rosenquist et al., 2006). On the farm, if possible, the infected flocks must be destroyed, or producers need to sanitize the hatching eggs and reduce the colonization so that flocks can be free of *Campylobacter* (Hermans et al., 2011; Rosenquist et al., 2006). In general terms, good animal husbandry practices must be followed (Hafez, 1999). In the processing plants, interventions must be as simple as staff training to reduce the percentage of visceral rupture and physical removal of faecal contamination, or as complicated as the redesign or replacement of the hygienic equipment (Berrang et al., 2000; Boysen and Rosenquist, 2009). Son et al. (2007) suggested that well-applied control measures for one bacteria should be useful enough to control others. Since the late 1990s, rapid diagnostic techniques (e.g. immunomagnetic separation, different PCR protocols, mass spectrometry) have been suggested to identify microorganisms (Mandrell and Wachtel, 1999); however, up to now, there has been no unification of methods.

Several techniques have been developed to control food-borne pathogens; however, consumer handling, especially improper refrigeration practices, are commonly linked to outbreaks and illnesses. Hence, it is important to develop guidelines to improve consumer storage and cooking operations (Bryan and Doyle, 1995; Roccato et al., 2015). Risk associated with bacterial contamination start on the farm (Marotta et al., 2015), and throughout the slaughtered process, the main critical points include scalding, defeathering, head pulling and evisceration, chilling and frozen (Fries, 2002; Reiter et al., 2007). Hence, interventions include preventing faecal contamination during transport, increased scalding temperature and countercurrent flow rate, reducing contamination of feathers, keeping water cold and acidic and increasing the residual chlorine in the chilled water (Osiriphun et al., 2011; Reiter et al., 2007). Separation of zones (e.g. raw and post-heat treatment) is also recommended to reduce contamination and cross-contamination (Lunden et al., 2003).

5 Summary and conclusions

The increment in poultry demand is forcing the industry to redesign its methods and technologies to fulfil the current demands of consumers, who are more concerned about animal welfare, environmental responsibility and safety. From the safety point of view, it is necessary not only to control traditional microorganisms but emerging pathogens as well. Transfer of bacteria can be at any point of the production system, including raw ingredients, food contact and non-food contact surfaces, environment and finished products.

The processing plants have not been able to reduce the incidence of pathogenic bacteria effectively. Various methods have been suggested to control the growth of microorganisms in poultry and poultry products. During processing, *Pseudomonas*, total aerobic mesophiles and presumptive coliforms are significantly reduced with air chilling. Water sprays or immersion baths reduce pathogens when PAA or ozone (in immersion bath) is added. Electrolysed water reduces the number of *Campylobacter jejuni* up to 3 logs and also prevents cross-contamination. The level of water in the chilled-immersion bath can be reduced to half with the same effect against bacteria. Water can also be saved when it is treated with ozone and UV light allowing 90% recirculation.

Post-processing food handling is an important factor for reducing food-borne-linked infections and outbreaks. Pasteurization in hot water with or without additives is favourable for reducing total aerobic counts and pathogens such as *Campylobacter*, *Salmonella*, *C. jejuni* and *E. coli*. Freezing by itself does not kill all bacteria. For example, *Campylobacter* does not grow below 30°C, but it can survive freezing temperatures. Irradiation reduces pathogens such as *Staphylococcus aureus*, *E. coli* and *S*. Typhimurium and the effect might remain for up to 28 days in refrigeration. Also, the application of high-power pulsed light is an effective technique against *S. enterica* serovar Typhimurium and *Listeria monocytogenes*, without causing changes in the sensorial characteristics of meat. The cold tolerance of *C. jejuni* has attracted the attention of many researchers, and studies include the presence of specific genes and the modification of the lipid composition of the cell membrane.

In conclusion, ensuring safety in chilling and freezing of poultry meat remains one of the major concerns of the poultry industry. Traditional microorganisms and emerging pathogens need to be controlled. Preventive measures have to be taken at the processing plants to reduce incidence of pathogenic bacteria. It is recommended that good animal husbandry practices must be followed in order to have flocks free of pathogens, and these practices should include the development of guidelines to improve consumer storage and cooking operations. Single or combined physical and chemical interventions are recommended to limit the growth of food-borne and spoilage bacteria. Most of the interventions recommended nowadays are environmentally friendly techniques in the poultry processing system, to focus on this, it is important to mention the reduction of the water level in the chilled-immersion bath.

6 Future directions and research needs

Animal welfare has a strong role in the choice of consumers; furthermore, stressed animals might yield defective meat. In poultry, handling prior to slaughter (e.g. harvesting, feed and water withdrawal, transport, etc.) will cause extreme stress to the animals if not done properly. Allostatic modulation is a strategy that intends to tackle antemortem stress (Stilborn et al., 1988). This modulation has been tried successfully in cattle and more recently in poultry (Pena et al., 2008). A mix of ascorbic acid, acetyl salicylic acid, chlorine, sodium and potassium improved the microbial quality of poultry carcasses and decreased carcass-processing defects (Rubio-Garcia et al., 2015). This strategy is not expensive and might be easily administrated in drinking water. Another treatment used at the slaughterhouse point is the spray and immersion baths with natural antimicrobials such as lysozyme, rosemary, oregano oil and their combinations, as they have had the same desirable effects in finished products as those of chemical additives (Ntzimani et al., 2010; Piskernik et al., 2011).

Food manufacturers need to rely on mild preservation techniques because consumers today look for minimally processed highly nutritious edibles apart from demanding that animals be raised under strict welfare standards. The more natural appearance might not be possible when using more astringent preservation methods. Products prepared with chemical preservatives might have a longer shelf-life, but today consumers are rejecting them and as a result they are being banned increasingly. Hence, producers need more natural alternatives that achieve the same safety and sensorial characteristics as those of the chemicals (Smid and Gorris, 2007). New techniques rely more on the hurdle effect to control food spoilage. Natural antimicrobials can be obtained from plants (phytoalexins, organic acids (citric, sorbic and propionic acid), phenolic compounds and essential oils) or from microorganisms (phages (Lone et al., 2016), enzymes (peroxidase and lysozyme), bacteriocins; mainly nisin, pediocin and sakacin). These compounds are then added to techniques such as modified atmosphere packaging, active packaging, intelligent packaging and degradable biopolymers (Azlin-Hasim et al., 2015; Gouvea et al., 2015; Higueras et al., 2013; Huang et al., 2015; Lone et al., 2016; Meredith et al., 2014; Otoni et al., 2016; Realini and Marcos, 2014).

These preservation additives might also be incorporated into edible coatings (Gennadios et al., 1997; Jiang et al., 2011; Sanchez-Ortega et al., 2014). Regularly used coatings such as wax, proteins, polysaccharides and fats have been replaced by essential oils (oregano, clover, rosemary (Fernandez-Pan et al., 2014)) and nanoparticles (silver and zinc oxide (Morsy et al., 2014), benzoic and sorbic acid (Cruz-Romero et al., 2013)). These additives might increase the safety of refrigerated meat and poultry products as they are effective against pathogens such as *S. aureus*, *L. monocytogenes*, *E. coli* O157:H7, *Bacillus cereus*, *Pseudomonas fluorescens* and *S.* Typhimurium (Cruz-Romero et al., 2013; Morsy et al., 2014).

Nano-food packaging is an emergent low environmental impact technology based on polymer nanocomposites (Huang et al., 2015). Nanofillers with antimicrobials may help to avoid the loss of moisture, reduce lipid oxidation and discolouration, and inhibit microbial growth. 'Intelligent' food packaging incorporating nanosensors are aimed to improve detection and tracing of food condition through the commercial chain (Chaudhry et al., 2008). For example, it has been proven that zinc oxide nanoparticles *in vitro* and *in vivo* (RTE poultry products) can reduce the number of *S. Typhimurium* and *Staph. aureus* (Akbar and Anal, 2014).

Still under investigation is the combination of gases for MAP. For example, CO2:O2:N2 (Meredith et al., 2014) and argon (Heinrich et al., 2016) seem to be a suitable option for ready-to-eat- and raw products, not only to achieve a longer shelf-life but also to inhibit the growth of pathogenic bacteria.

All these techniques by themselves are effective; however, the combinations of these have had very interesting results, as they not only reduce the number of bacteria but they might also improve the shelf-life and sensory characteristics. The use of combination of gases for MAP results in lower thiobarbituric acid-reactive substances, enhance oxidative stability and control colour changes in meat products extending its shelf-life. Most of the combinations are based on MAP and low-density polyethylene (LDPE) films containing nanoparticles of antimicrobials such as silver (Azlin-Hasim et al., 2015), nisin, chitosan (Higueras et al., 2013; Soysal et al., 2015), bacteriophages (Gouvea et al., 2015), zinc oxide (Panea et al., 2014), allyl isothiocyanate (Dias et al., 2013), essential oils (Stratakos et al., 2015) and so on. Among bacteria, MAP + LDPE + antimicrobial nanoparticles have been effective against *S. Typhimurium*, *L. monocytogenes*, mesophiles, psychrophiles, *Pseudomonas* spp., lactic acid bacteria, yeasts and moulds. The use of nanosensors in food packaging it is another option for minimally processed foods, as it only detects changes in time and temperature, freshness (Realini and Marcos, 2014) but also identifies microbial end products such as hydrogen sulphide (Koskela et al., 2015).

The popularity of these methods is growing, but the concerns of the consumer and governments as well as the safety or residual effects have not yet been well understood (Huang et al., 2011). For example, it is known that nanomaterials can migrate from the pack to the food (Cushen et al., 2014), but the consequences of this are still under investigation. Research will continue until the full consequences that some of the technologies might have and the use of natural antimicrobials might be restricted due to individual legislations (Cushen et al., 2014; Smid and Gorris, 2007).

7 Where to look for further information

Toldrá, F. and Reig, M. (2012). *Analytical Tools for Assessing the Chemical Safety of Meat and Poultry* (1;2012; ed.). Dordrecht: Springer New York.

Toldrá, F. (2009). *Safety of Meat and Processed Meat* (1. Aufl. ed.). DE: Springer-Verlag.

Gurtler, J. B., Doyle, M. P. and Kornacki, J. L. (2014). *The Microbiological Safety of Low Water Activity Foods and Spices* (2014th ed.). New York, NY: Springer.

Farber, J., Crichton, J. and Snyder, J., Peter, O. (2014). *Retail Food Safety* (2014th ed.). New York, NY: Springer.

Christian, E. W. and Vaclavik, V. A. (2014). *Essentials of Food Science (4th Edition)* (4th 2014 ed.). New York, NY: Springer New York.

Erginkaya, Z., Ahmad, S., Malik, A. and Erten, H. (2014). *Food Processing: Strategies for Quality Assessment* (2014th ed.). New York, NY: Springer Verlag.

Lima, G. P. P. and Vianello, F. (2013). *Food Quality, Safety and Technology* (2014;1; ed.). DE: Springer Verlag.

International Commission on Microbiological Specifications for Foods (ICMSF) (2011). *Microorganisms in Foods 8 : Use of Data for Assessing Process Control and Product Acceptance* (1st ed.). Boston, MA: Springer.

King, H. (2013). *Food Safety Management: Implementing a Food Safety Program in a Food Retail Business* (1. Aufl.;1; ed.). Dordrecht: Springer-Verlag.

O'Donnell, C. P., Fagan, C. and Cullen, P. J. (2014). *Process Analytical Technology for the Food Industry* (2014th ed.). New York, NY: Springer Verlag.

Yanniotis, S., Stoforos, N. G., Taoukis, P. and Karathanos, V. T. (2013). *Advances in Food Process Engineering Research and Applications* (2013;1; ed.). DE: Springer Verlag.

Weidenbörner, M. (2013, 2014). *Mycotoxins in Foodstuffs* (2nd;2;2nd;2nd 2013; ed.). Boston, MA: Springer Verlag.

Daszak, P., Richt, J. A., Jeggo, M. and Mackenzie, J. S. (2013). *One Health: The Human-Animal-Environment Interfaces in Emerging Infectious Diseases : Food Safety and Security, and International and National Plans for Implementation of One Health Activities* (2013th ed.). Berlin, Heidelberg: Springer Berlin Heidelberg.

Xu, S. (2014). *Proceedings of 2013 World Agricultural Outlook Conference* (1;2014; ed.). Dordrecht: Springer Berlin Heidelberg.

Mortimore, S. and Wallace, C. (2013). *HACCP: A Practical Approach* (3. Aufl.;3;3rd;3rd 2013; ed.). Boston, MA: Springer-Verlag.

Poultry Science Association
Poultry Science Journal
International Poultry Science
Journal of Applied Poultry Research
International Journal of Food Microbiology

8 References

Akbar, A., and A. Anal, 2014, Zinc oxide nanoparticles loaded active packaging, a challenge study against *Salmonella typhimurium* and *Staphylococcus aureus* in ready-to-eat poultry meat, *Food Control*, 38, 88–95.

Allen, V., S. Bull, J. Corry, G. Domingue, F. Jorgensen, J. Frost, R. Whyte, A. Gonzalez, N. Elviss and T. Humphrey, 2007, *Campylobacter* spp. contamination of chicken carcasses during processing in relation to flock colonisation, *International Journal of Food Microbiology*, 113, 54–61.

Allen, V., C. Burton, J. Corry, G. Mead and D. Tinker, 2000a, Investigation of hygiene aspects during air chilling of poultry carcases using a model rig, *British Poultry Science*, 41, 575–83.

Allen, V., J. Corry, C. Burton, R. Whyte and G. Mead, 2000b, Hygiene aspects of modern poultry chilling, *International Journal of Food Microbiology*, 58, 39–48.

Andersen, S., P. Saadbye, N. Shukri, H. Rosenquist, N. Nielsen and J. Boel, 2006, Antimicrobial resistance among *Campylobacter jejuni* isolated from raw poultry meat at retail level in Denmark, *International Journal of Food Microbiology*, 107, 250–5.

Archer, D., 2004, Freezing: An underutilized food safety technology?, *International Journal of Food Microbiology*, 90, 127–38.

Avens, J., S. Albright, A. Morton, B. Prewitt, P. Kendall and J. Sofos, 2002, Destruction of microorganisms on chicken carcasses by steam and boiling water immersion, *Food Control*, 13, 445–50.

Azlin-Hasim, S., M. Cruz-Romero, C., M. Morris, A., E. Cummins and J. Kerry, 2015, Effects of a combination of antimicrobial silves low density polyethylene nanocomposite films and modified atmosphere packaging on the shelf life of chicken breast fillets, *Food packaging and shelf life*, 4, 26–35.

Bashor, M., P. Curtis, K. Keener, B. Sheldon, S. Kathariou and J. Osborne, 2004, Effects of carcass washers on *Campylobacter* contamination in large broiler processing plants, *Poultry Science*, 83, 1232–9.

Berrang, M., C. Lyon, D. Smith and J. Northcutt, 2000, Incidence of *Listeria monocytogenes* on pre-scald and post-chill chicken, *Journal of Applied Poultry Research*, 9, 546–50.

Boyle, D., J. Sofos and J. Maga, 1988, Inhibition of spoilage and pathogenic microorganisms by liquid smoke from various woods, *Lebensmittel-Wissenschaft & Technologie*, 21, 54–8.

Boysen, L., and H. Rosenquist, 2009, Reduction of thermotolerant *Campylobacter* species on broiler carcasses following physical decontamination at slaughter, *Journal of Food Protection*, 72, 497–502.

Bryan, F., and M. Doyle, 1995, Health risks and consequences of *salmonella* and *Campylobacter jejuni* in raw poultry, *Journal of Food Protection*, 58, 326–44.

Chaudhry, Q., M. Scotter, J. Blackburn, B. Ross, A. Boxall, L. Castle, R. Aitken and R. Watkins, 2008, Applications and implications of nanotechnologies for the food sector, *Food Additives and Contaminants*. Part A. *Chemistry, Analysis, Control, Exposure & Risk Assessment*, 25, 241–58.

Chaves, B., I. Han, P. Dawson and J. Northcutt, 2011, Survival of artificially inoculated *Escherichia coli* and *Salmonella* Typhimurium on the surface of raw poultry products subjected to crust freezing, *Poultry Science*, 90, 2874–78.

Corry, J., S. James, G. Purnell, C. Barbedo-Pinto, Y. Chochois, M. Howell and C. James, 2007, Surface pasteurisation of chicken carcasses using hot water, *Journal of Food Engineering*, 79, 913–19.

Cruz-Romero, M., T. Murphy, M. Morris, E. Cummins and J. Kerry, 2013, Antimicrobial activity of chitosan, organic acids and nano-sized solubilisates for potential use in smart antimicrobially-active packaging for potential food applications, *Food Control*, 34, 393–7.

Culotti, A., and A. Packman, 2014, *Pseudomonas aeruginosa* promotes *Escherichia coli* biofilm formation in nutrient-limited medium, *Plos One*, 9, e107186. doi: 10.1371/journal.pone.0107186.

Culotti, A., and A. Packman, 2015, *Pseudomonas aeruginosa* facilitates *Campylobacter jejuni* growth in biofilms under oxic flow conditions, *FEMS Microbiology Ecology*, 91. doi: 10.1093/femsec/fiv136.

Cushen, M., J. Kerry, M. Morris, M. Cruz-Romero and E. Cummins, 2014, Silver migration from nanosilver and a commercially available zeolite filler polyethylene composites to food simulants, *Food Additives and Contaminants*. Part A. *Chemistry, Analysis, Control, Exposure & Risk Assessment*, 31, 1132–40.

del Rio, E., R. Muriente, M. Prieto, C. Alonso-Calleja and R. Capita, 2007a, Effectiveness of trisodium phosphate, acidified sodium chlorite, citric acid, and peroxyacids against pathogenic bacteria on poultry during refrigerated storage, *Journal of Food Protection*, 70, 2063–71.

del Rio, E., M. Panizo-Moran, M. Prieto, C. Alonso-Calleja and R. Capita, 2007b, Effect of various chemical decontamination treatments on natural microflora and sensory characteristics of poultry, *International Journal of Food Microbiology*, 115, 268–80.

Dias, M., N. Soares, S. Borges, M. de Sousa, C. Nunes, I. de Oliveira and E. Medeiros, 2013, Use of allyl isothiocyanate and carbon nanotubes in an antimicrobial film to package shredded, cooked chicken meat, *Food Chemistry*, 141, 3160–6.

Diaz, M., S. Law and J. Frank, 2001, Control of pathogenic microorganisms and turbidity in poultry-processing chiller water using UV-enhanced ozonation, *Ozone-Science & Engineering*, 23, 53–64.

FAO, 2014, Meat Consumption, Meat and Meat Products, Food and Agriculture Organization of the United Nations.

Fernandez-Pan, I., X. Carrion-Granda and J. Mate, 2014, Antimicrobial efficiency of edible coatings on the preservation of chicken breast fillets, *Food Control*, 36, 69–75.

Filet, S. M. H., G. C. Rosa, A. A. Ricardo, G. M. S. Amaral do, L. Blumer, W. J. M. Melges and R. E. Domarco, 2000, Gamma irradiation in the control of pathogenic bacteria in refrigerated chicken meat, *Sci Agric*, 57, 389–94.

Fox, E., P. Wall and S. Fanning, 2015, Control of *Listeria* species food safety at a poultry food production facility, *Food Microbiology*, 51, 81–6.

Fries, R., 2002, Reducing *Salmonella* transfer during industrial poultry meat production, *World's Poultry Science Journal*, 58, 527–40.

Gellynck, X., W. Messens, D. Halet, K. Grijspeerdt, E. Hartnett and J. Viaenei, 2008, Economics of reducing *Campylobacter* at different levels within the Belgian poultry meat, *Journal of Food Protection*, 71, 479–85.

Gennadios, A., M. Hanna and L. Kurth, 1997, Application of edible coatings on meats, poultry and seafoods: A review, *Food Science and Technology-Lebensmittel-Wissenschaft & Technologie*, 30, 337–50.

Georgsson, F., A. Porkelsson, M. Geirsdottir, J. Reiersen and N. Stern, 2006, The influence of freezing and duration of storage on *Campylobacter* and indicator bacteria in broiler carcasses, *Food Microbiology*, 23, 677–83.

Gomes, H., and E. da Silva, 2006, Effects of ionizing radiation on mechanically deboned chicken meat during frozen storage, *Journal of Radioanalytical and Nuclear Chemistry*, 270, 225–9.

Gonzalez-Miret, M., M. Escudero-Gilete and F. Heredia, 2006, The establishment of critical control points at the washing and air chilling stages in poultry meat production using multivariate statistics, *Food Control*, 17, 935–41.

Gouvea, D., R. Mendonca, M. Soto and R. Cruz, 2015, Acetate cellulose film with bacteriophages for potential antimicrobial use in food packaging, *Lwt-Food Science and Technology*, 63, 85–91.

Griekspoor, P., E. Engvall, B. Olsen and J. Waldenstrom, 2010, Multilocus sequence typing of *Campylobacter jejuni* from broilers, *Veterinary Microbiology*, 140, 180–5.

Haddad, N., C. Burns, J. Bolla, H. Prevost, M. Federighi, D. Drider and J. Cappelier, 2009, Long-Term survival of *Campylobacter jejuni* at low temperatures is dependent on polynucleotide phosphorylase activity, *Applied and Environmental Microbiology*, 75, 7310–18.

Hafez, H., 1999, Poultry meat and food safety: Pre- and post-harvest approaches to reduce foodborne pathogens, *World's Poultry Science Journal*, 55, 269–80.

Hanel, C., and V. Atanassova, 2007, Impact of different storage factors on the survivability of *Campylobacter jejuni* in turkey meat, *FEMS Immunology and Medical Microbiology*, 49, 146–8.

Heinrich, V., M. Zunabovic, L. Nehm, J. Bergmair and W. Kneifel, 2016, Influence of argon modified atmosphere packaging on the growth potential of strains of Listeria monocytogenes and Escherichia coli, *Food Control*, 59, 513–23.

Hermans, D., K. Van Deun, W. Messens, A. Martel, F. Van Immerseel, F. Haesebrouck, G. Rasschaert, M. Heyndrickx and F. Pasmans, 2011, *Campylobacter* control in poultry by current intervention measures ineffective: Urgent need for intensified fundamental research, *Veterinary Microbiology*, 152, 219–28.

Higueras, L., G. Lopez-Carballo, P. Hernandez-Munoz, R. Gavara and M. Rollini, 2013, Development of a novel antimicrobial film based on chitosan with LAE (ethyl-N alpha-dodecanoyl-L-arginate) and its application to fresh chicken, *International Journal of Food Microbiology*, 165, 339–45.

Huang, J., X. Li and W. Zhou, 2015, Safety assessment of nanocomposite for food packaging application, *Trends in Food Science and Technology*, 45, 187–99.

Huang, Y., S. Chen, X. Bing, C. Gao, T. Wang and B. Yuan, 2011, Nanosilver migrated into food-simulating solutions from commercially available food fresh containers, *Packaging Technology and Science*, 24, 291–7.

Huezo, R., J. Northcutt, D. Smith, D. Fletcher and K. Ingram, 2007, Effect of dry air or immersion chilling on recovery of bacteria from broiler carcasses, *Journal of Food Protection*, 70, 1829–34.

Hughes, R., K. Hallett, T. Cogan, M. Enser and T. Humphrey, 2009, The response of *Campylobacter jejuni* to low temperature differs from that of *Escherichia coli*, *Applied and Environmental Microbiology*, 75, 6292–8.

Hwang, S., B. Jeon, J. Yun and S. Ryu, 2011, Roles of *rpoN* in the resistance of *Campylobacter jejuni* under various stress conditions, *BMC Microbiology*, 11, 207.

James, C., S. James, N. Hannay, G. Purnell, C. Barbedo-Pinto, H. Yaman, M. Araujo, M. Gonzalez, J. Calvo, M. Howell and J. Corry, 2007, Decontamination of poultry carcasses using steam or hot water in combination with rapid cooling, chilling or freezing of carcass surfaces, *International Journal of Food Microbiology*, 114, 195–203.

James, C., C. Vincent, T. Lima and S. James, 2006, The primary chilling of poultry carcasses – a review, *International Journal of Refrigeration-Revue Internationale Du Froid*, 29, 847–62.

Jarquin, C., D. Alvarez, O. Morales, A. Morales, B. Lopez, P. Donado, M. Valencia, A. Arevalo, F. Munoz, I. Walls, M. Doyle and W. Alali, 2015, *Salmonella* on raw poultry in retail markets in guatemala: Levels, antibiotic susceptibility, and serovar distribution, *Journal of Food Protection*, 78, 1642–50.

Jiang, Z., H. Neetoo and H. Chen, 2011, Efficacy of freezing, frozen storage and edible antimicrobial coatings used in combination for control of *Listeria monocytogenes* on roasted turkey stored at chiller temperatures, *Food Microbiology*, 28, 1394–1401.

Koolman, L., P. Whyte, J. Meade, J. Lyng and D. Bolton, 2014, A combination of chemical and ultrasonication treatments to reduce *Campylobacter jejuni* on raw poultry, *Food and Bioprocess Technology*, 7, 3602–7.

Kosa, K., S. Cates, S. Bradley, E. Chambers and S. Godwin, 2015, Consumer-reported handling of raw poultry products at home: Results from a national survey, *Journal of Food Protection*, 78, 180–6.

Koskela, J., J. Sarfraz, P. Ihalainen, A. Maattanen, P. Pulkkinen, H. Tenhu, T. Nieminen, A. Kilpela and J. Peltonen, 2015, Monitoring the quality of raw poultry by detecting hydrogen sulfide with printed sensors, *Sensors and Actuators B:Chemical*, 218, 89–96.

Ligowska, M., M. Cohn, R. Stabler, B. Wren and L. Brondsted, 2011, Effect of chicken meat environment on gene expression of *Campylobacter jejuni* and its relevance to survival in food, *International Journal of Food Microbiology*, 145, S111–15.

Logue, C., J. Sherwood, P. Olah, L. Elijah and M. Dockter, 2003, The incidence of antimicrobial-resistant *Salmonella* spp. on freshly processed poultry from US Midwestern processing plants, *Journal of Applied Microbiology*, 94, 16–24.

Lone, A., H. Anany, M. Hakeem, L. Aguis, A. Avdjian, M. Bouget, A. Atashi, L. Brovko, D. Rochefort and M. Griffiths, 2016, Development of prototypes of bioactive packaging materials based on immobilized bacteriophages for control of growth of bacterial pathogens in foods, *International Journal of Food Microbiology*, 217, 49–58.

Loretz, M., R. Stephan and C. Zweifel, 2010, Antimicrobial activity of decontamination treatments for poultry carcasses: A literature survey, *Food Control*, 21, 791–804.

Lunden, J., T. Autio, A. Sjoberg and H. Korkeala, 2003, Persistent and nonpersistent *Listeria monocytogenes* contamination in meat and poultry processing plants, *Journal of Food Protection*, 66, 2062–9.

Mandrell, R., and M. Wachtel, 1999, Novel detection techniques for human pathogens that contaminate poultry, *Current Opinion in Biotechnology*, 10, 273–8.

Marotta, F., G. Garofolo, G. Di Donato, G. Aprea, I. Platone, S. Cianciavicchia, A. Alessiani and E. Di Giannatale, 2015, Population diversity of *Campylobacter jejuni* in Poultry and its dynamic of contamination in chicken meat, *Biomed Research International*, 859845. http://doi.org/10.1155/2015/859845.

Mead, G., 2004, Current trends in the microbiological safety of poultry meat, *World's Poultry Science Journal*, 60, 112–18.

Mead, G., Allen, C. Burton and J. Corry, 2000, Microbial cross-contamination during air chilling of poultry, *British Poultry Science*, 41, 158–62.

Mercogliano, R., A. De Felice, N. Murru, S. Santonicola and M. L. Cortesi, 2014, Ozone decontamination of poultry meat and biogenic amines as quality index, *Journal of Food Process Technology*, 5, 305–9.

Meredith, H., V. Valdramidis, B. Rotabakk, M. Sivertsvik, D. McDowell and D. Bolton, 2014, Effect of different modified atmospheric packaging (MAP) gaseous combinations on Campylobacter and the shelf-life of chilled poultry fillets, *Food Microbiology*, 44, 196–203.

Mor-Mur, M., and J. Yuste, 2010, Emerging bacterial pathogens in meat and poultry: An overview, *Food and Bioprocess Technology*, 3, 24–35.

Morsy, M., H. Khalaf, A. Sharoba, H. El-Tanahi and C. Cutter, 2014, Incorporation of essential oils and nanoparticles in pullulan films to control foodborne pathogens on meat and poultry products, *Journal of Food Science*, 79, M675–84.

Nauta, M., A. Hill, H. Rosenquist, S. Brynestad, A. Fetsch, P. van der Logt, A. Fazil, B. Christensen, E. Katsma, B. Borck and A. Havelaar, 2009, A comparison of risk assessments on *Campylobacter* in broiler meat, *International Journal of Food Microbiology*, 129, 107–23.

Northcutt, J., D. Smith, N. Cox, K. Ingram, R. Buhr, L. Richardson, J. Cason and A. Hinton, 2008a, Effects of chlorine or chlorine dioxide during immersion chilling on recovery of bacteria from broiler carcasses and chiller water, *Poultry Science*, 87(Suppl 1), 40.

Northcutt, J., D. Smith, R. Huezo and K. Ingram, 2008b, Microbiology of broiler carcasses and chemistry of chiller water as affected by water reuse, *Poultry Science*, 87, 1458–63.

Ntzimani, A., V. Giatrakou, and I. Savvaidis, 2010, Combined natural antimicrobial treatments (EDTA, lysozyme, rosemary and oregano oil) on semi cooked coated chicken meat stored in vacuum packages at 4 degrees C: Microbiological and sensory evaluation, *Innovative Food Science & Emerging Technologies*, 11, 187–96.

Okolocha, E., and L. Ellerbroek, 2005, The influence of acid and alkaline treatments on pathogens and the shelf life of poultry meat, *Food Control*, 16, 217–25.

Osiriphun, S., P. Iamtaweejaloen, P. Kooprasertying, W. Koetsinchai, K. Tuitemwong, L. Erickson and P. Tuitemwong, 2011, Exposure assessment and process sensitivity analysis of the contamination of *Campylobacter* in poultry products, *Poultry Science*, 90, 1562–73.

Otoni, G., P. Espitia, R. Avena-Bustillos and T. McHugh, 2016, Trends in antimicrobial food packaging systems: Emitting sachets and absorbent pads, *Food Research International*, 83, 60–73.

Panea, B., G. Ripoll, J. Gonzalez, A. Fernandez-Cuello and P. Alberti, 2014, Effect of nanocomposite packaging containing different proportions of ZnO and Ag on chicken breast meat quality, *Journal of Food Engineering*, 123, 104–12.

Parikh, P., R. Williams, P. Mallikarjunan, J. Eifert and J. Marcy, 2012, Control of listeria monocytogenes in recycled chill brine using ultraviolet light and antimicrobial agents, *Journal of Food Safety*, 32, 169–75.

Park, H., Y. Hung and R. Brackett, 2002, Antimicrobial effect of electrolyzed water for inactivating Campylobacter jejuni during poultry washing, *International Journal of Food Microbiology*, 72, 77–83.

Park, S., 2002, The physiology of Campylobacter species and its relevance to their role as foodborne pathogens, *International Journal of Food Microbiology*, 74, 177–88.

Paskeviciute, E., I. Buchovec and Z. Luksiene, 2011, High-power pulsed light for decontamination of chicken from food pathogens: A study on antimicrobial efficiency and organoleptic properties, *Journal of Food Safety*, 31, 61–8.

Pena, J., S. Vieira, J. Lopez, R. Reis, R. Barros, F. Furtado and P. Silva, 2008, Ascorbic acid and citric flavonoids for broilers under heat stress: Effects on performance and meat quality, *Brazilian Journal of Poultry Science*, 10, 125–30.

Piskernik, S., A. Klancnik, C. Riedel, L. Brondsted and S. Mozina, 2011, Reduction of *Campylobacter jejuni* by natural antimicrobials in chicken meat-related conditions, *Food Control*, 22, 718–24.

Purnell, G., K. Mattick and T. Humphrey, 2004, The use of 'hot wash' treatments to reduce the number of pathogenic and spoilage bacteria on raw retail poultry, *Journal of Food Engineering*, 62, 29–36.

Realini, C., and B. Marcos, 2014, Active and intelligent packaging systems for a modern society, *Meat Science*, 98, 404–19.

Reiter, M., M. Fiorese, G. Morett, M. Lopez and R. Jordano, 2007, Prevalence of *Salmonella* in a poultry slaughterhouse, *Journal of Food Protection*, 70, 1723–5.

Roccato, A., M. Uyttendaele, V. Cibin, F. Barrucci, V. Cappa, P. Zavagnin, A. Longo, P. Catellani and A. Ricci, 2015, Effects of domestic storage and thawing practices on *Salmonella* in poultry-based meat preparations, *Journal of Food Protection*, 78, 2117–25.

Rorvik, L., B. Aase, T. Alvestad and D. Caugant, 2003, Molecular epidemiological survey of *Listeria monocytogenes* in broilers and poultry products, *Journal of Applied Microbiology*, 94, 633–40.

Rosenquist, H., H. Sommer, N. Nielsen and B. Christensen, 2006, The effect of slaughter operations on the contamination of chicken carcasses with thermotolerant *Campylobacter*, *International Journal of Food Microbiology*, 108, 226–32.

Rubio-Garcia, M., M. Rubio-Lozano, E. Ponce-Alquicira, C. Rosario-Cortes, G. Nava and M. Castaneda-Serrano, 2015, Improving appearance and microbiologic quality of broiler carcasses with an allostatic modulator, *Poultry Science*, 94, 1957–63.

Sagarzazu, N., G. Cebrian, S. Condon, B. Mackey and P. Manas, 2010, High hydrostatic pressure resistance of *Campylobacter jejuni* after different sublethal stresses, *Journal of Applied Microbiology*, 109, 146–55.

Sampers, I., I. Habib, L. De Zutter, A. Dumoulin and M. Uyttendaele, 2010, Survival of *Campylobacter* spp. in poultry meat preparations subjected to freezing, refrigeration, minor salt concentration, and heat treatment, *International Journal of Food Microbiology*, 137, 147–53.

Sanchez-Ortega, I., B. Garica-Almendarez, E. Santos-Lopez, A. Amaro-Reyes, J. Barboza-Corona and C. Regalado, 2014, Antimicrobial edible films and coatings for meat and meat products preservation, *Scientific World Journal*, 248935. doi.org/10.1155/2014/248935.

Sarjeant, K., S. Williams and A. Hinton, 2005, The effect of electron beam irradiation on the survival of Salmonella enterica serovar *typhimurium* and psychrotrophic bacteria on raw chicken breasts stored at four degrees celsius for fourteen days, *Poultry Science*, 84, 955–8.

Shin, J., B. Harte, S. Selke and Y. Lee, 2011, Use of a controlled chlorine dioxide (clo2) release system in combination with modified atmosphere packaging (map) to control the growth of pathogens, *Journal of Food Quality*, 34, 220–8.

Smid, E. J., and L. G. M. Gorris, 2007, Natural Antimicrobials for Food Preservation, in M. S. Rahman (ed.), *Handbook of Food Preservation*, 2nd Ed., *Food Science and Technology*, Boca Raton (USA), CRC Press, 285–308.

Smith, J., S. Corkran, S. McKee, S. Bilgili and M. Singh, 2015, Evaluation of post-chill applications of antimicrobials against *Campylobacter jejuni* on poultry carcasses, *Journal of Applied Poultry Research*, 24, 451–6.

Sofos, J., 2008, Challenges to meat safety in the 21st century, *Meat Science*, 78, 3–13.

Son, I., M. Englen, M. Berrang, P. Fedorka-Cray and M. Harrison, 2007, Prevalence of *Arcobacter* and *Campylobacter* on broiler carcasses during processing, *International Journal of Food Microbiology*, 113, 16–22.

Soysal, C., H. Bozkurt, E. Dirican, M. Guclu, E. Bozhuyuk, A. Uslu and S. Kaya, 2015, Effect of antimicrobial packaging on physicochemical and microbial quality of chicken drumsticks, *Food Control*, 54, 294–9.

Spoto, M. H. F., C. R. Gallo, A. R. Alcarde, M. S. A. Gurgel, L. Blumer, J. M. M. Walder and R. E. Domarco, 2000, Gamma irradiation in the control of pathogenic bacteria in refrigerated chicken meat, *Scientia Agricola*, 57, 389–94.

Stilborn, H., G. Harris, W. Bottje and P. Waldroup, 1988, Ascorbic-acid and acetylsalicylic-acid (aspirin) in the diet of broilers maintained under heat-stress conditions, *Poultry Science*, 67, 1183–7.

Stintzi, A., 2003, Gene expression profile of *Campylobacter jejuni* in response to growth temperature variation, *Journal of Bacteriology*, 185, 2009–16.

Stratakos, A., G. Delgado-Pando, M. Linton, M. Patterson and A. Koidis, 2015, Synergism between high-pressure processing and active packaging against Listeria monocytogenes in ready-to-eat chicken breast, *Innovative Food Science & Emerging Technologies*, 27, 41–7.

The National Chicken Council, 2016, *Per Capita Consumption of Poultry and Livestock, 1965 to Estimated 2016, in Pounds – The National Chicken Council.* [online] Available at http://www.nationalchickencouncil.org/about-the-industry/statistics/per-capita-consumption-of-poultry-and-livestock-1965-to-estimated-2012-in-pounds/ (accessed 22 May 2016).

Tholozan, J., M. Federighi, J. Cappelier, N. Chihib, F. Jugiau, G. Delattre and R. Rosset, 1998, Physiological response of food spoilage or pathogenic microorganisms to cold treatment or to combined stabilization treatments including rapid cooling: Institut international du froid. Commission C2 & E1. Conference, Nantes, FRANCE (16/09/1998) Hygiene, Quality & Safety in the Cold Chain and Air Conditioning, 96–105.

Tompkins, N., J. Avens, P. Kendall and M. Salman, 2008, Effect of boiling water carcass immersion on aerobic bacteria counts of poultry skin and processed ground poultry meat, *Zoonoses and Public Health*, 55, 235–41.

Trindade, M., M. Kushida, N. Villanueva, D. Pereira and C. de Oliveira, 2012, Comparison of ozone and chlorine in low concentrations as sanitizing agents of chicken carcasses in the water immersion chiller, *Journal of Food Protection*, 75, 1139–43.

Tsola, E., E. Drosinos and P. Zoiopoulos, 2008, Impact of poultry slaughter house modernisation and updating of food safety management systems on the microbiological quality and safety of products, *Food Control*, 19, 423–31.

Vadhanasin, S., A. Bangtrakulnonth and T. Chidkrau, 2004, Critical control points for monitoring salmonellae reduction in Thai commercial frozen broiler processing, *Journal of Food Protection*, 67, 1480–3.

Voidarou, C., D. Vassos, T. Kegos, A. Koutsotoli, A. Tsiotsias, J. Skoufos, A. Tzora, Maipa A. Alexopoulos and E. Bezirtzoglou, 2007, Aerobic and anaerobic microbiology of the immersion chilling procedure during poultry processing, *Poultry Science*, 86, 1218–22.

Wieczorek, K., I. Kania and J. Osek, 2013, Prevalence and antimicrobial resistance of *Campylobacter* spp. isolated from poultry carcasses in Poland, *Journal of Food Protection*, 76, 1451–5.

Wieczorek, K., and J. Osek, 2015, Poultry flocks as a source of *Campylobacter* contamination of broiler carcasses, *Polish Journal of Veterinary Sciences*, 18, 101–6.

Zhao, T., G. Ezeike, M. Doyle, Y. Hung and R. Howell, 2003, Reduction of *Campylobacter jejuni* on poultry by low-temperature treatment, *Journal of Food Protection*, 66, 652–5.

Chapter 9

Case studies in food safety control in the production of fresh poultry meat: effective control of *Salmonella* in Sweden

Ivar Vågsholm, Swedish University of Agricultural Sciences, Sweden

1 Introduction

This chapter presents the Swedish *Salmonella* control programme in poultry with a view to highlight insights that could be helpful for poultry *Salmonella* control programmes in other countries. During the past 60 years the *Salmonella* control programme in poultry meat production has evolved, and the programme has been frequently reviewed and revised in light of new evidence and changed market conditions. Some of these elements will be outlined in this chapter, and hopefully some of the lessons learnt might be useful when designing or revising *Salmonella* control programmes along the poultry meat food chain in other countries. Perhaps it is good to keep in mind that any food business operator, operating all along the food chain, is responsible for the safety of the food, including *Salmonella* freedom.

According to the Swedish Board of Agriculture (SBA, 2015), in 2014, nearly 95% of poultry meat originated from broilers. The remaining poultry meat originated from spent layers, ducks, geese and turkeys. Therefore, I will mostly focus the presentation on the production on poultry meat from broilers (*Gallus gallus*). The achievements of the programme are illustrated by the fact that it is very unusual that salmonellosis in people can be traced back

http://dx.doi.org/10.19103/AS.2016.0010.12

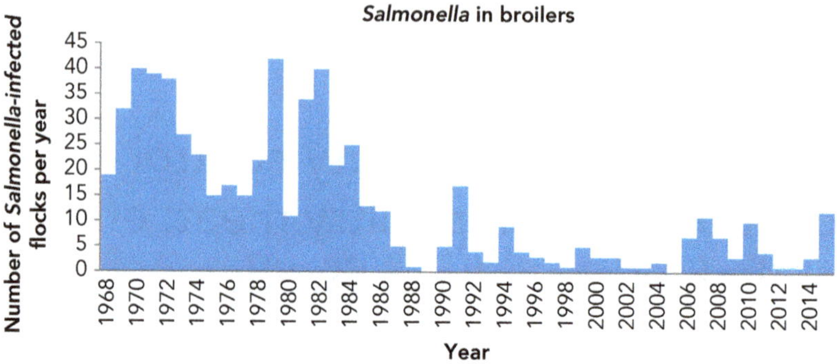

Figure 1 Number of reported outbreaks of *Salmonella* in broiler flocks 1968–2015 in Sweden. Both production and breeding flocks were included. Source: SVA (National Veterinary Institute), 2016a.

to *Salmonella* contamination of poultry meat produced in Sweden (SVA, 2016b), whilst some outbreaks have been suspected to be linked to imported poultry meat. The chapter begins by reviewing the history of *Salmonella* control in Sweden, from the Alvesta *Salmonella* outbreak of 1953 through to the national, comprehensive *Salmonella* control programme introduced in 2010. It then turns to the current *Salmonella* control and prevention system.

Salmonella cannot be eradicated except in compartments; hence, one unique feature of the Swedish *Salmonella* control from the onset is the aim to deliver *Salmonella*-free food to the consumers (Wierup et al., 1995), thus the practical goal of *Salmonella* freedom in the poultry meat compartment instead of a goal of national *Salmonella* eradication. All broilers (and other poultry) sent for slaughter should be without *Salmonella* and consequently the focus on pre-harvest control. This was achieved by applying the following strategies: (1) preventing and eliminating *Salmonella* contamination in feed given to broiler flocks; (2) preventing infection of the breeding pyramid (grandparent and parent flocks including hatcheries); (3) surveillance of the production chain at critical control points to detect *Salmonella* contamination; (4) enforcing strict biosecurity on the production site and; (5) whenever *Salmonella* is found, taking action to remove it from the food chain. Today, the pre-harvest *Salmonella* control of poultry meat production consists of a compulsory surveillance programme, with compulsory action if *Salmonella* is found and a voluntary preventive and biosecurity programme. Nearly all broiler producers participate in the voluntary programme. The feed control is compulsory for the poultry feed producers. In Fig. 1, the number of *Salmonella* outbreaks per year is shown, and hence illustrating that the goal of *Salmonella* freedom at slaughter has been practically achievable. For example, during 1988–2015, in 19 out of 28 years, the annual number of *Salmonella*-positive flocks was less than 5. The peaks in 2006–2008 and 2010 were associated with outbreaks in breeding flocks.

2 The history of *Salmonella* control in Sweden

2.1 The Alvesta outbreak and its consequences

The starting point was the large Alvesta outbreak in 1953 (Bengtsson et al., 1955; Ekstam, 1953; Olin, 1956; Silverstolpe and Wranne, 1955). The outbreak was unprecedented in Sweden with more than 9000 cases reported, of which approximately 4000 persons required hospital care and 90 people died. The outbreak lasted from June to December 1953, affecting major regions of Sweden. At the epidemic peak during July 1953, the daily attack rate was 220 persons. The investigation into this matter revealed the following: (1) lack of cooling capacity both at slaughter and during transport; (2) infected animals were sent to slaughter; and (3) breakdowns of biosecurity and hygiene resulted in massive cross-contaminations, all of which were made worse by an ongoing industrial action and an unusually hot summer. As a direct consequence, the Public Health Institute of Sweden created a department for epidemiology to investigate human outbreaks of foodborne diseases. On the poultry side a voluntary programme for the control of fowl typhoid (*Salmonella* Gallinarum/Pullorum) had been running since 1941 (SMPA, 2016). However, in 1961, the voluntary fowl typhoid was extended to an official control programme of all *Salmonella* serovars of public health importance (Wierup et al., 1992 and 1995). Moreover, *Salmonella* infections in poultry became a notifiable disease.

2.2 History and evolution of the control programme

The political objective for the poultry meat production in the 1960s and 1970s was to produce high-quality meat and at affordable prices (KSLA, 2003). Poultry meat was seen as the good value, more modern and indeed the healthier food. Therefore, poultry meat production expanded rapidly and became more industrialized. Food safety, including *Salmonella* control, was part of this high-quality objective, thereby being the main driver for evolution of the control programme.

In 1984 the previous voluntary *Salmonella* control programme became compulsory in poultry meat production. Consequently, the poultry industry had to find insurance solutions to accommodate the *Salmonella* clean-up costs of infected flocks. This change created stronger incentives for poultry producers to ensure *Salmonella* freedom amongst breeders supplying day-old chicks for meat production as well as the absence of *Salmonella* in poultry feed. Another consequence was the enhanced biosecurity on broiler farms.

2.3 The European Union and *Salmonella* protection guarantees

In 1995 Sweden joined the European Union (EU). Consequently, the closed Swedish market for poultry meat was now opened for imports from EU and 3rd countries. As a part of the accession process, the EU approved the Swedish *Salmonella* control programmes both pre- and post-harvest for poultry, egg, cattle and pig production. The approval enabled Sweden to demand equivalent *Salmonella* protection on imported fresh poultry meat and live poultry from other EU and 3rd countries. This equivalent protection could either be afforded by equivalent control programmes such as in Finland or that all imported consignments of live birds or fresh poultry meat had to be tested for *Salmonella*. This microbiological testing was designed to ensure that the prevalence of *Salmonella* was less than 5% with 95% confidence level – usually meant that 60 samples had to be taken and analysed with negative results. This has become a practical example for the acceptable level of *Salmonella* protection. For live poultry all individual birds imported had to be tested with negative results. This policy was successful in terms of food safety. Few cases of foodborne *Salmonella* illness have been traced back to contaminated poultry meat. However, one loophole was that the *Salmonella* guarantees only applied to fresh poultry meat. That meant that, for example, marinated poultry meat was exempt from the *Salmonella* guarantees.

2.4 Increasing imports and increasing control costs

In recent years (2006–15) the consumption poultry meat increased from 16 to 22 kg per person per year (SBA, 2016), but imported poultry meat constituted most of the increase. In 2015 domestic poultry meat production was around two-thirds of the poultry meat consumed in Sweden. This implied that the compulsory *Salmonella* testing of all poultry meat imported (*Salmonella* guarantees) to Sweden since 1995 did not function as a trade barrier. Compulsory *Salmonella* testing of imported poultry meat being a trade barrier was a concern for other EU countries when Sweden joined the EU. However, the poultry industry's profits were challenged due to increases in the direct and indirect costs of *Salmonella* control as more sampling was required and cheaper imports were available in the market.

2.5 Evaluations of the Swedish *Salmonella* control system

A perceived lack of cost efficiency and the suspicion that the *Salmonella* control programme crowded out resources for control of other foodborne zoonoses (e.g. campylobacter) were justifications for repeated critical appraisals. Accordingly, the *Salmonella* control programme has been evaluated in seven studies during the past 25 years, and some of the conclusions might be pertinent for a broader audience.

1. Engvall et al. (1993) concluded that the benefits were larger than the costs of the Swedish *Salmonella* control programme. This was done to show that the Swedish *Salmonella* control programme protected public health and was proportionate. This piece of information was important in the negotiations on Swedish membership to the EU during 1994.
2. Whether the Swedish *Salmonella* control programme crowded out available resources for control of other foodborne zoonoses was the aim of a public health perspective review in 1999 (NBHW, 1999). One noteworthy conclusion from this review was the recommendation to include the one-health perspective and thereby the public health profession when revising poultry *Salmonella* control programmes.
3. After the millennium, several *Salmonella* outbreaks were linked to contaminated feedstuffs (Österberg et al., 2006), mostly affecting pig farms. Wierup (2006) reviewed possible sources and production practices that could be linked to *Salmonella* contamination causing large outbreaks. He recommended that all feed should be heat treated and that the temperature at heat treatment should increase from 75 to 82°C. No indemnities should be paid for clean-up costs for farmers feeding their animals non-heat-treated feed. Furthermore, the broiler feed production should serve as an example to follow in terms of acceptable *Salmonella* risk.
4. In 2007 a road map for the future *Salmonella* control programme (SBA, 2007c) suggested several improvements. Most interesting from a poultry perspective was the introduction of a cost-sharing system for *Salmonella* clean-up costs in poultry flocks. In addition, the need to earmark government funds for improving prevention and surveillance of *Salmonella*, the need for a more efficient market for *Salmonella* clean-up services and the introduction of strict liability for feed producers were also

suggested. Thus, if *Salmonella* found in poultry flocks could be traced back to a feed producer, that producer would be liable for the total clean-up costs. The road map's overall aim was to enhance the incentives for stakeholders to prevent and contain *Salmonella* in animal production. The correlation between incentives (own insurance) and lower *Salmonella* risk for poultry meat was noted in contrast to pork and beef production. Perhaps the most important recommendation was that there should be regular reviews and evidence-based revisions of the *Salmonella* control programmes. Three-year intervals were suggested for the revisions.

5 A Swedish government enquiry (SOU, 2010) proposed a new division of tasks between the animal industry and the government. Of interest for poultry meat production was the emphasis on biosecurity, greater use of industry guidelines as in the food processing industry. Public funding should go to monitoring and surveillance, as well as for prevention of *Salmonella*. The enquiry proposed legislation in which *Salmonella* would be included and treated as any other foodborne zoonosis.
6 To evaluate the putative crowding out of resources for control of other foodborne zoonoses, Wahlstrom et al. (2014) studied a hypothetical relaxation of the *Salmonella* control and its impact on benefits and costs. It was concluded that a significant increase in number of *Salmonella* cases could be foreseen if the programme was relaxed. For example, a ten-fold increase in domestically acquired *Salmonella* cases was foreseen – from 800 to 8000 cases.
7 Moreover, Sundström et al. (2011) studied the economic impact of terminating the compulsory *Salmonella* control programme. While annual savings up to 10 million SEK (8.50 SEK = 1 USD) were foreseen, the increased putative costs to the society for salmonellosis ranged from 100 to 350 million SEK.

2.6 The 2010 national comprehensive *Salmonella* control programme

In 2010, all different control programmes in different parts of the food chain were amalgamated and a *Salmonella* control steering committee was created (SVA, 2011). In line with the recommendation of regular reviews, a national *Salmonella* control strategy was published, in which an explicit one-health perspective was applied (NBHW et al., 2013). The programme is from farm and feed producer to the slaughter house and meat processor, but with an emphasis on pre-harvest measures. We will present the current *Salmonella* programme and some of the thinking behind it subsequently.

3 The current *Salmonella* control system: preventing and eliminating feed contamination

3.1 *Salmonella* originating from feed – a driver for HACCP-based controls

From 1968 until 1984, *Salmonella* was detected in one or more slaughter groups on 20–40 poultry farms per year. After the introduction of the compulsory *Salmonella* control programme in 1984 in broiler production, the number of *Salmonella*-infected broiler farms decreased to 0–15 infected farms per year. In some years (e.g. 1989 and 2004) no flock was found infected with *Salmonella* (SVA, 2016a).

One recurrent problem in this period was that feed given to poultry were often suspected being contaminated with *Salmonella*. However, the samples taken from the feed seldom revealed *Salmonella*. Nevertheless, a few samples of 100 grams may not reveal much of the *Salmonella* status in a batch of hundreds of tonnes feed. This situation was unsatisfactory as farmers with broilers suffering from *Salmonella* could only claim compensation from feed producers if there were firm evidence that the feed were the source of *Salmonella*. The courts required to award compensation from the feed producer, finding the same *Salmonella* serovar and if possible type, in infected broilers and in the contaminated feed.

3.2 HACCP-based *Salmonella* control in feed

The introduction of a HACCP-based control of feed production in 1993 resulted in major improvements of the *Salmonella* control in poultry (Wierup and Häggblom, 2010). Perhaps most important was the compulsory heat treatment (75°C) of all poultry feed. Today, the feed producers are responsible for producing *Salmonella*-free feed, and liable for costs of *Salmonella* clean-up and poultry flocks. The SBA is the competent authority for feed control. The SBA (2011) has issued detailed rules that implement EU legislation on feed production, providing practical guidelines for the feed control in (SJVFS 2011: 40). In addition, it carries out on-the-spot inspections of feed producers to check their compliance. The HACCP includes both control of feed raw materials and process control when producing compound feeds.

Raw feed material

The focus is on control of the raw material used in the feed production and process control with an extensive testing for *Salmonella*. The feed raw material control sampling for *Salmonella* and risk classification is shown in Table 1.

The risk class protein feed materials of animal origin, that is, the risk class S1 is currently limited to fish meal because of EU legislation aimed at control prion diseases (e.g. mad cow disease). Thus currently the focus is on high- and low-risk vegetable feed raw materials.

Table 1 The Swedish *Salmonella* control of feed raw materials for animal feed production

	Risk class and examples of feed raw materials		
Number of samples in batches of	S1 animal raw material high risk (e.g. fish meal)	S2 vegetable high risk (e.g. soyabeans or canola meal, corn gluten extracts)	S3 vegetable low risk (e.g. rice)
<100 tonnes	40 samples	40 samples	20 samples
101–10 000 tonnes	80 samples	80 samples	40 samples
>100 000 tonnes	390 samples	390 samples	160 samples

Source: SBA, 2011.

The samples of 2.5 grams feed raw material are pooled 10 and 10 into 25 gram samples that are analysed to save on laboratory resources and reduce costs. In smaller batches of imported composite feed (less than 100 tonnes) then 80 samples shall be taken, while for larger batches (more than 100 tonnes) then while 160 samples shall be taken. These samples can be taken and analysed at the place of origin or after import but before use. All *Salmonella* shall be analysed according to MSRV (ISO 6579: 2002) or equivalent methods (SVA, 2016a). If *Salmonella* is found in feed raw material, the findings have to be notified to the SBA and also to the EU rapid alert system for food and feed (RASFF). Before being used to produce feed, the contaminated raw feed material batch has to be treated with organic acids (usually propionic acid) and be retested with negative results. In parallel, the *Salmonella* test results must be negative before feeding poultry imported composite feeds.

In 2015, the results of the control of raw feed materials indicated that *Salmonella* was detected in 18 samples out of 1187 samples (nearly 2%) from feed material of vegetable origin classified as S2 high risk (SVA, 2016a). This was from oil seeds such as rape seed (canola), soyabean and sunflower as well as palm kernels. The dominant serovar was *Salmonella* Senftenberg that was found in 10 samples. In raw feed materials of grain origin no *Salmonella* was found in 314 samples, that is, low-risk materials (S3). In raw feed material of animal origin *Salmonella* was detected in 1 out of 1157 samples (S1).

The process control of feed production

The Regulation (SJVFS 2011: 40) has detailed minimum rules for the process control of feed production intended for poultry (SBA, 2011). Weekly samples are taken from the specified spots before and after heat treatment (of at least 75°C) and analysed. The feed manufacturers have to pay for this control which is one point of controversy and thought by some conferring a competitive disadvantage for Swedish poultry production. One weekly environmental sample should be taken from (1) the intake pit or the bottom part of the elevator for feed raw materials, (2) dust from the aspiration system. That is the side before heat treatment. On the side after heat treatment one environmental weekly sample should be taken from (3) the top of the pellet cooler, (4) the dust in the room for pellet coolers and (5) top of the bin for final feed. In all 5 weekly samples are taken all year of operations.

Most feed mills produce compound feed for several animal species – poultry (broilers and layers) swine and cattle. In general, the type of risk mitigation action would depend on where the *Salmonella* was found, the specific outline of the production lines of the particular feed mill and for which species feed is produced (Binter et al., 2011). For poultry feed mills, the following applies: if *Salmonella* is found on the dirty side, that is, before heat treatment, the contaminated part of the production line is cleaned and disinfected. If *Salmonella* is found on the clean side, that is, after heat treatment, the whole feed mill is closed down for cleaning and disinfection. Before commencing feed production again environmental samples have to verify *Salmonella* absence. All contaminated feed has to be recalled and usually heat treated again if used as animal feed. The rules have been stricter for poultry feed production than for other food animal species, for example, five weekly environmental samples instead of two, respectively. Moreover, most feed mills have additional *Salmonella* sampling as part of their proprietary internal control systems.

If *Salmonella* is detected the isolates shall be serotyped by the National Veterinary Institute (SVA). Moreover, if *Salmonella* is found on the clean side after heat treatment the competent authority (SBA) shall be notified.

The surveillance results have been stable in recent years. The weekly surveillance of 11 large feed mills, representing 95% of the Swedish feed production, found 0.3% or 31 *Salmonella*-positive samples out of 9526 samples taken in 2014, and 0.2% or 17 out of 8960 samples, in 2014 and 2015, respectively (SVA, 2015 and 2016a).

Future challenges to *Salmonella* control in feed production

The animal feed production is a high-volume low-margin industry. This means that cost control or rather cost cutting is an ever-present question at feed mills. This might mean purchasing cheaper feed raw materials of that might present *Salmonella* risks. The costs of *Salmonella* control in feed are an ever-present topic and have been the driver for several reviews and studies, for example, Wierup (2006) and Wierup and Häggblom (2010).

Another issue is that larger farms now tend to use their own produced cereals as animal feed. By feeding their own grains, the farmers get added value of their cereal production and minimize their ecological footprint by fewer transports of feed. The farmers mix the feed themselves and only buy protein as well as mineral and vitamins to supplement their own grains. One question could then be how to assure *Salmonella* freedom. That should be easy for the protein supplements, but the question is how safe is the heat treatment of their own cereals and their process control. So far this seems be the case with other animal species.

4 The current *Salmonella* control system: *Salmonella* control in broilers

The control of *Salmonella* in poultry consists of three elements: (1) compulsory *Salmonella* surveillance, (2) compulsory eradication of *Salmonella* in the compartment and clean-up, and (3) a voluntary programme with focus on biosecurity.

The programme's achievement has been illustrated by the EU baseline studies of *Salmonella* prevalence in broilers and broiler carcasses. The EU average prevalence of *Salmonella*-positive broiler flocks was 24%, while it was 0% in Sweden and Finland (EFSA, 2007). Noteworthy was the large variation of *Salmonella* prevalence in broiler flocks amongst the EU member states ranging from 0 to 68.2% (EFSA, 2007). Three years later EFSA (2010) published a baseline study of *Salmonella* prevalence in broiler carcasses sampled at the abattoirs in the EU. The EU prevalence was 16% and ranging from 0 to 87% and the prevalence in Sweden was 0.3%. As the baseline studies were done the same way all over EU, that is, methods for sampling and microbiological analyses, these results were comparable within the EU.

4.1 Compulsory *Salmonella* surveillance programme

The Swedish Board of Agriculture (SBA, 2007a) has issued a Regulation for the compulsory surveillance programme (SJVFS 2007: 19) that gives the legal base for the actions to be taken. For broilers the breeding birds are imported as grandparents (GP). For other poultry meat-producing birds such as turkeys, geese and ducks, the breeding birds are imported

Table 2 Swedish *Salmonella* surveillance in poultry according to SVA (2016a)

Stages of control	Outline of testing
Breeders in rearing including grandparent (GP) and parent (P) birds	Two pairs of sock samples, that is pooled into one sample, to be taken at day 1, after 4 weeks and 2 weeks before moving
Breeders in production including grandparent (GP) and parent (P) birds	Five pairs of sock samples, that is pooled into two samples, to be taken at every second week of production
All birds (e.g. broilers, spent breeding birds, turkeys, geese, spent hens) intended for meat production	Two pairs of sock samples (or 2 × 75 grams of composited faecal samples) that are pooled into one sample, to be taken 2 weeks before intended slaughter

Table 3 Results of the *Salmonella* surveillance in poultry during 2015. This is in regard to poultry flocks tested (SVA, 2016a)

Type of production	No positives/flocks tested	Prevalence (%)
Grandparent broiler	0/20	0.0
Parent broiler	1/115	0.9
Meat production broiler	13/3390	0.38
Turkey parent	0/4	0
Turkey production	0/144	0
Geese	0/37	0
Ducks	1/23	4.4

as parents. All breeding flocks with more than 250 birds, and production flocks sending more than 500 birds per year to slaughter, are included. Table 2 outlines the current surveillance for *Salmonella* and some of the results are shown in Table 3.

The compulsory samples consist of sock samples from all parts of the house where broiler parent or grandparent birds are kept (SBA, 2007a). From parent flocks under rearing, two pairs of sock samples are taken, while for parent flocks in production five pairs of sock samples are taken (Table 2). The sock samples from rearing flocks and flocks in production are pooled into one and two, respectively. The samples should be analysed according to ISO 6579:2002 (MSRV) method. The costs for the sampling transport to the laboratory and microbiological analysis shall be paid by the producer. The laboratory doing the analyses shall have in place an officially recognized quality assurance and participate in *Salmonella* ring test trials organized by the national reference laboratory for *Salmonella* (The National Veterinary Institute – SVA). The sampling can be done by the producer, but at least once a year the samples shall be taken by an official veterinarian.

For poultry flocks going to slaughter the *Salmonella* samples shall be taken by an officially appointed veterinarian. In general sock samples are preferred as they are easier to take, are more representative than selecting droppings and appear to be more sensitive – that is, detect a larger proportion of truly positive flocks. A negative test result is required before the flock is permitted going to slaughter.

Perhaps noteworthy is that the *Salmonella* prevalence during the past five years ranged between 0.03 and 0.38% amongst meat production broilers, indicating a stable situation (SVA, 2016, 2015, 2014, 2013 and 2012). Hence, the annual numbers of *Salmonella*-positive flocks are few, thus enabling more stringent measures to clean-up those found positive.

Anyone who has sampled 30 droppings in a poultry house usually starts thinking about more efficient sampling methods after sampling a number of poultry houses. Questions were asked about representative the sampling of faecal droppings actually were – from a statistical sampling viewpoint, it was very close to convenience sampling. Hence, a major improvement was the introduction sock or boot swab sampling (Skov et al., 1999; Carrique-Mas and Davies, 2008) to take composite samples from the flocks before slaughter instead of individual faecal samples.

4.2 Measures implemented when *Salmonella* is found

If *Salmonella* (regardless of serovar) is found, the laboratory has to notify the Swedish Board of Agriculture and the County Veterinary Officer (SBA, 2004) as laid down in the Regulation for *Salmonella* eradication (SJVFS 2004: 2). The laboratory has to send the suspected sample(s) to the National Veterinary Institute for confirmation, resistance testing and serotyping. The birds are not allowed to leave the premises. Once the *Salmonella* infection is confirmed, the flock is destroyed (stamping out) regardless of the *Salmonella* serotype. The site premises and all possible contaminated areas have to be cleaned up and disinfected. Before restocking the site, the environmental samples from the site have to be negative for *Salmonella*. In addition, an investigation is done on putative sources for the *Salmonella* infection, for example, feed, breeding flocks or lapses in biosecurity.

Clean-up costs – a driver for the evolution of the programme

The costs for *Salmonella* clean-up could be high and has been since 1984 carried by the poultry producer in contrast to other food animal productions where the government subsidized 50–70% of the clean-up costs (SBA, 2007c). This has been a long-held grievance of Swedish poultry meat farmers that the government favoured pork and beef production (SPMA, 2016). Most poultry farmers have preferred insurance solutions, meaning that 90% of the clean-up costs are covered by their insurance. Both insurance and cost sharing were viable alternatives while the public resources should not be used to pay for clean-up costs. Cost sharing was discussed in a review by OECD (2012) and appears to be at least as efficient as private insurance solutions. However, this discussion has triggered searches for alternatives to the current Swedish policy of stamping out of *Salmonella*-infected flocks.

Alternatives to stamping out

Antibiotic treatment, vaccination and/or competitive exclusion are possible alternative control strategies to stamping out. The use of antibiotic treatments of *Salmonella*-infected birds as a clean-up appears to be ruled out for the foreseeable future by following arguments – the efficacy as elimination method is uncertain, and there are risks for development, selection and spread of resistant bacteria (EFSA, 2004a). Moreover, use of antibiotics for *Salmonella* clean-up is in conflict with the Swedish policy of restricting the use of antibiotics in animal production (NBHW and SBA, 2015).

Salmonella vaccination is prohibited in Sweden. However, the SBA can by derogation give permission to vaccinate against *Salmonella*. However, this approach has not been used in poultry to the author's knowledge for many years. EFSA (2004b) reviewed the use of vaccines as a control option in poultry production. In situations where *Salmonella* is common and with control programmes targeting *Salmonella* Typhimurium and Enteriditis then vaccination could be helpful. For example, the positive Belgian experiences from *Salmonella* vaccination of poultry seem to support this notion (Collard et al., 2008). However, in low prevalence situations or while targeting all *Salmonella* serovars, vaccination might not suffice as the only control option.

Competitive exclusion (CE) was used from 1981 for *Salmonella* control in broilers (Wierup et al., 1992). The rule was that that if finding *Salmonella* in a flock, then after cleaning and disinfection CE was used for the next two consecutive flocks to be raised on the same site. Wierup et al. (1988) concluded that CE has appeared to prevent *Salmonella* infection in broilers under Swedish field conditions without any adverse effects. Hirn et al. (1992) reported similar results of CE in Finland. After Sweden joined the EU in 1995, CE was used to a much lesser extent. The increasing prevalence of ESBL in Swedish poultry meat (Nilsson et al., 2014) is thought to originate from the breeding pyramid. Nilsson et al. (2014) reported that 6 out 8 GP flocks were positive for ESBL. Hence, there is a need for practical control options. In this regard Nutio et al. (2013) suggested that CE might be helpful for controlling ESBL in broiler flocks. Furthermore, in a meta-analysis Kerr et al. (2013) concluded that CE was a helpful control option for controlling *Salmonella* in broilers. Hence if the control policy changes to control *Salmonella* and/or ESBL non-destructively, the CE might be the best alternative. Furthermore, CE might be a practical alternative to ensure that the breeding pyramid remains free of ESBL and *Salmonella*.

4.3 The voluntary biosecurity programme for poultry flocks

The legal basis of voluntary programme is laid down in regulation SJVFS 2007:78 issued by the (SBA, 2007b). The poultry farmer shall only purchase live birds from breeding flocks with *Salmonella*-free status, and only use batch production (all in–all out) with depopulation of the whole house, before new birds arrive. Although no *Salmonella* has been found in the previous flock, the house always must be cleaned and disinfected and only *Salmonella*-free litter must be used before restocking. Moreover, the biosecurity programme includes (1) measures to prevent domestic or wild animals entering the premises, (2) compulsory provision of protective clothing for all visiting the clean parts of the farm where broilers are kept, (3) manure or slurry should not be spread nearby the production sites and (4) an officially appointed veterinary officer supervises the programme and its implementation for each farm. The official veterinarian shall visit the farm at least annually, and during the visit take official samples, inspect the biosecurity and the compliance to the programme, and write a report to the Swedish Poultry Meat Association.

The voluntary *Salmonella* control programme in poultry meat flocks (broilers) is operated by the Swedish Poultry Meat Association under the supervision of the SBA. This programme has been operating for more than 40 years and most likely been a key to the successful *Salmonella* control in poultry. Although the programme is voluntary, more than 99% of the broilers slaughtered originate from participating farms. An incentive to participate has been that participating poultry farmers receive higher compensation in the case of a *Salmonella* outbreak on their farms from their insurance.

5 The current *Salmonella* control system: *Salmonella* monitoring and control during and after harvest

5.1 *Salmonella* monitoring of abattoirs and cutting plants

After Sweden joined EU in 1995 a monitoring programme for *Salmonella* at slaughter and cutting plants was initiated. The aim was verifying the absence of *Salmonella* at poultry slaughter (Table 4). The National Food Administration is the competent authority for the harvest and post-harvest part of the poultry food chain.

Annually around 3000–6000 neck skin samples and around 900 meat scrap samples are taken annually (SVA, 2016a). The prevalence of *Salmonella* has been less than 0.1% during 1996–2016 in these samples. For example, in 2014 and 2015, no *Salmonella* was detected in samples of neck skins and in meat scrapings. The samples were analysed using the NMKL (nr 71: 1991) method. The consistent low prevalence for nearly 20 years indicate the successful control of *Salmonella* in poultry meat produced in Sweden and a low *Salmonella* burden entering the food chain. The meat processing industry has to pay for the *Salmonella* sampling from neck skins and meat scraps according to Table 4. These added costs has been controversial and seen as an additional burden on the Swedish meat industry (SBA, 2007c).

Hence, an important driver for the repeated reviews of the control programme has been to assess whether the *Salmonella* control is cost-effective and do not place an undue burden on the Swedish poultry and other food animal industries. On the other hand, any food business operator importing fresh poultry meat to Sweden has to pay for the testing of *Salmonella* – 60 samples to be taken, analysed without *Salmonella* detection for each batch that are allowed imported.

5.2 *Salmonella* control in food post-harvest

The focus of the Swedish *Salmonella* control has been pre-harvest and harvest. Thus, *Salmonella* has been less of a priority in the post-harvest part of the food chain. The local municipalities are the competent authority for food control post-harvest – wholesale, distribution, retail, catering and consumption.

The basic tenet is that the food business operator is responsible for the safety of the food, including absence of *Salmonella*. The internal audit and controls are keys to safe food, while local municipalities verify that such checks are in place. The food business shall have regular monitoring that their programme ensures this absence of *Salmonella*.

Table 4 Number of *Salmonella* samples to be taken and costs of sampling at cutting plants in the Swedish *Salmonella* control

Meat production per week in tonnes	Number of samples to be taken	Sampling occasions per year	Hours per year (1 hour costs 900 SEK)	Costs per year (8.5 SEK = 1 USD)
>100	260	52	156	140400 SEK
20–100	52	12	36	32400 SEK
5–19	12	6	12	10800 SEK
<5	2	2	4	3600 SEK

Any foodstuff including poultry meat found contaminated with *Salmonella* is deemed unfit for human consumption regardless of serovar. What is absence of *Salmonella* in poultry meat is a frequently discussed question at a practical level. How many samples should be taken to ensure freedom and to enable food business operators verify they have shown due diligence in their food production? The microbiological criteria as laid down in the Regulation 2073/2005 guides the control of poultry meat and poultry meat products (European Commission, 2005). In general, 5 samples of 25 grams each should be analysed with EN/ISO 6579 method without finding *Salmonella*, to deem the foodstuff acceptable. If a food business operator finds *Salmonella* during their routine monitoring, it is customary to recall the foodstuff from the market. If the foodstuff originates from imported poultry meat, then it might be possible to alert the EU through the RASFF system.

6 Summary and critical comments

The starting point was the huge Alvesta outbreak in 1953, which was unprecedented in Sweden, with more than 9000 cases reported. Approximately 4000 persons required hospital care and 90 persons died. The outbreak lasted from June to December 1953, affecting major regions of Sweden. Consequently, several measures were enacted. The Public Health Institute of Sweden created a department of epidemiology to investigate outbreaks; doctors were recommended when seeing patients with diarrhoea to take faecal samples for bacteriological investigations. In 1961, official control of all *Salmonella* serovars of public health importance on the animal side commenced. Moreover, *Salmonella* infection in animals became a notifiable disease.

One unique feature of the Swedish *Salmonella* control is the aim to deliver *Salmonella*-free food to consumers. Therefore, all broilers sent for slaughter should be without *Salmonella*. This was achieved by the following strategies: (1) preventing contamination of the breeding pyramid (grandparent and parent flocks including hatcheries); (2) preventing and eliminating *Salmonella* contamination in feed given to broiler flocks; (3) surveillance of the production chain at critical control points to detect *Salmonella* contamination; (4) biosecurity on the production site; and (5) whenever *Salmonella* is found, taking action to remove it from the food chain.

The *Salmonella* control of poultry has been successful, with rare findings of *Salmonella* in meat and almost no human cases linked to Swedish poultry meat. The baseline studies organized by EFSA (2007 and 2010) found *Salmonella* in 0% of the broiler flocks and 0.3% of the broiler carcasses.

Such a large programme as the Swedish *Salmonella* control programme in poultry and other food animals has been a reliable source for discussions over the years. The costs and the distribution of costs and benefits have been contentious topics. Examples of critical comments were as follows: (1) the programme missed vegetable sources of salmonellosis – does not help to have poultry meat without *Salmonella* if the salad was contaminated, (2) the programme had a bias to pre-harvest controls which meant that post-harvest parts of the food chain were ill prepared for handling *Salmonella* contaminated poultry meat and (3) the programme was too expensive and crowded out resources for control of other public health hazards in poultry meat such as campylobacter and ESBL.

The high costs of the *Salmonella* control were borne by the poultry industry of Sweden. Wierup et al. (1995) suggested that the overall control costs for both feed, breeding pyramid amounted to 0.90 SEK (0.11 USD) per broiler produced. Although Wahlström et al. (2014) found the control resulted in larger benefits than costs, their distribution was a sore point. The benefits accrued to the consumers of poultry meat and the health care system – in terms of lower costs while the costs where paid by the industry. In recent years the focus has been to make the control more cost-effective without harming animal or public health and food safety (SBA, 2007c).

7 Future trends in research

The future trends of research in *Salmonella* control are complex and dynamic. Lots of papers are published on *Salmonella* (around 10 per working day) and the readers should carefully update themselves on the latest evidence. Perhaps telling is on 26 August 2016 when the second draft of this chapter was finished, there were 81863 papers on PUBMED using the search term *Salmonella*.

One challenging research question has been – Can we differentiate between *Salmonella* serovars found in poultry or poultry meat in terms of food safety risk? If yes, on which scientific evidence and criteria should one differentiate? Possible criteria could be prevalence in poultry and incidence and disease burden in persons, the invasiveness of *Salmonella* infections. In the EU the criterion was that the Typhimurium and Enteriditis was the most frequent serovars associated with human illness. For example, the EU has set a broiler flock prevalence target of 1% of Typhimurium (*including strains with the antigenic formula 1,4,[5],12:i:-*) and Enteriditis (European Commission, 2012). Currently, the Swedish policy has been that all *Salmonella* serovars if found in poultry present a public health risk that should be mitigated. It might be that current Swedish approach is more practical in low prevalence situations. You do not have to wait for typing results to start clean-up procedures.

Another trend is more rapid diagnostics for *Salmonella*. Is it possible to detect and type *Salmonella* in hours instead of days? What could be the role of whole genome sequencing of isolates? Yoshida et al. (2016) noted that most public health laboratories are switching to whole genome sequencing, but that the methods are still three times more expensive than traditional typing. The great advantage is being able in outbreaks to rapidly assess the relationship between cases and *Salmonella* findings earlier in the food chain.

A challenge for the poultry industry is to be sustainable and address concerns of food safety and food security simultaneously. From a food safety perspective the future poultry meat control programmes should cover *Salmonella* but also campylobacter and ESBL. These pathogens were identified as the main foodborne risks in the poultry food chain (EFSA, 2012). The challenges are to design a control programme addressing all three pathogens concurrently and determine where along the food chain they can be controlled cost-efficiently.

A further challenge is to integrate biological evidence with insights from the behavioural sciences. Would it be more efficient to use soft law, that is, to nudge farmers or consumer into lower risk behaviours or should one focus on hard law issuing binding regulations?

8 Where to look for further information

The readers wishing to inform themselves on the Swedish *Salmonella* control would be well advised to look at the homepages of the Swedish Poultry Meat Association (SMPA) and the National Veterinary Institute (SVA). The industry perspective is important and too often ignored. A good place to start is to look at the Swedish Poultry Meat Association's homepage on *Salmonella* control (http://www.svenskfagel.se/?p=2216). For the regulatory and risk assessment perspective, the National Veterinary Institute (SVA) has a *Salmonella* homepage translated to English that is updated frequently and is informative on the latest developments on *Salmonella* control (http://www.sva.se/en/animal-health/zoonoses/*Salmonella*-as-a-zoonosis/general-facts-about-*Salmonella*).

One conference that is very good for catching up on the latest developments on *Salmonella* control is the I3S conference in France (http://www.i3s2016.com/content.php?PAGE=27). The conference in June 2016 includes sessions on following topics: (1) host–pathogen interactions, (2) ecology and epidemiology along the food chain, (3) risk assessment and control, (4) antimicrobial resistance, (5) whole genome sequencing, (6) detection and characterization and (7) epidemiology and public health, all on *Salmonella*.

In the EU a lot of work has been done on *Salmonella* control and the number of reported human cases has decreased (nearly halved) while EU has expanded (doubled). To catch up with policy and risk management developments on *Salmonella* control the European Commission's DG SANTE homepage is very helpful (http://ec.europa.eu/food/safety/biosafety/food_borne_diseases/*Salmonella*/index_en.htm). Perhaps the key message is the control programmes for foodborne *Salmonella* in particular in poultry have been a major success story in the EU. Before 2004, more than 200 000 human cases were reported annually in the 15 member states. Note the contrast to 2014 where the reported number of *Salmonella* cases has dropped to less than 90 000 cases. In the meantime EU has expanded to 28 member states.

The European Food Safety Authority (EFSA) has issued several opinions on *Salmonella* control. These have delivered scientific evidence upon which EU control policy for *Salmonella* has been developed. Of particular importance was the opinion on public health risks originating from poultry meat (EFSA, 2012) and the baseline studies done on *Salmonella* prevalence in broiler flocks (EFSA, 2007) and on broiler carcasses (EFSA, 2010). EFSA has a *Salmonella* homepage which is very helpful for those wishing to update themselves on the ongoing and completed work concerning risk assessment and control of *Salmonella* (http://www.efsa.europa.eu/en/topics/topic/*Salmonella*).

The public health situation in the EU is reported annually through the reports on trends and sources of zoonoses (ECDC and EFSA, 2015). This reporting has been ongoing for 20 years and gives a good and indeed revealing picture of the state of foodborne diseases, including findings of *Salmonella* in the poultry food chain. In 2014 the report included the results of the zoonoses monitoring activities carried out in 2014 in 32 European countries: 28 member states and four non-member states.

9 References

Bengtsson, E., Hedlund, P., Nisell, A. and Nordenstam, H., 1955. An epidemic due to *Salmonella* typhimurium (Breslau) occurring in Sweden in 1953; with special reference to clinical complications, bacteriology, serology, antibiotic treatment, and morbid anatomy. *Acta. Med. Scand.* 16 December 1955, 153(1): 1–20.

Binter, C., Straver, J. M., Häggblom, P., Bruggeman, G., Lindqvist, P. A., Zentek, J. and Andersson, M. G., 2011. Transmission and control of *Salmonella* in the pig feed chain: a conceptual model. *Int. J. Food Microbiol.* 145(Suppl 1): S7–17. doi: 10.1016/j.ijfoodmicro.2010.09.001.

Carrique-Mas, J. J. and Davies, R. H., 2008. Sampling and bacteriological detection of *Salmonella* in poultry and poultry premises: a review. *Rev. Sci. Tech.* 27(3): 665–77. Review.

Collard, J. M., Bertrand, S., Dierick, K., Godard, C., Wildemauwe, C., Vermeersch, K., Duculot, J., Van Immerseel, F., Pasmans, F., Imberechts, H. and Quinet, C., 2008. Drastic decrease of *Salmonella* Enteritidis isolated from humans in Belgium in 2005, shift in phage types and influence on foodborne outbreaks. *Epidemiol Infect.* 136(6): 771–81.

European Food Safety Authority (EFSA), 2004a. Opinion of the Scientific Panel on Biological Hazards on a request from the Commission related to the use of antimicrobials for the control of *Salmonella* in poultry. *The EFSA Journal* 115: 1–76. doi: 10.2903/j.efsa.2004.115.

European Food Safety Authority (EFSA), 2004b. Opinion of the Scientific Panel on Biological Hazards on a request from the Commission related to the use of vaccines for the control of *Salmonella* in poultry. *EFSA J.* 114: 1–74. doi: 10.2903/j.efsa.2004.114.

European Food Safety Authority (EFSA), 2007. Report of the Task Force on Zoonoses Data Collection on the Analysis of the baseline survey on the prevalence of *Salmonella* in broiler flocks of *Gallus gallus*, Part A. *EFSA J.* 98: 85. doi: 10.2903/j.efsa.2007.98r.

European Food Safety Authority (EFSA), 2010. Analysis of the baseline survey on the prevalence of Campylobacter in broiler batches and of Campylobacter and *Salmonella* on broiler carcasses in the EU, 2008, Part A: Campylobacter and *Salmonella* prevalence estimates. *EFSA J.* 8(3):1503 [100 p.] doi:10.2903/j.efsa.2010.1503.

European Food Safety Authority (EFSA), 2012. EFSA Panels on Biological Hazards (BIOHAZ), on Contaminants in the Food Chain (CONTAM), and on Animal Health and Welfare (AHAW); Scientific Opinion on the public health hazards to be covered by inspection of meat (poultry). *EFSA J.* 10(6): 2741 [179 p.] doi:10.2903/j.efsa.2012.2741.

EFSA (European Food Safety Authority) and ECDC (European Centre for Disease Prevention and Control), 2015. The European Union summary report on trends and sources of zoonoses, zoonotic agents and food-borne outbreaks in 2014. *EFSA J.* 13(12): 4329 [191 p.] doi:10.2903/j.efsa.2015.4329.

Ekstam, M., 1953. Paratyfusepidemien. Medlemsblad för Sveriges Veterinär Förbund; årg. 5, nr 20, sida 271–80 (in Swedish).

Engvall, A., Andersson, Y. and Cerenius, F., 1993. The economics of the Swedish *Salmonella* control. A cost/benefit analysis. In *Proceedings International Course on* Salmonella *Control in Animal Production*. Arranged by the National Veterinary Institute of Sweden and the World Health Organization, pp. 221–37.

European Commission, 2005. Commission Regulation (EU) No2073/2005 of 15 November 2005. On microbiological criteria for foodsuffs. OJ L 22.12.2005, pp. 1–25. Available at http://eur-lex.europa.eu/legal-content/EN/TXT/?qid=1453305238121&uri=CELEX:02005R2073-20140601 (Accessed 25 August 2016).

European Commission, 2012. Commission Regulation (EU) No 200/2012 of 8 March 2012 concerning a Union target for the reduction of *Salmonella* enteritidis and *Salmonella* typhimurium in flocks of broilers, as provided for in Regulation (EC) No 2160/2003 of the European Parliament and of the Council Text with EEA relevance. OJ L 71, 9.3.2012, pp. 31–6. Available at http://data.europa.eu/eli/reg/2012/200/oj (Accessed 6 June 2016).

Hirn, J., Nurmi, E., Johansson, T. and Nuotio, L., 1992. Long-term experience with competitive exclusion and *Salmonella*s in Finland. *Int. J. Food Microbiol.* 15(3–4): 281–5. Review.

Kerr, A. K., Farrar, A. M., Waddell, L. A., Wilkins, W., Wilhelm, B. J., Bucher, O., Wills, R. W., Bailey, R. H., Varga, C., McEwen, S. A. and Raji , A., 2013. A systematic review-meta-analysis and meta-regression on the effect of selected competitive exclusion products on *Salmonella* spp. prevalence and concentration in broiler chickens. *Prev. Vet. Med.* 111(1–2): 112–25. doi: 10.1016/j.prevetmed.2013.04.005.

KSLA (Royal Swedish Forest and Agriculture Academy), 2003. Chapter 13, Agriculture policy – a classical drama. (Jordbrukspolitiken – ett klassisk drama). Lantbrukets historia. KSLA, Stockholm, Sweden, pp. 227–57. Available at http://www.ksla.se/anh/files/2012/06/13.-Jordbrukspolitiken-ett-klassiskt-drama-sid-227-257.pdf (Accessed 1 June 2016, in Swedish).

NBHW (National Board of Health and Welfare – Socialstyrelsen), 1999. *Salmonella* strategy (*Salmonella* – Strategidokument.) Socialstyrelsen, Smittskyddsinstitutet, Smittskyddsläkarföreningen. Socialstyrelsen, Stockholm, Sweden, p. 82 (in Swedish).

NBHW (National Board of Health and Welfare), NFA (National Food Agency), National Veterinary Institute (SVA) SBA (Swedish Board of Agriculture), SIIDS (Swedish Institute of Infectious Disease Control), 2013. *Salmonella* – a National Strategy (*Salmonella* ett nationelt strategidokument). Socialstyrelsen, Stockholm, Sweden, p. 54. Available at https://www.folkhalsomyndigheten.se/publicerat-material/publikationer/Salmonella--ett-nationellt-strategidokument/ (Accessed 3 June 2016, in Swedish).

NBHW (National Board of Health and Welfare) and SBA (Swedish Board of Agriculture), 2015. Action plan against antibiotic resistance and health care related infections. (Handlingsplan mot antibiotikaresistens och vårdrelaterade infektioner). Socialstyrelsen, Stockholm, Sweden, p. 67. Available at https://www.folkhalsomyndigheten.se/publicerat-material/publikationer/Handlingsplan-mot-antibiotikaresistens-och-vardrelaterade-infektioner--Underlag-for-myndigheternas-fortsatta-arbete/ (Accessed 2 June 2016, in Swedish).

Nilsson, O., Börjesson, S., Landén, A. and Bengtsson, B., 2014. Vertical transmission of Escherichia coli carrying plasmid-mediated AmpC (pAmpC) through the broiler production pyramid. *J. Antimicrob. Chemother.* 69(6): 1497–500. doi:10.1093/jac/dku030.

Nuotio, L., Schneitz, C. and Nilsson, O., 2013. Effect of competitive exclusion in reducing the occurrence of Escherichia coli producing extended-spectrum β-lactamases in the ceca of broiler chicks. *Poult. Sci.* 92(1): 250–4. doi: 10.3382/ps.2012-02575.

OECD, 2012. *Livestock Diseases: Prevention, Control and Compensation Schemes*, OECD Publishing, Paris. doi: http://dx.doi.org/10.1787/9789264178762-en

Olin, G., 1956. *Salmonella* epidemics of 1953 in Sweden and the problems raised by them. *Nord. Med.*, 26 April 1956, 55(17): 577–85 (In Swedish).

Osterberg, J., Vågsholm, I., Boqvist, S. and Lewerin, S. S., 2006. Feed-borne outbreak of *Salmonella* cubana in Swedish pig farms: risk factors and factors affecting the restriction period in infected farms. *Acta. Vet. Scand.* 47: 13–21.

SBA (Swedish Board of Agriculture), 2004. SBA regulation on *Salmonella* eradication and clean-up in live animals (Statens jordbruksverks föreskrifter om bekämpande av *Salmonella* hos djur). SJVFS 2004:2. Available at http://www.jordbruksverket.se/download/18.26424bf71212ecc74b080001416/1370040585494/2004-002.pdf (Accessed 2 June 2016, in Swedish).

SBA (Swedish Board of Agriculture), 2007a. SBA regulation of compulsory *Salmonella* control in poultry (Statens jordbruksverks föreskrifter om obligatorisk *Salmonella*kontroll av fjäderfän). SJVFS 2007:19. Available at http://www.jordbruksverket.se/download/18.26424bf71212ecc74b080001110/1370040518386/2007-019.pdf (Accessed 2 June 2016, in Swedish).

SBA (Swedish Board of Agriculture), 2007b. SBA regulation on voluntary and preventive *Salmonella* control in poultry (Statens jordbruksverks föreskrifter om frivillig och förebyggande hälsokontroll avseende *Salmonella* hos fjäderfä). SJVFS 2007:78. Available at http://www.jordbruksverket.se/download/18.26424bf71212ecc74b080001168/1370040518436/2007-078.pdf (Accessed 2 June 2016, in Swedish).

SBA (Swedish Board of Agriculture), 2007c. Report on the road map for the Swedish *Salmonella* control program. SJV rapport 2007: 10. Jordbruksverket, Jönköping, Sweden. Available at http://www2.jordbruksverket.se/webdav/files/SJV/trycksaker/Pdf_rapporter/ra07_10.pdf (Accessed 2 April 2016, in Swedish with English summary).

SBA (Swedish Board of Agriculture), 2011. SBA regulation on animal feedingstuffs and voluntary guidelines (Statens jordbruksverks föreskrifter och allmänna råd om foder). SJVFS 2011:40. Available at http://www.jordbruksverket.se/download/18.6f9b86741329df6fab480007681/1370040540758/2011-040.pdf (Accessed 26 August 2016, in Swedish).

SBA (Swedish Board of Agriculture), 2015. Official statistics for domestic poultry meat production at slaughter 2014 (Slakt av fjäderfä vid slakteri, 2014). Available at http://statistik.sjv.se/PXWeb/

pxweb/sv/Jordbruksverkets%20statistikdatabas/Jordbruksverkets%20statistikdatabas__Animalieproduktion__Slakt/JO0604A6.px/table/tableViewLayout2/?rxid=5adf4929-f548-4f27-9bc9-78e127837625 (Accessed 19 May 2016, in Swedish).

SBA (Swedish Board of Agriculture), 2016. Trade balance for poultry meat (Marknaden för matfågel). Jordbruksverket, Jönköping Sweden. Available at http://www.jordbruksverket.se/amnesomraden/handelmarknad/kottmjolkochagg/handelmedkottmjolkochagg/marknadenformatfagel.4.3a3862f81373bf24eab80001808.html (Accessed 19 May 2016, in Swedish).

Silverstolpe, L. and Wranne, N., 1955. *Salmonella* outbreaks in Sweden in the county of Ostergötland in 1953. *Nord. Hyg. Tidskr.* 36(11–12): 213–29 (in Swedish).

Skov, M. N., Carstensen, B., Tornøe, N. and Madsen, M., 1999. Evaluation of sampling methods for the detection of *Salmonella* in broiler flocks. *J. Appl. Microbiol.* 86(4): 695–700.

SMPA, 2016. Swedish Poultry Meat Association (Svensk Fågel) homepage on *Salmonella* control in poultry. Available at http://www.svenskfagel.se/sida/om-svensk-fagel/in-english (Accessed 7 June 2016).

SOU (Statens offentliga utredningar), 2010. Folkhälsa – Djurhälsa: Ny ansvarsfördelning mellan stat och näring (SOU 2010:106). Available at http://www.regeringen.se/rattsdokument/statens-offentliga-utredningar/2011/01/sou-2010106/ (Accessed10 June 2016, in Swedish).

Sundström, K., Ivarsson, S., Sternberg Lewerin, S. and Wahlstrom, H., 2011. Economic impact of termination of the compulsory Swedish *Salmonella* control (Ekonomiska effekter av att avveckla det obligatoriska svenska *Salmonella*kontrollprogrammet). Report to the Swedish *Salmonella* control steering committee (*Salmonella* kontroll nämnden). Agrifood Economics, Lund, Sweden. Summary accessible at http://www.agrifood.se/publication.aspx?fKeyID=646 (Accessed 20 May 2016, in Swedish).

SVA (National Veterinary Institute), 2011. SVA – 100 years of knowledge 1911-2011 (SVA – 100 år av kunskap 1911-2011). SVA, Uppsala, Sweden, p. 81. Downloadable at http://www.sva.se/globalassets/redesign2011/pdf/om_sva/publikationer/sva_100_webb.pdf (Accessed 19 May 2016).

SVA, 2012. Surveillance of infectious diseases in animals and humans in Sweden 2011, National Veterinary Institute (SVA), Uppsala, Sweden. SVA:s rapportserie 25 ISSN 1654-7098. 120 pp.

SVA, 2013. Surveillance of infectious diseases in animals and humans in Sweden 2012, National Veterinary Institute (SVA), Uppsala, Sweden. SVA:s rapportserie 26 ISSN 1654-7098.124 pp

SVA, 2014. Surveillance of infectious diseases in animals and humans in Sweden 2013. National Veterinary Institute (SVA), Uppsala, Sweden. SVA:s rapportserie 28 ISBN 1654-7098. 136 pp.

SVA, 2015. Surveillance of infectious diseases in animals and humans in Sweden 2014. National Veterinary Institute (SVA), Uppsala, Sweden. SVA:s rapportserie 31 ISSN 1654-7098. 128 pp.

SVA, 2016a. Surveillance of infectious diseases in animals and humans in Sweden 2015. National Veterinary Institute (SVA), Uppsala, Sweden. SVA:s rapportserie 34 ISSN 1654-7098. 131 pp.

SVA, 2016b. *Salmonella* as a zoonosis. Available at http://www.sva.se/en/animal-health/zoonoses/*Salmonella*-as-a-zoonosis (Accessed 10 June 2016).

Wahlstrom, H., Sternberg Lewerin, S., Sundstrom, K. and Ivarsson, S., 2014. Estimation of the Expected Change in Domestic Human *Salmonella* Cases in Sweden in 2010, Given a Hypothetical Relaxation of the Current *Salmonella* Control Programme. *PLoS ONE* 9(3): e89833. doi:10.1371/journal.pone.0089833.

Wierup, M., 2006. *Salmonella* i foder – en utredning på uppdrag av Jordbruksverket om orsaker och risker samt förslag till åtgärder. Jordbruksverket Jönköping, Sweden, p. 132. ISBN 91 88 264-32-7 (in Swedish).

Wierup, M., Engström, B., Engvall, A. and Wahlström, H., 1995. Control of *Salmonella* enteritidis in Sweden. *Int. J. Food Microbiol.*, May 1995, 25(3): 219–26. Review.

Wierup, M. and Häggblom, P., 2010. An assessment of soybeans and other vegetable proteins as source of *Salmonella* contamination in pig production. *Acta. Vet. Scand.* 17 February 2010, 52: 15. doi: 10.1186/1751-0147-52-15.

Wierup, M., Wahlström, H. and Engström, B., 1992. Experience of a 10-year use of competitive exclusion treatment as part of the *Salmonella* control programme in Sweden. *Int. J. Food Microbiol.*, March to April 1992,15(3–4): 287–91.

Wierup, M., Wold-Troell, M., Nurmi, E. and Häkkinen, M., 1988. Epidemiological evaluation of the *Salmonella*-controlling effect of a nationwide use of a competitive exclusion culture in poultry. *Poult. Sci.* 67(7): 1026–33.

Yoshida, C., Gurnik, S., Ahmad, A., Blimkie, T., Murphy, S. A., Kropinski, A. M. and Nash, J. H., 2016. Evaluation of molecular methods for the identification of *Salmonella* serovars. *J. Clin. Microbiol.*, 54(8): 1992–8. doi: 10.1128/JCM.00262-16. Epub 18 May 2016.

Chapter 10

Food safety control on poultry farms: effective control of *Campylobacter*

Xiang Liu, University of Tennessee, USA; Irene Hanning, Lincoln International Academy, Nicaragua; Sandra Diaz-Sanchez, SaBio IREC, Spain; and Jun Lin, University of Tennessee, USA

1 Introduction

The zoonotic pathogen, *Campylobacter*, is the leading cause of foodborne illness worldwide. *Campylobacter* colonizes the intestinal tract of chickens, which is the primary host for the bacterium. In chickens, this pathogen is considered commensal and does not cause illness or have adverse impacts on production performance. The farm is the first point in the food production chain; for this reason, control in the farm could reduce the risk of campylobacteriosis. A number of methods aim at controlling *Campylobacter* in the farm. To date, no single method is truly effective. Using multiple methods along with a good biosecurity programme is currently the most effective control method. However, many other methods are being designed and optimized to further reduce the risk of campylobacteriosis to the public. The objective of this chapter is to summarize and discuss the literature available regarding the current methods that are available or being developed to control *Campylobacter spp.* in the farm.

http://dx.doi.org/10.19103/AS.2016.0010.13

2 *Campylobacter*: general features and epidemiology

Campylobacter, a genus of Gram-negative microaerophilic bacteria, was first observed in the colons of infants who died of 'cholera infantum' in 1886 (Samie et al., 2007) and then successfully isolated from human faeces in 1972 (Dekeyser et al., 1972). Ever since the discovery, *Campylobacter* has been recognized as an important human enteric pathogen worldwide. According to the Foodborne Illness Risk Ranking Model (FIRRM) generated by the Emerging Pathogens Institute, *Campylobacter* remains the most common foodborne pathogen causing the greatest burden to the public health in the United States (Batz et al., 2011, 2014). Among a large and diverse group of species, *Campylobacter jejuni* causes most of the cases, while *Campylobacter coli* causes 1–25% of the *Campylobacter*-related diseases (Kaakoush et al., 2015; Man, 2011; Sahin et al., 2015). The susceptible populations to *Campylobacter* were reported to be concentrated in the United States and European countries, likely due to infrequent exposure to this pathogen (Friedman, 2000). The seasonality of campylobacteriosis has also been observed with a peak during the summer months, which might be related to the increase in flies and other vectors (Gölz et al., 2014; Nichols, 2005).

The majority of human infections are caused by the consumption of undercooked, contaminated animal products, such as meat and milk, especially chicken meat (Gölz et al., 2014). Nonetheless, people can also be infected by consumption of contaminated water, and contact with animals and other environmental sources (Gölz et al., 2014). The typical clinical symptom of campylobacteriosis is acute diarrhoea, often accompanied by abdominal cramping, headache and fever (Blaser, 1997). Being considered a self-limiting disease, the latent period of campylobacteriosis is usually two to five days, while the disease can last for up to two weeks (Young et al., 2007). In some cases, the symptom may still last for several weeks and medical intervention is required for 10% of reported cases (Lee and Newell, 2006). *C. jejuni* infection is also associated with Guillain-Barre syndrome (GBS), a life-threatening autoimmune disease which leads to peripheral nervous damage with a mortality of 2–7% (Mawla et al., 2014).

In 2013, FoodNet identified 6621 cases of laboratory-confirmed infection, 1010 hospitalizations, and 12 deaths caused by *C. jejuni* (Crim et al., 2014). However, it has been widely considered that *C. jejuni* leads to 400–500 million cases annually due primarily to underreporting (Ruiz-Palacios, 2007). Approximately 11% of the population acquires the infection every year in the United States (Ruiz-Palacios, 2007). Moreover, *C. jejuni* has been estimated to cause 200 deaths per year in that country (Ruiz-Palacios, 2007) and the annual cost attributed to this disease is estimated to be between $1.2 and 4 billion per year (Batz et al., 2014; Frenzen, 2008).

Molecular epidemiologic studies have suggested that chicken is the primary source of human campylobacteriosis worldwide (Agunos et al., 2014; Sheppard et al., 2009; Wilson et al., 2008). Importantly, poultry is one of the world's most significant animal protein sources, due to the higher feeding efficiency and faster growth rate compared to pork and beef (Smil, 2002). It is expected that chicken will become the largest meat product worldwide in 2020, as a result of changing food preferences, household income and population growth (OECD/FAO, 2012).

Epidemiological simulation has shown that the reduction of *C. jejuni* contamination in poultry carcasses by two log units can result in a reduction in the incidence of human infection by 30-fold (Rosenquist et al., 2003). Since the farm is the first point in the production chain where chickens become colonized, on-farm control of *C. jejuni* would make a significant impact on the reduction of campylobacteriosis in humans.

Multiple approaches have been developed to control *Campylobacter* on the broiler farm level. Those approaches include (1) reduction of environmental exposure, for example, biosecurity measures; (2) antimicrobial alternatives, such as bacteriophage therapy and bacteriocin treatment; (3) increase in host resistance, such as host genetics selection; and (4) vaccinations (Lin, 2009). However, many of these proposed intervention strategies still have significant disadvantages and may not be practically available (Lin, 2009).

3 Improving biosecurity in the farm

Once the first bird in a flock becomes colonized, the infection spreads to the entire flock in just a few days (Gölz et al., 2014; Katsma et al., 2005). This rapid spread of *Campylobacter* throughout the flock is likely a result of faecal-oral transmission, compounded by communal water and feed (Lee and Newell, 2006). Based on the visualization of the confocal scanning laser microscopy, *C. jejuni* can also survive in feather follicles and the pores of chicken skin at a depth of 20–30 μm upon contacting a poultry host, which provides *C. jejuni* with a microenvironment with little exposure to oxygen, appropriate humidity and temperature to survive stress conditions (Chantarapanont et al., 2003).

Colonization of *C. jejuni* can persist throughout the lifetime of a broiler, which may lead to carcass contamination at the slaughter facility. Although *Campylobacter* can be isolated from most intestinal sites of broiler chickens, it is mainly found in the cecal and cloacal crypts where it does not adhere to epithelial cells but colonizes the mucous layer (Lee and Newell, 2006). Broiler chickens can be colonized with *C. jejuni* as high as 10^6–10^{10} colony-forming units (CFU) per gram of faeces (Lee and Newell, 2006; Rice et al., 1997). The detection of *C. jejuni* in tissues other than the intestinal tract, such as the spleen, lung, heart and liver, suggests that this pathogen can translocate to intestinal epithelial cells and become systemic (Knudsen et al., 2006).

Biosecurity measures on the farm are employed to reduce the risk of chickens coming in contact with *Campylobacter* and hence, infection. Typical security measures include limiting the number of personnel and the number of times that the personnel can come in contact with the birds. Additionally, reducing the chances of vectors contacting birds may also reduce the risk (Huneau-Salaun et al., 2007). Proper rodent and insect control programmes are important, as these are significant vectors of *Campylobacter* (Lyngstad et al., 2008).

Currently, there is no strong evidence for vertical transmission of *Campylobacter*. Although the organism can be cultured from the oviduct of hens, internal infection of eggs occurring prior to the shell deposition significantly reduces viability of the egg (Clark and Bueschkens, 1985; King et al., 1993). Further, evidence exists of *C. jejuni*-specific maternal antibodies that may be vertically transferred from layers to the offspring (Sahin et al., 2001). Thus, if vertical transmission did occur, some protection is afforded by maternal immune factors. Other hatchery debris may be more likely sources of *Campylobacter*. Fluff from feathers, faecal contamination and liners that were evaluated in studies have been, in some cases, culture positive (Hiett et al., 2002; Byrd et al., 2007; Rasschaert et al., 2007). For this reason, taking care to prevent *Campylobacter* introduction on the farm includes careful inspection of material associated with chick transportation boxes.

As noted, once a single bird is infected, horizontal transmission occurs rapidly. Control of insects, rodents and wild birds can prevent infection. Bates et al. (2004) reported isolating *Campylobacter* from the darkling beetle with the same short variable region (SVR) genotype as isolates obtained from faecal samples. Flies have also been reported to be a source of *Campylobacter* (Rosef and Kapperud, 1983; Gregory et al., 1997). Although *Campylobacter* have been isolated from flies, consumption of contaminated flies has not been reported to lead to infection. Experiments which involved feeding chickens flies contaminated with *Salmonella* showed that infection did not occur from this source (Holt et al., 2007), but no data are available for *Campylobacter*. Rodents are also reported as vectors for *Campylobacter*. In this case, faecal pellets of rodents in the chicken houses are the most likely sources of contamination. However, like flies, there is no evidence that consumption of infected rodent faecal pellets leads to infection in chickens.

Farm labour can also be a biosecurity issue. The shoes, clothing and personal items of workers can become contaminated and thus, workers entering the bird houses are a risk factor for *Campylobacter* transmission (Agunos et al., 2014). It has been demonstrated that the risk of *Campylobacter* infection in the birds is directly proportional to the number of workers and the number of times they enter the house (Huneau-Salaun et al., 2007). On multi-commodity farms that rear livestock in addition to chickens, the risk of transmission is increased (Neubauer et al., 2005; Newell et al., 2011).

Infected birds can be considered as vectors to other birds and therefore control of horizontal transmission is an important factor. Antimicrobials may be delivered to the birds as a proactive method to reduce the risk of infection. Therefore, although one bird may become colonized, horizontal transmission can be reduced or potentially slowed.

4 Antimicrobial therapy and increased host resistance

Antimicrobials used in poultry production may include natural and synthetic antibiotics, plant extracts and herbs, probiotics and prebiotics, and organic acids. Due to concerns of the use of antibiotics in animal production, many farms are now using alternatives to traditional antibiotics. Public pressure has also driven alternatives due to concerns of antibiotic residues in meat and the emergence of antibiotic-resistant zoonotic bacteria.

Probiotics are more commonly used than other antibiotic alternatives (Morishita et al., 1997; Willis and Reid, 2008; Santini et al., 2010). One advantage of using probiotics is the increase in weight gain that occurs in the first two weeks after hatch (Higgins et al., 2007). Probiotic bacteria hasten the histological and immune maturation of the intestines. A mature intestinal tract contains elongated villi which dramatically increase the amount of surface area available for nutrient uptake. Thus, the probiotic bacteria indirectly increase weight gain.

Most importantly, the probiotic bacteria defend the host from infection and zoonotic bacteria including *Salmonella*. The effectiveness of probiotics for protection against *Campylobacter* has largely failed. This is because *Campylobacter* thrives in a mature intestinal tract for two main reasons: (1) it cannot use six-carbon sugars and instead depends on the bacterial community to produce the secondary metabolites it needs; and (2) *Campylobacter* requires a low oxygen environment which is provided only when the intestinal tract is mature. For these reasons, it can be inferred that probiotics may actually support *Campylobacter* in the intestinal tract, with the assumption that the probiotic treatment did reach the ceca and established colonization.

Herbs, spices and plant extracts are gaining some popularity especially among the organic and pasture flock producers. The mechanism of action by which these botanicals work against zoonotic bacteria is not fully understood. Some reports of weight gain and improved feed conversion after supplementing the feed with botanicals are available (Diaz-Sanchez et al., 2015). As with the protective effects of botanicals, how these substances improve production performance is not fully understood.

Acidifying the drinking water has also been demonstrated to protect the host from colonization with *Salmonella* and *Campylobacter.* The protective effect is due to the acid sensitivity of these pathogens (Ganguly, 2013). In addition, beneficial bacteria, including *Lactobacillus*, thrive in an acidic pH environment. These commensals produce antimicrobial substances and inhibit colonization of zoonotic bacteria by occupying niches and reducing the nutrient availability for other bacteria.

Methods to increase host resistance to *Campylobacter* are being explored. Methods including selective breeding and vaccination are attractive to producers because production parameters do not need to be adjusted. No additives are required in the feed or water and, in the case of vaccinations, only one inoculation is required and this can be delivered at the hatchery.

The resistance of inbred chicken lines to *Salmonella* and *Campylobacter* infection has been described. For both pathogens, the major histocompatibility complex (MHC) is not linked to resistance (Wigley et al., 2002; Boyd et al., 2005). Additionally, for *Salmonella*, resistance is not a function of sex or the gene Slc11a1. It appears that the mononuclear phagocyte system is responsible for *Salmonella* clearance mediated by the gene *sal1*, but it is unknown what product this gene expresses and how it facilitates *Salmonella* clearance (Wigley et al., 2002).

5 Boosting the immune response

An early study by Stern et al. (1990) showed that the *C. jejuni* dose needed for colonization in chickens increased by 50% if chickens were pre-incubated with *Campylobacter*-specific immunoglobulins. This suggests that immune intervention, either vaccination or passive immunity, can be an effective strategy to reduce *Campylobacter* colonization in chickens.

An effective chicken *Campylobacter* vaccine should meet the following standards:

(1) the vaccine should prevent colonization or reduce bacteria numbers for more than two log units in chickens; (2) the vaccine must induce a rapid immune response as chickens usually come in contact with *Campylobacter* a few days post-hatch; (3) immunity should be cross-protective against different *Campylobacter* isolates; (4) the vaccine should be cost-effective and easy to deliver; and (5) the vaccine should be safe for both chickens and humans (de Zoete et al., 2007). Various vaccines have been developed and evaluated for protection against *Campylobacter* in avian, including live attenuated vaccines, killed whole-cell vaccines (WCV), live attenuated *Salmonella*-vectored vaccines and subunit vaccines.

Many studies have reported that *C. jejuni* infection could lead to both mucosal and systemic immune response in chickens (Cawthraw et al., 1994; de Zoete et al., 2007; Widders et al., 1998). Cawthraw et al. (1994) have shown that levels of *C. jejuni*-specific serum antibodies (IgG, IgM and IgA) rose two weeks after infection with a peak at five to seven weeks after infection, and then declined. Intestinal mucosal IgA against *C. jejuni* also rose three to four weeks after challenge (Cawthraw et al., 1994). Furthermore, the elevated levels of *C. jejuni*-specific antibodies are associated with the reduction of *C. jejuni* colonization level, suggesting an important role of the humoral immunity in controlling *C. jejuni* infection in chickens (Lin, 2009). Together, these findings indicate that *C. jejuni*-specific antibodies may reduce *Campylobacter* colonization in birds. *C. jejuni* has been demonstrated to be recognized by the Toll-like receptor 4 (TLR4) as well as TLR21 in chickens, which then leads to the innate immune responses in the intestinal tract that cause an influx of inflammatory cells, including heterophils, a functional equivalent of the neutrophil (de Zoete et al., 2010; Meade et al., 2009; Smith et al., 2008).

Killed WCV have been the most widely studied for various pathogens and have been demonstrated to be effective in controlling intestinal pathogens (Pace et al., 1998). In chickens, several WCV against *Campylobacter* have also been tested. In one study, birds were administered with 10^9 formalin-killed *C. jejuni* through oral route multiple times within 16 days before challenge. The cecal *C. jejuni* level in the vaccine groups was 1.5 log units lower than the control (Rice et al., 1997). However, in another formalin-killed *C. jejuni* trial, no serum or mucosal immunoglobulins could be detected and the mean cecal colonization level was only 1 log unit lower than the control (Cawthraw et al., 1998).

Outer membrane proteins (OMPs) are attractive antigens for immune intervention in Gram-negative bacteria (Lin et al., 2002a), and various recombinant subunit vaccines as well as live attenuated *Salmonella*-vectored vaccines expressing *Campylobacter* OMPs were exploited. These vaccines include flagellum vaccines (Meeusen et al., 2007; Widders et al., 1996, 1998); a CmeC subunit vaccine (Zeng et al., 2010); a CadF-FlaA-FlpA fusion vaccine (Neal-McKinney et al., 2014); an attenuated *Salmonella*-vectored CjaA vaccine (Wyszyńska et al., 2004); an attenuated *Salmonella*-vectored Peb1 vaccine (Sizemore et al., 2006); and a live *Salmonella* vaccine expressing Cj0113 (Omp18/CjaD), Cj0982c (CjaA) and Cj0420 (ACE393) (Layton et al., 2011).

As the major component of the flagellum, Fla is one of the best-studied antigenic proteins in *Campylobacter*. In two flagellin vaccine trials, birds were vaccinated with flagellin alone or in combination with heat-killed bacteria intraperitoneally (IP), followed by an IP or oral booster. The vaccine with the flagellin/whole-cell combination administered IP/IP provided higher serum and mucosal antibody levels and resulted in a 10–100 fold reduction of cecal CFU level after challenge (Widders et al., 1996, 1998). Another flagellum vaccine developed by Meeusen et al. (2007) also induced partial protection. The reason for the partial success may be that the flagellin is not surface-exposed in the flagellar structure and can be modified by glycosylation leading to antigenic variation (Doig et al., 1996; Widders et al., 1998). In a CmeC subunit vaccination trial, the vaccine triggered systemic immune responses, while the intestinal secretory IgA response was not significantly stimulated (Zeng et al., 2010). Therefore, this CmeC vaccine did not confer protection of broilers against *C. jejuni* colonization with that specific vaccination regimen. Recently, a CadF, FlaA and FlpA peptide vaccine has been reported to reduce cecal *C. jejuni* colonization by three log units after challenge (Neal-McKinney et al., 2014).

Attenuated *Salmonella*-based vaccines are another attractive approach for *C. jejuni* control in poultry. In one study, an attenuated *Salmonella* CjaA vaccine triggered *C. jejuni*-specific serum IgG and intestinal IgA and reduced wild-type *C. jejuni* infection in the chicken cecum (Wyszyńska et al., 2004). However, the attenuated *Salmonella* strain expressing Peb1 failed to induce protection against *C. jejuni* in chickens (Sizemore et al.,

2006). Recently, Layton et al. (2011) developed live *Salmonella* vaccines expressing Cj0113 (Omp18/CjaD), Cj0982c (CjaA) or Cj0420 (ACE393) and administered them to chickens orally on the first day post-hatching. All three candidates induced serum and intestinal mucosal immune responses and decreased *C. jejuni* recovery from the ileum, with the best response (4.8 log unit reduction) from the Cj0113 group.

Passive immunization has also been studied to reduce *C. jejuni* colonization in broilers. In a study by Al-Adwani et al. (2013), specific-pathogen-free laying hens were hyper-immunized with one of the five *C. jejuni* colonization-associated proteins or CAPs (CadF, FlaA, MOMP, FlpA or CmeC) and egg-yolk-derived antibodies (IgY) were then obtained from egg-yolk powder (EYP). Indirect enzyme-linked immunosorbent assays (ELISA) showed that *C. jejuni* CAP-specific IgY levels were significantly ($P < 0.05$) higher in both serum and EYP obtained from immunized hens as compared with the non-immunized hens. The study also showed that CadF-, MOMP- and CmeC-specific IgY greatly reduced *C. jejuni* colonization in the chicken hepatocellular carcinoma cells. In a follow-up study, laying hens were immunized with seven *C. jejuni* CAPs (CadF, FlaA, MOMP, FlpA, CmeC, Peb1A and JlpA), respectively, or as a cocktail containing equal parts of each EYP. Nevertheless, no significant differences in the cecal colonization of *C. jejuni* were observed between the treated chickens and the control (Paul et al., 2014). In another study, laying hens were immunized with either a whole-cell lysate or the hydrophobic protein fraction of *C. jejuni* (Hermans et al., 2014). Results showed that preventive administration of hyper-immune egg yolk significantly reduced *C. jejuni* level in chickens by approximately four log units. Western blot analysis in combination with mass spectrometry also revealed that the immunodominant antigens (AtpA, EF-Tu, GroEL, CtpA, etc.) were highly conserved and were involved in a variety of cell functions of *C. jejuni*.

6 DNA vaccines

In 1990, the plasmid DNA encoding luciferase was reported to transfect muscle cells *in vivo* after an intramuscular injection (Wolff et al., 1990). The luciferase activity was present in the muscle for over two months. This finding led to the development of the first DNA vaccine reported (Ulmer et al., 1993). Since then, various DNA vaccines have been developed against infectious diseases, cancer, autoimmunity and allergies in human and animals (Liu, 2011; Restifo and Rosenberg, 1999). The principle of DNA vaccine is straightforward: the gene encoding foreign antigen is cloned into an appropriate eukaryotic expression plasmid that can be replicated in a bacterial host; then the purified recombinant plasmid can be directly used as a vaccine with or without adjuvant (Shah et al., 2014). DNA vaccination has a variety of advantages, including the ability to induce both cellular and humoral immune responses, lack of risk for infection, long-term persistence of immunogen and the ability to be easily manufactured by standard molecular biology techniques (Restifo and Rosenberg, 1999; Wahren and Liu, 2014). DNA vaccines are also more stable for storage and shipping because of the structural and chemical characteristics when compared to traditional vaccines (Shah et al., 2014).

Many veterinary DNA vaccines have been licensed, including an equine vaccine against West Nile virus, a fish vaccine against infectious haematopoietic necrosis virus, a pig vaccine expressing growth hormone releasing hormone, and a therapeutic canine vaccine for melanoma (Redding and Weiner, 2009; Wahren and Liu, 2014). In chickens, several

DNA vaccines have also been developed, most of which are viral vaccines. A recombinant plasmid encoding the VP2 gene fragment of infectious bursal disease virus was tested in chickens through the intramuscular route and the pVP2 vaccine group developed a higher titre of anti-VP2 antibodies than the control (Pradhan et al., 2014). Moreover, splenocytes from the vaccine group showed a significantly higher proliferation to the whole viral and recombinant antigen, which implies that the DNA vaccine elicited both B and T cell responses. In another trial, the DNA vaccine encoding chicken infectious anaemia virus (CIAV) VP1/VP2 genes were co-administered with a truncated chicken high-mobility group box 1 (HMGB1ΔC) as adjuvant in chickens with a booster and a strong immunity for CIAV was observed (Sawant et al., 2015). Scientists have also studied different DNA vaccines against coccidiosis in poultry. In one study by Shah et al. (2011), five different DNA vaccines were tested in chickens intramuscularly. The result indicates that vaccination groups greatly ($P < 0.05$) alleviated intestinal lesions, body weight loss and oocyst count. Similarly, in another DNA vaccination study by Xu et al. (2008), the vaccinated chickens also showed significant ($P < 0.05$) lower weight gain loss and higher oocyst decrease ratio, imparting partial protection against homologous challenge.

DNA vaccines against *Campylobacter* have also been developed recently (Huang et al., 2010; Liu et al., 2014). Intranasal vaccination of chickens with a DNA vaccine expressing the flagellin gene *flaA* induced *C. jejuni*-specific serum IgG and intestinal IgA and reduced *C. jejuni* colonization by 2 to 3 log units within the cecum (Huang et al., 2010). In a mouse study for evaluating a DNA vaccine expressing *C. jejuni* PEB1 permease, the stimulation index of lymphocytes, serum IgG, IL-4, IFN-γ, and intestinal IgA were significantly higher in the mice immunized with DNA vaccines with a protein boost via the intranasal route (Huang et al., 2010; Liu et al., 2014).

In ovo delivery, an attractive vaccination route for chickens (de Zoete et al., 2007), offers various advantages when combined with DNA vaccine (Haygreen et al., 2005). In 1997, a plasmid encoding β-galactosidase delivered into the breast muscle via *in ovo* route achieved successful gene transfer and expression, which showed the potential for the development of *in ovo* DNA vaccines (Johnston et al., 1997). The immune system of chickens is well developed by Day 18 of inoculation, indicating the feasibility for stimulating immune response via *in ovo* vaccination (Ricks et al., 1999). In addition, *in ovo* vaccination is a fully automatic method to vaccinate massive numbers of eggs (20 000–30 000 per hour), and has been applied to various vaccines for viral, bacterial and protozoal diseases in broilers, without compromising embryo viability (Johnston et al., 1997; Ricks et al., 1999). This method also reduces egg handling, improves hatchery manageability and reduces cost and labour (Johnston et al., 1997). To date, several poultry vaccines have been approved by the United States department of agriculture (USDA) for *in ovo* administration (Johnston et al., 1997).

7 Conclusions

Campylobacter can be introduced to a farm and subsequently colonize chickens by numerous routes. Biosecurity is crucial to prevent introduction of this zoonotic pathogen. A proper insect and rodent control programme, education programme for farm workers and secure infrastructure are keys to a good biosecurity programme. Even with a proper biosecurity programme, *Campylobacter* may eventually be present on the farm. Thus, along with biosecurity, preventative treatments for the birds can act synergistically to significantly reduce the risk of contamination. Probiotics, prebiotics, antimicrobials and vaccines remain ineffective, but are being improved, and promising therapies may be available to effectively control this pathogen. Future needs include optimizing control methods and further development of methods that have only been moderately explored. In this way, control on the farm will be more effective and the reduction in the number of human campylobacteriosis cases can be reduced.

8 Acknowledgements

This work was supported by the Agriculture and Food Research Initiative (AFRI) Competitive Grants No. 2012-68003-19679.

9 Where to look for further information

Much of the peer reviewed literature contains up to date information regarding this subject. Databases, including Medline and Google Scholar can be searched to find information. Extension agents at local Universities can also provide local information.

10 References

Agunos, A., Waddell, L., Léger, D. and Taboada, E. (2014). A systematic review characterizing on-farm sources of *Campylobacter spp*. for broiler chickens. *PLoS ONE* 9, e104905.

Al-Adwani, S. R., Crespo, R. and Shah, D. H. (2013). Production and evaluation of chicken egg-yolk-derived antibodies against *Campylobacter jejuni* colonization- associated proteins. *Foodborne Pathogens and Disease* 10, 624–31.

Bates, C., Hiett, K. L. and Stern, N. J. (2004). Relationship of Campylobacter isolated from poultry and from darkling beetles in New Zealand. *Avian diseases* 48 (1), 138–47.

Batz, B., Morris, Jr., J. G. and Hoffmann, S. (2011). Ranking the risks: the 10 pathogen- food combinations with the greatest burden on public health. Available at http://epi.ufl.edu/wp-content/uploads/2011/04/RankingTheRisksREPORT.pdf.

Batz, M., Hoffmann, S. and Morris, Jr., J. G. (2014). Disease-outcome trees, EQ-5D scores, and estimated annual losses of quality-adjusted life years (QALYs) for 14 foodborne pathogens in the United States. *Foodborne Pathogens and Disease* 11, 395–402.

Blaser, M. J. (1997). Epidemiologic and clinical features of *Campylobacter jejuni* infections. *Journal of Infectious Diseases* 176, S103–5.

Boyd, Y., Herbert, E. G., Marston, K. L., Jones, M. A. and Barrow, P. A. (2005). Host genes affect intestinal colonisation of newly hatched chickens by *Campylobacter jejuni*. *Immunogenetics*, 57 (3–4), 248–53.

Byrd, J., Bailey, R. H., Wills, R. and Nisbet, D. (2007). Recovery of *Campylobacter* from commercial broiler hatchery trayliners. *Poultry Science* 86, 26–29.

Cawthraw, S., Ayling, R., Nuijten, P., Wassenaar, T. and Newell, D. (1994). Isotype, specificity, and kinetics of systemic and mucosal antibodies to *Campylobacter jejuni* antigens, including flagellin, during experimental oral infections of chickens. *Avian Diseases*, 341–9.

Cawthraw, S., Gorringe, C. and Newell, D. (1998). Prior infection, but not a killed vaccine, reduces colonization of chickens by *Campylobacter jejuni*. *Campylobacter, Helicobacter and Related Organisms Workshop*, 364–72.

Chantarapanont, W., Berrang, M. and Frank, J. F. (2003). Direct microscopic observation and viability determination of *Campylobacter jejuni* on chicken skin. *Journal of Food Protection®* 66, 2222–30.

Clark, A. G. and Bueschkens, D. H. (1985). Laboratory infection of chicken eggs with *Campylobacter jejuni* by using temperature or pressure differentials. *Applied Environmental Microbiology* 49, 1467–71.

Crim, S. M., Iwamoto, M., Huang, J. Y., Griffin, P. M., Gilliss, D., Cronquist, A. B., Cartter, M., Tobin-D'Angelo, M., Blythe, D. and Smith, K. (2014). Incidence and trends of infection with pathogens transmitted commonly through food – Foodborne Diseases Active Surveillance Network, 10 US Sites, 2006–2013. *MMWR. Morbidity and Mortality Weekly Report* 63, 328–32.

de Zoete, M. R., van Putten, J. P. and Wagenaar, J. A. (2007). Vaccination of chickens against *Campylobacter*. *Vaccine* 25, 5548–57.

de Zoete, M. R., Keestra, A. M., Roszczenko, P. and van Putten, J. P. (2010). Activation of human and chicken toll-like receptors by *Campylobacter spp*. *Infection and Immunity* 78, 1229–38.

Dekeyser, P., Gossuin-Detrain, M., Butzler, J. and Sternon, J. (1972). Acute enteritis due to related vibrio: first positive stool cultures. *Journal of Infectious Diseases* 125, 390–92.

Diaz-Sanchez, S., Hawkins, R., Okimoto, R., Layton, A., Saxton, A., Blakeley-Ruiz, A. and Hanning, I. (15–18 April 2015). *Correlation of Microbiome Populations with Performance Measurements and Gender in Broiler Breeders*. Boston, MA: Congress on Gastrointestinal Function.

Ding, X., Lillehoj, H. S., Dalloul, R. A., Min, W., Sato, T., Yasuda, A. and Lillehoj, E. P. (2005). *In ovo* vaccination with the *Eimeria tenella* EtMIC2 gene induces protective immunity against coccidiosis. *Vaccine* 23, 3733–40.

Doig, P., Kinsella, N., Guerry, P. and Trust, T. J. (1996). Characterization of a post-translational modification of *Campylobacter* flagellin: identification of a sero-specific glycosyl moiety. *Molecular Microbiology* 19, 379–87.

Frenzen, P. (2008). Economic cost of Guillain-Barré syndrome in the United States. *Neurology* 71, 21–7.

Friedman, C. (2000). Epidemiology of *Campylobacter jejuni* infections in the United States and other industrialized nations. *Campylobacter*, 121–38.

Ganguly, S. (2013). Supplementation of prebiotics, probiotics and acids on immunity in poultry feed: a brief review. *World's Poultry Science Journal* 69 (03), 639–48.

Gölz, G., Rosner, B., Hofreuter, D., Josenhans, C., Kreienbrock, L., Löwenstein, A., Schielke, A., Stark, K., Suerbaum, S. and Wieler, L. H. (2014). Relevance of *Campylobacter* to public health—The need for a One Health approach. *International Journal of Medical Microbiology* 304, 817–23.

Gregory, E., Barnhart, H., Dreesen, D. W., Stern, N. J. and Corn, J. L. (1997). Epidemiological study of Campylobacter spp. in broilers: source, time of colonization, and prevalence. *Avian diseases*, 890–8.

Haygreen, L., Davison, F. and Kaiser, P. (2005). DNA vaccines for poultry: the jump from theory to practice. *Expert Review of Vaccines* 4, 51–62.

Hermans, D., Van Steendam, K., Verbrugghe, E., Verlinden, M., Martel, A., Seliwiorstow, T., Heyndrickx, M., Haesebrouck, F., De Zutter, L. and Deforce, D. (2014). Passive immunization to reduce *Campylobacter jejuni* colonization and transmission in broiler chickens. *Veterinary Research* 45, 27.

Hiett, K. L., Cox, N. A. and Stern, N. J. (2002). Direct polymerase chain reaction detection of *Campylobacter* spp. in poultry hatchery samples. *Avian Diseases* 46, 219–23.

Higgins, J. P., Higgins, S. E., Vicente, J. L., Wolfenden, A. D., Tellez, G. and Hargis, B. M. (2007). Temporal effects of lactic acid bacteria probiotic culture on *Salmonella* in neonatal broilers. *Poultry Science* 86(8), 1662–66.

Holt, P. S., Geden, C. J., Moore, R. W. and Gast, R. K. (2007). Isolation of *Salmonella enterica* serovar Enteritidis from houseflies (*Musca domestica*) found in rooms containing *Salmonella* enterica serovar Enteritidis-challenged hens. *Applied Environmental Microbiology* 73, 6030–5.

Huang, J.-L., Yin, Y.-X., Pan, Z.-M., Zhang, G., Zhu, A.-P., Liu, X.-F. and Jiao, X.-A. (2010). Intranasal immunization with chitosan/pCAGGS-flaA nanoparticles inhibits *Campylobacter jejuni* in a White Leghorn model. *BioMed Research International*.

Huneau-Salaun, A., Denis, M., Balaine, L. and Salvat, G. (2007). Risk factors for *Campylobacter* spp. colonization in French free-range broiler-chicken flocks at the end of the indoor rearing period. *Preventive Veterinary Medicine* 80, 34–48.

Johnston, P., Liu, H., O'Connell, T., Phelps, P., Bland, M., Tyczkowski, J., Kemper, A., Harding, T., Avakian, A. and Haddad, E. (1997). Applications in *in ovo* technology. *Poultry Science* 76, 165–78.

Kaakoush, N. O., Castaño-Rodríguez, N., Mitchell, H. M. and Man, S. M. (2015). Global epidemiology of *Campylobacter* infection. *Clinical Microbiology Reviews* 28, 687–720.

Katsma, W., De Koeijer, A., Fischer, E., Jacobs Reitsma, W. and Wagenaar, J. (2005). *Campylobacter* prevalence in broiler flocks in the Netherlands Modelling transmission within and between flocks and efficacy of interventions. Lelystad: Animal Science Group. ASG05.

King, V., Bavetsia, A. and Bumstead, N. (1993). Effect of host lineage on the virulence of *Campylobacter jejuni/coli* in the chicken embryo model. *FEMS Microbiology Letters* 106, 271–4.

Knudsen, K. N., Bang, D. D., Andresen, L. O. and Madsen, M. (2006). *Campylobacter jejuni* strains of human and chicken origin are invasive in chickens after oral challenge. *Avian Diseases* 50, 10–14.

Layton, S., Morgan, M., Cole, K., Kwon, Y., Donoghue, D., Hargis, B. and Pumford, N. (2011). Evaluation of *Salmonella*-vectored *Campylobacter* peptide epitopes for reduction of *Campylobacter jejuni* in broiler chickens. *Clinical and Vaccine Immunology* 18, 449–54.

Lee, M. D. and Newell, D. G. (2006). *Campylobacter* in poultry: filling an ecological niche. *Avian Diseases* 50, 1–9.

Lin, J. (2009). Novel approaches for *Campylobacter* control in poultry. *Foodborne Pathogens and Disease* 6, 755–65.

Lin, J., Huang, S. and Zhang, Q. (2002a). Outer membrane proteins: key players for bacterial adaptation in host niches. *Microbes and Infection* 4, 325–31.

Liu, L., Lai, W., Hu, N., Zhang, W., Chen, T., Wang, F. and Gu, R. (2014). Evaluation of immunological efficiency induced by *Campylobacter jejuni* PEB1 DNA combined with PEB1 protein in mice. *Xi bao yu fen zi mian yi xue za zhi (Chinese Journal of Cellular and Molecular Immunology)* 30, 576–80.

Liu, M. A. (2011). DNA vaccines: an historical perspective and view to the future. *Immunological Reviews* 239, 62–84.

Lyngstad, T. M., Jonsson, M. E., Hofshagen, M. and Heier, B. T. (2008). Risk factors associated with the presence of *Campylobacter* species in Norwegian broiler flocks. *Poultry Science* 87, 1987–94.

Man, S. M. (2011). The clinical importance of emerging *Campylobacter* species. *Nature Reviews Gastroenterology and Hepatology* 8, 669–85.

Mawla, N. N., Sultana, S. and Akhter, N. (2014). Guillain-Barré syndrome and *Campylobacter jejuni* infection: a review. *Delta Medical College Journal* 2, 28–35.

Meade, K. G., Narciandi, F., Cahalane, S., Reiman, C., Allan, B. and O'Farrelly, C. (2009). Comparative *in vivo* infection models yield insights on early host immune response to *Campylobacter* in chickens. *Immunogenetics* 61, 101–10.

Meeusen, E. N., Walker, J., Peters, A., Pastoret, P.-P. and Jungersen, G. (2007). Current status of veterinary vaccines. *Clinical Microbiology Reviews* 20, 489–510.

Morishita, T. Y., Aye, P. P., Harr, B. S., Cobb, C. W. and Clifford, J. R. (1997). Evaluation of an avian-specific probiotic to reduce the colonization and shedding of *Campylobacter jejuni* in broilers. *Avian Diseases* 41, 850–5.

Neal-McKinney, J. M., Samuelson, D. R., Eucker, T. P., Nissen, M. S., Crespo, R. and Konkel, M. E. (2014). Reducing *Campylobacter jejuni* colonization of poultry via vaccination. *PLoS ONE* 9, e114254.

Neubauer, C., Bibl, D., Szölgyenyi, W., Jauk, V., Schmid, M., Gabler, C. and Vasicek, L. (2005). Epidemiological investigation of *Campylobacter* spp. in Austrian broiler flocks: prevalence and risk factors. *Wiener Tierärztliche Monatsschrift* 92, 4–10.

Newell, D. G., Elvers, K. T., Dopfer, D., Hansson, I., Jones, P., James, S. and Allen, V. M. (2011). Biosecurity-Based interventions and strategies to reduce *Campylobacter spp.* on poultry farms. *Applied and Environmental Microbiology* 77, 8605–14.

Nichols, G. L. (2005). Fly transmission of *Campylobacter*. *Emerging Infectious Diseases* 11, 361–4.

OECD/FAO (2012). OECD-FAO Agricultural Outlook 2012–21, OECD Publishing and FAO. http://dx.doi.org/10.1787/agr_outlook-2012-en.

Pace, J. L., Rossi, H. A., Esposito, V. M., Frey, S. M., Tucker, K. D. and Walker, R. I. (1998). Inactivated whole-cell bacterial vaccines: current status and novel strategies. *Vaccine* 16, 1563–74.

Paul, N. C., Al-Adwani, S., Crespo, R. and Shah, D. H. (2014). Evaluation of passive immunotherapeutic efficacy of hyperimmunized egg yolk powder against intestinal colonization of *Campylobacter jejuni* in chickens. *Poultry Science* 93, 2779–87.

Pinkerton, H. and Bessey, O. A. (1939). The loss of resistance to murine typhus infection resulting from riboflavin deficiency in rats. *Science* 89, 368–70.

Pradhan, S. N., Prince, P. R., Madhumathi, J., Arunkumar, C., Roy, P., Narayanan, R. B. and Antony, U. (2014). DNA vaccination with VP2 gene fragment confers protection against Infectious Bursal Disease Virus in Chickens. *Veterinary Microbiology* 171, 13–22.

Redding, L. and Weiner, D. B. (2009). DNA vaccines in veterinary use. *Expert Review of Vaccines* 8, 1251–76.

Rasschaert, G., Houf, K. and De Zutter, L. (2007). External contamination of *Campylobacter*-free flocks after transport in cleaned and disinfected containers. *Journal of Food Protection* 70, 40–6.

Restifo, N. P. and Rosenberg, S. A. (1999). Developing recombinant and synthetic vaccines for the treatment of melanoma. *Current Opinion in Oncology* 11, 50.

Rice, B. E., Rollins, D. M., Mallinson, E. T., Carr, L. and Joseph, S. W. (1997). *Campylobacter jejuni* in broiler chickens: colonization and humoral immunity following oral vaccination and experimental infection. *Vaccine* 15, 1922–32.

Ricks, C., Avakian, A., Bryan, T., Gildersleeve, R., Haddad, E., Ilich, R., King, S., Murray, L., Phelps, P. and Poston, R. (1999). *In ovo* vaccination technology. *Advances in Veterinary Medicine* 41, 495–515.

Rosef, O. and G. Kapperud. (1983). House flies (*Musca domestica*) as possible vectors of *Campylobacter fetus* subsp I. *Applied Environmental Microbiology* 45, 381–3.

Rosenquist, H., Nielsen, N. L., Sommer, H. M., Nørrung, B. and Christensen, B. B. (2003). Quantitative risk assessment of human campylobacteriosis associated with thermophilic *Campylobacter* species in chickens. *International Journal of Food Microbiology* 83, 87–103.

Ruiz-Palacios, G. M. (2007). The health burden of *Campylobacter* infection and the impact of antimicrobial resistance: playing chicken. *Clinical Infectious Diseases* 44, 701–3.

Sahin, O., Zhang, Q., Meitzler, J. C., Harr, B. S., Morishita, T. Y. and Mohan, R. (2001). Prevalence, antigenic specificity, and bactericidal activity of poultry anti- *Campylobacter* maternal antibodies. *Applied and Environmental Microbiology* 67, 3951–7.

Sahin, O., Kassem, I. I., Shen, Z., Lin, J., Rajashekara, G. and Zhang, Q. (2015). *Campylobacter* in poultry: ecology and potential interventions. *Avian Diseases* 59(2), 185–200.

Salahi, A., Khabisi, M. M. and Mousavi, S. N. (2011). Optimum time of *in ovo* injection in eggs of young broiler breeder flock. In *18th European Symposium on Poultry Nutrition*, 557–9. Izmir, Turkey.

Samie, A., Obi, C., Barrett, L., Powell, S. and Guerrant, R. (2007). Prevalence of *Campylobacter* species, *Helicobacter pylori* and *Arcobacter* species in stool samples from the Venda region, Limpopo, South Africa: Studies using molecular diagnostic methods. *Journal of Infection* 54, 558–66.

Santini, S., Baffoni, L., Gaggia, F., Granata, M., Gasbarri, G., Di Gioia, D. and Biavati, B. (2010). Characterization of probiotic strains: an application as feed additives in poultry against *Campylobacter jejuni*. *International Journal of Food Microbiology* 141, S98–S108, http://dx.doi.org/10.1016/j.ijfoodmicro.2010.03.039.

Sawant, P. M., Dhama, K., Rawool, D. B., Wani, M. Y., Tiwari, R., Singh, S. D. and Singh, R. K. (2015). Development of a DNA vaccine for chicken infectious anemia and its immunogenicity studies using high mobility group box 1 protein as a novel immunoadjuvant indicated induction of promising protective immune responses. *Vaccine* 33, 333–40.

Shah, M. A. A., Song, X., Xu, L., Yan, R. and Li, X. (2011). Construction of DNA vaccines encoding *Eimeria acervulina* cSZ-2 with chicken IL-2 and IFN-γ and their efficacy against poultry coccidiosis. *Research in Veterinary Science* 90, 72–7.

Shah, M. A. A., Umar, S., Iqbal, M., Rehman, F., Qadri, I. and He, N. (2014). Recent developments in DNA vaccination approaches against poultry coccidiosis and its future endeavours. *World's Poultry Science Journal* 70, 315–28.

Shane, S. M. (2000). *Campylobacter* infection of commercial poultry. *Revue scientifique et technique (International Office of Epizootics)* 19, 376–95.

Sheppard, S. K., Dallas, J. F., Strachan, N. J., MacRae, M., McCarthy, N. D., Wilson, D. J., Gormley, F. J., Falush, D., Ogden, I. D. and Maiden, M. C. (2009). *Campylobacter* genotyping to determine the source of human infection. *Clinical Infectious Diseases* 48, 1072–8.

Sizemore, D. R., Warner, B., Lawrence, J., Jones, A. and Killeen, K. P. (2006). Live, attenuated *Salmonella* typhimurium vectoring *Campylobacter* antigens. *Vaccine* 24, 3793–803.

Smil, V. (2002). Eating meat: evolution, patterns, and consequences. *Population and Development Review* 28, 599–639.

Smith, C. K., AbuOun, M., Cawthraw, S. A., Humphrey, T. J., Rothwell, L., Kaiser, P., Barrow, P. A. and Jones, M. A. (2008). *Campylobacter* colonization of the chicken induces a proinflammatory response in mucosal tissues. *FEMS Immunology & Medical Microbiology* 54, 114–21.

Stern, N. J., Meinersmann, R. J. and Dickerson, H. W. (1990). Influence of antibody treatment of *Campylobacter jejuni* on the dose required to colonize chicks. *Avian Diseases*, 595–601.

Ulmer, J. B., Donnelly, J. J., Parker, S. E., Rhodes, G. H., Felgner, P. L., Dwarki, V., Gromkowski, S. H., Deck, R. R., Dewitt, C. M. and Friedman, A. (1993). Heterologous protection against influenza by injection of DNA encoding a viral protein. *Science* 259, 1745–9.

Wahren, B. and Liu, M. A. (2014). DNA vaccines: recent developments and the future. *Vaccines* 2, 785–96.

Widders, P., Perry, R., Muir, W., Husband, A. and Long, K. (1996). Immunisation of chickens to reduce intestinal colonisation with *Campylobacter jejuni*. *British Poultry Science* 37, 765–78.

Widders, P., Thomas, L., Long, K., Tokhi, M., Panaccio, M. and Apos, E. (1998). The specificity of antibody in chickens immunised to reduce intestinal colonisation with *Campylobacter jejuni*. *Veterinary Microbiology* 64, 39–50.

Wigley, P., Hulme, S. D., Bumstead, N. and Barrow, P. A. (2002). *In vivo* and *in vitro* studies of genetic resistance to systemic salmonellosis in the chicken encoded by the SAL1 locus. *Microbes and Infection* 4 (11), 1111–20.

Wilson, D. J., Gabriel, E., Leatherbarrow, A. J., Cheesbrough, J., Gee, S., Bolton, E., Fox, A., Fearnhead, P., Hart, C. A. and Diggle, P. J. (2008). Tracing the source of campylobacteriosis. *PLoS Genetics* 4, e1000203, doi:10.1371/journal.pgen.1000203.

Willis, W. L. and Reid, L. (2008). Investigating the effects of dietary probiotic feeding regimens on broiler chicken production and *Campylobacter jejuni* presence. *Poultry Science* 87, 606–11.

Wolff, J. A., Malone, R. W., Williams, P., Chong, W., Acsadi, G., Jani, A. and Felgner, P. L. (1990). Direct gene transfer into mouse muscle *in vivo*. *Science* 247, 1465–8.

Wyszyńska, A., Raczko, A., Lis, M. and Jagusztyn-Krynicka, E. K. (2004). Oral immunization of chickens with avirulent *Salmonella* vaccine strain carrying *C. jejuni* 72Dz/92 *cjaA* gene elicits specific humoral immune response associated with protection against challenge with wild-type *Campylobacter*. *Vaccine* 22, 1379–89.

Xu, Q., Song, X., Xu, L., Yan, R., Shah, M. A. A. and Li, X. (2008). Vaccination of chickens with a chimeric DNA vaccine encoding Eimeria tenella TA4 and chicken IL-2 induces protective immunity against coccidiosis. *Veterinary Parasitology* 156, 319–23.

Young, K. T., Davis, L. M. and DiRita, V. J. (2007). *Campylobacter jejuni*: molecular biology and pathogenesis. *Nature Reviews Microbiology* 5, 665–79.

Zeng, X., Xu, F. and Lin, J. (2010). Development and evaluation of CmeC subunit vaccine against *Campylobacter jejuni*. *Journal of Vaccines & Vaccination* 1, 3.

Part 2

Poultry meat quality

Chapter 11

Poultry meat quality: an overview

Michael A. Grashorn, University of Hohenheim, Germany

1 Introduction

Consumers have defined expectations on the quality of their food. This especially holds for meat as meat is assumed to be a high-quality product. In contrast to production of vegetables, consumers have also specific expectations on how meat is produced, equating their expectations with quality features. In general, poultry meat is produced in large houses under very intensive conditions, raising animal welfare concerns in consumers. Therefore, the attitude of the consumer has to be considered as a new quality attribute of poultry meat.

The product quality of meat-type poultry is determined by carcass quality (mainly, appearance and meatiness), meat quality and fat quality (after Kallweit, 2000). Separation of meat and fat quality is indicated for poultry meat due to the low intramuscular fat content of poultry meat. In general, fat content is a component of the nutritive value of the meat and the total fat content is rather low in chickens, guinea fowls, pheasants and turkeys. Only waterfowl (ducks, geese) possess a higher fat content. Lipids are mainly concentrated in depots (subcutaneous fat, organ fat, abdominal fat) which are partly removed during slaughtering. In the present overview, fat quality will be considered together with the nutrient content of the meat.

Meat quality is determined by (1) ingredients (all-included components), by (2) nutritive-physiological aspects (nutrients such as protein/amino acids, fat/fatty acids, vitamins, minerals, etc.) with regard to human requirements, by (3) hygienic and toxicological aspects (microbial load, residues, contaminants, anti-nutritive factors, etc.), by (4) physical and technological aspects (colour, water-holding capacity (WHC), texture, etc.) and by (5) sensory attributes (juiciness, tenderness, flavour, etc.) (Kallweit, 2000). The sensory attributes

http://dx.doi.org/10.19103/AS.2016.0010.14

summarize the subjective assessments of the consumers, including their emotions. In contrast, the other four categories describe meat quality in an objective way. The five categories of meat quality are influenced by genotype of bird (fast vs. slow growing), composition of feed (nutrient density, fatty acids profile), husbandry conditions (intensive vs. extensive), hygienic conditions (partly confounded with husbandry conditions), meat production (slaughter conditions, meat maturation) and marketing of meat (offer form, packaging, storage conditions).

In the following sections, meat quality will be defined, described and probable factors of influence named. These sections will comprise the nutritive value, physical properties, sensory attributes, durability and functional properties. Here also appropriate methods will be indicated and/or described briefly, which are suitable and/or even recommended for determining the different quality features. In 2011 a paper of the working group 5 (Poultry Meat Quality) of the European Federation of the World's Poultry Science Association (WPSA) was published, summarizing recommended methods for determining quality features of poultry meat (Petracci and Baeza, 2011). This paper was intended to harmonize the applied methods in the laboratories making results from different studies more comparable.

2 The nutritive value of poultry meat: description, assessment and improvement

The nutritive value of meat is described by the content of main nutrients like water (moisture), protein and fat. Appropriate methods are given by AOAC (1990). Typically, the moisture content is about 75% in poultry meat. Table 1 shows that the protein and fat content is quite variable within and between species. Chicken meat has a moderate protein content with a fat content comparable to veal. Pig, beef and lamb meat show a higher fat content, whereas rabbit meat is the lowest. The fat in chicken and rabbit is mainly found in depots. The other species show both a considerable fat depot and a high content of intramuscular fat. Marbling does not occur in poultry meat, in contrast to beef and pork meat. Contents of the main nutrients vary distinctly in broiler meat, depending on the cut and on the existence of skin (Table 2). In general, breast meat has the highest protein and the lowest fat content. In turkey breast meat fat content is lower and protein content is higher than in broiler breast meat (Table 3). Animal fat also contains distinct amounts of the lipid concomitant substance cholesterol. The cholesterol content does not differ much between chicken and turkey breast meat and amounts to 60–70 mg/100 g raw meat. Although, it has been proven that dietary cholesterol is not related to the incidence of atherosclerosis, humans with a disturbed cholesterol metabolism have to consider this when consuming poultry meat. Purines are also included in poultry meat, but the content is somewhat lower than in pork meat (120–170 mg/100 g).

The protein fraction consists of 11–12% structural proteins (actin and myosin), 5–6% soluble sarcoplasmic proteins and 2% connective tissue (collagen and elastin) (Karlsson and Font-i-Furnols, 2015). Methods to determine the amino acids and collagen contents are described by Petracci and Baeza (2011). The content of connective tissue mainly influences the tenderness of the meat. According to Fernandez et al. (2001) slow-growing turkey breeds have a higher content of soluble collagen, which may be

Table 1 Nutritive value of meat from different species (Kallweit et al., 1988)

	Life weight (kg)	Protein (%)	Fat (%)
Veal	150–200	14–20	8–10
Beef	200–300	15–21	12–19
Pork	70–80	12–16	30–38
Lamb	5–10	11–16	20–25
Rabbit	1.0–1.3	19–25	3–6
Chicken	1.3–1.5	12–18	9–10

Table 2 Nutritive value of broiler meat (Smith, 1997)

Per 100 g	Energy Kcal	Protein (g)	Total fat (g)
Breast meat with skin	172	20.9	9.3
Breast meat without skin	110	23.1	1.2
Thigh meat with skin	211	17.3	15.3
Thigh meat without skin	119	19.7	3.9
Drumstick meat with skin	161	19.3	8.7
Drumstick meat without skin	119	20.6	3.4

Table 3 Comparison of the nutritive value of chicken and turkey breast meat (per 100 g fresh matter; Heseker and Heseker, 1999; Ternes et al., 2005)

Nutrient		Chicken	Turkey
Protein	g	22.2	24.1
Fat	g	1.2	1.0
Cholesterol	mg	66	60
Purine	mg	115	150

responsible for the higher texture values in this meat (Table 4). Poultry meat also contains distinct amounts of essential amino acids, contributing significantly to meeting human requirements (Table 5). The recommended daily intake for humans for methionine can meet a quarter of the requirement, for lysine by 100% when consuming 100 g meat. Change in the profile of amino acids is rarely possible.

Besides amino acids, chicken meat contains high amounts of fatty acids such as linoleic acid (omega-6 and/or n-6), whereas there are low amounts of n-3 fatty acids (α-linolenic acid) (Table 6). But, in contrast to amino acids, the contents and profiles of fatty acids can be easily changed by feeding birds. Determination of fatty acid profiles and/or contents in tissues is mainly done by gas chromatography (recommendation

Table 4 Chemical composition of *M. pectoralis major* in turkeys with different growth capacity (Fernandez et al., 2001)

	Growth capacity		
	Slow	Medium	Fast
Collagen (%)	0.58	0.58	0.69
Soluble collagen (%)	24.9	22.5	23.2
Myoglobin (μg/g)	0.80	0.76	0.77
Dry matter (%)	26.0	26.5	26.2
Ash (%)	1.31	1.30	1.33

Table 5 Content of water, protein and major amino acids in broiler breast and thigh meat (g/100 g fresh matter)

	Breast	Thigh	RDI
Water	75.0	74.7	
Crude protein	22.8	20.6	
Isoleucine	1.21	1.17	1.4
Leucine	1.57	1.48	2.2
Lysine	1.71	1.79	1.6
Methionine	0.62	0.58	2.2
Phenylalanine	0.87	0.82	2.2
Threonine	0.89	0.78	1.0
Tryptophan	0.27	0.19	0.5
Valine	1.07	0.95	1.6

RDI: recommended daily intake (Ternes et al., 2005).

Table 6 Content of fatty acids and n-6:n-3 ratio in chicken and turkey cuts (mg/100 g fresh matter; after Hahn, 2016)

	Oleic acid	Linoleic acid	Linolenic acid	n-6:n-3
	C18:1	C 18:2	C 18:3	
Chicken breast meat with skin	234	142	6	23.7
Chicken thigh meat with skin	2832	2036	78	26.6
Turkey breast meat without skin	210	180	9	20.0
Duck meat with skin	7322	2054	171	12.0

given by Petracci and Baeza (2011)). In the past numerous attempts have been undertaken to enrich broiler meat with n-3 fatty acids and conjugated linoleic acid (CLA). The example in Table 7 shows that using linseed oil (rich in α-linolenic acid) results in a significant increase in the contents of n-3 fatty acids in muscle tissues, especially in thigh muscle. Assuming a recommended daily intake level for n-3 fatty acids of 400 mg indicates that 100 g breast and thigh meat can meet about 250% and 850% of this RDI, respectively. According to nutritionists, the intake of n-3 fatty acids has to be increased in most Western countries, as a diet based on n-6 fatty acids has been proven to distinctly increase the probability of cardiovascular and inflammatory diseases (Farrell, 1995). Meanwhile, this led to the development of many n-3 enriched poultry products (eggs and meat), which have reached a significant market share in some countries. Besides n-3 fatty acids, CLA also may be enriched by feeding birds with CLA-enriched diets. CLA has proven anti-oxidative and anti-cancerogenic effects (Aletor et al., 2003). Sirri et al. (2003) enriched broiler thigh meat by 177 mg CLA/100 g muscle which is about 9% of the RDI for humans (2 g; Table 8). Like fatty

Table 7 Contents of n-6 and n-3 fatty acids in chicken breast and thigh meat using different dietary fat sources (Crespo and Esteve-Garcia, 2001)

	Tallow	Olive oil	Sunflower oil	Linseed oil
Breast (1.2% fat)				
Σn-6 (mg/g fat)	167.8	155.3	350.4	160.6
Σn-3 (mg/g fat)	22.1	22.6	12.0	204.7
n-6:n-3	7.8	7.0	30.0	0.8
mg n-3/ 100 g tissue	26.5	27.1	14.4	245.6
Thigh (3.5% fat)				
Σn-6 (mg/g fat)	179.9	159.4	370.5	185.9
Σn-3 (mg/g fat)	19.9	19.3	9.0	244.0
n-6:n-3	9.1	8.4	43.3	0.8
mg n-3/ 100 g tissue	69.7	67.6	31.5	854.0

Table 8 Content of saturated, mono-unsaturated, poly-unsaturated (without CLA) and conjugated linoleic acids (CLA) in thigh meat depending on the dietary CLA content (mg/100 tissue; Sirri et al., 2003)

Fatty acid	0% CLA	2% CLA	4% CLA
SAT	593	596	625
MUFA	670	425	463
PUFA (without CLA)	830	679	675
CLA	4.7	95.5	177
n-3	64.7	50.8	51.0

acids, fat-soluble substances can be easily enriched in poultry tissues. The most interesting group is antioxidants such as α-tocopherol or selenium. Control of lipid oxidation in tissues is important as some metabolites from this process, such as peroxides and free radicals (also named TBARS = thiobarbituric reactive substances), may damage cell membranes, resulting in cancer. The level of lipid oxidation is expressed by the concentration of malondialdehyde (MDA), which is generated during the reaction of peroxides and free radicals with thiobarbituric acid. It is recommended to determine the contents of TBARS by thin-layer chromatography (Baeza, 2004).

According to Surai (2002), the oxidative defence system has three levels (I, control of oxidative processes; II, neutralizing free radicals; III, enzymatic repair of damaged components). On the first level, the extent of oxidation is controlled by glutathione peroxidase, which has selenium as a primary cofactor. The most important antioxidant is α-tocopherol, which acts on the second level of the oxidative defence system by scavenging and neutralizing free radicals (Surai, 2002). Deposition of α-tocopherol in broiler thigh muscle directly depends on the content of polyunsaturated fatty acids (PUFAs) and α-tocopherol in the diet. Cortinas et al. (2003) showed that even when using diets with a high PUFA content, distinct amounts of α-tocopherol can be deposited in the muscle (Table 9). In the same way, selenium may be enriched in chicken muscle (Table 10). The achieved content amounts up to 114 and 170% of the RDI (70 and/or 47 μg/d) for UK and German consumers, respectively. Besides this enrichment, poultry meat contains distinct amounts of several minerals, trace elements and vitamins (Table 11).

Table 9 Effect of dietary unsaturation and α-tocopherol acetate supplementation on α-tocopherol content in thigh meat (mg/kg muscle; Cortinas et al., 2003)

Dietary PUFA (mg/kg)	Global mean	α-tocopheryl acetate (mg(kg)			
		0	100	200	400
27	23.2[x]	0.33[g]	12.6[ef]	24.3[cd]	55.5[a]
38	18.5[y]	0.43[g]	12.9[ef]	19.1[de]	41.6[b]
48	14.6[z]	0.10[g]	10.3[ef]	17.6[de]	30.5[c]
59	12.9[z]	0.04[g]	6.05[fg]	15.0[ef]	10.6[c]
Global mean		0.22[z]	10.5[y]	19.0[x]	39.6[w]

[a,b,c,d,e,f,g] indicate significant differences between single means (P < 0.05).

[w,x,y,z] indicate significant differences between global means (P < 0.05).

Table 10 Examples of selenium-enriched broiler meat

Diet (μg/kg)	μg/100 g	% of RDI	Reference
840	39	83	Haug et al. (2008)
390	80	170	Cvrtila et al. (2005)
800	28	60	Yaroshenko et al. (2004)

RDI: recommended daily intake.

Table 11 Contents of minerals, trace elements and vitamins in chicken and turkey breast meat without skin (Heseker and Heseker, 1999)

Nutrient	per 100 g	Chicken	Turkey
Mg	mg	27	20
Ca	mg	14	13
Na	mg	66	46
J	µg	0.4	1.5
Vitamin A	µg	25	1.0
Vitamin E	mg	0.3	0.9

3 Important physical and functional properties of poultry meat

3.1 Physical properties

Colour is one of the most important sensory attributes. Customers have distinct expectations on the colour of their food. In poultry meat, skin colour is the first visible quality attribute. It mainly depends on the feeding of birds, but it may also be influenced by genotype. Lipid-soluble carotenoids from the feed are transported and deposited to subcutaneous and to depot fat. Corn and alfalfa meal contain distinct amounts of lutein and zeaxanthin which result in a yellow skin and fat colour (EFSA, 2009). These properties are used for producing the 'corn chicken' in France according to EU Directives 91/1538 and 90/1906, which prescribe at least 50% corn in the feed. Sometimes, artificial carotenoids approved as feed additives (EU Directive 70/524) are also used. In most countries, poultry carcasses with pigmented skin are not requested by customers, with the exception of France (already mentioned), Austria or Mexico (Grashorn, 2016). The yellowness of the skin/fat can be measured by photometry, as described below. In chicken, skin and fat can be pigmented quite easily and pigmentation of slow-growing genotypes is more efficient than of fast-growing ones. This is especially used in organic chicken meat production to make this product distinguishable from the commercial one (Grashorn and Serini, 2006). In contrast to chicken, pigments are rarely deposited in fat tissues of turkeys.

Poultry breast meat consists mainly of white muscle fibres (fast twitched, type IIb), whereas in thigh muscles intermediate muscle fibres (fast twitched, type IIa) also occur (Remignon et al., 1995). Therefore, poultry meat is referred to as 'white' meat in contrast to the 'red' beef and pork meat. As the intramuscular fat content is rather low in poultry meat, meat colour depends mainly on the contents of deoxymoyglobin (Mb), oxymyoglobin (MbO_2) and metmyoglobin (MetMb; Lebret et al., 2015). Oxidation of the purple red Mb results in the formation of brown MetMb. Relevant methods for determining the myoglobin content are given by Petracci and Baeza (2011) and Muth (2015). Redness can be measured by photometry, where the a^* value indicates the red colour intensity. According to Berri et al. (2001) meat of slow-growing chicken genotypes is more red than the meat of commercial fast-growing ones. Thigh meat is somewhat darker and more red than breast meat. In contrast to chicken and turkey, meat of waterfowl (ducks, geese), pheasants, guinea fowl and quails is more dark and has a more intense red colour. Meat colour in chicken is not affected by feeding. Chilling and cold storage during slaughtering retain an intensive red colour, whereas heat treatment during processing results in a decrease in redness due to the destruction of myoglobin (Lebret et al., 2015). In contrast, use of modified atmosphere packaging (e.g. with oxygen) improves and retains redness due to oxidation (Orkusz et al., 2011).

Most physical properties of poultry meat depend on the species, the genotype within the species and on the specific tissue. The properties vary between chicken, turkey and waterfowl; between fast-and slow-growing genotypes and between breast and thigh muscle. Conditions before and during slaughtering and conditions and survey of meat maturation are fundamental for the physical properties of the final meat. It is well known that stress before and during slaughter impairs the maturation process by either leaving a too low amount of glycogen or a too high content of lactic acid in the muscles at the time of killing. A low glycogen content will result in a decreased rate of pH decline, whereas a high lactic acid content will result in a very rapid pH decline. There exist several methods for determining the primary conditions before the onset of the maturation process. A quite common method is the measurement of pH post-mortem. In poultry, pH is measured at the cranial end of the breast muscle. In the living bird pH is approximate neutral (7.0; Prevolnik Povše et al., 2015). As the slaughtering process is very fast in poultry this measurement has to be done within the first 15 min post-mortem. Unfortunately, the pH decline is not sufficient at that time to differentiate between different courses of the breakdown of energy sources (Fig. 1). A normal pH 15 min *post-mortem* in poultry is about 6.0. Complementary methods may be the measurement of conductivity, brightness and drip loss. Conductivity is just another way to determine the amount of H^+ ions in the muscle and does not result in a better prediction than measurement of pH. Brightness is normally measured with a photometer according to CIE (1976). Here, the colour components are arranged in a

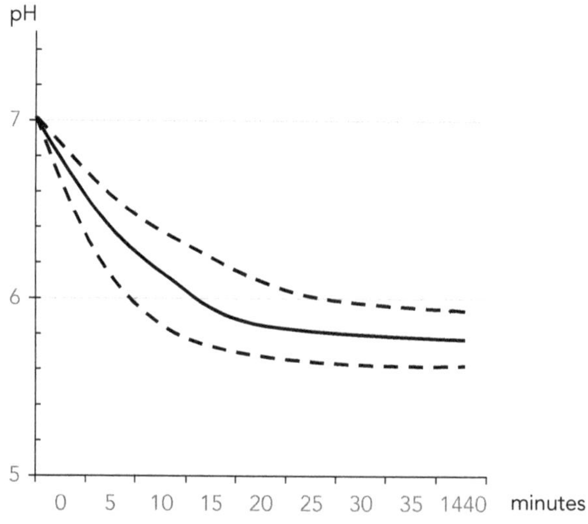

Figure 1 Course of pH decline between killing and 24 h post-mortem.

three-dimensional system of coordinates. *L** stands for brightness (0 = black, 100 = white), *a** for redness and *b** for yellowness. It was found that higher brightness of poultry breast meat correlates with a reduced WHC and a higher texture of the meat. These changes are typical for PSE meat (pale, soft, exudative) which is quite common in pigs. According to Petracci et al. (2015), PSE in poultry may be assigned to two classes: (1) fast acidifying (pH fall to <6.0 within 1 h post-mortem) and (2) acid meat (normal pH decline, but ultimate pH (pHu) <5.8). These changes may be caused by an altered cation regulation and result in paler meat with lower WHC. The nutritive value is not impaired. It is estimated that 40–60% of all broiler and turkey flocks show signs of PSE (Barbut, 1997a,b). In turkey breast meat there is a clear correlation between pH, *L** value and drip loss 14 min post-mortem (Table 12). Therefore, it is generally acknowledged that the *L** value is a good indicator for PSE condition both in broiler and turkey breast meat (Barbut, 1997a,b).

Further meat quality aberrations are deep pectoralis myopathy (DPM), white striping (WS) and wooden breast (WB). DPM was already reported in the 1970s (Jones et al., 1974) and is characterized by a degeneration of the fillet, resulting in a green and red colour. This aberration is mainly found in heavy birds and may be the result of the interruption of blood supply, probably due to the bird falling on the breast or receiving a hit to the breast. The heritability is estimated as <0.1 (Bailey et al., 2015). WS is characterized by 'superficial white striations' (Mudalal et al., 2015) and is also found predominantly in thick breast muscle pieces. In contrast to WS, WB is characterized by 'pale and bulged areas with substantial hardness' (Mudalal et al., 2015). WS and WB are especially found in breeds with a high breast muscle proportion (Table 13) and the heritability is estimated as 0.34 and <0.1, respectively (Bailey et al., 2015). Both aberrations result in impaired processing characteristics and may be indicated by the

Table 12 Drip loss and colour of turkey breast meat according to pH class (14 min p.m.; Wynveen et al., 1999)

pH	Drip loss (%)	*L**
<5.70	1.75	57.8
5.70–6.18	0.81	56.2
>6.18	0.75	54.0

Table 13 Live weight, breast meat yield and incidences of deep pectoralis myopathy, wooden breast and white striping in two broiler lines with different growth capacity (Bailey et al., 2015)

Trait	Line A	Line B
Body weight (kg)	2.33	1.91
Breast yield (%)	29.4	21.7
Deep pectoralis myopathy (%)	6.96	0.41
Wooden breast (%)	3.19	0.16
White striping (%)	49.6	14.5

texture of the caudal part of the muscle (Table 14). WS meat contains less protein and more fat and collagen (Petracci et al., 2015).

The *R*-value is another indicator for assessing meat quality early post-mortem. It is the relation between the content of Adenosin-Tri-Phosphate (ATP) and inosine monophosphate (IMP) in the muscle and reflects the degradation of energy sources. For determining this value, a meat sample has to be taken from the muscle, immediately frozen in liquid nitrogen and analysed in the laboratory. Thus, this method is too laborious for routine measurements.

The quality of the mature meat can be assessed more accurately by different methods. Measurement of the pH gives an indication whether the maturation was run normally. In 'normal' poultry meat the ultimate pH (pHu, 24 h p.m.) should be around 5.8. Again, determination of conductivity may be an alternative. Measurement of brightness will also give further information as to whether a PSE condition is to be expected. The most important quality attributes are WHC and texture of the meat.

The WHC may be determined by drip and press methods or by cooking or grilling. WHC is defined as the capacity of a given sample to retain internal or external water under standardized pressure. The lowest WHC values are attained at the isoelectric point of soluble proteins (Prevolnik-Povse et al., 2015). This means that WHC increases with the difference between the pH at the isoelectric point (5.1–5.3) and the actual pH in the muscle tissue. The drip loss is based on the free exudate. According to Honikel (1997), a piece of meat is hanged in a plastic bag for 1–3 days and the water loss is determined. In general, this method is less suited for poultry meat due to the quite low free water loss. Despite this low free water loss, WS breast meat seems to have lower and WB meat seems to have higher dripping losses than normal breast meat (Table 15).

The original press method for determining WHC was developed by Grau and Hamm (1957). Here, exactly 0.3 g of meat are taken from the sample and put on a filter paper with a given soaking capacity. The filter paper is laid between two plastic plates in a pressing tool, exhibiting a pressure of about 10 N. After 5 min of pressing, the filter paper is removed and the liquid area is marked with a pencil. The area is then determined by planimetry. Areas can then be compared directly. As this method is too costly for routine use, a modified system, called the Braunschweiger model (Fig. 2), was developed. Here, the amount of meat may vary as both the areas of the meat sample and of the liquid are determined. The quotient between the meat area and the liquid area directly reflects the

Table 14 Relations between weight and dimension of chicken breast fillets with normal and insufficient quality and texture (Mudalal et al., 2015)

Trait	Normal (N)	White striped (WS)	Wooden breast (WB)
Weight (g)	145	306	299
Top height (mm)[1]	38.1	45.7	43.9
Middle height (mm)[2]	24.7	31.2	30.5
Bottom height (mm)[3]	8.2	8.7	11.0
Compression test (kg)	2.02	2.28	4.02

[1] measured at the thickest point in the cranial part.
[2] measured at half distance of the breast length.
[3] measured from the far end of the caudal part by 1 cm towards a dorsal direction.

Table 15 Effect of breast meat abnormalities on quality traits of non-marinated chicken meat with normal and insufficient quality (Mudalal et al., 2015)

Trait	Normal	White striped	Wooden breast
Raw meat			
pH	5.80	5.90	5.87
Drip loss (%)	0.93	0.72	1.19
Cooked meat			
Cooking loss (%)	21.6	24.7	28.0
Shear force (kg)	2.37	2.35	2.19
Yellowness (b^*)	8.01	8.30	8.43

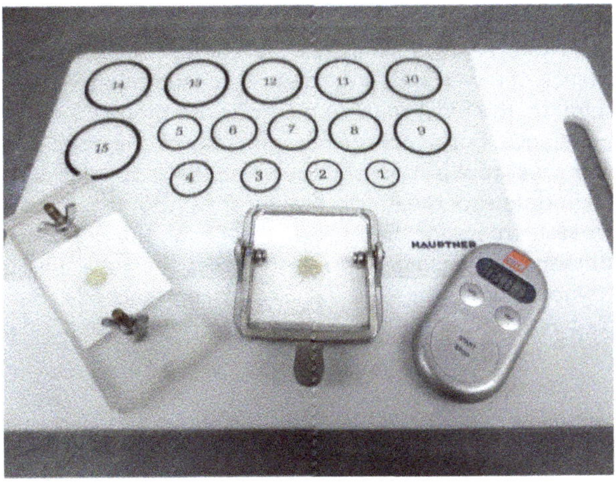

Figure 2 Press tool Braunschweiger model to determine water-holding capacity.

WHC. A quotient <0.5 indicates a reduced WHC, whereas a quotient of >0.65 indicates rather dry meat (ARMH, 1986). WHC mainly depends on genetics (fast vs. slow growth), the age (young vs. old), pre-slaughter handling (stress vs. no stress; starvation vs. no starvation), pre-slaughter energy resources (low vs. high glycogen store), slaughter conditions (electric vs. gas stunning; high temperature vs. low temperature scalding; mechanical vs. hand plucking) and post-slaughter conditions (slow vs. fast chilling, early vs. late deboning). All these factors mainly influence the meat maturation process, that is, the pH decline.

Using thermal force (cooking and grilling) for determining WHC simulates the preparation process in the kitchen. For measuring cooking losses meat samples are put in a plastic bag and are cooked until a defined core temperature is reached. This temperature has to be at least 72°C (Honikel, 1997). At this temperature, proteins are denatured. The temperature course is controlled by an insertion probe. Determination of cooking losses is less common for poultry meat, as poultry meat is mainly roasted. To determine grilling losses, the meat samples are wrapped in aluminium foil and put into a grill. Again, the temperature course has to be controlled. After cooking or grilling, meat samples have to cool down. The samples are weighed before and after cooking/grilling to calculate the losses. Normal cooking or grilling losses are in the range of 15–25% for both broiler and turkey meat.

Tenderness and/or hardness of the meat are the key sensory characteristics during meat consumption. The firmness of meat can be measured by texture analysis. According to Honikel (1998) a wide variety of methods exist for assessing meat tenderness, each having its benefits and limitations. These may be distinguished between invasive and non-destructive methods. Ampuero Kragten and Gil (2015) summarized as invasive methods for texture analysis texture profile analysis (TPA), Volodkevich Bite Tenderometer (VBT), slice shear force (SSF), Kramer shear press (KPS), MIRINZ Tenderometer, NIP Tenderometer, Meullenet–Owen razor blade (MORBS), Warner–Bratzler shear force (WB), Armour Meat Tenderometer and Tensipresser. Here, the meat sample is destroyed during measurement. But, there also exist some non-destructive methods such as spectroscopy (NIR, Raman), ultrasonics, image analysis, dual-energy x-ray (DEXA), electromyography and elastography (Ampuero Kragten and Gil, 2015). The choice of the appropriate method depends on the cut to be investigated.

The destructive texture analysis simulates the eating process of the consumer who connects the resistance of the food during chewing consumption with its tenderness and/or toughness. Therefore, the most suitable methods use thermally denaturated meat for measurement of resistance. For samples consisting only of one muscle (e.g. breast meat), the Warner–Bratzler shear tool is most suited, whereas for samples consisting of several muscles (e.g. thigh or drumstick meat), the Kramer shear tool should be used. To improve the exactness of measurement for the Kramer shear tool, the prepared meat should be minced. As for determining grilling or cooking losses the meat has to be heated up until an internal temperature of at least 72°C has been reached. This can be done by cooking, grilling or baking. After the sample is cooled down, a defined piece is taken for

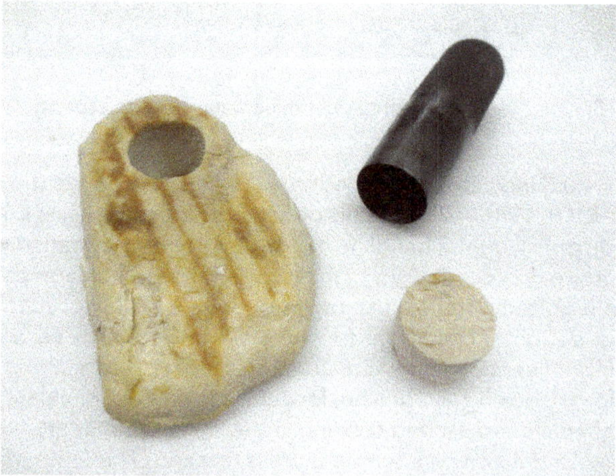

Figure 3 Punched cylinder of chicken breast meat for determining texture with the Warner–Bratzler shear tool.

texture analysis. For the Warner–Bratzler shear tool normally a cylinder with a diameter of 20 mm is punched out of the sample (Fig. 3). The upper and the lower surface are cut away as they may result in greater shear forces. Cylinders should be of similar height. The measurement is then done in a rectangular fashion to the muscle fibre direction (Fig. 4). Using the Kramer shear tool requires a distinct amount of sample. As Fig. 5 shows, about 50–60 g meat are optimal for texture measurement. The crosshead speed is important for the repeatability of the measurement. In most research studies a crosshead speed of 200–250 mm/min is applied (Ampuero Kragten and Gil, 2015), but it may vary between research facilities. Several manufacturers of texture analysers exist. The most known one is Instron (Instron Worldwide Headquarters, Norwood, MA) or Zwick (Zwick Roell Ltd., Ulm, Germany).

In general, shear values are good indicators of tenderness assessed by sensory test. Solomon et al. (2010) reported correlation coefficients between shear value and tenderness (from sensory test) ranging from 0.7 to 0.9. Texture values are highest when reaching rigor mortis and are continuously declining within 24 h post-mortem (Fig. 6). Normal values for broiler breast meat are between 15 and 25 N (Newton). The time point of deboning post-mortem is of significant importance (Fig. 7). Deboning early in the slaughter process (called hot boning) results in a high loss of intercellular fluid and an impairment of meat maturation. The result is quite tough meat. Therefore, if broiler meat should be deboned, the maturation process should either be completed after cooling down carcasses to a core

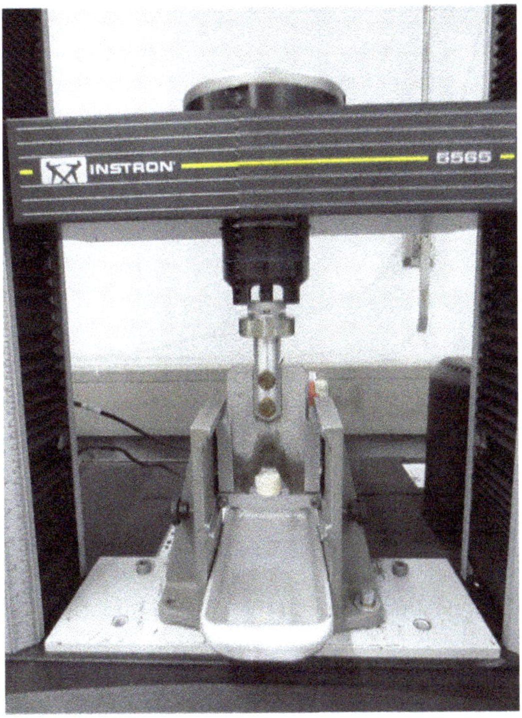

Figure 4 Texture analyzer Instron model 5565 equipped with a Warner–Bratzler shear tool.

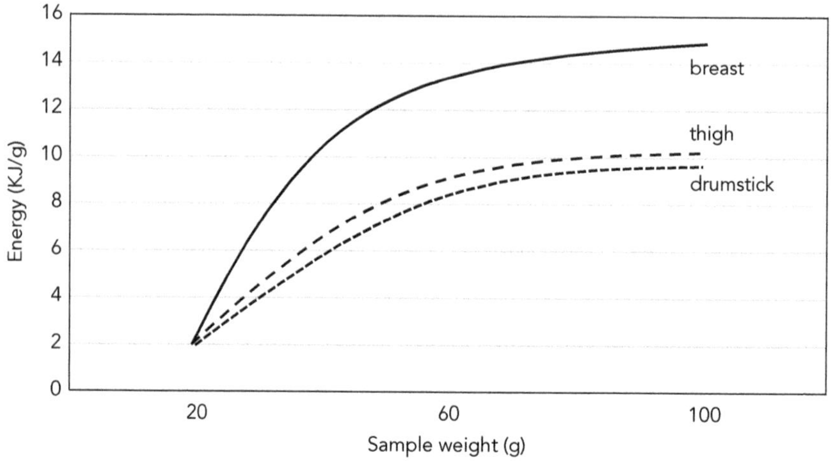

Figure 5 Relation between sample amount in the Kramer Lee shear tool and the necessary energy (after Scholtyssek, 1985).

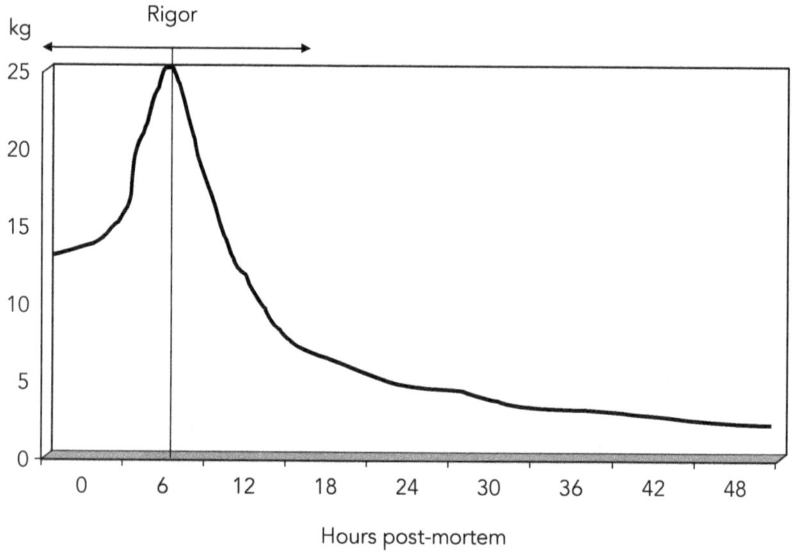

Figure 6 Typical course of breast muscle texture during meat maturation (Schreurs, 2000).

temperature of <4°C, or electrostimulation should be introduced. Carcasses have to pass a contact bar where electricity provokes contractions of muscles, by this accelerating the meat maturation process.

Tenderness of meat may be influenced by many factors. The genotype of the bird has the greatest effect on the farm. In organic production slower-growing breeds are used, which have significantly tougher breast meat than fast-growing breeds,

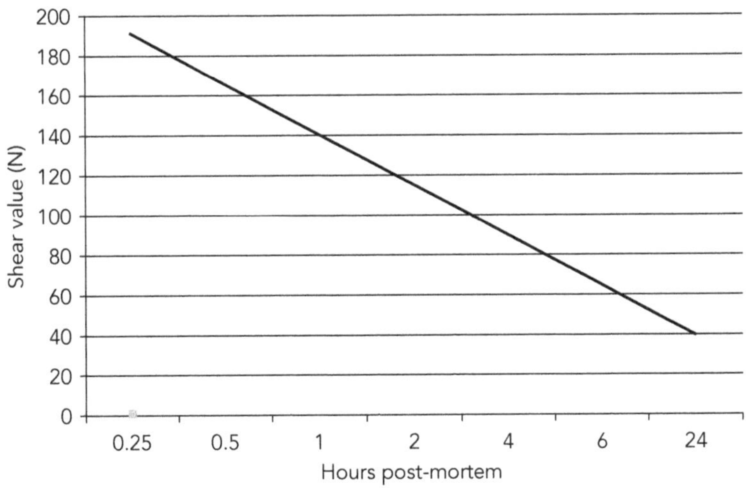

Figure 7 Shear force values of broiler breast meat after 24 h storage depending on time point of deboning post-mortem (N; Kramer Shear Tool; after Smith et al., 1992).

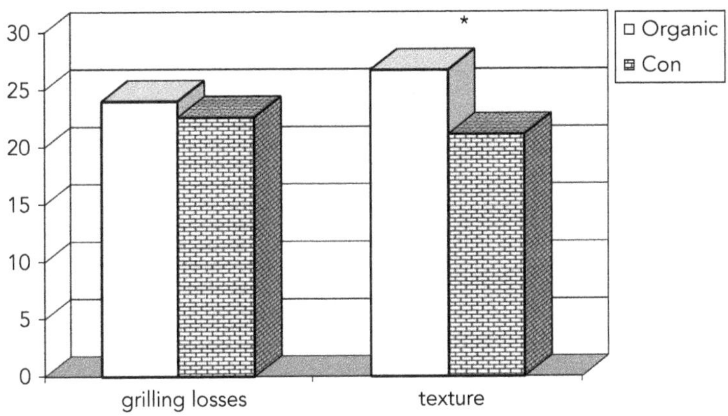

Figure 8 Grilling losses (%) and texture (N) of conventional and organic broiler breast meat (* = $P < 0.05$; Grashorn and Serini, 2006).

although grilling losses are about on the same level (Fig. 8). This is mainly caused by the higher content of soluble collagen in muscle tissues of slow-growing breeds. Meat of older birds, for example, of spent hens, is tougher than meat of young birds, and a higher intra- and inter-muscular fat content results in more tender meat. Stress affects the course of meat maturation and may be one of the factors causing meat quality aberrations such as PSE. Stunning with electricity results in excessive muscle contractions, on the one side using a distinct level of energy resources and on the other side tearing cell membranes and increasing leakage of intercellular fluid. Both side effects result in a retardation of the maturation process and a tougher meat (Fig. 9).

High-temperature scalding impairs texture of meat in the same way as mechanical plucking of feathers. Fresh meat is more tender than deeply frozen meat as during thawing of meat ice crystals are built up rupturing the cells and increasing the leakage of inter-cellular fluid.

Sarcomere length is also an indicator for tenderness of the meat. A recommended method for determination is described by Petracci and Baeza (2011).

3.2 Functional properties

Functional properties of poultry meat are important for processing. Poultry meat is widely used for making sausages and the quality level of the raw meat has to be known. Froning (1991) summarizes functional properties like water binding and absorption, viscosity,

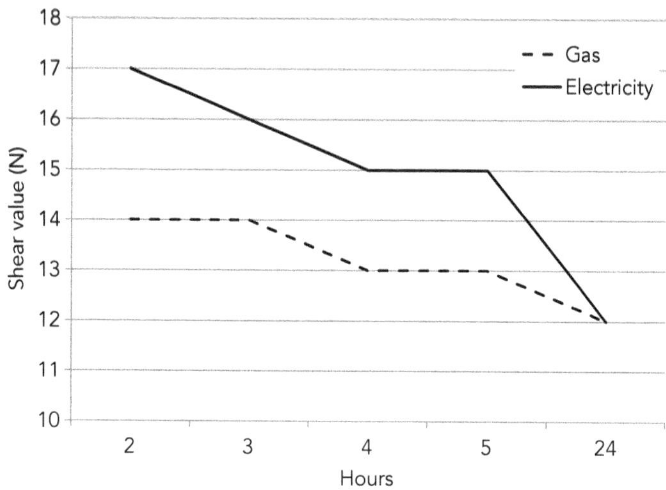

Figure 9 Texture of broiler breast meat 2–24 h post-mortem depending on stunning method (Raj et al., 1997).

Table 16 Functional properties of chicken breast meat with normal and insufficient quality (Zhang and Barbut, 2005)

	Normal	PSE	DFD
Texture (N)	28.1	42.1	60.0
Elasticity (cm)	0.91	0.93	0.96
Chewiness (N/cm)	7.92	9.72	11.82
Gumminess (N/cm^2)	8.68	10.48	12.47

PSE: pale, soft, exudative.
DFD: dark, firm dry.

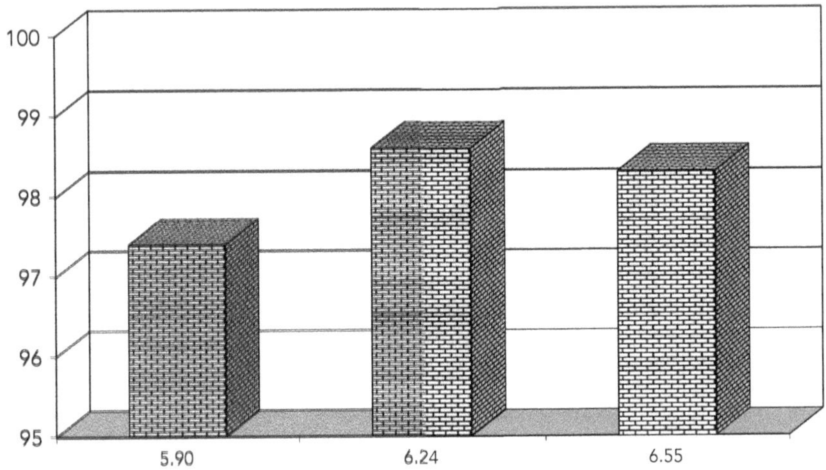

Figure 10 Technological gain (%) of cured and cooked turkey breast meat as influenced by pH value (Fernandez et al., 2002).

gelling, co/adhesion, elasticity, emulsifying, fat absorption and flavour binding. Properties related to eating are hardness, springiness, cohesiveness, gumminess and chewiness (Bourne, 1978). Whereas PSE condition has only a minor impact on consumption quality of the meat, physical measures indicate differences between normal, PSE and DFD chicken breast meat (Table 16). Fernandez et al. (2002) showed that PSE meat (indicated by low pH) has a lower technological gain and thus should not be used for scalded sausage (Fig. 10).

4 Poultry meat shelf-life

Durability of poultry meat depends on the microbial load and the content of oxidizable substances (e.g. fatty acids, cholesterol). In fresh slaughtered poultry microbial counts are lower than $10^5/cm^2$ surface or $10^6/g$ meat (Table 17). Values inside of the muscle are lower than on edible organs. Spoilage of the meat will start when microbial counts exceed 10^7. TBARS values also increase continuously during storage, mainly depending on the content of polyunsaturated fatty acids. Normally, microbial counts are the first limiting changes in storage of poultry meat.

Fresh chicken, waterfowl or turkey meat may be stored at 2°C for 8 and 14 days, respectively (Table 18). The longer storage period for turkey meat depends on the bigger size of the muscles and the lower relative surface. Deep frozen chicken or turkey meat may be stored more than 15 months as multiplication of microorganisms is nearly stopped. In contrast, meat of waterfowl may only be stored deep frozen for up to 6 months, due to its higher fat content.

Table 17 Microbial ratios for poultry meat (Hahn, 2016)

Location	Microbial counts
Standard	10^4–10^5/cm² surface
	Max. 1.0 × 10^6/g or max. 1.0 × 10^5 Enterobacteriaceae/g
Inside the muscle	Max. 10^4/g
Edible giblets	10^5–10^6/g
Start of spoilage	10^7–10^8/cm² surface (neck skin)

Table 18 Storability of poultry meat

Species	Way of offer	Temperature (°C)	Duration
Chicken	Fresh	2	8 days
	Deep frozen	−18	15 months
Turkey	Fresh	2	14 days
	Deep frozen	−18	24 months
Waterfowl	Fresh	As chicken	
	Deep frozen	−18	6 months

5 Sensory assessment of poultry meat

Sensory assessment of food is a complex procedure. Different sensations (optical, olfactory, gustatory, haptic, acoustic) sum up the general impression. Sensations are subjective as

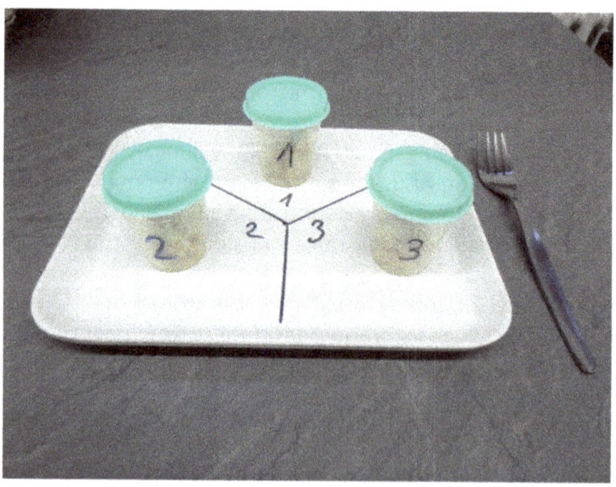

Figure 11 Arrangement of meat samples for a triangle test.

they are based on the experience and attitude of humans. Besides optical properties, the smell of the food especially gives a first indication of the quality. Aroma is defined as the total olfactory impression and results in combination with haptic sensation and taste in the overall oral sensation (= flavour; Fliedner and Wilhelmi, 1993). Taste comprises the attributes sweet, salty, bitter, sour, astringent, burning, cool and umami. Thus, panellists doing sensory tests have to be selected for their ability to differentiate between these attributes according to EN DIN 10961. Depending on the objectives of the planned sensory test, different test procedures are available. For poultry meat there exists a general recommendation of the World Poultry Science Association (WPSA, 1987). This recommendation is designed to elaborate differences between two given samples. For example, the objective may be to find out whether there exists a difference in sensory between conventional and organic poultry meat. For this, the recommendation describes the design of the test station, preparation of the samples, organization of the panel and the test conditions, the test design and the statistical evaluation of the test results. The basic principle is the use of a triangle test (Fig. 11). The test plate presents three samples, from which two are identical. The panellist has to find the single probe. As the probability to give the correct probe number is one-third, the test is repeated with a different combination of samples. The test is only accepted as correct if the panellist has identified the single probe in both tests. Depending on the number of panellists (at least six) and the number of correct identifications, the difference between the two tested samples can be tested for significance. For example, if three out of six panellists identified the single probe correctly, the difference between the two samples is significant with a statistical probability of $P < 0.05$. This decision test can be combined with an assessment test. Here the single probe is assessed whether the general taste is better than for the double probe. For this, a rank scale of up to 9 points may be applied (9 – typical chicken breast meat, 1 – strong negative deviation from the expectation).

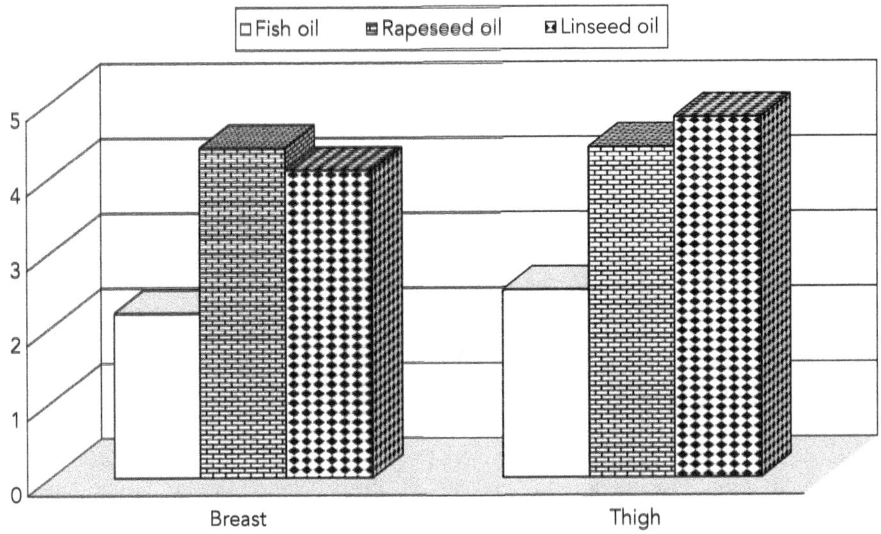

Figure 12 Effect of dietary fat source on flavour of chicken breast and thigh meat (Score 0 = strong negative deviation, 5 = typical poultry; Lopez-Ferrer et al., 1999).

This WPSA sensory test has been applied in several research studies. Comparing conventional broiler meat with Label Rouge broiler meat in France, Culioli et al. (1990) found that the Label Rouge meat is less tender but has a better flavour and overall preference. In the same way, Grashorn and Serini (2006) found that breast and thigh meat from organic broilers is less tender than the meat from conventional broilers. Using fish oil, linseed oil or rapeseed oil in broiler diets resulted in a better sensory assessment of the meat from the linseed oil and rapeseed oil groups (Lopez-Ferrer et al., 1999) (Fig. 12).

6 Conclusions

Poultry meat is a highly valuable food with many important nutrients for human nutrition. Enriching poultry meat with health-promoting substances such as n-3 fatty acids or antioxidants is important for improving its nutritive value. In the past, poultry breeding focussed mainly on increasing growth and meat yield, whereas meat quality was given less importance. Today, broilers reach their market weight within 30–35 days and turkey toms need 20 weeks to reach a market weight of 20 kg. Consumption quality deteriorated and the incidence of meat quality aberrations increased. Meat quality aberrations such as PSE (pale, soft, exudative) condition, WB and WS are quite common today and lead to a reduced processing suitability of the meat. Prolonging the fattening duration by using breeds with a reduced growth capacity reduces the incidence of this problem. This results in an improved eating quality of the meat and causes less problems during processing. Another important quality feature is microbial contamination of the meat, which still needs further improvement. Contamination with *Salmonella* is a minor risk today, whereas contamination with *Campylobacter* is still a challenge. Here, further efforts are required to further improve the microbiological quality of poultry meat. Anyway, the quality of poultry meat has already reached a high level, but this level can still be improved.

7 Where to look for further information

Searching literature data bases with the keywords 'poultry', 'meat' and 'quality' results in many hits, but only particular aspects of poultry meat quality are covered mostly. Therefore, books can provide a more extended overview of this topic. Unfortunately, most recent books written in English are mainly focusing on red meat (beef, goat, sheep, pork), with minor information on poultry meat. Specific books on poultry meat are concentrated on poultry meat processing, although these books contain basic information on raw meat quality.

The book *Poultry Meat Science* of R. I. Richardson and G. C. Mead, published by CABI Publishing in 1999, contains a lot of basic information, although some information is obsolete.

Two books on poultry meat processing have been published more recently: G. C. Mead's (2004), *Poultry Meat Processing and Quality*, CRC Press (Woodhead Publishing Ltd.) and C. M. Owens's (2010), *Poultry Meat Processing*, CRC. Both books summarise mainly poultry slaughter and processing technology, but contain also basic information on raw meat quality.

People interested in receiving more information on methods used for determining meat quality criteria are requested to download the Final Report of COST (European Cooperation in Science and Technology) Action FA1102 in cooperation with FAIM (Farm Animal Imaging) (http://www.teknologisk.dk/_/media/63061_faim%20handbook%20 ref%20methods%20for%20mq.pdf) – M. Font-i-Furnols, M. Čandek-Potokar, C. Maltin and M. Prevlovik-Povše's (2015), *A Handbook of Reference Methods for Meat Quality Assessment*. This was a project supported by the European Union Framework Programme 'Horizon 2020'. For poultry meat quality the suggestion of Working Group 5 'Poultry Meat Quality' of European Federation of World's Poultry Science Association (WPSA) is recommended: M. Petracci and E. Baeza's (2011), 'Harmonization methodologies for the assessment of poultry meat quality features', *World's Poult. Sci. J.*, 67. doi: 10.1017/ S0043933911000122. This is probably the most comprehensive and informative review on methods for determining poultry meat quality criteria.

Mainly two associations are incorporating the topic poultry meat quality in their activities: WPSA (www.wpsa.com) and Poultry Science Association (PSA, www.poultryscience.org). WPSA is running its own journal (*World's Poultry Science Journal*, more details on the homepage), which publishes mainly reviews. Furthermore, in the name of WPSA, several regular conferences are organised at changing locations. Especially, Working Group 5 'Poultry Meat Quality' of the European Federation of WPSA branches is organising its symposia on this topic in biannual intervals in odd years (for details visit www.wpsa.com). PSA is running two journals, *Poultry Science* and *Applied Poultry Science*, of which *Poultry Science* frequently publishes papers on poultry meat quality (more details on the homepage). Another journal, independent of both associations, is *British Poultry Science* (http://www.tandfonline.com/loi/cbps20#.V1EX2ZGLSUI), which has a long history in publishing papers on this topic.

In many countries, research centres have been established to deal with poultry meat quality. One of the biggest and probably the most active research centres is the National Institute for Agricultural Research – INRA (http://www.val-de-loire.inra.fr/en) in France. The most comprehensive database is maintained by the United States Department of Agriculture – USDA (http://www.usda.gov/wps/portal/usda/usdahome), which is mainly focusing on food safety.

8 References

Aletor, V. A., Eder, K., Becker, K., Paulicks, B. R., Roth, F. X. and Roth-Maier, D. A. (2003), 'The effects of conjugated linoleic acids or an a-glucosidase inhibitor on tissue lipid concentrations and fatty acid composition of broiler chicks fed a low-protein diet', *Poult. Sci.*, 82, 796–804.

Ampuero Kragten, S. and Gil, M. (2015), 'Instrumental tenderness – shear force', in M. Font-i-Furnols, M. Čandek-Potokar, C. Maltin and M. Prevolnik Povše (eds), 'A handbook of reference methods for meat quality', COST – European Cooperation in Science and Technology.

AOAC (1990), *Official Method of Analysis* (15th Ed.), Assoc. Offic. Anal. Chem., Washington, DC, USA.

ARMH (1986), Verwaltungsvorschrift Fleischhygienegesetz, 11 December 1986.

BAEZA, E. (2004), 'Measuring quality parameters', in G. C. Mead (ed.), 'Poultry meat processing and quality', Cambridge, Woodhead Publishing Limited, UK.

Bailey, R. A., Watson, K. A., Bilgili, S. F. and Avendano, S. (2015), 'The genetic basis of pectoralis myopathies in modern broiler chicken lines', *Poult. Sci.*, 94, 2870–9.

Barbut, S. (1997a), 'Problem of pale soft exudative meat in broiler chickens', *Br. Poult. Sci.*, 38, 355–8.

Barbut, S. (1997b), 'Occurrence of pale soft exudative meat in mature turkey hens', *Br. Poult. Sci.*, 38, 74–7.

Berri, C., Wacrenier, N., Millet, N. and Le Bihan-Duval, E. (2001), 'Effect of selection for improved body composition on muscle and meat characteristics of broilers from experimental and commercial lines', *Poult. Sci.*, 80, 833–8.

Bourne, M. C. (1978), 'Texture profile analysis', *Food Tech.*, 32, 62–6.

CIE (1976), 'Recommendations on uniform color spaces-color difference equations, Psychometric Color Terms', Supplement No. 2 to CIE Publication No. 15 (E-1.3.1.) 1976, 1971/(TC-1-3). Commission Internationale de l'Eclairage, Paris, France.

Cortinas, L., Villaverde, C., Baucells, M. D., Guardiola, F. and Barroeta, A. C. (2003), 'Interaction between dietary unsaturation and alpha-tocopherol levels. Vitamin E content in thigh meat', XVIth European Symposium on the Quality of Poultry Meat, St. Brieuc, France, 23–26 September.

Crespo, N. and Esteve-Garcia, E. (2001), 'Dietary fatty acid profile modifies abdominal fat deposition in broiler chickens', *Poult. Sci.*, 80, 71–8.

Culioli, J., Touraille, C., Bordes, P. and Girard, J. P. (1990), 'Characteristiques des carcasses et de la viande du poulet label fermier', *Arch. Geflügelk.*, 53, 237–45.

Cvrtila, Z., Kozačinski, L., Hadziosmanovic, M., Milinovic-Tur, S. and Filipovic, I. (2005), 'Importance of selenium in poultry meat', *Stočarstvo*, 59, 281–7.

EFSA (2009), 'Scientific Opinion on the "Safety of use of colouring agents in animal nutrition. Part III: β-apo-8'-carotenal, ethyl ester of β-apo-8'-carotenoic acid, lutein, zeaxanthin and concluding remarks"', *The EFSA Journal*, 1098, 1–48.

Farrell, D. J. (1995), 'The hearty egg is good for you', *World Poultry Misset*, 11(4), 27–9.

Fernandez, X., Sante, V., Baeza, E., Lebihan-Duval, E., Berri, C., Remignon, H., Babile, R., Le Pottier, G., Millet, N., Berge, P. and Astruc, T. (2001), 'Post mortem muscle metabolism and meat quality in three genetic types of turkey', *Br. Poult. Sci.*, 42, 462–9.

Fernandez, X., Santé, V., Baeza, E., Lebihan-Duval, E., Berri, C., Rémignon, H., Babilé, R., Le Pottier, G. and Astruc, T. (2002), 'Effects of the rate of muscle post mortem pH fall on the technological quality of turkey meat', *Br. Poult. Sci.*, 43, 245–52.

Fliedner, I. and Wilhelmi, F. (1993), 'Grundlagen und Prüfverfahren der Lebensmittelsensorik', Behr's Verlag, Hamburg, Germany.

Froning, G. W. (1991), 'Methods for measuring functional properties of poultry meat', 10th European Symposium on the Quality of Poultry Meat, Doorwerth, The Netherlands, 12–17 May.

Grashorn, M. A. and Serini, C. (2006), 'Quality of chicken meat from conventional and organic production', XII European Poultry Conference, Verona, Italy, 10–14 September.

Grashorn, M. A. (2016), 'Feed additives for influencing chicken meat and egg yolk colour', in R. Carle and R. M. Schweiggert (eds), *Handbook on Natural Pigments in Food and Beverages*, Elsevier, Oxford, UK.

Grau, R. and Hamm, R. (1957), 'Über das Wasserbindungsvermögen des Säugetiermuskels II', *Lebensmittel-Intersuchung und-Forschung*, 105, 446–60.

Hahn, G. (2016), 'Faustzahlen über Schlachtgeflügel', in K. Damme and F. Muth (eds), *Geflügeljahrbuch 2016*, Eugen Ulmer, Stuttgart, Germany.

Haug, A., Eich-Greatorex, S., Bernhoft, A., Hetland, H. and Sogn, T. (2008), 'Selenium bioavailability in chicken fed selenium-fertilized wheat', *Acta Agric. Scand. Sec. A.* 68, 65–70.

Heseker, B. and Heseker, H. (1999), 'Nährstoffe in Lebensmitteln', Umschau-Zeitschrift-Verlag, Frankfurt a.M., Germany, 2nd Edition.

Honikel, K. O. (1997), 'Reference methods supported by OECD and their use in Mediterranean meat products', *Food Chem.*, 59, 573–82.

Honikel, K. O. (1998), 'Reference methods for the assessment of physical characteristics of meat', *Meat Sci.*, 49, 447–57.

Jones, J. K., King, N. R. and Mulliner, M. M. (1974), 'Degenerative myopathy in turkey breeder hens: a comparative study of normal and affected muscle', *Br. Poult. Sci.*, 15, 191–6.

Kallweit, E., Fries, R., Kielwein, G. and Scholtyssek, S. (1988), 'Qualität tierischer Nahrungsmittel', Eugen Ulmer, Stuttgart, Germany.

Kallweit, E. (2000), 'Was ist Fleischqualität?', *DGfZ-Schriftenreihe*, Qualitätssicherung bei Landwirtschaftlichen Nutztieren', issue 18.

Karlsson, A. H. and Font-i-Furnols, M. (2015), 'Protein, fat, moisture and ash', in M. Font-i-Furnols, M. Čandek-Potokar, C. Maltin and M. Prevolnik Povše (eds), 'A handbook of reference methods for meat quality', COST – European Cooperation in Science and Technology.

Lebret, B., Prevolnik Povše, M. and Čandek-Potokar, M. (2015), 'Muscle and fat colour', in M. Font-i-Furnols, M. Čandek-Potokar, C. Maltin and M. Prevolnik Povše (eds), 'A handbook of reference methods for meat quality', COST – European Cooperation in Science and Technology.

Lopez-Ferrer, S., Baucells, M. D., Barroeta, A. C. and Grashorn, M. A. (1999), 'n-3 enrichment of chicken meat using fish oil: alternative substitution with rapeseed and linseed oils', *Poult. Sci.*, 78, 356–65.

Mudalal, S., Lorenzi, M., Soglia, F., Cavani, C. and Petracci, M. (2015), 'Implications of white striping and wooden breast abnormalities on quality traits of raw and marinated chicken meat', *Animal*, 9, 728–34.

Muth, P. (2015), 'Development of high quality niche products from local chicken and pig genetic resources', PhD thesis, University of Hohenheim, Germany.

Orkusz, A., Wołoszyn, J. and Okruszek, A. (2011), 'Characteristics of the thigh muscles colour from turkey packaged under modified atmosphere in different film types', *Arch. Geflügelk.*, 75, 196–203.

Petracci, M. and Baeza, E. (2011), 'Harmonization methodologies for the assessment of poultry meat quality features', *World's Poult. Sci. J.*, 67. doi:10.1017/S0043933911000122.

Petracci, M., Mudalal, S., Soglia, F. and Cavani, C. (2015), 'Meat quality in fast-growing broiler chickens', *World's Poult. Sci. J.*, 71, 363–73.

Prevolnik Povše, M., Čandek-Potokar, M., Gispert, M. and Lebret, B. (2015), 'pH and water holding capacity', in M. Font-i-Furnols, M. Čandek-Potokar, C. Maltin and M. Prevolnik Povše (eds), 'A handbook of reference methods for meat quality', COST – European Cooperation in Science and Technology.

Raj, A. B. M., Wilkins, L. J., Richardson, R. I., Johnson, S. P. and Wotton, S. B. (1997), 'Carcase and meat quality in broilers either killed with a gas mixture or stunned with an electric current under commercial processing conditions', *Br. Poult. Sci.*, 38, 169–74.

Remignon, H., Gardaheut, M. F., Marche, G. and Ricard, F. H. (1995), 'Selection for rapid growth increases the number and size of muscle fibres without changing their typing in chickens', *J. Muscle Res. Cell Mot.*, 16, 95–102.

Scholtyssek, S. (1985), 'Methods for tenderness measurements in poultry meat', VII European Symposium on Poultry Meat Quality, Vejle, Denmark.

Schreurs, F. J. G. (2000), 'Post-mortem changes in chicken muscle', *World's Poult. Sci. J.*, 56, 319–46.

Sirri, F., Tallarico, N., Meluzzi, A. and Franchini, A. (2003), 'Fatty acid composition and productive traits of broiler fed diets containing conjugated linoleic acid', *Poult. Sci.*, 82, 1356–61.

Smith, D. P., Fletcher, D. L. and Papa, C. M. (1992), 'Duckling and chicken processing yields and breast meat tenderness', *Poult. Sci.*, 71, 197–202.

Smith, D. M., (1997), 'Low fat and low salt poultry products', in A. M. Perarson and T. R. Dutson (eds), *Production and Processing of Healthy Meat, Poultry and Fish Products*, Black Academic & Professional, London, UK.

Solomon, M. B., Eastridge, J. S., Paroczay, E. W. and Bowker, B. C. (2010), 'Measuring meat texture', in L. M. L. Nollet and F. Toldra (eds), *Sensory Analysis of Foods of Animal Origin*, CRC Press, Boca Raton, USA.

Surai, P. F. (2002), *Natural Antioxidants in Avian Nutrition and Reproduction*, Nottingham University Press, UK.

Ternes, W., Täufel, A., Tunger, L. and Zobel, M. (2005), *Lebensmittellexikon*, Behr's Verlag Hamburg, Germany.

WPSA (1987), 'Working Group No. 5. Recommendation for a standardized method of sensory analysis for broilers', *World's Poult. Sci. J.*, 43, 64–8.

Wynveen, E. J., Bowker, B. C., Grant, A. L., Demos, B. P. and Gerrard, D. E. (1999), 'Effects of muscle pH and chilling on development of PSE-like turkey breast meat', *Br. Poult. Sci.*, 40, 253–60.

Yaroshenko, F. O., Surai, P. F., Yaroshenko, Y. F., Karadas, F. and Sparks, N. H. C. (2004), 'Theoretical background and commercial application of production of Se-enriched chicken', XXII World's Poultry Congress, Istanbul (Turkey), 8–13 June.

Zhang, L. and Barbut, S. (2005), 'Rheological characteristics of fresh and frozen PSE, normal and DFD chicken breast meat', *Br. Poult. Sci.*, 46, 687–93.

Chapter 12

Enhancing the nutritional quality of poultry meat

Michael S. Lilburn, Ohio State University, USA

1 Introduction

The nature and form of poultry meat consumption has changed considerably over the last 40 years. There has been a significant increase in the consumption of poultry meat products that are easily purchased and/or prepared (Barbut, 2012), and as such, these types of products are often referred as 'convenience foods' (Petracci and Cavani, 2012). The market demands for increased poultry meat have been largely met through genetic selection for achieving growth rate and improved protein accretion, particularly with respect to the breast muscle or white meat portions of the carcass (Havenstein et al., 2003a,b; Gaya et al., 2012). The article by Cruikshank et al. (1934) is often cited as one of the first reports establishing the link between dietary intake and yolk/lipid depot fatty acid profiles and this was later shown to be particularly true for unsaturated fatty acids (Naber, 1979). The growing awareness of the cardiovascular benefits of omega-3 fatty acids (Leaf and Weber, 1988; Barlow et al., 1990) and the potential for manipulating essential fatty acid profiles with fish oil were already shown to be feasible in earlier studies to this effect (Murty and Reiser, 1961, Navarro et al., 1972). Hargis et al. (1991) subsequently showed that total $n-3$ enrichment with fish oil was accompanied by parallel increases in the longer chain $n-3$ fatty acids eicosapentaenoic (EPA) and docosahexaenoic (DHA), which was not the case for $n-3$ enrichment of poultry tissues from plant source oils (Phettaplace and Watkins, 1989).

Much of the aforementioned research was being done at the same time that basic research on lipid metabolism in poultry was also being conducted. The objective of this chapter is

http://dx.doi.org/10.19103/AS.2016.0010.15

to review the basics of poultry carcass growth and development and the fundamentals of lipid metabolism and at the same time discuss new research developments in the matter of nutrient enhancement of poultry meat.

2 Selection for growth and feed efficiency

The continued growth of the broiler and turkey industries has been largely fuelled by significant improvements in body weight, feed efficiency and increases in breast muscle size and yield. These genetic changes have largely been due to the short generation intervals in commercial poultry versus other meat animal species and the relative ease of non-invasive assessment of differences in the size and proportion of the breast muscles.

Fairfull and Chambers (1984) had suggested that improvements in feed efficiency could be achieved by decreasing maintenance energy requirements and this has been accomplished largely by improved growth and a concomitant reduction in days to market

Figure 1 Age-related changes in size of University of Alberta Meat Control strains unselected since 1957 and 1978 and Ross 308 broilers (2005).

Within each strain, images are of the same bird at 0, 28 and 56 days of age.
Source: Zuidhof et al. (2005) University of Alberta.

weight. Zuidhof et al. (2014) summarized the various performance differences of 1957-era versus 2005 broilers (Fig. 1). They reported 3.30% yearly improvements in body weight coupled with a 2.55% reduction in feed efficiency to similar age. The positive relationship between improved feed efficiency and genetic reductions in carcass fat described by Washburn et al. (1975) and Fairfull and Chambers (1984) supports numerous studies which showed both the variability in carcass fat among commercial broiler crosses (Griffiths et al., 1977) and the genetic relationships between abdominal and carcass fat accretion and feed efficiency in poultry (Pym and Solvyns, 1979; Leclercq et al., 1980; Lin et al., 1980; Lilburn et al., 1982; Becker et al., 1984; Whitehead and Griffin, 1984). These studies were accompanied by studies in which biochemical and physiological relationships underlying fat accretion were also described (Griffin and Whitehead, 1982; Hood and Pym, 1982). Genetic improvements in live production and carcass traits, however, have been accompanied by negatively correlated muscle or meat quality traits due to changes in both the anatomical and metabolic characteristics of individual breast muscles (Dransfield and Sosnicki, 1999).

3 Understanding lipid metabolism and carcass lipid deposition

When considering the lipid portion of a poultry carcass and the potential for nutritional modulation, some consideration needs to be given to the unique aspects of *de novo* lipid biosynthesis and its effect on carcass lipid (fatty acid) composition and deposition. A series of *in vitro* and *in vivo* studies in pigeons by Goodridge and Ball (1966, 1967) and Goodridge (1968) suggested that approximately 96% of total *de novo* lipogenesis occurred in the liver with negligible contribution from the adipose tissue itself. In subsequent studies with chickens, Leveille (1969) and O'Hea and Leveille (1968, 1969) showed that like the pigeon the liver was also the primary site of *de novo* lipogenesis, and upon fasting, liver lipid accumulation significantly declined, unlike what was observed in rats (McDonald and Johnson, 1965). This was also an era in which there was considerable research interest in the effects of dietary nutrient effects on lipid metabolism and the enzymatic mechanisms underlying these responses (Pearce, 1968; Yeh and Leveille, 1969; Yeh et al., 1970). Brady et al. (1976) went on to show that radiolabelled ^{3}H from tritium was uniformly present in both liver and carcass lipids, but the label found in the adipose tissue was primarily a result of fatty acid elongation and not of *de novo* lipogenesis.

While the basic mechanisms and tissue sites associated with lipogenesis were clearly defined during the period described above, the differential contribution of dietary fat versus *de novo* synthesized fatty acids to stored carcass lipids had been first observed 30 years earlier. Cruikshank (1934) was among the first to show that 'egg fat and depot fat' could be influenced by the source of dietary fat, while Reiser and Gibson (1950) later contributed to our understanding of essential fatty acids in poultry diets, particularly linoleic acid. This was followed a decade later by numerous reports that confirmed the relationship between dietary fat sources and tissue fatty acid profiles in poultry (Feigenbaum and Fisher, 1959; Machlin and Gordon, 1961; Machlin et al., 1962; Marion and Woodroof, 1963).

Two distinct aspects of protein and lipid deposition in a poultry carcass are the different fibre types that make up white (breast) and dark (drum/thigh) muscles and the deposition of carcass fat in distinct adipose depots. The relative proportions of total lipid, triglycerides

Table 1 Total lipid content and lipid class composition of white and dark meats and skin from chickens fed a standard diet

		Lipid class (% wt total lipid)	
	Total lipid (% wt issue)	TG	PL
Breast (white meat)	0.9	43	55
Thigh (dark meat)	2.2	83	16
Skin	30.3	100	–

TG – triacylglycerol; PL – phospholipid.
Source: Adapted from Ratnayake et al. (1989).

and phospholipids within breast and thigh muscles compared with skin are shown in Table 1. The significant increase in thigh total lipid compared with the breast, as well as the higher proportion of triglycerides, supports the hypothesis that dark meat is more amenable to dietary lipid modulation and potential enrichment with selected lipid-soluble nutrients. It also explains why dark meat may be more susceptible to factors that could decrease fresh and frozen poultry meat quality via increased lipid oxidation.

As alluded to above, there is a selective deposition of fat in distinct depots in poultry and one of these depots is the easily quantified accumulation of abdominal fat. The high correlation between abdominal fat and total carcass lipid has resulted in the 'abdominal fat pad' becoming the most commonly used metric for studying dietary and genetic influences on total carcass lipid. Fraps (1943) was among the first to observe that the ratio of energy to protein in the diet (calorie: protein ratio) had an effect on carcass fat deposition and this was later confirmed by numerous other authors (Donaldson et al., 1956; Summers et al., 1965; Boomgaardt and Baker, 1973). Summers and Leeson (1985) showed that supplementing amino acids to otherwise low protein diets can partially offset some of the observed increase in carcass lipid accumulation. The data in Table 2 illustrate that dietary energy density alone is not a primary driver of carcass lipid deposition; the response is to the total protein and not to the source of the dietary protein.

4 Nutritional enrichment of poultry meat to increase polyunsaturated fat content

With respect to all animal products, there has been a major change in the concentration and relative proportions of different polyunsaturated fatty acids (PUFAs) and more recently, concern for the dietary intake of key essential polyunsaturated fatty acids (PUFA; *n*–3) and the balance between *n*–6 and *n*–3 PUFA. For an excellent introduction to the relationship between *n*–3 intake and human health, we would refer readers to Leskanich and Noble (1997). There has also been a significant increase in the attention paid to PUFA that have been reconfigured via hydrogenation and the generation of *trans* fatty acids in the process of cooking (Simopoulos, 1999). While the emphasis of this chapter will be on indirect enrichment of poultry meat products through diet, there are also excellent reviews on direct enrichment of meat products during processing and

Table 2 Effect of energy level and calorie:protein ratio on mean 4 to 8 week body weight gain, caloric intake, feed:body weight gain, abdominal fat pad size and abdominal fat pad moisture content

Contrasts	BW Gain (4–8 wk)	Feed Intake	Caloric Intake (Kcal ME)	Feed:Gain	Abdominal Fat Pad (g/100 g BW)
M.E (kcal/kg)					
2970	1448	3503[a]	10,404	2.42[a]	2.57
3190	1454	3268[b]	10,425	2.25[b]	2.55
Calorie: Protein					
187.6	1451	3503[a]	10,789[a]	2.41[a]	3.19[a]
159.5	1463	3352[b]	10,324[b]	2.29[b]	2.61[b]
138.7	1440	3296[b]	10,152[b]	2.29[b]	2.22[c]
138.7 (feathermeal)	1450	3389[ab]	10,426[ab]	2.35[ab]	2.22[c]

[a–c] Different superscripts represent significant differences, $P < 0.05$.

Source: Adapted from Griffiths et al. (1977).

fabrication of further processed products (Decker and Park, 2010; Zhang et al., 2010). As discussed by Zdunczyk and Jankowski (2013), there are species and country/regional constraints on the direct fortification of muscle products with *n*–3 PUFA resulting in 'functional' food products.

With respect to poultry products, the manipulation of yolk lipid fatty acid profiles through diet modification has received considerable research attention (Navarro et al., 1972; Hargis et al., 1991; Leskanich and Noble, 1997). 'Enriched' eggs with high levels of omega-3 fatty acids are now commonplace in many grocery stores. After the initial observations on the relationship between dietary and tissue fatty acid profiles by Cruikshank (1934), data on dietary lipid modulation with respect to improving *n*–3 fatty acid levels in edible tissues have lagged behind the research done with eggs. As with egg yolks, the modulation of tissue (muscle) fatty acids can be accomplished without effecting any major changes in the total lipid content of these tissues (Hulan et al., 1989; Phetteplace and Watkins, 1990). A seminal study by Marion and Woodroof (1963) studied the effects of different sources and combinations of dietary fat on fatty acid profiles in broiler thigh muscle, breast muscle and skin (Table 3a–c). Not surprisingly, skin lipids showed the greatest response to differences in dietary sources, while breast muscle lipids had the lowest response. This was particularly true when corn oil was the sole source of lipid and linoleic acid ($18{:}2_{n-6}$) levels increased substantially in thigh and skin lipids.

The concentrations of linolenic acid (ALA; $18{:}3_{n-3}$) were uniformly lower in breast meat than in thigh meat and the greatest *n*–3 response to supplemental menhaden fish oil was observed in skin lipids. After the initial report presented by Marion and Woodroof (1963), the appreciation for the importance of total *n*–3 fatty acid intake and more specifically, for dietary *n*–6: *n*–3 ratios relative to optimizing human health, has increased significantly.

The status of *n*–3 nutrition now includes the longer chain *n*–3 PUFA, eicosapentaenoic acid (EPA; $20{:}5_{n-3}$) and docosahexaenoic acid (DHA; $20{:}6_{n-3}$). EPA and DHA can be synthesized by elongation and desaturation of ALA, but the conversion process itself is

Table 3 Fatty acid composition of (a) breast meat, (b) thigh meat and (c) skin as influenced by diet

Added dietary fat	None	6% CO	6% BT	5% CO, 1% MO	5% BT, 1% MO
(a) Breast meat					
18:0	8.0	7.5	10.6	7.3	11.3
18:1	26.5	22.1	29.9	21.1	27.7
18:2	22.3	29.9	14.0	26.4	14.6
18:3	0.9	1.0	1.0	0.8	1.3
20:4	4.1	5.6	3.9	3.2	1.7
20:5	0.5	0.7	0.5	1.4	2.2
22:6	2.5	0.6	2.6	0.7	3.4
(b) Thigh meat					
18:0	8.4	6.0	10.4	5.2	10.6
18:1	27.7	24.7	34.5	28.7	29.8
18:2	23.6	38.7	13.8	33.8	15.3
18:3	1.0	1.8	1.6	1.0	1.9
$20{:}4_{n-6}$	3.7	3.1	2.0	1.6	1.1
$20{:}5_{n-3}$	0.5	0.8	0.3	1.0	1.7
$22{:}6_{n-3}$	1.9	0.7	1.4	1.6	2.4
(c) Skin					
18:0	5.2	4.6	8.6	3.6	9.0
18:1	33.9	28.8	39.2	28.4	32.6
18:2	26.0	44.0	13.8	39.0	16.4
18:3	1.8	1.6	2.0	2.2	2.4
$20{:}4_{n-6}$	0.3	0.6	0.2	0.4	0.3
$20{:}5_{n-3}$	0.0	0.0	0.0	1.2	1.9

CO – corn oil; BT – beef tallow; MO – menhaden oil.

Source: Adapted from Marion and Woodroof (1963).

very low (Emken et al., 1994; Lopez-Ferrer et al., 2001; Rymer and Givens, 2005). Thus, there is a considerable interest in the direct supplementation of EPA and DHA (Ribeiro et al., 2013, 2014).

When fishmeal and fish oil were easily obtained and the sustainability of the industry was not of concern, dietary supplementation with the longer chain *n*–3 PUFA was primarily through some form of marine product (i.e. fishmeal, fish oil; Tables 4, 5). As shown by Ajuyah et al. (1991; Table 6), the enhancement of poultry meat with longer chain *n*–3 can also be accomplished with plant-based ingredients (i.e. full-fat flaxseed), although the

Table 4 Potential sources of omega-3 fatty acids for poultry diets

Source	$18:3_{n-3}$	$20:5_{n-3}$	$22:5_{n-3}$	$22:6_{n-3}$	Total_{n-3}	Total_{n-6}
Flaxseed oil	53.3	–	–	–	53.3	12.7
Menhaden oil	0.3	11.0	1.9	9.1	25.5	1.5
Marine algae	–	–	3.8	7.4	11.2	–
Canola oil	12.0	–	–	–	12.0	20.2

Source: Adapted from Gonzalez-Esquerra and Leeson (2001).

performance and economics of high levels of inclusion may be an issue. The reduction in performance could be the result of a number of anti-nutritional factors present in ground flaxseed that become problematic when it is incorporated at even moderately high levels (Madhusudhan et al., 1986). More recent data on feeding high levels of full-fat flaxseed to enhance *n*–3 deposition in breast meat confirmed the previously observed negative growth effects. Zuidhof et al. (2009) found that feeding 10% ground flaxseed reduced feed intake and body weight and this was exacerbated when the levels were increased to 17%. The Canadian government threshold for omega-3-enriched poultry (300 mg/ 100 g meat) was attained in breast meat at the 10% level of flaxseed after 24 days on the enriched diets (11–35 days; 316 mg) and 4–8 days for thigh meat. In the breast meat, there was also a concomitant increase in EPA ($C20:5_{n-3}$), which led the authors to suggest that ALA elongation and desaturation mechanisms may be greater than previously thought. The largest proportion of the total *n*–3 PUFA, however, was still ALA and it was primarily deposited in the triacylglycerol fraction of both muscle types (Betti et al., 2009a). However, there were potential meat quality trade-offs associated with PUFA enrichment (Betti et al., 2009b). Drip loss percentage increased by 24% (2.5 to 3.2%) and 70% (2.5 to 4.3%) when the supplemental flaxseed was fed for 4–12- and 16–20-day periods, respectively. The same breast and thigh meat samples had a far greater oxidative potential (thiobarbituric acid reactive substances (TBARS); nmol malondialdehyde/mg meat) and the treatment effects were already evident at the onset of the experimental incubation period. The treatment effects became even greater with increasing length of the incubation period and all values were higher in thigh versus breast meat samples. These results with fresh ground meat confirmed earlier studies by Rahimi et al. (2001) with frozen breast and thigh meat from broilers fed diets with 7.5% or 15% flaxseed and canola.

Far fewer studies have been conducted on turkeys partly because of the very lean nature of turkey breast meat and the reduced potential for *n*–3 enhancement. This has also been observed in very lean, slow-growing broiler strains (Baeza et al., 2013). Jankowski et al. (2012) fed three different sources of vegetable oil (soy, rapeseed, linseed) to turkeys and the linseed oil resulted in the highest levels of PUFA in breast meat and the lowest total *n*–6/*n*–3 ratio (Table 7). Retinol (vitamin A) and a-tocopherol (vitamin E) levels, however, were highest with supplemental rapeseed oil. In another long-term study with turkeys (16 weeks; Delezie et al., 2010), fish oil supplementation was far more successful than soybean oil or linseed oil at increasing the levels of EPA and DHA, as had been previously observed in chickens.

As alluded to earlier in this report, there are longer-term concerns with the sourcing of *n*–3 fatty acids from marine sources (Pauly et al., 2005). Algae have long been known to be a potential source of nutrients for poultry (Combs, 1952). More recent studies have demonstrated

Table 5 Fatty acid composition of broiler and turkey meat if birds are fed 40–50 g/kg fishmeal

			FA composition (% FA mg/100 g meat)			
Muscle	Species (male)	Lipid content (g/kg)	ALA	EPA+DHA	ALA	EPA +DHA
White	Chicken	9	0.4	5.4	3.6	48.5
	Turkey	7	0.8	7.3	5.6	54.4
Dark	Chicken	19	0.6	2.6	11.4	49.5
	Turkey	22	1.5	3.9	33.4	85.8

Source: Adapted from Ratnayake et al. (1989) and Komprda et al. (2002).

its effectiveness as a source of pigment, lutein and omega-3 fatty acids in yolks when fed to laying hens (An et al., 2014; Anderson et al., 1991; Bruneel et al., 2013) and a potential source of dietary protein for broilers (Austic et al., 2013; Evans et al., 2015). There has been renewed interest in utilizing the defatted algal by-products resulting from the production of biofuels (Pyle et al., 2008; Kalogeropoulos et al., 2010). In the latter report, dietary supplementation with a DHA-enriched microalgae increased breast and leg DHA content by three- to five-fold in raw meat concomitant with a significant increase in carotenoids.

Dried biomass from the blue green microalgae *Spirulina* and a fresh water green microalgae, *Chlorella*, have both been shown to be high in total carotenoids and other essential nutrients, and this has recently been reviewed by Swiatkiewicz et al. (2015). These products have been shown to have significant pigmentation value in poultry diets and can be incorporated at fairly high concentrations without any negative effects (Grau and Klein, 1957; Evans, 2015). *Chlorella* was first described as a potential ingredient by Combs (1952) who reported it to be high in carotene, vitamin B_{12} and riboflavin. While algal defatted biomass could be an excellent protein source for poultry, the cost of high inclusion levels may restrict its use to pigmentation and as a source of micronutrients (Kotrbacek et al., 2015). With respect to micronutrients, *Chlorella* has the potential to accumulate selenium from its growing environment and could be a potential source of tissue enrichment (Dlouha et al., 2008). In the latter report, Se-enriched *Chlorella* was effective at decreasing the level of oxidation of broiler breast meat during refrigerated storage via increased glutathione peroxidase activity resulting from enhanced tissue accumulation. This corroborated an earlier report by Downs et al. (2000) on the enhanced incorporation of Se into breast meat when organic versus inorganic Se was incorporated into a diet at the same concentration. It has long been recognized that Se and vitamin E are two essential components of tissue defence against free radical oxidative challenges (Combs, 1981) and organic forms of supplemental Se would be more physiologically available than similar quantities of Se from inorganic sources (i.e. sodium selenite; Cantor et al., 1982).

5 Reducing tissue lipid oxidation

A central theme of this chapter has been the potential for enriching poultry (meat) products with healthful PUFA, particularly long chain $n-3$. This enrichment, however, is accompanied

Table 6 Effect of full-fat flax seeds and canola oil on fatty acid profiles in broiler meat lipid

	White meat					
	Fatty acids (%)					
	C18:2$_{n-6}$	C18:3$_{n-3}$	C20:5$_{n-3}$	C22:6$_{n-3}$	Total$_{n-3}$	Total$_{n-6}$
Control (%)	18.4	1.2	0.8	2.5	6.4	27.1
FFS 10	20.6	4.1	2.1	4.8	14.1	26.3
FFS 20	23.8	7.0	3.6	3.9	19.2	29.2
CO+FFS 10	19.8	2.5	1.6	4.5	11.4	25.3
CO+FFS 20	21.6	3.4	1.0	3.6	10.6	29.1
Dark meat						
Control (%)	18.5	1.2	0.5	2.3	5.4	24.8
FFS 10	22.1	6.9	1.3	4.7	15.6	27.6
FFS 20	26.9	10.3	2.2	2.8	18.2	31.3
CO+FFS 10	21.7	3.6	0.6	2.9	9.1	26.8
CO+FFS 20	24.4	4.8	0.5	3.1	10.2	30.2

FFS – full-fat flax seed; CO – canola oil.

Source: Adapted from Ajuyah et al. (1991).

by the potential for increased oxidation of muscle lipids, which is the major contributor to the decline in meat quality during storage (Pearson et al., 1983). During the transition of muscle to meat immediately after harvest, one factor that can significantly influence meat quality is the rate of pH decline. As glycogen (carbohydrate) stores are metabolized via glycolysis, there is an accumulation of lactic acid resulting in a progressive decline in muscle pH. The rate of pH decline is faster in white (breast) than in dark (leg) muscles and can be significantly influenced by environmental or dietary factors that influence the concentration of pre-harvest muscle glycogen (Berri, 2000). Extremes in final muscle pH (i.e. >6.0 versus <5.5) can influence the final muscle quality, with effects ranging from poor water-holding capacity and reduced pigmentation (low pH) to reduced shelf life and dry, firm and dark colouration (high pH; Allen et al., 1997). These biochemical post-harvest changes can also increase the potential for muscle lipids, which are most often membrane-associated PUFA, to undergo oxidation challenges with the decline in cellular antioxidant systems (O'Neill et al., 1998). Further processing of muscle/meat further exacerbates the breakdown of cellular membranes in muscle and facilitates the potential interactions between pro-oxidants and these membrane-associated PUFA (Morrissey et al., 1998). Betti et al. (2009b) reported that there was a consistently positive relationship between the length of time that flaxseed was incorporated into broiler diets and the ultimate pH of breast meat. A feeding period of 16 days resulted in a significant reduction in the ultimate pH by almost 30 units (5.65 versus 5.93) and the authors concluded that flaxseed supplementation and its augmentation of membrane PUFA with *n*–3 fatty acids could influence post-harvest muscle metabolism. As noted by Morrisey et al. (1994), the societal increase in pre-cooked convenience foods and

Table 7 Fatty acid composition of diets supplemented with different fatty acid sources and fatty acid composition of turkey breast meat

	Experimental treatments		
	SO	RO	LO
n–3 PUFA in diet (% total FA)	7.02	7.96	30.6
n–6/*n*–3 PUFA ratios in diet	7.3	4.4	1.0
PUFA in turkey breast meat (% FA)	29.4	27.1	37.3
n–6/*n*–3 PUFA ratios	5.6	5.1	1.2

SO – soybean oil; RO – rapeseed oil; LO – linseed oil.
Source: Adapted from Jankowski et al. (2012).

fabricated or restructured meat products has increased the predisposition for muscle lipid oxidation. Among the quality factors that have been associated with lipid oxidation are nutritive value, flavour and colour (Batifoulier et al., 2002).

From the perspective of lipid stabilization, there have been demonstrated benefits from supplemental vitamin E to enhance oxidative protection, ostensibly by increasing tissue membrane incorporation for antioxidant protection post-harvest. Sklan et al. (1983) fed diets with high (60 mg/kg) and low (5 mg/kg) all-rac *a*-tocopherol levels to turkeys and the higher level of supplementation increased tissue levels and decreased muscle lipid oxidation upon storage. Sheldon et al. (1997) later showed that 10 × and 25 × levels of DL-alpha-tocopheryl acetate resulted in the lowest incidence of off flavours in fresh and frozen samples of turkey breast meat when compared with samples taken from birds fed with control diets (1994 Poultry NRC recommended levels). These authors also reported a time and concentration effect of vitamin E on muscle colour, although this observation is not consistent with what was later reported by Mercier et al. (1998). Morrissey et al. (1997) reported that diets containing 200 mg/kg α-tocopherol acetate needed to be fed at least four weeks prior to slaughter for muscle concentrations to be optimized against post-harvest lipid oxidation in broilers.

While the antioxidant value of supplemental vitamin E has been well established, there have been studies that have looked at other types of 'natural' antioxidants that can be used as dietary supplements. The Labiatae family of plants has been reported to have antioxidant activity, with oregano, rosemary and sage having received the most attention (Economou et al., 1991). The antioxidant activity in the essential oils from these plants is attributed to the phenols within each species (Yanishlieva et al., 1999). While it is clear that both rosemary and sage have antioxidant properties when supplemented in poultry diets, the magnitude of the post-harvest response in both refrigerated and frozen poultry meat products has been variable, particularly when compared with the gold standard, α-tocopherol (Lopez-Bote et al., 1998; Yesilbag et al., 2011; Loetscher et al., 2013).

6 Summary

As alluded to throughout this chapter, consumer interest and subsequent demand for more healthful foods is driving the poultry industry to adopt dietary practices that can enhance the nutritional value of their final products, whether it be eggs or poultry meat products. The poultry industry has largely transitioned away from the use of marine products to plant-based lipids for fatty acid enrichment, and a move to an increased use of modified algae may be a next step for direct supplementation of DHA and EPA (Rymer and Givens, 2005). As with many practices in agriculture that involve dietary changes, however, the move to product enhancement comes with potential economic, processing and product trade-offs. The latter two categories may present the most immediate challenge. The body weight and the size of breast muscles in modern broilers have already resulted in visual, and sometimes physical, muscle anomalies that involve a lipid component (white striping; Kuttappan et al., 2013; Petracci et al., 2013). A move to changing the fatty acid profile associated with dietary $n-3$ enrichment would be proportionally greater in leg muscles and could present potentially increased oxidative stability questions to processors and end users as demography and politics (exports) result in a more domestic consumption of leg products. The poultry industry, however, has shown a tremendous resilience over time in its ability to address the ever-changing production and product challenges. The volume of data that has already been published and future research will no doubt allow the industry to continue the production of poultry meat products with an optimized nutrient content.

7 Where to look for further information

Fats and Fatty Acids in Poultry Nutrition and Health.
Edited by Dr. Gita Cherian and Dr. Reza Poureslami, 2012.
Nottingham University Press.

8 References

Ajuyah, A. O., K. H. Lee, R. T. Hardin, and J. S. Sim (1991). Changes in the yield and in the fatty acid composition of whole carcass and selected meat portions of broiler chickens fed full-fat oil seeds. *Poult. Sci.* 70: 2304–14.

Allen, C. D., S. M. Russell, and D. L. Fletcher (1997). The relationship of broiler breast meat color and pH to shelf-life and color development. *Poult. Sci.* 76: 1042–6.

An, B. K., J. Y. Jeon, C. W. Kang, J. M. Kim, and J. K. Hwang (2014). The tissue distribution of lutein in laying hens fed lutein fortified Chlorella and production of chicken eggs enriched with lutein. *Korean J. Food Sci. Anim. Res.* 34: 172–7.

Anderson, D. W., C. S. Tang, and E. Ross (1991). The xanthophylls of Spirulina and their effect on egg yolk pigmentation. *Poult. Sci.* 70: 115–19.

Austic, R. E., A. Mustafa, B. Jung, S. Gatrell, and X. G. Lie (2013). Potential and limitations of a new defatted diatom microalgal biomass in replacing soybean meal and corn in diets for broiler chickens. *J. Agric. Food Chem.* 61: 7341–8.

Baeza, E., P. Chartrina, V. Gigaudb, S. Tautyc, K. Meteaud, M. Lessirea and C. Berria Baeza (2013). Effects of dietary enrichment with n–3 fatty acids on the quality of raw and processed breast meat of high and low growth rate chickens. *Br. Poult. Sci.* 54: 190–8.

Barbut, S. (2012). Convenience breaded poultry meat products – new developments. *Trends Food Sci. Tech.* 26: 14–20.

Barlow, S. M., F. V. K. Young, and I. F. Duthie (1990). Nutritional recommendations for omega-3 polyunsaturated fatty acids and the challenge to the food industry. *Proc. Nutr. Soc.* 49: 13–21.

Batifoulier, F., Y. Mercier, P. Gatellier, and M. Renerre (2002). Influence of vitamin E and protein oxidation induced by H_2O_2-activated MetMb in microsomal membranes from turkey muscle. *Meat Sci.* 61: 389–95.

Becker, W. A., J. V. Spencer, L. W. Mirosh, and J. A. Verstrate (1984). Genetic variation of abdominal fat, body weight, and carcass weight in a female broiler line. *Poult. Sci.* 63: 607–11.

Berri, C. (2000). Variability of sensory and processing qualities of poultry meat. *World's Poult. Sci. J.* 56: 209–24.

Betti, M., T. I. Perez, M. J. Zuidhof, and R. A. Renema (2009a). Omega-3-enriched broiler meat: 3. Fatty acid distribution between triacylglycerol and phospholipid classes. *Poult. Sci.* 88: 1740–54.

Betti, M., B. L. Schneider, W. V. Wismer, V. L. Carney, M. J. Zuidhof, and R. A. Renema (2009b). Omega-3-enriched broiler meat: 2. Functional properties, oxidative stability, and consumer acceptance. *Poult. Sci.* 88: 1085–95.

Boomgaardt, J. and D. H. Baker (1973). Effect of dietary energy concentration on sulfur amino acid requirements and body composition of young chicks. *J. Anim. Sci.* 36: 307–11.

Brady, L., D. R. Romsos, and G. A. Leveille (1976). In vivo estimation of fatty acid synthesis in the chicken (*Gallus domesticus*) utilizing H3O. *Comp. Biochem. Physiol. Pt. B. Comp. Biochem.* 54: 403–7.

Cantor, A. H., P. D. Moorhead, and M. A. Musser (1982). Comparative effects of sodium selenite and selenomethionine upon nutritional muscular dystrophy, selenium-dependent glutathionine peroxidase and tissue selenium concentrations of turkey poults. *Poult. Sci.* 61: 478–84.

Cohen, Z., A. Vonshak, and A. Richmond (1987). Fatty acid composition of *Spirulina* strains grown under various environmental conditions. *Phytochemistry* 26: 2255–8.

Combs, G. F., Jr. (1952). Algae (*Chlorella*) as a source of nutrients for the chick. *Science* 116: 453–4.

Combs, G. F., Jr. (1981). Influences of dietary vitamin E and selenium on the oxidant system of the chick. *Poult. Sci.* 60: 2098–105.

Cruikshank, E. M. (1934). Studies in fat metabolism in the fowl. 1. The composition of the egg fat and depot fat of the fowl as affected by the ingestion of large amounts of different fats. *Biochem. J.* 28: 965–77.

Decker, E. A. and Y. Park (2010). Healthier meat products as functional foods. *Meat Sci.* 86: 49–55.

Delezie, E., J. M. Aerts, L. Maertens, and G. Huyghebaert (2010). The efficiency of long chain *n*–3 fatty acid deposition of different oils in turkeys at different ages. *Arch. Geflugelk.* 74: 51–7.

Dlouha, G., S. Sevcikova, A. Dokoupilova, L. Zita, J. Heindl, and M. Skrivan (2008). Effect of dietary selenium sources on growth performance, breast muscle selenium, glutathionine peroxidase activity and oxidative stability in broilers. *Czech J. Anim. Sci.* 53: 265–9.

Donaldson, W. E., G. P. Combs, and G. L. Romoser (1956). Studies on energy levels in poultry ration. 1. The effect of calorie-protein ratio of the ration on growth, nutrient utilization, and body composition of chicks. *Poult. Sci.* 35: 101–5.

Downs, K. M., J. B. Hess, and S. F. Bilgili (2000). Selenium source effect on broiler carcass characteristics, meat quality, and drip loss. *J. Appl. Poult. Res.* 18: 61–72.

Dransfield, E. and A. A. Sosnicki (1999). Relationship between muscle growth and poultry meat quality. *Poult. Sci.* 78: 743–6.

Economou, K., V. Oreopoulou, and C. Thomopoulos (1991). Antioxidant activity of some plant extracts of the family *Labiatae*. *J. Am. Oil. Chem. Soc.* 68: 109–13.

Emken, E. A., R. O Adlof, and R. M. Gulley (1994). Dietary linoleic acid influences desaturation and acylation of deuterium labeled linoleic and linolenic acids in young adult males. *Biochim. Biophys. Acta* 1213: 277–88.

Evans, A. M., D. L. Smith, and J. S. Moritz (2015). Effects of algae incorporation into broiler starter diet formulations on nutrient digestibility and 3 to 21 d bird performance. *J. Appl. Poult. Res.* 24: 206–14.

Fairfull, R. W. and J. R. Chambers (1984). Breeding for feed efficiency: *Poultry. Can. J. Anim. Sci.* 64: 513–27.

Feigenbaum, A. S. and H. Fisher (1959). The influence of dietary fat on the incorporation of fatty acids into body and egg fat of the hen. *Arch. Biochem. Biophys.* 79: 302–6.

Fraps, G. S. (1943). Relation of the protein, fat and energy of the ration to the composition of chickens. *Poult. Sci.* 22: 421–4.

Gaya, L. G., J. B. S. Ferraz, F. M. Rezende, G. B. Mourão, E. C. Mattos, J. P. Eler, and T. Michelan Filho (2012). Heritability and genetic correlation estimates for performance and carcass and body composition traits in a Male Broiler line. *Poult. Sci.* 85: 837–43.

Gibbs, R. A., C. Rymer, and D. I. Givens (2010). Post graduate symposium long chain *n*–3 PUFA: intake in the UK and the potential of a chicken meat prototype to increase them. *Proc. Nutr. Soc.* 69: 144–55.

Gonzalez-Esquerra, R. and S. Leeson (2001). Alternatives for enrichment of eggs and chicken meat with omega-3 fatty acids. *Can. J. Anim. Sci.* 73: 177–81.

Goodridge, A. G. (1968). The effect of starvation and starvation followed by feeding on enzyme activity and the metabolism of [U–14C] glucose in the liver from growing chicks. *Biochem. J.* 108: 667.

Goodridge, A. G. and E. G. Ball (1966). Lipogenesis in the pigeon: in vitro studies. *Am. J. Physiol.* 211: 803–8.

Goodridge, A. G. and E. G. Ball (1967). Lipogenesis in the pigeon: in vivo studies. *Am. J. Physiol.* 213: 245–9.

Grau, C. R. and N. W. Klein (1957). Sewage-grown algae as a feedstuff for chicks. *Poult. Sci.* 36: 1046–51.

Griffin, H. D. and C. C. Whitehead (1982). Plasma lipoprotein concentration as an indicator of fatness in broilers: development and use of a simple assay for plasma very low density lipoproteins. *Br. Poult. Sci.* 23: 307–13.

Griffiths, L., S Leeson, and J. D. Summers (1977). Fat deposition in broilers: effect of dietary energy to protein balance, and early life caloric restriction on productive performance and abdominal fat pad size. *Poult. Sci.* 56: 638–46.

Hargis, P. S., M. E. Van Elswyk, and B. M. Hargis (1991). Dietary manipulation of yolk lipid with menhaden oil. *Poult. Sci.* 70: 874–83.

Havenstein, G. B., P. R. Ferket, and M. A. Qureshi (2003a). Growth, livability, and feed conversion of 1957 versus 2001 broilers when fed representative 1957 and 2001 broiler diets. *Poult. Sci.* 82: 1500–8.

Havenstein, G. B., P. R. Ferket, and M. A. Qureshi (2003b). Carcass composition and yield of 1957 versus 2001 broilers when fed representative 1957 and 2001 broiler diets. *Poult. Sci.* 82: 1509–18.

Hood, R. L. and R. A. E. Pym (1982). Correlated responses for lipogenesis and adipose tissue cellularity in chickens selected for body weight gain, food consumption, and food conversion efficiency. *Poult Sci.* 61: 122–7.

Hulan, H. W., R. G. Ackman, W. M. N. Ratnayake, and F. G. Proudfoot (1989). Omega-3 fatty acid levels and general performance of commercial broilers fed practical levels of redfish meal. *Poult. Sci.* 68: 153–62.

Jankowski, J., P. Zdunczyk, D. Mikulski, J. Juskiewicz, M. Mikulska, and Z. Zdunczyk (2012). Effects of dietary soyabean, rapeseed, and linseed oils on performance, slaughter yield, and fatty acid profile of breast meat in turkeys. *J. Anim. Feed Sci.* 21: 715–27.

Kalogeropoulos, N., A. C. Chiou, E. Gavala, M. Christea, and N. K. Andrikopoulos (2010). Nutritional evaluation and bioactive microconstituents (carotenoids, tocopherols, sterols and squalene) of raw and roasted chicken fed on DHA-rich microalgae. *Food Res. Int.* 43: 2006–13.

Komprda, T., J. Zelenka, P. Bakaj, D. Kladroba, E. Blazkova, and E. Fajmonova (2003). Cholesterol and fatty acid content in meat of turkeys fed diets with sunflower, linseed, or fish oil. *Arch Geflugel.* 67: 65–75.

Kotrbacek, V., J. Doubek, and J. Doucha (2015). The chlorococcalean alga Chlorella in animal nutrition: a review. *J. Appl. Phycol.* 27: 2173–80.

Kuttappan, V. A., V. B. Brewer, A. Mauromoustakos, S. R. McKee, J. L. Emmert, J. F. Meullenet, and C. M. Owens (2013). Estimation of factors associated with the occurrence of white striping in broiler breast fillets. *Poult. Sci.* 92: 811–19.

Leclercq, B., J. C. Blum, and J. P. Boyer (1980). Selecting broilers for low or high abdominal fat: Initial observations. *Br. Poult Sci.* 21: 107–13.

Leskanich, C. O. and R. C. Noble (1997). Manipulation of the *n*-3 polyunsaturated fatty acid composition of avian eggs and meat. *World's Poult. Sci. J.* 53: 155–83.

Leveille, G. A. (1969). In vivo fatty acid and cholesterol biosynthesis in fasted and fasted-refed chicks. *J. Nutr.* 98: 367–72.

Lilburn, M. S., R. M. Leach. Jr., E. G. Buss, and R. J. Martin (1982). The developmental characteristics of two strains of chickens selected for differences in mature abdominal fat pad size. *Growth* 46: 171–81.

Lin, C. Y., G. W. Friars, and E. T. Moran (1980). Genetic and environmental aspects of obesity in broilers. *W. Poul. Sci. Jour.* 36: 103–11.

Loetscher, Y., M. Kruezer, and R. E. Messikommer (2013). Oxidative stability of the meat of broilers supplemented with rosemary leaves, rosehip fruits, chokeberry pomace and entire nettle and effects on performance and meat quality. *Poult. Sci.* 92: 2938–48.

Lopez-Bote, C. J., J. I. Gray, E. A. Gomaa, and C. J. Flegal (1998). Effect of dietary administration of oil extracts from rosemary and sage on lipid oxidation in broiler meat. *Br. Poult. Sci.* 39: 235–40.

López-Ferrer, S., M. D. Baucells, A. C. Barroeta, J. Galobart, and M. A. Grashorn (2001). $n-3$ Enrichment of chicken meat. 2. Use of precursors of long-chain polyunsaturated fatty acids: Linseed oil. *Poult. Sci.* 80: 753–61.

Machlin, L. J. and R. S. Gordon (1961). Effect of dietary fatty acids and cholesterol on growth and fatty acid composition of the chick. *J. Nutr.* 75: 157–64.

Machlin, L. J., R. S. Gordon, J. Marr, and C. W. Pope (1962). Effect of dietary fat on the fatty acid composition of eggs and tissues of the hen. *Poult. Sci.* 41: 1340–3.

Marion, J. E. and J. G. Woodroof (1963). The fatty acid composition of breast, thigh, and skin tissues of chicken broilers as influenced by dietary fats. *Poult. Sci.* 48: 1202–7.

Madhusudhan, K. T., J. P. Ramesh, T. Ogawua, K. Sasaoka, and N. Singh (1986). Detoxification of commercial linseed meal for use in broiler rations. *Poult. Sci.* 65: 164–71.

McDonald, B. E. and B. C. Johnson (1965). Metabolic response to realimentation following chronic starvation in the adult male rat. *J. Nutr.* 87: 161–7

Mercier, Y., P. Gatellier, M. Viau, H. Remignon, and M. Renerre (1998). Effect of dietary fat and vitamin E on colour stability and on lipid and protein oxidation in turkey meat during storage. *Meat Sci.* 48: 301–18.

Morrissey, P. A., S. Brandon, D. J. Buckley, P. J. A. Sheehy, and M. Frigg (1997). Tissue content of α tocopherol and oxidative stability of broilers receiving dietary α tocopheryl acetate supplement for various periods pre slaughter. *Br. Poult. Sci.* 38: 84–8.

Morrissey, P. A., P. J. A. Sheehy, K. Galvin, J. P. Kerry, and D. J. Buckley (1998). Lipid stability in meat and meat products. *Meat Sci.* 49 (Suppl. 1): 73–86.

Muhling, M., A. Belay, and B. A. Whitton (2005). Variation in fatty acid composition of *Arthrospira* (Spirulina) strains. *J. Appl. Phycol.* 17: 137–46.

Murty, N. L. and R. Reiser (1961). Influence of graded levels of dietary linoleic and linolenic acids on the fatty acid composition of hen's eggs. *J. Nutr.* 75: 287–94.

Naber, E. C. (1979). The effect of nutrition on the composition of eggs. *Poult. Sci.* 58: 518–28.

Navarro, J. G., J. C. Saavedra, F. B. Borie, and M. M. Caiozzi (1972). Influence of dietary fish meal on egg fatty acid composition. *J. Sci. Food Agri.* 23: 1287–92.

O'Hea, E. K. and G. A. Leveille (1968). Lipogenesis in isolated adipose tissue of the domestic chick (Gallus domesticus). *Comp. Biochem. Physiol.* 26: 111–20.

O'Hea, E. K. and G. A. Leveille (1969). Lipid biosynthesis and transport in the domestic chick (*Gallus domesticus*). *Comp. Biochem. Physiol.* 30: 149–59.

O'Neill, L. M., K. Galvin, P. A. Morrissey, and D. J. Buckley (1998). Comparison of effects of dietary olive oil, tallow, and vitamin E on the quality of broiler meat and meat products. *Poult. Sci.* 39: 365–71.

Pauly, D., R. Watson, and J. Alder (2005). Global trends in world fisheries: impacts on marine ecosystems and food security. *Philos. Trans. R. Soc. B. Biol. Sci.* 360: 5–12.

Pearce, J. (1968). The effect of dietary fat on lipogenic enzymes in the liver of the domestic fowl. *Biochem. J.* 109: 702–4.

Pearson, A. M., J. I. Gray, A. M. Wolzak, and N. A. Horenstein (1983). Safety implications of oxidized lipids in muscle foods. *Food Tech.* 37: 121–9.

Petracci, M. and C. Cavani (2012). Muscle growth and meat quality issues. *Nutrients* 4: 1–12.

Petracci, M., S. Mudalal, A. Bonfiglio, and C. Cavani (2013). Occurrence of white striping under commercial conditions and its impact on breast meat quality in broiler chickens. *Poult. Sci.* 92: 1670–5.

Phetteplace, H. W. and B. A. Watkins (1990). Lipid measurements in chickens fed different combinations of chicken fat and menhaden oil. *J. Agric. Food Chem.* 38: 1848–53.

Pyle, D. J., R. A. Garcia, and Z. Wen (2008). Producing docosahexaenoic Acid (DHA)-rich algae from biodiesel-derived crude glycerol: effects of impurities on DHA production and algal biomass composition. *J. Agric. Food Chem.* 56: 3933–9.

Pym, R. A. E. and A. J. Solvyns (1979). Selection for food conversion in broilers: bodycomposition of birds selected for increased body weight gain, food consumption and food conversion ratio. *Br. Poult. Sci.* 20: 87–97.

Rahimi, S., S. A. Kamran, and M. A. Karimi Torshizi (2011). Omega-3 enrichment of broiler meat by using two oil seeds. *J. Agr. Sci. Tech.* 13: 353–65.

Reiser, R. and R. Gibson (1950). The metabolism of polyunsaturated fatty acids in growing chicks. *J. Nutr.* 42: 325–36.

Ribeiro, T., M. M. Lordelo, S. P. Alves, R. J. B. Bessa, P. Costa, J. P. C. Lemos, L. M. A. Ferreira, C. M. G. A. Fontes, and J. A. M. Prates (2013). Direct supplementation of diet is the most efficient way of enriching broiler meat with $n-3$ long-chain polyunsaturated fatty acids. *Br. Poult. Sci.* 54: 753–65.

Ribeiro, T., M. M. Lordelo, P. Costa, S. P. Alves, W. S. Benevides, R. J. B. Bessa, J. P. C. Lemos, R. M. A. Pinto, L. M. A. Ferreira, C. M. G. A. Fontes, and J. A. M. Prates (2014). Effect of reduced dietary protein and supplementation with a docosahexaenoic acid product on broiler performance and meat quality. *Br. Poult. Sci.* 55: 752–65.

Richard, F. H., B. Leclercq, and C. Touraillеb (1983). Selecting broilers for low or high abdominal fat: distribution of carcass fat and quality of meat. *Br. Poult. Sci.* 24: 511–16.

Rymer, C. and D. I. Givens (2005). $n-3$ Fatty acid enrichment of edible tissues of poultry: a review. *Lipids* 40: 121–30.

Sheldon, B. W., P. A. Curtis, P. L. Dawson, and P. R. Ferket (1997). Effect of dietary vitamin E on the oxidative stability, flavor, color, and volatile profiles of refrigerated and frozen turkey breast meat. *Poult. Sci.* 76: 634–41.

Simopoulos, A. P. (1999). New products from the agri-food industry: the return of *n*-3 fatty acids into the food supply. *Lipids* 34: S297–301.

Sklan, D., Z. Tenne, and P. Budowski (1983). The effect of dietary fat and tocopherol on lipolysis and oxidation in turkey meat stored at different temperatures. *Poult. Sci.* 62: 2017–21.

Summers, J. D. and S. Leeson (1985). Broiler carcass composition as affected by amino acid supplementation. *Can. J. Anim. Sci.* 65: 717–23.

Summers, J. D., S. J. Slinger, and G. C. Ashton (1965). The effect of dietary energy and protein on carcass composition with a note for estimating carcass composition. *Poult. Sci.* 44: 501–9.

Swiatkiewicz, S., A. Arczewska-Wlosek, and D. Jozefiak (2015). Application of microalgae biomass in poultry nutrition. *W. Poult. Sci. J.* 71: 663–72.

Washburn, K. W., R. A. Guill, and H. M. Edwards, Jr. (1975). Influence of genetic differences in feed efficiency on carcass composition of young chickens. *J. Nutr.* 105: 1311–17.

Whitehead, C. C. and H. D. Griffin (1984). Development of divergent lines of lean and fat broilers using plasma very low density lipoprotein concentration as selection criterion: the first three generations. *Br. Poult. Sci.* 25: 573–82.

Yanishlieva, V., M. Marinova, M. H. Gordon, and V. G. Raneva (1999). Antioxidant activity and mechanism of action of thymol and carvacrol in two lipid systems. *Food Chem.* 64: 59–66.

Yeh, Y. Y. and G. A. Leveille (1969). Effect of dietary protein on hepatic lipogenesis in the growing chick. *J. Nutr.* 98: 356–66.

Yeh, Y. Y., G. A. Leveille, and J. H. Wiley (1970). Influence of dietary lipid on lipogenesis and on the activity of malic enzyme and citrate cleavage enzyme in liver of the growing chick. *J. Nutr.* 100: 917–24.

Yesilbag, D., M. Eren, H. Agel, A. Kovanlikaya, and F. Balci (2011). Effects of dietary rosemary, rosemary volatile oil and vitamin E on broiler performance, meat quality and serum SOD activity. *Br. Poult. Sci.* 52: 472–82.

Zdunczyk, Z. and J. Jankowski (2013). Poultry meat as functional food: modification of the fatty acid profile – a review. *Ann. Anim. Sci.* 13: 463–80.

Zhang, W., S. Xiao, H. Samaraweera, E. J. Lee, and D. U. Ahn (2010). Improving functional value of meat products. *Meat Sci.* 86: 15–31.

Zuidhof, M. J., M. Betti, D. R. Korver, F. I. L. Hernandez, B. L. Schneider, V. L. Carney, and R. A. Renema (2009). Omega-3-enriched broiler meat. 1. Optimization of a production system. *Poult. Sci.* 88: 1108–20.

Zuidhof, M. J., B. L. Schneider, V. L. Carney, D. R. Korver, and F. E. Robinson (2014). Growth, efficiency, and yield of commercial broilers from 1957, 1978, and 2005. *Poult. Sci.* 93: 2970–82.

Chapter 13

Enhancing the flavour of poultry meat

Dinesh D. Jayasena, Uva Wellassa University, Sri Lanka; and Cheorun Jo, Seoul National University, Republic of Korea

1 Introduction

A tasty, nutritious and safe meat supply can ensure a continued consumption of meat and meat products by consumers (Joo and Kim, 2011). Of all types of meat consumed in the world, chicken continues to be the cheapest meat and its consumption is expected to increase by 34% by 2018 (Jung et al., 2011). Fast-growing commercial broiler strains and a few indigenous chicken breeds have fulfilled the global demand of chicken meat over the years (Jaturasitha et al., 2008). A consumer's decision to purchase meat, however, is influenced by many criteria, including appearance, taste, aroma and texture. Flavour, one of the most important factors affecting consumers' meat-buying behaviour and preferences, even before the meat is eaten, is comprised mainly of taste and aroma (Sitz et al., 2005; Shahidi, 1989).

Raw meat has only a blood-like taste with little or no aroma. Flavour development in cooked meat is a complex scenario. Aromatic notes and most of the characteristic flavours responsible for the development of meat flavour are primarily contributed by the volatile compounds originated through thermally induced reactions between non-volatile components of lean and fatty tissues during cooking (Mottram, 1998). Therefore, cooked meat flavour relies on several production and processing factors relating to the source animal, including age, breed, sex, nutritional status, chiller ageing, presence of free amino acids and nucleotides, antioxidants, pH and method of cooking (Liu et al., 2012; Spanier et al., 1997). These ante- and post-mortem factors can change the status of poultry meat flavour. Furthermore, poultry meat is prone to the development of off-flavours through lipid oxidation, which is considered to be one of the main problems

http://dx.doi.org/10.19103/AS.2016.0010.16

regarding the quality of the poultry meat. As a result, the producers and processors in poultry meat sector, and even consumers, try to avoid off-flavour development through various prevention mechanisms to preserve natural flavour and taste (Aymerich et al., 2008).

Understanding the chemistry and factors affecting poultry meat flavour is therefore vital in order to produce the most flavourful and consistent product possible by preserving and/or improving the flavour of poultry meat. The intent of this chapter is to highlight different mechanisms and chemical compounds responsible for poultry meat flavour and off-flavour development and to describe the main factors affecting poultry meat flavour, which can be manipulated to improve the flavour of poultry meat.

2 The chemistry of poultry meat flavour

Bloody, metallic and salty taste is generally a unique characteristic of fresh uncooked meat. Its aroma resembles blood serum (Joo and Kim, 2011; Wasserman, 1972). However, significant changes take place in the flavour of meat during cooking. The main reactions involved during cooking that are responsible for flavour development are Maillard reactions, thermal degradation of lipids and Maillard-lipid interactions (Brunton et al., 2002). Flavour gets developed during cooking through complex reactions between components found in raw meat during heating. More than 1,000 chemicals have so far been identified in the volatiles of different muscle foods (Shahidi, 1994). The majority of the volatile compounds identified in cooked poultry meat have been recognized in chicken (Brunton et al., 2002). However, many of these have little influence on flavour of meat and relatively few make a key contribution to the odour and flavour of cooked meat (Aliani and Farmer, 2005). Melton (1999) has named sweet, sour, salty, bitter and the 'umami' or savoury taste as the basic tastes of meat. Hydrocarbons, aldehydes, ketones, alcohols, furans, thriphenes, pyrroles, pyridines, pyrazines, oxazoles, thiazoles, sulphurous compounds and many others have been identified as the flavour and aroma compounds found in meat (Ho et al., 1994; MacLeod, 1994). Sulphurous and carbonyl compounds are considered to be the principal contributors to meat flavour (Perez-Alvarez et al., 2010).

With respect to poultry meats, in particular, chicken meats, many of their key flavour and odour compounds, together with the mechanisms for the formation, have been identified (Aliani and Farmer, 2005). According to Shi and Ho (1994), 16 primary odour components have been identified in chicken broth, of which 14 are structurally identified. It has been further demonstrated that 2-methyl-3-furanthiol, generated from the Maillard reaction and lipid oxidation, is the most vital chemical compound responsible for the meaty flavour of chicken broth. In addition, other volatile compounds originated from the above two reactions include 2-furfurylthiol, methionol, 2,4,5-trimethylthiazole, nonanol, 2-trans-nonenal, 2-formyl-5-methylthiophene, *p*-cresol, 2-trans-4-trans-nonadienal, 2-trans-4-trans-decadienal, 2-undecenal, β-ionone, γ-decalactone and γ-dodecalactone. These compounds are obviously the major sources of chicken flavour (Varavinit et al., 2000; Shi and Ho, 1994). Sasaki et al. (2007) showed that components responsible for umami characteristics also contribute to the brothy taste of meat. With respect to the primary odourants of the broth, 2-trans-4-trans-decadienal and γ-dodecalactone predominated in chicken broth compared to that of beef (Table 1).

Table 1 Flavour dilution factors of odourants identified in broths from chicken and beef

Compounds	Flavour dilution factor		Odour description
	Chicken	Beef	
2-Methyl-3-furanthiol	1024	512	Meat-like, sweet
bis (2-Methyl-3-furyl) disulphide	<16	2048	Meat-like
2-furfurylthiol	512	512	Roasty
2,5-dimethyl-3-furanthiol	256	<16	Meaty
3-mercapto-2-pentanone	128	32	Sulphurous
Methionol	128	512	Cooked potato
2,4,5-trimethylthiazole	128	<16	Earthy
2-formyl-5-methylthiophene	64	64	Sulphurous
Phenylacetaldehyde	16	64	Honey-like
2-trans-4-trans-decadienal	2048	<16	Fatty
2-trans-4-cis-decadienal	128	<16	Fatty, tallowy
2-undecenal	256	<16	Tallowy, sweet
γ-dodecalactone	512	<16	Tallowy, fruity
γ-decalactone	64	<16	Peach-like
Nonanol	64	<16	Tallowy, green
2-trans-nonenal	64	<16	Tallowy, fatty
2-trans-4-trans-nonadienal	64	<16	Fatty
β-ionone	64	<16	Violet-like
p-cresol	64	<16	Phenolic

Source: Adapted from Gasser and Grosch, 1990.

3 Flavour precursors of poultry meat

The major flavour precursors found in meat including poultry meat can be divided into two main groups: water-soluble components and lipids (Mottram, 1998). Differences in lipid-derived volatile components between species are mainly responsible for the species differences in flavour (Perez-Alvarez et al., 2010), whereas the precursors supplied by lean tissues generate the meaty flavour common to all cooked meats (Mottram, 1998). Koriyama et al. (2002) showed that eicosapentaenoic acid and docosahexaenoic acid suppressed sourness and bitterness but increased sweetness and umami characteristics. Takahashi et al. (2012) concluded that arachidonic acid can serve as a flavour enhancer by activating the transient receptor potential cation channel, subfamily M, member 5 (TRPM5), which is considered a component of the sweet, bitter and umami taste pathways of type II receptor cells.

Free sugars, sugar phosphates, nucleotide-bound sugars, free amino acids, peptides, nucleotides and other nitrogenous components such as thiamine are considered as the

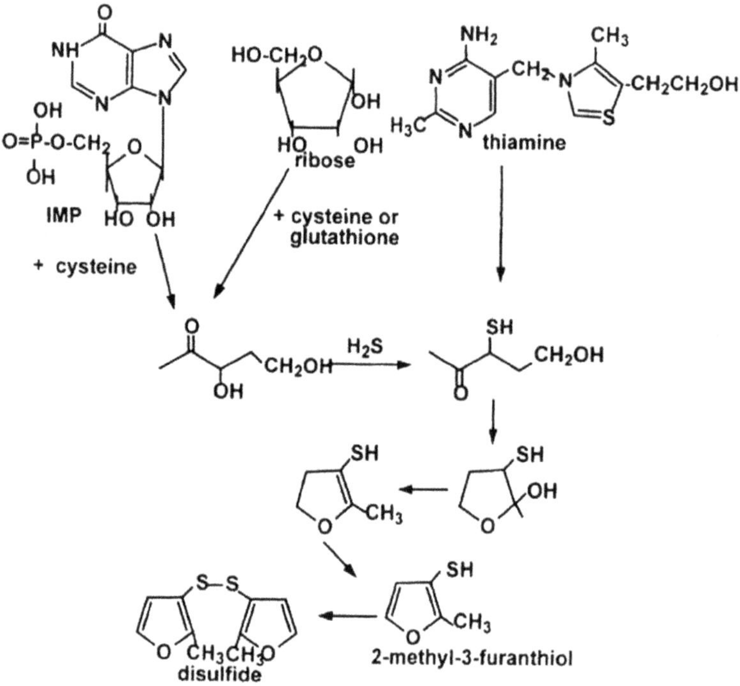

Figure 1 Mechanism for formation of 2-methyl-3-furanthiol and its disulphide by interaction of pentose sugars with cysteine or glutathione or by degradation of thiamine (Shi and Ho, 1994).

main water-soluble flavour precursors (Mottram, 1998). The reaction of cysteine and sugar can lead to the characteristic meat flavour, especially for chicken and pork (Varavinit et al., 2000). This was further confirmed by research where it was demonstrated that the quantities of carbohydrates (reducing sugars) and amino acids, in particular ribose and cysteine, are reduced during heating (Jayasena et al., 2014, 2015; Meinert et al., 2009). The main carbohydrates with flavour-forming potential include ribose, ribose-5-phosphate, glucose and glucose-6-phosphate (Meinert et al., 2009). Ribose, which is associated with the ribonucleotides in the muscle, is highly involved in flavour-producing reactions during heating of meat (Mottram, 1998).

Many flavour compounds, however, may be formed by two or more possible mechanisms. For instance, principal precursors for 2-methyl-3-furanthiol, which is responsible for the meaty flavour of chicken broth (Aliani and Farmer, 2005; Shi and Ho, 1994), are pentose sugars and cysteine/cysteine/glutathione or thiamine. As pentose sugars are mainly derived from ribonucleotides, in particular inosine-5′-monophosphate (IMP), 2-methyl-3-furanthiol and its disulphide can be formed from reaction of IMP or ribose with cysteine or cysteine or glutathione or by thiamine degradation as shown in Fig. 1 (Melton, 1999; Shi and Ho, 1994). Therefore, the relative importance of each precursor for flavour generation in cooked chicken meat is still unclear (Aliani and Farmer, 2005).

3.1 Lipid-derived flavour compounds

The role of lipid-derived carbonyl compounds in poultry flavour has been assessed by many researchers over the years. Desirable as well as undesirable flavours can result in meat due to lipid oxidation as lean meat contains intramuscular triglycerides and structural phospholipids. Mild thermal oxidative changes of lipids lead to the generation of desirable flavour compounds and aromas in cooked meats (Shahidi, 2002).

Similar to other meats, the flavour development of poultry meat is partly attributed to its lipids (Perez-Alvarez et al., 2010). Lipid degradation, primarily the oxidation of the fatty acid components of lipids, generates several hundred volatile compounds in cooked meat, including aliphatic hydrocarbons, aldehydes, alcohols, ketones, esters, carboxylic acids, some aromatic hydrocarbons and oxygenated heterocyclic compounds such as lactones and alkyl furans (Mottram, 1998). A total of 41 out of the 193 total compounds reported in the flavour of roasted chicken are lipid-derived aldehydes. Selected aldehydes identified in roasted and cooked chicken flavour are shown in Table 2 (Shi and Ho, 1994). According to Table 2, hexanal and 2,4-decadienal are the most abundant aldehydes identified in chicken flavour which are known to be the primary oxidation products of linoleic acid. However, 2,4-decadienal is considered a more important odourant for chicken flavour than hexanal due to its much lower odour threshold (Shi and Ho, 1994).

Phospholipids are much more important in the development of aroma volatiles during the cooking of meat than triglycerides (Mottram, 1998). This is attributed to the presence of a much higher proportion of unsaturated fatty acids, including significant amounts of PUFAs such as arachidonic acid (20:4) in phospholipids (Mottram and Edwards, 1983). Saturated and unsaturated aldehydes having green, fatty and tallowy aroma play a vital role in all cooked meat aroma (Mottram, 1998). The higher levels of unsaturated fatty acids compared with red meat generate more unsaturated volatile aldehydes, which are vital in the determination of specific aromas of chicken meat (Noleau and Toulemonde, 1987). Further, the aroma of fat-fried food is reported to be due to 2,4-decadienal (Mottram, 1998). Therefore, aliphatic aldehydes contribute to the fatty flavours of cooked meat, including poultry meat.

A different composition of volatiles that is responsible for the desirable flavours is formed due to rapid reactions taking place in cooked meat (Mottram, 1998) compared to those volatiles formed during long-term storage that lead to rancid off-flavours. In chicken meat, lack of α-tocopherol is, however, considered as the main reason for meat flavour deterioration and formation of undesirable 'warmed over flavour (WOF)' in chicken meat products (Shahidi, 2002). Lipid-derived compounds in meat volatiles have greater odour threshold values as opposed to sulphur- and nitrogen-containing heterocyclic compounds, which make them less significant as contributor to aroma. Even at relatively low concentrations, heterocyclic compounds possess a significant effect on aroma due to their low odour threshold values (Mottram, 1998).

3.2 Flavour compounds from the Maillard reaction

The Maillard reaction is one of the main chemical reactions that take place during cooking of meat, including poultry meat. This typically occurs between amino compounds and reducing sugars and eventually results in a large number of compounds responsible

Table 2 Selected aldehydes identified in roasted and cooked chicken flavour

	Concentration (mg per kg)	
Aldehyde	Roasted chicken	Cooked chicken meat
Butanal	0.133	
Pentanal	0.319	
Hexanal	1.804	25.6
Heptanal	0.212	2.1
Octanal	0.422	2.3
Nonanal	0.467	1.7
Decanal	0.052	0.3
Undecanal	0.058	
Dodecanal	0.022	
Tridecanal	0.151	
Tetradecanal	0.125	0.2
Pentadecanal	0.383	
Hexadecanal	19.788	1.4
Heptadecanal	0.276	0.1
Octadecanal	2.664	
trans-2-butenal	tr	
cis-2-pentenal	tr	
trans-2-pentenal	0.085	1.1
cis-2-hexenal	tr	
trans-2-hexenal	0.060	0.3
trans-2-heptenal	0.104	1.2
cis-2-octenal	0.004	
trans-2-octenal	0.195	3.7
trans-2-nonenal	0.084	
cis-2-decenal	0.003	
trans-2-decenal	0.139	1.0
cis-2-undecenal	0.002	
trans-2-undecenal	0.139	0.4
trans-dodecenal	0.002	0.3
trans,cis-2,4-nonadienal	tr	
trans,trans-2,4-nonadienal	tr	0.3
trans,cis-2,4-decadienal	0.051	1.0
trans,trans-2,4-decadienal	0.137	5.2
trans,trans-2,4-undecadienal	0.001	0.2

tr=trace

Source: Adapted from Shi and Ho, 1994.

for the flavour of any meat (Mottram, 1994a). During the initial stages of this reaction, Amadori products are formed via glycosylamine as a result of condensation of the carbonyl group of a reducing sugar with amino compounds. Various sugar dehydration and degradation products such as furfural and furanone derivatives, hydroxyketones and dicarbonyl compounds are formed by rearranging and dehydration of the resulting product via deoxyosones. Subsequently, these compounds interact with other reactive components such as amines, amino acids, aldehydes, hydrogen sulphide and ammonia and the corresponding aroma compounds are formed (Mottram, 1998; Calkins and Hodgen, 2007).

Strecker degradation of amino acids by Maillard reaction-derived dicarbonyl compounds is an essentially associated reaction, during which an aldehyde is formed due to the decarboxylation and deamination of an amino acid and an α-aminoketone or amino alcohol results from the dicarbonyl compound as well. Hydrogen sulphide, ammonia and acetaldehyde are also formed by Strecker degradation when cysteine is used as the amino acid (Mottram, 1998). Maillard reaction-derived carbonyl compounds interact with these compounds forming intermediates that further involve in flavour-forming reactions. This eventually produces many important classes of flavour compounds, including furans, pyrazines, pyrroles, oxazoles, thiophenes, thiazoles and other heterocyclic compounds (Melton, 1999). Shahidi (1989) and Mottram and Madruga (1994) reported that sulphur compounds derived from ribose and cysteine, and carbonyl compounds are the principal contributors to meat flavour. Ribose is considered as the most important flavour precursor in chicken (Meinert et al., 2009). In addition, thiamine has also been proved to be an important precursor that provides a wide range of sulphur compounds (Aliani and Farmer, 2005). Due to the degradation process of nucleotides such as IMP, ribose is formed and it is then involved in a number of secondary reactions yielding a large number of volatile compounds (Kavitha and Modi, 2007). Hence, IMP is generally considered as the major nucleotide in muscle that imparts flavour to the meat (Yamaguchi, 1991). In general, the IMP content in chicken meat is 75–122 mg/100 g meat (Maga, 1983), but some differences among poultry breeds exist (Jayasena et al., 2014; Jung et al., 2011).

Formation of 2-methyl-3-furanthiol in chicken broth via the Maillard reaction involves interaction between ribose and sulphur-containing amino acids (cysteine or cystine) or peptide (glutathione). Glutathione liberates hydrogen sulphide rapidly during the initial stages of cooking, whereas cysteine does upon prolonged heating. Another important odourant in chicken broth, 2-furfurylthiol, is formed from the reaction between furfural and cysteine (Shi and Ho, 1994). In addition, the Maillard reaction-derived volatiles are the major components in meat grilled under severe conditions (Mottram, 1998).

3.3 Flavour compounds from lipid-Maillard interactions

Interaction of lipid with the Maillard reaction leads to the formation of a number of volatiles that have been identified in meat (Mottram, 1994b; Whitfield, 1992). Several thiazoles with C_4–C_8 *n*-alkyl substituents in the 2-position and some other alkyl thiazoles with much longer 2-alkyl substituents (C_{13}–C_{15}) have been reported in fried chicken and heated chicken, respectively (Farmer and Mottram, 1994; Tang et al., 1983). Melton (1999) reported that 2-octyl-4,5-dimethylthiazole found in chicken meat is also a product from the interaction of Maillard reactions and lipids.

3.4 Heterocyclic compounds in poultry meat flavour

Heterocyclic compounds such as pyrazines, thiazoles and oxazoles are usually considered to be responsible for the roast flavours in foods including poultry meat. Melton (1999) reported that a large number of heterocyclic compounds are associated with roasted, grilled, fried or pressure-cooked meats, but not boiled meat, due to higher temperatures used in those cooking methods. Different alkyl pyrazines and two classes of bicyclic compounds, 6,7-dihydro-5(H)-cyclopentapyrazines and pyrrolopyrazines, were found in meat volatiles (Mottram, 1998). Both classes of compounds increase greatly with the increasing severity of heat treatment. However, Mottram (1991) reported that boiled meat contained higher levels of sulphur-containing heterocyclic compounds such as thiophenes, trithiolanes, and trithianes that have low odour thresholds with sulphurous, onion-like and, sometimes, meaty aromas (Fors, 1983). These compounds contributed to the overall flavour and aroma of boiled meat.

Thermal degradation of cysteine and glutathione results in two major volatile compounds of fried chicken. At frying temperature (180°C), cysteine produced 3,5-dimethyl-1,2,4-trithiolanes and 2,4,6-trimethylperhydro-1,3,5-dithiazines (thialdine), whereas only the former compound was produced by glutathione under the same conditions. These volatiles were identified in cooked chicken as well (Shi and Ho, 1994). In addition to 3,5-dimethyl-1,2,4-trithiolanes, three other alkyl-substituted trithiolanes, 3,5-diisobutyl-1,2,4-trithiolane, 3-methyl-5-butyl-1,2,4-trithiolane and 3-methyl-5-pentyl-1,2,4-trithiolane, were identified in fried chicken flavour. The formation of latter two volatiles involves thermal and oxidative degradation of lipids, pentanal and hexanal (Shi and Ho, 1994). According to the same authors, flavour of deep-fat-fried foods such as fried chicken is attributed to lipid-derived aldehyde, 2,4-decadienal. Being an oxidation product of 2,4-decadienal, trans-4,5-epoxy-trans-2-decenal having a low odour threshold plays a key role in the flavour of fried chicken.

Heterocyclic compounds, mainly pyrazines, pyridines, pyrroles and thiazoles, found in fried and roasted chicken are listed in Table 3. Alkylpyrazines have roasted, nut-like or toasted flavour notes and are present in the flavours of fried chicken and roasted chicken but not in chicken broth. This confirms that the generation of pyrazines requires high temperature and low moisture. Shi and Ho (1994) reported that fried chicken flavour was further intensified by 2-pentylpyridine (strong fatty and tallow-like odour), 2-isobutyl-3,5-diisopropylpyridine (roasted cocoa-like aroma), 2-pentyl-4-methyl-5-ethylthiazole (strong paprika pepper flavour), 2-heptyl-4,5-dimethylthiazole (strong spicy flavour) and 2-octyl-4,5-dimethylthiazole (sweet fatty aroma).

4 Factors influencing the formation of flavour and off-flavour in poultry meat

Research findings have shown that several pre- and post-mortem factors influence the flavour formation in poultry meat. Hence, producers and processors in the poultry meat sector, and even consumers, should have at least a general understanding on factors influencing the flavour formation in poultry meat in order to preserve and/or enhance the flavour compounds in poultry meat during production and processing.

Table 3 Heterocyclic compounds identified in fried and roasted chicken flavours

	Compound	Fried chicken	Roasted chicken
Pyrazines	Pyrazine	+	
	2-Methylpyrazine	+	+
	2,3-Dimethylpyrazine	+	+
	2,5-Dimethylpyrazine	+	
	2,6-Dimethylpyrazine	+	+
	Trimethylpyrazine	+	+
	2-Isopropylpyrazine	+	
	2-Methyl-3-ethylpyrazine	+	+
	2-Methyl-6(5)-ethylpyrazine	+	+
	2-Butylpyrazine	+	
	2,3-Dimethyl-5-ethylpyrazine	+	+
	2,5-Dimethyl-3-ethylpyrazine	+	
	2,6-Dimethyl-3-ethylpyrazine		+
	2,6-Diethylpyrazine	+	+
	2-Methyl-5,6-diethylpyrazine	+	
	2-Methyl-3,5-diethylpyrazine	+	
	2-Methyl-3-butylpyrazine	+	
	2-Methyl-5-vinylpyrazine	+	
	2-Methyl-6-vinylpyrazine	+	
	2-Isopropenylpyrazine	+	
	6,7-Dihydro-5*H*-cyclopentapyrazine	+	
	2-Methyl-6,7-dihydro-5*H*-cyclopentapyrazine	+	
Pyridines	Pyridine	+	+
	2-Methylpyridine	+	+
	3-Ethylpyridine	+	+
	4-Ethylpyridine	+	
	2-Methyl-5-ethylpyridine	+	+
	2-Ethyl-3-methylpyridine	+	
	2-Butylpyridine	+	
	2-Pentylpyridine	+	
	2-Isobutyl-3,5-dipropylpyridine	+	

(Continued)

Table 3 (*Continued*)

	Compound	Fried chicken	Roasted chicken
Pyrroles	Pyrrole	+	+
	N-methylpyrrole		+
	2-Methylpyrrole	+	
	2-Ethylpyrrole		+
	N-acetylpyrrole		+
	2-Acetylpyrrole	+	
	2-Isobutylpyrrole	+	
	N-isobutylpyrrole	+	
	N-(2-butanoyl)pyrrole	+	
	N-furfurylpyrrole		+
Thiazoles	Thiazole	+	+
	2-Methylthiazole	+	+
	2,4,5-Trimethylthiazole	+	
	2-Methyl-4-ethylthiazole	+	
	2-Methyl-5-ethylthiazole		+
	2,4-Dimethyl-5-ethylthiazole	+	
	2-Isopropyl-4,5-dimethylthiazole	+	
	2,5-Dimethyl-4-butylthiazole	+	
	2-Isopropyl-4-ethyl-5-methylthiazole	+	
	2-Butyl-4,5-dimethylthiazole	+	
	2-Butyl-4-methyl-5-ethylthiazole	+	
	2-Pentyl-4,5-dimethylthiazole	+	
	2-Hexyl-4,5-dimethylthiazole	+	
	2-Heptyl-4,5-dimethylthiazole	+	
	2-Heptyl-4-ethyl-5-methylthiazole	+	
	2-Octyl-4,5-dimethylthiazole	+	

+ = Present

Source: Adapted from Shi and Ho, 1994.

The main determinants influencing the flavour formation in poultry meat have been identified as breed/strain of the chicken, diet of the bird, production system, presence of free amino acids and nucleotides, irradiation, high pressure treatment, cooking, antioxidants, pH, sex and ageing (Jayasena et al., 2013; Farmer, 1999). Different breeds/strains contain different levels of flavour precursors leading to various types and concentrations of volatile compounds (Jayasena et al., 2013, 2014). Additionally, chicken results in more unsaturated volatile aldehydes as compared to beef or lamb because their muscle contains higher levels of PUFAs in the triglycerides than beef and lamb (Calkins and Hodgen, 2007). These compounds contribute to the specific aromas of chicken (Mottram, 1991; Noleau and Toulemonde, 1987), which confirm the effect of lipids on the flavour of chicken primarily via the differences in fatty acid composition and the resulting carbonyls (Perez-Alvarez et al., 2010). Furthermore, positive correlations between the flavour of chicken meat and the intramuscular contents of amino acids and nucleotides were demonstrated by Takahashi et al. (2012), Rikimaru and Takahashi (2010) and Kurihara (1987). Effect of sex on chicken meat flavour was demonstrated by many researchers, although the results were not consistent. Meat from male birds received higher scores for flavour as opposed to that from female birds (Farmer, 1999; Ramaswamy and Richards, 1982). However, it was also shown that the breast and leg meat of female birds were preferred to those of male birds. Many other researchers reported no significant relationship between the two parameters (Farmer, 1999).

Diet of the bird also plays a vital role towards the flavour of chicken meat (Perez-Alvarez et al., 2010; Fanatico et al., 2007) because the types of feed mainly affect the fatty acid profile of meat. In addition, the production system has a more varying effect on taste and flavour of chicken meat (Farmer, 1999). Many research studies claim that chicken meat from alternative production systems such as free-range, organic or some combination are characterized by a better taste than meat from conventional production systems (Lin et al., 2014; Horsted et al., 2012; Sogunle et al., 2012; Seabra et al., 2001; Culioli et al., 1990). Processing steps such as chiller ageing, cooking, irradiation and high pressure treatment also affect the flavour of chicken meat. Post-mortem ageing results in the generation of many chemical compounds responsible for poultry meat flavour, including sugars, organic acids, peptides, and free amino acids (Spanier et al., 2004; Yano et al., 1995) and thereby leads to increased flavour. Cooking is important in flavour development and it affects the acceptability and volatile flavour components of poultry meat (Sanudo et al., 2000). Cooking methods such as roasting, grilling, frying and pressure cooking generate many pyrazines, pyridines, pyrroles and thiazoles compared to boiling of chicken meat (Shi and Ho, 1994). Irradiation also affects flavour and aroma of chicken meat primarily through the production of free radicals. Aldehydes (hexanal, pentanal, heptanal, octanal and nonanal) and sulphur volatiles, mainly dimethyl trisulphide, generated during irradiation result in the associated off-odour (Ahn et al., 2016a,b; Perez-Alvarez et al., 2010; Patterson and Stevenson, 1995). Effect of high pressure treatment on flavour of meat, including chicken meat, had been variable over the years.

Lipid oxidation has been considered as the primary cause of flavour deterioration and development of off-flavour known as oxidized flavours or 'WOF' in poultry meat (Shi and Ho, 1994; Lillard, 1987). Generally, WOF is generated within 24 h of refrigerated storage in precooked poultry (Graf and Panter, 1991). WOF development is also related to rancid, and sulphur/rubber sensory notes and a parallel reduction in chicken 'meaty' flavour can also be observed. A remarkable off-flavour problem in mechanically deboned chicken meat was reported and it was attributed to rancidity of fat (Shi and Ho, 1994). It was further stated by these authors that haem pigments, being the catalyst of the above reaction, are largely responsible for the off-flavour formation. However, more recent studies indicated that raw poultry meats are highly resistant to oxidative changes due to various antioxidants present in chicken meat (Min et al., 2010). Meanwhile, it has also been reported that increasing cooking temperature is associated with increased roasted, toasted, and bitter sensory notes (Perez-Alvarez et al., 2010). Studies conducted to determine the critical changes in odour compounds of boiled chicken during refrigerated storage and reheating proved that refrigerated storage and reheating of boiled chicken leading to WOF were due to loss of meaty, chicken-like and sweet odour notes and the formation of green, cardboard-like and metallic off-odours by the secondary by-products of lipid oxidation. These changes were caused primarily by an increase in hexanal (seven-fold) and six-fold decrease in both (*E,E*)-2,4-decadienal and 2-furfurylthiol (Kerler and Grosch, 1997). Rhee et al. (2005) reported that 'cardboard' flavour and 'sour' flavour intensity increased with storage time for chicken breasts, while 'cooked chicken' intensity decreased and, 'cardboard' intensity increased for chicken thighs. Results further illustrated relative storage effects on intensity of the species-specific natural meat flavours and 'cardboard' flavour, with day-3 and day-6 scores expressed relative to day-0 scores. However, chicken exhibited the slowest rate of decrease in species-specific natural flavour intensity and the slowest rate of increase in 'cardboard' flavour intensity as compared to pork and beef.

5 Strategies for preserving or enhancing the flavour of poultry meat

Chemical compounds responsible for flavour of poultry meat can be preserved and/or enhanced through manipulating the factors discussed in the above section.

5.1 Breed/strain of chicken

The flavour of chicken meat can be improved by selecting the suitable breed/strain of chicken as different breeds/strains contain different levels of flavour precursors such as fat, IMP, amino acids, arachidonic acid, eicosapentaenoic acid, and docosahexaenoic acid. Increased palatability of indigenous chicken as compared to broilers is well documented by many authors. Superior flavours of Korean native/farm chicken, Hinai-jidori chicken (Japan), and Wenchang and Xianju (China) to broilers have been reported (Tang et al., 2009; Jung et al., 2011; Kiyohara et al., 2011). Further, it was reported that concentrations of IMP – the major nucleotide in muscle that imparts flavour to the cooked meat (Yamaguchi, 1991) – in chicken meat from the same indigenous and slow-growing genotypes were higher than those in broiler chicken lines (Table 4). In addition, a more recent study revealed that Korean native chicken contained significantly higher concentrations of taste-active compounds such as reducing sugars, glutamic acid, linoleic acid, arachidonic acid and docosahexaenoic acid than commercial broilers did (Jayasena et al., 2014; Lee et al., 2012). Recently, Liu et al. (2012) showed that several unique proteins or peptides, including phosphoglucomutase 1, NRXN3 96 kDa protein, HSPB1 22 kDa protein, TF 78 kDa protein, were present in higher levels in Korean native chickens than in broilers. Authors explained that these proteins were mostly attributed to degradation of muscle protein during the post-mortem ageing period and thereby contributed to the characteristic flavour in the cooked meat.

Table 4 Inosine-5′-monophosphate contents of different chicken breeds

Breed/strain	Inosine-5′-monophosphate (mg/g)
Broilers*	1.44
Hinai-jidori chicken*	1.57
Wenchang†	1.52
Xianju†	1.44
Avian†	0.95
Lingnanhuang†	0.97
Korean native chicken‡	2.26–2.31
Broilers‡	0.94–1.54

*Rikimaru and Takahashi (2010)
†Tang et al. (2009)
‡Jayasena et al. (2014) and Jung et al. (2011)

5.2 Lipid class and fatty acid composition

Aldehydes produced during lipid degradation are probably characteristic features of certain species. Especially, the 2-alkenals such as hexenal, heptenal, octenal, nonenal, undecenal and dodecenal as well as other aldehydes, including octanal, nonanal, decanal and decadienal, are related to both chicken-specific aroma and flavour (Ramarathnam et al., 1993). Poultry and pork muscle have higher levels of PUFAs in the triglycerides than lamb or beef (Calkins and Hodgen, 2007). Thus, more unsaturated volatile aldehydes are formed in chicken and pork than in beef or lamb. These compounds contribute to the specific aromas of chicken and pork (Noleau and Toulemonde, 1987; Mottram, 1991). Hence, better and characteristic flavour in poultry meat can be achieved through changing the fatty acid profile of meat, that is, increasing PUFA content such as linoleic acid, arachidonic acid, eicosapentaenoic acid and docosahexaenoic acid via different feeds.

5.3 Intramuscular contents of amino acids and nucleotides

Positive correlations between the flavour of chicken meat and the intramuscular contents of amino acids, including glutamic acid and nucleotides, such as IMP, were demonstrated by Kurihara (1987), Rikimaru and Takahashi (2010) and Takahashi et al. (2012). In addition, Fujimura et al. (1996) indicated that glutamate and IMP contributed to the taste of chicken meat. Glutamic acid is one of the most influential amino acids for chicken meat flavour, particularly on the umami sensation, where it plays an important role singly or in combination with other taste-related compounds (Kurihara, 1987). Nishimura et al. (1988) noted that an increase in free amino acids during conditioning affected the umami taste positively. A study that analysed the chemical components in meat has suggested that free amino acids are responsible for improving the taste of meat during storage (Rikimaru and Takahashi, 2010). In contrast, Fukunaga et al. (1989) has reported that phenylalanine content has a close association with bitter taste in meat. Aliani and Farmer (2005) have investigated the most important flavour precursors affecting the sensory attributes of consumers by monitoring a series of sensory experiments. The authors isolated precursors and added them to ground raw chicken before cooking, and the cooked products were then evaluated by trained sensory panellists. On using the thiamine precursor, the panellists described the flavour of the cooked product as 'roasted' and 'vegetable soup'. In the same study, the flavours of samples with added cysteine and ribose were described as 'chicken' and 'savoury'.

However, different volatile compounds were generated when different sugars were combined with amino groups under different conditions. Cysteine and glucose produced mainly sulphur compounds, whereas more pyrazines and furans were formed by the combination of cysteine with glucose under oxidation conditions (Tai and Ho, 1997). Maillard volatile compounds from glutathione/glucose systems at pH 6 and 8 produced sulphur-containing compounds such as thiophenes, thiazoles and cyclic polysulphides, whereas furans often became the end products at a more acidic pH. When glutathione is oxidized, it becomes glutathione sulphonic acid, which produces furans, carbonyls, pyrroles and pyrazines with glucose. Sulphur-containing compounds are not formed when glutathione is oxidized (Tai and Ho, 1998). Hence, suitable pH and oxidation conditions should be implemented in order to improve/preserve the flavour compounds.

5.4 Diet

Diet of the bird also plays a vital role towards the flavour of chicken meat (Perez-Alvarez et al., 2010; Fanatico et al., 2007). Although it was considered that poultry flavour could be improved by manipulating the diet (Fanatico et al., 2007), diet can either positively or negatively affect the flavour of chicken meat. Corn-fed chicken and arachidonic-acid-enriched diet-fed chickens have produced tastier meat (Takahashi et al., 2012; Kiyohara et al., 2011; Lyon et al., 2004), while diets supplemented with fish meal have negatively affected the flavour of chicken meat (Poste, 1990). After cooking, the off-flavour was less apparent but increased when the samples were maintained overnight at 4°C and re-evaluated 24 h later (Poste, 1990). Jang et al. (2008) indicated that taste of breast meat from broiler chickens which were fed a dietary medicinal herb extract mix (at 0.3% level) consisting of mulberry leaf, Japanese honeysuckle and goldthread at a ratio of 48.5:48.5:3.0 scored higher than control meat. Furthermore, feed supplements, including dietary fat source, *dl*-α-tocopheryl acetate and ascorbic acid, influenced the flavour of chicken meat positively (Bou et al., 2001). Sensory panellists noted that variations in sensory scores (rancid aroma, rancid flavour, and acceptability) were due to dietary fat sources and α-tocopheryl acetate content. However, ascorbic acid had no influence on these scores. No adverse effect on chicken flavour was observed when 1.2% of formic acid or up to 7.5% of shrimp by-catch composed mainly of brush tooth lizard fish and herring (Poste, 1990) was added to the feed.

Diet can also be used to increase the taste-active PUFA content of poultry meat. Kiyohara et al. (2011) showed that meat obtained from Hinai-jidori chicken-fed diets containing high levels of arachidonic acid displayed better sensory attributes, including total taste intensity, umami, kokumi, aftertaste and saltiness. It was further confirmed by a similar study conducted by Takahashi et al. (2012) using broiler chicken. Correlations between sensory data and arachidonic acid content in thigh meat of broiler chickens have shown a significant positive relationship. Huang et al. (1990) demonstrated that DHA in thigh flesh could be increased without causing fishy flavour by feeding up to 3% fish oil to chickens. Additionally, Oh et al. (2012) demonstrated that the addition of local ingredients, such as perilla meal with conventional sources (flaxseed and fish oil), could be used to obtain value-enhanced meat by enhancing ω-3 PUFA content in Korean native chicken meat.

5.5 Production system

Several studies have revealed that alternative production systems such as free-range, organic, or some combination positively affect the sensory quality of poultry meat, including flavour. Free-range chicken received significantly higher taste panel scores for flavour than the cage group (Lin et al., 2014). In addition, the production system influenced the taste and flavour of the grilled chicken with birds on free-range having higher values than those managed on deep litter (Sogunle et al., 2012). There was a huge difference in the sensory profile between the conventional standard and the organic niche products. The positive aroma attributes 'chicken' and 'bouillon' were significantly more pronounced and the negative attribute 'iron/liver' was significantly lower in the organic niche broilers than in the conventional standard broilers (Horsted et al., 2012). Similarly, the panellists classified meat from birds on free-range with the highest value for flavour (Seabra et al., 2001). Culioli et al. (1990) compared the meat characteristics of chicken raised according to the French 'label fermier' norms of production with those of standard type chickens raised intensively in confinement conditions. The sensory evaluation showed that the 'label' meat had a stronger flavour than standard meat.

Sogunle et al. (2012) stated that free-range production can reduce stress due to low bird density per unit area and increase comfort and bird welfare, thereby leading to products with better taste and flavour than those obtained from conventionally reared broiler chicken (Fanatico et al., 2006). Birds are reared for a longer period of time in the unconventional rearing systems, which allows their meat to attain more intense meat flavours (Michalczuk et al., 2014; Horsted et al., 2012; Farmer, 1999). Furthermore, breast meat from outdoor-reared chicken has shown higher PUFA levels and ω-3 fatty acid levels than those obtained from indoor-reared chicken (Michalczuk et al., 2014), which might contribute to the increased flavour of meat from outdoor-reared chicken.

Several prior studies, however, have demonstrated that production systems had no significant effect on the flavour of poultry meat. For instance, Cheng et al. (2008) and Husak et al. (2008) indicated that there were no significant differences in flavour and aroma of breast and thigh meats among conventionally reared, free-range, and organic chicken. Similarly, no difference was found for flavour attribute between conventional and Label Rouge-type chicken breast fillets (Smith et al., 2012). The production system alone did not have an effect on flavour of chicken meat reared in indoor or outdoor systems (Fanatico et al., 2007). Furthermore, the intake of a subterranean clover pasture had no impact on the flavour of broiler meat. However, the consumer panel classified the meat from free-range broilers with greater scores for overall acceptance compared to indoor-reared broilers (Ponte et al., 2008). In addition, several prior studies have compared conventionally and alternatively reared broiler strains for meat sensory attributes. Due to many variables involved in those studies, no obvious trend for conventional or alternative strains for taste evaluation was detected (Smith et al., 2012).

Several factors such as breed, age at slaughter, diet, outdoor access and housing conditions were very different between the alternative (free-range, organic, and Label Rouge-type) and standard broilers, which resulted in very different sensory profiles of these products as indicated above. Furthermore, Horsted et al. (2010) emphasized the significance of selecting appropriate birds to be suitable for alternative rearing systems as in such productions systems slow-growing birds have proven well due to their lower growth rate and lower body weight, which facilitate better health status and lesser mortality (Fanatico et al., 2008; Mikulski et al., 2011).

5.6 Ageing

The most important factor that determines the final flavour of meat is post-mortem ageing (Liu et al., 2012) due to the generation of many chemical flavour compounds during this process, including sugars, organic acids, peptides, free amino acids and metabolites of adenine nucleotide metabolism. Amino acids and peptide levels change with post-mortem ageing in muscle (Spanier et al., 2004) and are known to increase during ageing (Yano et al., 1995). Nishimura et al. (1988) also noted that an increase in free amino acids during conditioning is responsible for the meaty taste. These components serve either directly as flavour components or as a pool of reactive flavour intermediates that form many of the characteristic meat flavours after cooking (Spanier et al., 1997). Because the content of free amino acids is interrelated with the taste of chicken meat, ageing can improve the flavour of poultry meat.

5.7 Cooking

Type and concentration of flavour compounds generated during thermal treatment depend on the method of cooking. For instance, cooking methods such as roasting, grilling, frying and pressure cooking generates many more pyrazines, pyridines, pyrroles and thiazoles compared to boiling of chicken meat (Shi and Ho, 1994). Many flavour compounds isolated more recently undergo alterations throughout storage and cooking (Perez-Alvarez et al., 2010). Chikuni et al. (2002) investigated and confirmed a relationship between raw and cooked meat in the contents of free amino acids, IMP and peptides. They found that intramuscular content of these umami-relevant components decreased on cooking and the decline may be caused by leaching of these substances from meat samples.

The flavour attributes of chicken meat were slightly, but positively, affected by the low temperature prolonged time (53 and 58°C for 30 h) cooking (Christensen et al., 2012), where time had the main impact on the increasing intensity of cooked meat flavour. This confirms the fact that intensities of flavour compounds at low temperatures are in low quantities because they are mainly generated at higher temperatures. Similarly, Vani et al. (2006) suggested that cooking of Indian country hen meat at high temperatures resulted in improved flavour, primarily because of the formation of high levels of IMP degradation compounds.

As discussed earlier, cooking methods such as roasting, grilling, frying and pressure cooking compared to boiling involve heat treatments in which the temperature exceeds 100°C. Such a high temperature favours the formation of a vast number of heterocyclic compounds found in the aroma of cooked meats (Melton, 1999). In particular, the generation of pyrazines requires high temperature and low moisture conditions. The occurrence of heterocyclic compounds generated at higher temperatures and low moisture conditions during cooking of meats are responsible for the flavours of chicken meat prepared by roasting, grilling, frying or pressure cooking. It is further governed by the concentration and characteristics of the compounds formed via the Maillard reaction, lipids degradation and Maillard-lipid interactions discussed earlier (Melton, 1999). Hence, the flavour of poultry meat can be improved by using the suitable cooking method with suitable conditions.

5.8 Irradiation

It has been reported that irradiated raw chicken meat gave off a bloody and sweet aroma (Heath et al., 1990). Increases in irradiation-induced sulphur-containing volatiles appear to be the result of radiolytic degradation of sulphur-containing amino acids (Ahn, 2002) and lipid oxidation (Jo and Ahn, 2000), which produces cabbage-like or rotten vegetable-like putrid odours. However, Hashim et al. (1995) reported that irradiated raw chicken gave off more intense 'fresh chickeny', bloody and sweet aromatic odours than did non-irradiated samples. They also determined that cooked irradiated frozen dark chicken meat had a more intense chicken flavour than did non-irradiated samples. However, very little negative effect was expressed by electron beam irradiation on the flavour of preheated chicken breast meat (Rababah et al., 2006). Further, modified atmosphere packaging, reducing temperature (freezing) before irradiation and addition of antioxidants could decrease the irradiation-associated odours (Ahn et al., 1999; Jo et al., 2001). The detrimental effects of irradiation can be reduced or eliminated by ensuring oxygen-permeable packaging after irradiation. Vacuum packaging retains irradiation-generated sulphur-containing compounds, but re-packaging meat in oxygen-permeable materials allows for dissipation of these compounds (Nam and Ahn, 2003). Rababah et al. (2006) demonstrated that plant extracts, grape seed and green tea extracts infused into the chicken meat could decrease lipid oxidation in both irradiated and non-irradiated chicken meat.

5.9 High pressure treatment

Sensory quality of various meat products was not affected by high pressure treatment (Hayman et al., 2004). However, Kruk et al. (2011) demonstrated that exposure of chicken meat to a pressure of 300 MPa led to a better flavour and taste as opposed to a 450 MPa treatment. Pressure treatments at approximately 300 MPa at room temperature initiated the changes which eventually led to catalysis of lipid oxidation in pressure-processed meat (Cheah and Ledward, 1996). Additionally, Rivas-Canedo et al. (2008) showed that high pressure treatment of chicken breast at 400 MPa significantly changed the levels of some volatile compounds. The amount of alcohols and aldehydes decreased in this study, while other compounds, such as 2,3-butanedione and 2-butanone, were more abundant in high-pressure-treated meats. This could have an impact on flavour, especially on the aroma strength. High pressure treatment may also accelerate other reactions such as denaturation of muscle proteins and pressure-induced increases in proteolytic enzyme activity in the muscle that have a bearing on food flavour.

5.10 Additives

Plant extracts are being incorporated into several meat products to reduce/arrest lipid oxidation mainly because of toxicological concerns over synthetic antioxidants (Ahn et al., 1998). Plant extracts having high content of antioxidants such as grape seed extract, green tea extracts, and their combinations, were effective in minimizing lipid oxidation products such as pentanal and hexanal from chicken meat (Rababah et al., 2006). Therefore, these additives may reduce the formation of off-flavours in chicken meat. A relationship among α-tocopherol, total content of PUFAs and omega-3 fatty acids and flavour components from cooked chicken was identified by Jahan et al. (2005).

PUFAs and α-tocopherol contents were strongly related (r^2 = 0.96 and 0.99, respectively) to total flavour component abundance. However, there were no relationships between PUFAs and abundance of nitrogen-rich components, such as pyrazines and pyridines. They further reported that two antioxidant enzymes, glutathione peroxidase and glutathione reductase, were related to chicken flavour components. Therefore, suitable antioxidants may reduce the formation of off-flavours in poultry meat, thus preserving and/or improving the existing flavour.

6 Summary

The Maillard reaction, thermal degradation of lipids and Maillard-lipid interactions are the main pathways by which a large number of flavour and aroma compounds responsible for poultry meat flavour are formed during cooking. Volatile compounds generated from the Maillard reaction and lipid oxidation, including 2-methyl-3-furanthiol, 2-furfurylthiol, methionol, 2,4,5-trimethylthiazole, nonanol, 2-trans-nonenal, 2-formyl-5-methylthiophene, *p*-cresol, 2-trans-4-trans-nonadienal, 2-trans-4-trans-decadienal, 2-undecenal, β-ionone, γ-decalactone and γ-dodecalactone, are obviously the major sources of poultry meat flavour. However, 2-methyl-3-furanthiol that is produced from the reaction between ribose and cysteine or cystine, and from degradation of thiamin, is considered as the most important compound in chicken flavour. A large number of heterocyclic compounds are formed during roasting, grilling, frying or pressure cooking of chicken meat due to higher temperature and low moisture conditions used in those cooking methods, but not during the boiling of the meat. Flavour of cooked poultry meat is affected by the breed/strain, diet, intramuscular content of free amino acids and nucleotides, classes of the lipids and fatty acid composition, irradiation, high pressure treatment, cooking, antioxidants and ageing. Flavour of poultry meat can be preserved and/or improved by effective manipulation of the above-mentioned factors.

7 Where to look for further information

1. *Poultry Science* (http://ps.oxfordjournals.org) is an international journal publishing original papers, research notes, symposium papers and reviews of basic science as applied to poultry. *World's Poultry Science Journal* (http://journals.cambridge.org/WPS) provides authoritative reviews in poultry science and an international forum for the exchange and dissemination of information including research, education and industry structure. Both journals are of interest to academics, researchers, students, extension workers and commercial poultry producers.
2. Khan, M. I., Jo, C. and Tariq, M. R. (2015), 'Meat flavor precursors and factors influencing flavor precursors – A systematic review', *Meat Sci.*, 110, 278–84.
3. World's Poultry Congress organized annually by *World's Poultry Science Association.*
4. Shahidi, F. (1994), *Flavour of meat and meat products,* Blackie Academic and Professional.
5. Jayasena, D. D., Ahn, D. U., Nam, K. C. and Jo, C. (2013), 'Flavour chemistry of chicken meat: a review', *Asian-Australas. J. Anim. Sci.*, 26(5), 732–42.

8 References

Ahn, D. U. (2002), 'Production of volatiles from amino acid homopolymers by irradiation', *J. Food Sci.*, 67, 419–26.

Ahn, D. U., Feng, X., Lee, E. J., Zhang, W., Lee, J. W., Jo, C. and Nam, K. C. (2016), 'Mechanisms of volatile production from non-sulfur amino acids by irradiation', *Radiat. Phys. Chem.*, 119, 64–73, 80–4.

Ahn, D. U., Olson, D. G., Jo, C., Chen, X., Wu, C. and Lee, J. I. (1998), 'Effect of muscle type, packaging, and irradiation on lipid oxidation, volatile production, and color in raw pork patties', *Meat Sci.*, 47, 27–39.

Ahn, D. U., Olson, D. G., Jo, C., Love, J. and Jin, S. K. (1999), 'Volatiles production and lipid oxidation of irradiated cooked sausage with different fat content and packaging during storage', *Meat Sci.*, 51, 355–61.

Aliani, M. and Farmer, L. J. (2005), 'Precursors of chicken flavour II: Identification of key flavour precursors using sensory methods', *J. Agric. Food Chem.*, 53, 6455–62.

Aymerich, T., Picouet, P. A. and Monfort, J. M. (2008), 'Decontamination technologies for meat products', *Meat Sci.*, 78, 114–29.

Bou, R., Guardiola, F., Grau, A., Grimpa, S., Manich, A., Barroeta, A. and Codony, R. (2001), 'Influence of dietary fat source, alpha-tocopherol, and ascorbic acid supplementation on sensory quality of dark chicken meat', *Poultry Sci.*, 80, 800–7.

Brunton, N. P., Cronin, D. A. and Monahan, F. J. (2002), 'Volatile components associated with freshly cooked and oxidized off-flavours in turkey breast meat', *Flavour Frag. J.*, 17, 327–34.

Calkins, C. R. and Hodgen, J. M. (2007), 'A fresh look at meat flavour', *Meat Sci.*, 77, 63–80.

Cheah, P. B. and Ledward, D. A. (1996), 'High pressure effects on lipid oxidation in minced pork', *Meat Sci.*, 43, 123–34.

Cheng, F. Y., Huang, C. W., Wan, T. C., Liu, Y. T., Lin, L. C. and Lou Chyr, C. Y. (2008), 'Effect of free-range farming on carcass and meat qualities of black-feathered Taiwan native chicken', *Asian Australas. J. Anim. Sci.*, 21, 1201–6.

Chikuni, K., Sasaki, K., Emori, T., Iwaki, F., Tani, F., Nakajima, I., Muroya, S. and Mitsumoto, M. (2002), 'Effect of cooking on the taste- and flavor-related compounds in pork', *Jpn. J. Swine Sci.*, 39, 191–9.

Christensen, L., Gunvig, A., Tørngren, M. A., Aaslyng, M. D., Knøchel, S. and Christensen, M. (2012), 'Sensory characteristics of meat cooked for prolonged times at low temperature', *Meat Sci.*, 90, 485–9.

Culioli, J., Touraille, C., Bordes, P. and Giraud, J. P. (1990), 'Caracteristiques des carcasses et de la viande du poulet label fermier', *Arch. Geflugelk.*, 53, 237–45.

Fanatico, A. C., Pillai, P. B., Cavitt, L. C., Emmert, J. L., Meullenet, J. F. and Owens, C. M. (2006), 'Evaluation of slow-growing broiler genotypes grown with and without outdoor access: sensory attributes', *Poultry Sci.*, 85, 337–43.

Fanatico, A. C., Pillai, P. B., Emmert, J. L., Gbur, E. E., Meullenet, J. F. and Owens, C. M. (2007), 'Sensory attributes of slow- and fast-growing chicken genotypes raised indoors or with outdoor access', *Poultry Sci.*, 86, 2441–9.

Fanatico, A. C., Pillai, P. B., Hester, P. Y., Falcone, C., Mench, J. A., Owens, C. M. and Emmert, J. L. (2008), 'Performance, livability, and carcass yield of slow- and fast-growing chicken genotypes fed low-nutrient or standard diets and raised indoors or with outdoor access', *Poultry Sci.*, 87, 1012–21.

Farmer, L. J. (1999), 'Poultry meat flavour'. In *Poultry Meat Science: Poultry Science Symposium Series Vol. 25*, ed. R. I. Richardson and G. C. Mead, 127–58. CABI Publishing, Oxon.

Farmer, L. J. and Mottram, D. S. (1994), 'Lipid-Maillard interactions in the formation of volatile aroma compounds'. In *Trends in Flavour Research*, ed. H. Maarse and D. G. vander Heij, 313–26. Elsevier, Amsterdam.

Fors, S. (1983), 'Sensory properties of volatile Maillard reaction products and related compounds'. In *The Maillard Reaction in Foods and Nutrition*, ed. G. R. Waller and M. S. Feather, 185–286. American Chemical Society, Washington.

Fujimura, S., Koga, H., Takeda, H., Tone, N., Kadowaki, M. and Ishibashi, T. (1996), 'Role of taste-active components, glutamic acid, 5'-inosinic acid and potassium ion in taste of chicken meat extract', *Anim. Sci. Technol.*, 67, 423–9.

Fukunaga, T., Koga, K., Maita, Y. and Matsuoka, S. (1989), 'Free amino acid, carnosine and 5'-inosinic acid contents in the breast and leg meats from the cross and tripe-cross chickens of Satsuma native fowl', *Bulletin Fac. Agric., Kagoshima University*, 39, 223–32.

Gasser, U. and Grosch, W. (1990), 'Primary odorants of chicken broth', *Z. Lebensm. Unters. Forsch.*, 190, 3–8.

Graf, E. and Panter, S. S. (1991), 'Inhibition of warmed-over flavour development by polyvalent cations', *J. Food Sci.*, 56(4), 1055–8.

Hashim, I. B., Resurreccíon, A. V. A. and McWatters, K. H. (1995), 'Descriptive sensory analysis of irradiated frozen or refrigerated chicken', *J. Food Sci.*, 60, 664–6.

Hayman, M. M., Baxter, I., O'riordan, P. J. and Stewart, C. M. (2004), 'Effects of high pressure processing on the safety, quality, and shelf life of ready-to-eat meats', *J. Food Protect.*, 67, 1709–18.

Heath, J. L., Owens, S. L., Tesch, S. and Hannah, K.W. (1990), 'Effect of high-energy electron irradiation of chicken on thiobarbituric acid values, shear values, odor, and cook yield', *Poultry Sci.*, 69, 313–19.

Ho, C. T., Oh, Y. C. and Bae-Lee, M. (1994), 'The flavour of pork'. In *Flavour of Meat and Meat Products*, ed. F. Shahidi, 38–51. Blackie Academic and Professional, London.

Horsted, K., Allesen-Holm, B. H. and Hermansen, J. E. (2010), 'The effect of breed and feed-type on the sensory profile on breast meat in male broilers reared in an organic free-range system', *J. Sci. Food Agric.*, 92, 258–65.

Horsted, K., Allesen-Holm, B. H., Hermansen, J. E. and Kongsted, A. G. (2012), 'Sensory profiles of breast meat from broilers reared in an organic niche production system and conventional standard broilers', *Brit. Poultry Sci.*, 4, 515–24.

Huang, Z. B., Leibovitz, H., Lee, C. M. and Millar, R. (1990), 'Effect of dietary fish oil on ω-3 fatty acid levels in chicken eggs and thigh flesh', *J. Agric. Food Chem.*, 38, 743–7.

Husak, R. L., Sebranek, J. G. and Bregendahl, K. (2008), 'A survey of commercially available broilers marketed as organic, free-range, and conventional broilers for cooked meat yields, meat composition, and relative value', *Poultry Sci.*, 87, 2367–76.

Jahan, K., Paterson, A., Piggott, J. and Spickett, C. (2005), 'Chemometric modeling to relate antioxidants, neutral lipid fatty acids, and flavor components in chicken breasts', *Poultry Sci.*, 84, 158–66.

Jang, A., Liu, X. D., Shin, M. H., Lee, B. D., Lee, S. K., Lee, J. H. and Jo, C. (2008), 'Antioxidative potential of raw breast meat from broiler chicks fed a dietary medicinal herb extract mix', *Poultry Sci.*, 87, 2382–9.

Jaturasitha, S., Srikanchai, T., Kreuzer M. and Wicke, M. (2008), 'Difference in carcass and meat characteristics between chicken indigenous to northern Thailand (blackboned and Thai native) and imported extensive breeds (Bresse and Rhode Island Red)', *Poultry Sci.* 87, 160–9.

Jayasena, D. D., Ahn, D. U., Nam, K. C. and Jo, C. (2013), 'Factors affecting cooked chicken meat flavor: A review', *Worlds Poultry Sci. J.*, 69, 515–26.

Jayasena, D. D., Jung, S., Park, H. B., Lee, J. H., Nam, K. C., Lee, K. H. and Jo, C. (2015), 'Taste-active compound levels in Korean native chicken meat: The effects of bird age and the cooking process', *Poultry Sci.*, 94, 1964–72.

Jayasena, D. D., Kim, S. H., Lee, H. J., Jung, S., Lee, J. H., Park, H. B. and Jo, C. (2014), 'Comparison of the amounts of taste-related compounds in raw and cooked meats from broilers and Korean native chickens', *Poultry Sci.*, 93, 3163–70.

Jo, C. and Ahn, D. U. (2000), 'Production of volatile compounds from irradiated oil emulsion containing amino acids and proteins', *J. Food Sci.*, 65, 612–16.

Jo, C., Ahn, D. U. and Byun, M. W. (2001), 'Lipid oxidation and volatile production in irradiated raw pork batters prepared with commercial soybean oil containing vitamin E', *Radiat. Phys. Chem.*, 61, 41–7.

Joo, S. T. and Kim, G. D. (2011), 'Meat quality traits and control technologies'. In *Control of Meat Quality,* ed. S.T. Joo, 1–20. Research Signpost, Kerala.

Jung, Y., Jeon, H. J., Jung, S., Choe, J. H., Lee, J. H., Heo, K. N., Kang, B. S. and Jo, C. (2011), 'Comparison of quality traits of thigh meat from Korean native chickens and broilers', *Korean J. Food Sci. An.*, 31(5), 684–92.

Kavitha, S. and Modi, V. K. (2007), 'Effect of water activity and temperature on degradation of 5'-inosine monophosphate in a meat model system', *LWT-Food Sci. Technol.* 40, 1280–6.

Kerler, J. and Grosch, W. (1997), 'Character impact odourants of boiled chicken: changes during refrigerated storage and reheating', *Food Res. Technol.*, 205(3), 232–8.

Kiyohara, R., Yamaguchi, S., Rikimaru, K. and Takahashi, H. (2011), 'Supplemental arachidonic acid-enriched oil improves the taste of thigh meat of Hinai-jidori chickens', *Poultry Sci.*, 90, 1817–22.

Koriyama, T., Wongso, S., Watanabe, K. and Abe, H. (2002), 'Fatty acid compositions of oil species affect the 5 basic taste perceptions', *J. Food Sci.* 67, 868–73.

Kruk, Z. A., Yun, H., Rutley, D. L., Lee, E. J., Kim, Y. J. and Jo, C. (2011), 'The effect of high pressure on microbial population, meat quality and sensory characteristics of chicken breast fillet', *Food Control*, 22, 6–12.

Kurihara, K. (1987), 'Recent progress in the taste receptor'. In *Umami: A Basic Taste,* ed. Y. Kawamura and M. R. Kare, 3–39. Marcel Dekker, New York.

Lee, K. H., Kim, H. J., Lee, H. J., Kang, M. and Jo, C. (2012), 'A study on components related to flavor and taste in commercial broiler and Korean native chicken meat', *Korean J. Food Preserv.*, 19, 385–92.

Lillard, D. A. (1987), 'Oxidative deterioration in meat, poultry, and fish'. In *Warmed-Over Flavour of Meat,* ed. A. J. St. Angelo and M. E. Bailey, 41–67. Academic Press, Orlando.

Lin, C.-Y., Kuo, H.-Y. and Wan, T.-C. (2014), 'Effect of free-range rearing on meat composition, physical properties and sensory evaluation in Taiwan game hens', *Asian Australas. J. Anim. Sci.*, 27, 880–5.

Liu, X. D., Jayasena, D. D., Jung, Y., Jung, S., Kang, B. S., Heo, K. N., Lee, J. H. and Jo, C. (2012), 'Differential proteome analysis of breast and thigh muscles between Korean native chickens and commercial broilers', *Asian-Australas. J. Anim. Sci.,* 25(6), 895–902.

Lyon, B. G., Smith, D. P., Lyon, C. E. and Savage, E. M. (2004), 'Effects of diet and feed withdrawal on the sensory descriptive and instrumental profiles of broiler breast fillets', *Poultry Sci.*, 83, 275–81.

MacLeod, G. (1994), 'The flavour of beef'. In *Flavour of Meat and Meat Products,* ed. F. Shahidi, 4–37. Blackie Academic and Professional, Glasgow.

Maga, J. A. (1983), 'Flavour potentiator', *Crit. Rev. Food Sci. Nutr.*, 18, 231–312.

Meinert, L., Schäfer, A., Bjergegaard, C., Aaslyng M. D. and Bredie, W. L. P. (2009), 'Comparison of glucose, glucose 6-phosphate, ribose, and mannose as flavour precursors in pork; the effect of monosaccharide addition on flavour generation', *Meat Sci.*, 81(3), 419–25.

Melton, S. L. (1999), 'Current status of meat flavour'. In *Quality Attributes of Muscle Foods,* ed. Y. L. Xiong, C. T. Ho and F. Shahidi, 115–30. Kluwer Academic/ Plenum Publishers, New York.

Michalczuk, M., Lukasiewicz, M., Zdanowska-Sqsiadek, Z. and Niemiec, J. (2014), 'Comparison of selected quality attributes of chicken meat as affected by rearing systems', *Pol. J. Food Nutr. Sci.* 64, 121–6.

Mikulski, D., Celej, J., Jankowski, J., Majewska, T. and Mikulska, M. (2011), 'Growth performance, carcass traits and meat quality of slower-growing and fast-growing chickens reared with and without outdoor access', *Asian Australas. J. Anim. Sci.*, 24, 1407–16.

Min, B. R., Cordray, J. C. and Ahn, D. U. (2010), 'Effect of NaCl, myoglobin, Fe(II), and Fe(III) on lipid oxidation of raw and cooked chicken breast and beef loin', *J. Agric. Food Chem.*, 58, 600–5.

Mottram, D. S. (1991), 'Meat'. In *Volatile Compounds in Foods and Beverages*, ed. H. Maarse, 107–77. Marcel Dekker, New York.

Mottram, D. S. (1994a), 'Flavour compounds formed during the Maillard reaction'. In *Thermally Generated Flavours. Maillard, Microwave, and Extrusion Processes*, ed. T. H. Parliament, M. J. Morello and R. J. McGorrin, 104–26. American Chemical Society, Washington.

Mottram, D. S. (1994b), 'Some aspects of the chemistry of meat flavour'. In *The Flavour of Meat and Meat Products*, ed. F. Shahidi, 210–30. Chapman and Hall, London.

Mottram, D. S. (1998), 'Flavour formation in meat and meat products: a review', *Food Chem.*, 62(4), 415–24.

Mottram, D. S. and Edwards, R. A. (1983), 'The role of triglycerides and phospholipids in the aroma of cooked beef', *J. Sci. Food Agric.*, 34, 517–22.

Mottram, D. S. and Madruga, M. S. (1994), 'Important sulfur containing aroma volatiles in meat'. In *Sulfur Compounds in Foods*, ed. C. J. Mussinan and M. E. Keelan, 180–7. American Chemical Society, Washington.

Nam, K. C. and Ahn, D. U. (2003), 'Use of double-packaging and antioxidant combinations to improve color, lipid oxidation, and volatiles of irradiated raw and cooked turkey breast patties', *Poultry Sci.*, 82, 850–7.

Nishimura, T., Rhue, M. R., Okitani, A. and Kato, H. (1988), 'Components contributing to the improvement of meat taste during storage', *Agric. Biol. Chem.* 52, 2323–30.

Noleau, I. and Toulemonde, B. (1987), 'Volatile components of roasted chicken fat', *Lebensm. Wiss. Technol.*, 20, 37–41.

Oh, S.-T., Jhun, H. K., Park, J.-M., Kim, J.-M., Kang, C.-W. and An, B.-K. (2012), 'Effects of dietary sources containing ω-3 fatty acids on the fatty acid composition of meats in Korean native chickens', 32(4), 476–82.

Patterson, R. L. S. and Stevenson, M. H. (1995), 'Irradiation-induced off-odor in chicken and its possible control', *Br. Poultry Sci.*, 36, 425–41.

Perez-Alvarez, J. A., Sendra-Nadal, E., Sanchez-Zapata, E. J. and Viuda-Martos, M. (2010), 'Poultry flavour: General aspects and applications'. In *Handbook of Poultry Science and Technology Volume 2: Secondary Processing*, ed. I. Guerrero-Legarreta and Y. H. Hui, 339–57. John Wiley and Sons Inc, New Jersey.

Ponte, P. I. P., Rosado, C. M. C., Crespo, J. P., Crespo, D. G., Mourao, J. L., Chaveiro-Soares, M. A., Bras, J. L. A., Mendes, I., Gama, L. T., Prates, J. A. M., Ferreira, L. M. A. and Fontes, C. M. G. A. (2008), 'Pasture intake improves the performance and meat sensory attributes of free-range broilers', *Poultry Sci.*, 87, 71–9.

Poste, L. M. (1990), 'A sensory perspective of effect of feeds on flavor in meats: poultry meats', *J. Anim. Sci.*, 68, 4414–20.

Rababah, T., Hettiarachchy, N. S., Horax, R., Cho, M. J., Davis, B. and Dickson, J. (2006), 'Thiobarbituric acid reactive substances and volatile compounds in chicken breast meat infused with plant extracts and subjected to electron beam irradiation', *Poultry Sci.*, 85, 1107–13.

Ramarathnam, N., Rubin, L. J. and Diosady, L. L. (1993), 'Studies on meat flavor: 4. Fractionation, characterization, and quantitation of volatiles from uncured and cured beef and chicken', *J. Agric. Food Chem.* 41, 939–45.

Ramaswamy, H. S. and Richards, J. F. (1982), 'Flavour of poultry meat- a review', *Can. Inst. Food Sci. Technol. J.*, 15, 7–18.

Rhee, K. S., Anderson, L. M. and Sams, A. R. (2005), 'Comparison of flavour changes in cooked–refrigerated beef, pork and chicken meat patties', *Meat Sci.*, 71, 392–6.

Rikimaru, K. and Takahashi, H. (2010), 'Evaluation of the meat from Hinai-jidori chickens and broilers: Analysis of general biochemical components, free amino acids, inosine 5′-monophosphate, and fatty acids', *J. Appl. Poultry Res.*, 19, 327–33.

Rivas-Canedo, A., Fernandez-Garcia, E. and Nunez, M. (2008), 'Volatile compounds in fresh meats subjected to high pressure processing: effect of the packaging material', *Meat Sci.*, 81, 321–8.

Sanudo, C., Enser, M. E., Campo, M. M., Nute, G. R., Maria, G., Sierra I. and Wood J. D. (2000), 'Fatty acid composition and sensory characteristics of lamb carcasses from Britain and Spain', *Meat Sci.*, 54, 339–46.

Sasaki, K., Motoyama, M. and Mitsumoto, M. (2007), 'Changes in the amounts of water-soluble umami-related substances in porcine longissimus and biceps femoris muscles during moist heat cooking', *Meat Sci.*, 77, 167–72.

Seabra, L. M., Zapata, J. F., Fuentes, M. F., Aguiar, C. M., Freitas, E. R. and Rodrigues, M. C. (2001), 'Effect of deboning time, muscle tensioning and calcium chloride marination on texture characteristics of chicken breast meat', *Poultry Sci.*, 80, 109–12.

Shahidi, F. (1989), 'Flavour of cooked meats'. In *Flavour Chemistry: Trends and Developments,* ed. R. Teranishi, R. G. Buttery and F. Shahidi, 188–201. American Chemical Society, Washington.

Shahidi, F. (1994), 'Flavour of meat and meat products-an overview'. In *Flavour of Meat and Meat Products,* ed. F. Shahidi, 1–3. Blackie Academic and Professional, Glasgow.

Shahidi, F. (2002), 'Lipid derived flavours in meat products'. In *Meat Processing: Improving Quality,* ed. J. Kerry, J. Kerry and D. Ledward, 105–21. Woodhead Publishing Ltd, Cambridge.

Shi, H. and Ho, C. T. (1994), 'The flavour of poultry meat'. In *Flavour of Meat and Meat Products,* ed. F. Shahidi, 52–69. Blackie Academic and Professional, Glasgow.

Sitz, B. M., Calkins, C. R., Feuz, D. M., Umberger W. J. and Eskridge K. M. (2005), 'Consumer sensory acceptance and value of domestic, Canadian, and Australian grass-fed beef steaks', *J. Anim. Sci.*, 83(12), 2863–8.

Smith, D. P., Northcutt, J. K. and Steinberg, E. L. (2012), 'Meat quality and sensory attributes of a conventional and a Label Rouge-type broiler strain obtained at retail', *Poultry Sci.*, 91, 1489–95.

Sogunle, O. M., Olaniyi, O. A., Shittu, T. A. and Abiola, S. S. (2012), 'Performance and meat attributes of chickens reared on deep litter and free-range', *Arch. Zootec.*, 236, 569–76.

Spanier, A. M., Flores, M., Mcmillin K. W. and Bidner T. D. (1997), 'The effect of postmortem aging on meat flavour quality. Correlation of treatment, sensory, instrumental, and chemical descriptors', *Food Chem.*, 59, 531–8.

Spanier A. M., Flores, M., Toldrá, F., Aristoy, M. C., Bett, K. L., Bystricky P. and Bland J. M. (2004), 'Meat flavor: contribution of proteins and peptides to the flavor of beef', *Adv. Exp. Med. Biol.*, 542, 33–49.

Tai, C. Y. and Ho, C. T. (1997), 'Influence of cysteine oxidation on thermal formation of Maillard aromas', *J. Agric. Food Chem.*, 45, 3586–9.

Tai, C. Y. and Ho, C. T. (1998), 'Influence of glutathione oxidation and pH on thermal formation of Maillard-type volatile compounds', *J. Agric. Food Chem.*, 46, 2260–5.

Takahashi, H., Rikimaru, K., Kiyohara, R. and Yamaguchi, S. (2012), 'Effect of arachidonic acid-enriched oil diet supplementation on the taste of broiler meat', *Asian-Australas. J. Anim. Sci.*, 25, 845–51.

Tang, H., Gong, Y. Z., Wu, C. X., Jiang, J., Wang Y. and Li, K. (2009), 'Variation of meat quality traits among five genotypes of chicken', *Poultry Sci.*, 88, 2212–18.

Tang, J., Jin, Q. Z., Shen, G. H., Ho, C. T. and Chang S. S. (1983), 'Isolation and identification of volatile compounds from fried chicken', *J. Agric. Food Chem.*, 31, 1287–92.

Vani, N. D., Modi, V. K., Kavitha, S., Sachindra, N. M. and Mahendrakar, N. S. (2006), 'Degradation of inosine-5'-monophosphate in aqueous and in laying chicken muscle fibre system: effect of pH and temperature', *LWT-J. Food Sci. Technol.*, 39, 627–32.

Varavinit, S., Shobsngob, S., Bhidyachakorawat, M. and Suphantharika, M. (2000), 'Production of meat-like flavour', *ScienceAsia*, 26, 219–24.

Wasserman, A. E. (1972), 'Thermally produced flavour components in the aroma of meat and poultry', *J. Agric. Food Chem.*, 20, 737.

Whitfield, F. B. (1992), 'Volatiles from interactions of Maillard reactions and lipids', *Crit. Rev. Food Sci. Nutr.*, 31, 1–58.
Yamaguchi, S. (1991), 'Roles and efficacy of sensory evaluation in studies of taste', *J. Japan Soc. Food Sci. Technol.*, 38, 972–8.
Yano, T., Kataho, N., Watanabe, M., Nakamura, T. and Asano, Y. (1995), 'Evaluation of beef aging by determination of hypoxanthine and xanthine contents: application of a xanthine sensor', *Food Chem.*, 52, 439–45.

Chapter 14

The colour of poultry meat: understanding, measuring and maintaining product quality

KiChang Nam, Sunchon National University, Republic of Korea; Eun Joo Lee, University of Wisconsin-Stout, USA; and Dong Uk Ahn, Iowa State University, USA

1 Introduction

Colour is a crucial criterion that determines the consumer acceptance of meat products. Consumers can rapidly assess the visual appearance of fresh meat, and colour causes an immediate positive or negative psychological response (Nanke et al., 1998). Consumers know what fresh or processed meat colour should look like and relate the colour to the product quality, safety and storage history (Kropf, 1980; Seideman et al., 1984; Allen et al., 1997, 1998). In case of U.S. beef industry, about 15% of retail beef is discounted in price due to discolouration, which can be estimated to annual revenue losses of 1 billion dollars (Liu et al., 1995; Mancini and Hunt, 2005). Despite no available estimation of economic losses for poultry meat, the cost reduction caused by surface discolouration in poultry meat is likely to be considerable (Kuttappan et al., 2012). This chapter reviews the fundamentals of meat pigments; the colour of fresh, cooked, cured and irradiated poultry meats; the mechanism and prevention of discolouration and methods for colour measurement.

http://dx.doi.org/10.19103/AS.2016.0010.17

2 Meat pigments

2.1 Overview

Meat colour, which is one of the most important organoleptic factors that determines consumers' acceptability, can be different depending upon the chemical status of haem pigments, muscle type (red or white) and observation conditions such as light sources. However, the most crucial attribute is the concentration and chemical status of meat pigments.

There are three types of representative haem pigments (myoglobin, haemoglobin and cytochrome c), which are responsible for the colour of poultry meat. In well-bled red muscle tissues, myoglobin is the main haem pigment, and it constitutes about 70 to 95% of the total pigment (Rickansrud and Henrickson, 1967; Fox, 1987; Judge et al., 1989). Some oxygen-utilizing enzyme such as catalase can also act as a pigment, but its contribution to meat colour is not considerable (Bandman, 1987; Pikul et al., 1986).

The concentration of haem pigments is also an important factor (Nocito et al., 1973). Poultry meat, especially breast meat, has very low myoglobin content compared with the red meats such as beef, pork and goat meat. The concentration of haem pigment varies with species, strain and age of the animal, and muscle types (Potthast, 1987): pigment concentration increases with the age of poultry. The meats of ducks and geese, particularly in the breast muscle, contain many times more haem pigments than those of chickens or turkeys. The leg and thigh muscles of chickens or turkeys have 5 to 10 times more haem pigments than their breast muscles (Saffle, 1973). The myoglobin content in the red muscles of poultry meat is 2 to 4 mg per g, whereas the content in white muscles is less than 0.5 mg per g of wet tissues (Romans et al., 1994; Pearson and Young, 1989). Thus, the phenomena of colour changes in poultry breast meat are different from those in thigh meat.

2.2 Chemistry of haem pigments

Haem pigments consist of two components: one is haem ring, a non-protein portion that is the key to colour changes, and globin, a protein portion. Myoglobin and

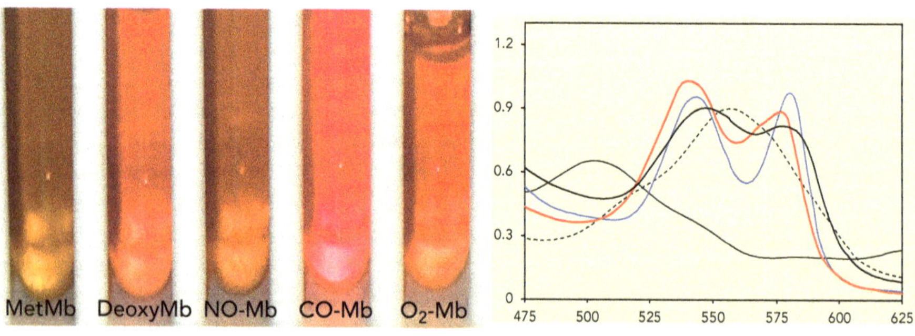

Figure 1 Colour and the absorption maxima of myoglobin with six different ligand compounds (pigments and absorption maxima): Metmyoglobin (505 and 630 nm); oxymyoglobin (543 and 580 nm); deoxymyoglobin (540 nm, not shown); carbon monoxide myoglobin (541 and 577 nm) and nitric oxide myoglobin (547 and 578 nm).

cytochrome c molecule have a globin and a haem ring, whereas haemoglobin is made of four myoglobin subunits. The amino acid residues of globin are oriented so that their non-polar portion points inward. The only polar amino acids inside the myoglobin are two histidines, which have critical functions at the haem-binding site (Bandman, 1987). The haem portion of the pigment is of special interest because the characteristics meat colour and the formation of the sixth ligand depend on the state (valency) of the iron within the haem ring (Judge et al., 1989). The fifth ligand of haem ring is linked to proximal histidine of haem pigments.

The reflectance and absorption spectra of haem pigments differ depending on the status of haem iron and the ligand molecule attached to the sixth coordinate of the haem ring. The ability of haem iron to coordinate with a sixth ligand is changed by the chemical state of haem iron. When haem iron is in oxidized form (ferric state, Fe^{3+}), it cannot combine with any other molecule. Only when the haem iron is in reduced form (ferrous state, Fe^{2+}), it can combine with a compound to make a sixth ligand (Judge et al., 1989). A few gaseous molecules (oxygen, nitric oxide and carbon monoxide) are permitted to access the reduced haem iron (Fig. 1). Various conditions of haem iron and sixth ligand determine the colour properties of myoglobin (Table 1).

Oxymyoglobin has a reduced iron bound with oxygen as a sixth ligand. This molecule imparts bright-red colour desirable for fresh meat colour. During the conversion of muscle to meat, oxygen is mainly present at the outer surface of meat. Therefore, the myoglobin in the centre of meat is usually in the reduced form and is weakly combined with a water molecule or stabilized by the postal histidine of globin (Lehninger, 1982). The colour of the reduced haem pigment is purple, and is called deoxymyoglobin. The purple colour can be turned into bright-red colour by exposing the meat to air (Suman and Joseph, 2014). If oxygen partial pressure in meat is low, haem iron becomes oxidized to form brown

Table 1 The colour properties of myoglobin depending on the chemical state of iron atom and a colour-generating ligand

Haem protein	Haem iron	Ligand	Colour	Name
Globin	Ferric	H_2O	Brown	Metmyoglobin
Globin	Ferrous	O_2	Bright red	Oxymyoglobin
Globin	Ferrous	–	Purple	Deoxymyoglobin
Globin	Ferrous	NO	Pink	NO-myoglobin
Globin	Ferric	NO	Brown	NO-metmyoglobin
Globin	Ferrous	CO	Pink	CO-myoglobin
Globin	Ferrous	S	Green	Sulphmyoglobin
–	Ferrous	NO	Pink	Haemochrome
–	Ferric	NO	Brown	Haemichrome
Denatured globin	Ferrous	NO	Pink	Nitrosyl haemochrome
Denatured globin	Ferric	NO	Brown	Denatured haemichrome
Denatured globin	Ferrous	CO	Pink	Denatured CO-haemochrome

Source: Nam and Ahn (*Food Sci. Biotechnol.*, 2003).

colour, and the pigment is called metmyoglobin. Formation of high-level metmyoglobin in fresh meat is not desirable because consumers can misunderstand the meat as old or low-quality meat.

The oxymyoglobin is considerably stable and is not oxidized to metmyoglobin under reduced conditions (Judge et al., 1989). However, once the meat is oxidized by exposing the meat to the open air for long time, it is hard to revert the meat under reduced conditions. Therefore, along with the haem complex formation, the oxidation–reduction potential (ORP) of meat is also critical for colour changes because the colour intensity of ferrous haem pigment is stronger than that of the ferric state. The reduction of metmyoglobin involves an enzyme system, metmyoglobin reductase that require NADH and NADPH coenzymes (Hagler et al., 1979), and the reducing activity increases with the rise of meat pH (Walter and Taylor, 1965).

3 The colour of fresh poultry meat

3.1 Fresh meat pigment

A bright-red colour is considered as a desirable fresh meat colour except for poultry breast meat. Bluish-white is the normal colour of skinned poultry breast meat due to its low pigment concentration. Although all three common forms (deoxy-, oxy- and met-) of myoglobin exist in fresh meat, the colour of fresh meat is mainly imparted by the bright-red oxymyoglobin and purple deoxymyoglobin (Ghorpade and Cornforth, 1993). The bright-red colour of fresh meat depends on the proportion of oxymyoglobin and the rate of oxygen diffusion, oxygen consumption rate (OCR), and partial pressure of oxygen at the meat surface (Giddings, 1977). The colour of fresh poultry meat is influenced by sex (Smith et al., 2015), ante-mortem factors such as diet, bird management and preslaughter stress (Schneider et al., 2012). The stress level reflects a decrease of pH in post-mortem muscle. The ultimate pH is strongly correlated with meat quality parameters such as tenderness, water-holding capacity (Zhu et al., 2012), cooking loss, juiciness and meat colour (Guidi and Castigliego, 2010; Bianchi and Fletcher, 2002; Qiao et al., 2001). Poultry meat with lower ultimate pH has higher L-values, imparting a paler appearance with lower consumers' acceptance (Fletcher, 1999). Heat stress and struggling during the slaughtering process are also factors influencing the pH decline of poultry meat (Ngoka and Froning, 1982).

At the beginning stage of storage, the degree of oxygen penetration in meat is low due to the decreased oxygen consumption by mitochondria (Bendall and Taylor, 1972). At a few millimetres below the meat surface, there is a region where the oxygen partial pressure is in the optimal range for the formation of a metmyoglobin layer (Ledward, 1970). Thus, packaging with high oxygen partial pressure can be beneficial to extend the fresh meat colour (Taylor and MacDougall, 1973), even though the conditions can trigger lipid oxidation during the extended storage. Vacuum-packaged meats have mainly purple deoxymyoglobin. Some approaches have utilized anoxic atmospheres in master packs for shipment to retail (Chouliara et al., 2007; Petrou et al., 2012), and meat cuts are subsequently displayed in traditional oxygen-permeable packaging to bloom. The use of modified atmosphere packaging can discolour fresh meat because the inner gases such as carbon dioxide or nitrogen reduce the pH or partial pressure of oxygen, resulting in brown colour (Seideman et al., 1984).

When fresh meat is cut, the deoxymyoglobin predominated before cutting changes to bright-red colour because the reduced myoglobin contacts with the oxygen of air and forms oxymyoglobin (Renerre, 1999). However, the stability of formed oxymyoglobin can vary depending upon the continued supply of oxygen and reducing environments for pigments. The OCR of tissue is different depending on the species and muscle types. A high OCR decreases oxymyoglobin formation and maintains myoglobin in a reduced state. Because the deoxymyoglobin is less stable than oxymyoglobin, the pigment can be oxidized easily to metmyoglobin. If meat is continuously exposed to oxygen, myoglobin is gradually oxidized to metmyoglobin especially at the low oxygen partial pressure, high temperature, high ionic strength and low pH (Renerre, 1999). Since the oxygen solubility in water increases as the temperature decreases, the maintenance of low temperature is highly desirable. The activity of oxygen-utilizing enzymes will increase with increasing storage temperature, which contributes to lowering the oxygen tension at the meat surface. After long-term storage, metmyoglobin becomes visible at the surface, and the meat colour turns to complete brown.

Aerobic microbial flora in their logarithmic growth phase have a high oxygen demand and may reduce the oxygen partial pressure to increase the formation of metmyoglobin causing brown discolouration (Andersen and Skibsted, 1992). Raw meat pigment is susceptible to bacterial discolouration, in which case green colour develops on product surfaces. Hydrogen sulphide produced by microbial contamination can form a green pigment called sulphmyoglobin. This greening usually results from poor sanitation or improper storage conditions where products are contaminated.

The loss of moisture from meat surfaces concentrates pigments and reduces light reflection due to the loss of intracellular water, both of which cause dark colour. During frozen storage, an excessive loss of moisture from meat surfaces will result in dehydration and discolouration of localized areas. Upon thawing, meat colour is improved just a little (Judge et al., 1989).

3.2 Metmyoglobin reduction

Specific attention is given to metmyoglobin reduction and antioxidant approaches for minimizing oxymyoglobin oxidation (Faustman et al., 1996; Reddy and Carpenter, 1991). The metmyoglobin-reducing enzyme (metmyoglobin reductase) has been considered to minimize the oxidation of oxymyoglobin (Faustman et al., 1988). α-Tocopherol maintained oxymyoglobin by the enhancement of cytochrome b_5-mediated reduction of metmyoglobin (Lynch et al., 1998; Sheldon et al., 1997; Rey et al., 2015). Meats contain negligible amounts of ascorbic acid, and the addition of ascorbic acid to meats may be beneficial for inhibiting metmyoglobin formation (Lee et al., 1999). The reduction in meat is related with NADH as a coenzyme, which facilitates the conversion of ferrimyoglobin to its ferrous form (Andersen and Skibsted, 1992). Metmyoglobin-reducing activity increases with decreasing pH, although to a lesser extent than for oxymyoglobin autoxidation. Reducing activity can be indirectly measured using ORP. Vacuum-packaged meat has lower ORP values than aerobically packaged one, and vacuum conditions maintained haem pigments of poultry breast with strongly reduced conditions (Nam and Ahn, 2002a). Irradiation can also provide poultry meat with reduced conditions, and irradiation dose was negatively correlated with ORP values (Table 2). Once reducing equivalents in the meat are exhausted, a complete metmyoglobin formation will occur (Ledward, 1984).

Table 2 Pearson correlation coefficients between colour values, irradiation dose, storage time, carbon monoxide (CO), redox potential and TBARS of vacuum-packaged turkey breast

	*a**-value	*b**-value	IR[a]	Storage	CO	ORP[b]	TBARS[c]
*L**-value	0.05	−0.76*	0.20	−0.21	0.23	−0.09	−0.30
*a**-value	–	0.33	0.88**	0.26	0.79**	−0.39	0.58
*b**-value		–	0.17	0.03	0.23	−0.23	0.73*
Irradiation			–	0.00	0.89**	−0.44	0.47
Storage				–	−0.23	0.37	−0.19
CO					–	−0.74*	0.63
ORP						–	−0.42

[a]Irradiation dose; [b]Oxidation–reduction potential; [c]2-Thiobarbituric acid-reactive substances.
*Value with significant correlation ($p < 0.05$), $n = 18$.
**Value with significant correlation ($p < 0.01$).
Source: Nam and Ahn (*Meat Sci*., 2002).

3.3 Discolouration

Discolouration in fresh meat is mainly caused by oxidation of haem pigment to metmyoglobin, which has an unattractive brown colour. Oxymyoglobin is more stable than deoxymyoglobin because of the hydrogen bonding between the bound oxygen and a distal histidine (Giddings, 1977; Renerre, 1999). The differences between species and muscles in their ability to form metmyoglobin can be related to the inherent differences in the concentration of mitochondria, activity of mitochondrial enzymes and content of accessory factors (Renerre, 1999). Therefore, meat has different partial oxygen pressure, pH, reducing ability and glycolytic rate (Pedrão et al., 2015).

At high oxygen tension, oxymyoglobin can persist for several days before discolouration occurs. When the oxygen partial pressure is reduced and the oxidation to metmyoglobin is favoured, brown discolouration develops (Fox, 1987). If the oxygen pressure is reduced to partially vacuum conditions, myoglobin will be oxidized to metmyoglobin. However, if the oxygen partial pressure reaches zero, deoxymyoglobin will be formed (Lawrie, 1983). The autoxidation rate was more dependent on the ligand accessibility of the myoglobin. Partial or complete loss of tertiary structure of globin results in an increased rate of oxymyoglobin oxidation (Livingston and Brown, 1981). The brown discolouration of fresh meat can be found in predominantly spoiled or after prolonged storage. However, in even fresh meat, the discolouration phenomenon can occur because of several reasons. Meat illuminated with light has relatively high rate constant of autoxidation (Kropt, 1980).

There is a strong relationship between broiler breast meat colour and muscle pH (Fletcher, 1999; Van Laack et al., 2000). The oxygen partial pressure also depends on temperature and ultimate pH of meat (Petracci et al., 2004). A high ultimate pH nearing neutral maintains respiratory activity and causes dark meat where deoxymyoglobin predominates (Livingston and Brown, 1981). An acidic ultimate pH enhances myoglobin oxidation and may lead to browning or fading of colour (Alnahhas et al., 2014). Low pH reduces the stability constant for the haem–globin linkage and increases the autoxidation rate (Chan et al., 2011). At pH values below 5, protonation of bound oxygen is accelerated, and release of superoxide anion is favoured. The unfolding of the globin moiety allows much easier attack of water molecules or hydroxyl radicals (Faustman et al., 1996). Low pH will also cause myoglobin to be more readily oxidized to metmyoglobin (Seideman et al., 1984). Guardia et al. (2014) reported that short-term nutritional strategies such as changing amino acid profile before slaughter can alter the pH of breast meat with limited impact on broiler growth and carcass composition.

Unusual discolouration can be found in an unexpected pH range of meat (Li et al., 2015). The abnormal paleness of PSE (pale, soft and exudative) meat is due to the high proportion of extracellular free water (Popp et al., 2013). The high free water in tissues increases the reflecting surface and L-value, and decreases the colour intensity greatly (Judge et al., 1989; Barbut, 1996). Tissues with high pH also can have high OCR due to high enzyme activities. Zhu et al. (2011) reported that PSE is generated due to elevated temperature during early post-mortem period, resulting in rapid glycolysis, which consequently induces phosphorylase denaturation. In addition, high temperature at early post-mortem could activate AMP-activated protein kinase (AMPK) and affect protein solubility (Zhu et al., 2013; Van Laack et al., 2000). Pedrão et al. (2015) reported that refrigeration treatment was effective in retarding the post-mortem glycolytic rate and preventing PSE conditions in poultry.

3.4 Lipid and myoglobin oxidation

The presence of oxidized unsaturated fatty acids may affect the colour of fresh meat (Qiao et al., 2002b). Primary changes during lipid oxidation are the production of free radicals and peroxides such as superoxide anion, hydrogen peroxide, hydroxyl radical and ferryl myoglobin radical. The oxidation of fatty acids influences not only flavour but also colour and texture of meat (Arroyo et al., 2015). Colour deterioration is essentially dependent on the rates of myoglobin autoxidation induced by lipid oxidation (Yin et al., 1973). Consequently, the oxidation of fatty acids or pigments can accelerate the oxidation of the other, which is called co-oxidation (Faustman and Cassens, 1990; Gatellier et al., 1993).

Free metal ions such as copper, iron, zinc and aluminium can catalyse the oxidation of oxymyoglobin (Kanner, 1994). Processing of raw meat such as deboning, mechanical separation, restructuring and grinding can also increase lipid and pigment oxidation (Renerre, 1999). The disruption of muscle membrane promotes the formation of free radicals through the interaction of iron with oxygen. In frozen meat, light in conjunction with oxygen increased lipid and colour oxidation (Berthelsen and Skibsted, 1987). Superoxide anion radicals generated from membrane electron transport systems may autoxidize oxymyoglobin into metmyoglobin. Ferrous iron can be converted to ferric iron and produces free radicals by oxidative stress. Therefore, lipid and pigment oxidation was closely coupled. Lipid oxidation can promote pigment oxidation and vice versa. Lipid oxidation may be a promoter of myoglobin oxidation, but it was not always possible to deduce whether or not the pigment oxidation caused lipid oxidation (Renerre and Labadie, 1993).

3.5 Blood spots and white striping

Improper slaughtering, which generates blood spot, affects the colour of meat. In fresh meat, a bloody appearance can reduce consumers' acceptability, and the pigments in the blood spots can be oxidized and converted to undesirable grey-brown or darker colour after cooking (Lyon et al., 1986; Schilling et al., 2015). The blood spots on breast and thigh can be minimized through optimal stunning and handling conditions (Hoen and Lankhaar, 1999). The migration of haemoglobin-rich bone marrow from the thigh femur to meat tissue in frozen meat can also result in undesirable colour changes (black colour) after cooking (Guidi and Castigliego, 2010). Bone discolouration can occur around the leg and thigh bones, knees joints and wing by freezing and thawing young chickens (Judge et al., 1989). White striping in broiler breast fillets is a defect resulting in the reduction of consumer acceptance. It is characterized grossly by the occurrence of white striations seen parallel to the direction of muscle fibres and could be associated with increased growth rate in birds (Kuttappan et al., 2013).

4 The colour of cooked poultry meat

4.1 Cooked meat pigments

The colour of cooked meat must be grey-brown and is caused by protein (globin) denaturation during heat processing under aerobic conditions. The brown pigments of cooked meat are denatured metmyoglobin due to the formation of denatured globin haemichrome (ferrihaemochrome). When meat is cooked under the anaerobic conditions such as vacuum-packaging conditions, denatured globin haemochrome (ferrohaemochrome, pink/red colour) is formed. However, it is easily oxidized to ferrihaemochrome (brown colour) (Suman and Joseph, 2014). The cooked brown pigment could be partially reversible to red colour during refrigerated storage (Froning et al., 1968). When a nitrogen compound (nicotinamide) reacted with the haem during cooking, pink colour was produced (Ahn and Maurer, 1989c). For the pink colour formation in cooked turkey breast meat, ORP of meat was an important factor (Cornforth et al., 1986). The pH was also important for the colour of cooked meat as the heat stability of the metmyoglobin increased with a rise in pH of the pigment solution (Satterlee and Zachariah, 1972; Ahn and Maurer, 1989b). In some cases, pink colour, which will be explained later, can be found in even fully cooked meat. Complete disappearance of pinkness in cooked meat does not require intensive cooking (Howe et al., 1982).

4.2 Discolouration during cooking

The brown colour of cooked meat is detectable when approximately more than 60% of the myoglobin is in metmyoglobin form. Metmyoglobin formation is accelerated by the conditions that cause denaturation of protein portion of myoglobin, and by oxidation under low oxygen tension. Denaturation increases by lowering the pH of meat (Fletcher et al., 2000). Low-pH muscles have open muscle fibre structure, allowing for greater discolouration. The addition of phosphates to the meat reduces the amount of pigment denaturation due to pH increase. The addition of NaCl increased the rate of heat-induced denaturation of myoglobin. The reverse effect of NaCl and phosphates was observed with cytochrome c (Ahn and Maurer, 1989a). Cytochrome c is very stable against heat denaturation and can resist denaturation at 105°C (Cornish and Froning, 1974).

The effect of fat contents to the discolouration of cooked meat is very low, but the surface browning is occasionally the result of fat decomposition and polymerization with carbohydrate and protein decomposition products (Judge et al., 1989).

4.3 Pink colour defect

Pink discolouration in cooked, uncured poultry meat, especially breast meat (white muscle), is considered a quality defect, which can cause consumer's complaint because consumers may perceive it as an indicator of undercooked or contaminated meat (Holownia et al., 2003). Therefore, pink discolouration is a negative factor for cooked meat colour with an economic concern to the processors. Pink defect has been generated by the undenatured myoglobin, contamination with nitrite or nitrate (Froning et al., 1969; Ahn and Maurer, 1987), severe stress at pre-slaughtering (Ngoka and Froning, 1982), absorption of combustion gases such as nitric oxide or carbon monoxide (Froning et al., 1969) or use of non-thermal pasteurization by irradiation (Nam and Ahn, 2002). Poultry breast meat was more susceptible than highly pigmented beef to pinking in the presence of sodium nitrate in soy isolates (Heaton et al., 2000).

Although the usual concentration of nitrate and nitrite in turkey breast meat was not high enough to cause a pink colour defect, it is possible under certain conditions such as high nitrate levels in feed or water supplies, high microbial load and long storage conditions (Ahn and Maurer, 1987). While cooking in a gas oven, a small amount (*ca.* 0.4 ppm) of nitrogen dioxide gas causes pinking of turkey roll, but the solubility of nitrogen monoxide gas at meat surfaces was lower than nitrogen dioxide (Cornforth et al., 1998). The pink colour generated by nitrite can be identified easily by acetone–HCl extraction (Hornsey, 1956; Ahn and Maurer, 1989a). Irradiation also increases the pink colour in cooked poultry breast meat (Nam and Ahn, 2002b).

The haem-complex forming ligands such as pyridine and its derivatives were suggested as possible nitrite substitutes to fix meat colour. The haem complexes have a strong effect on the redness of the cooked meat, especially under reduced conditions (Ahn and Maurer, 1990a,c). ORP measurements of cooked turkey rolls showed that haemochrome formation was promoted by reducing conditions and prevented by oxidizing conditions (Cornforth et al., 1986; Ahn and Maurer, 1989a).

Cytochrome c has been suggested as a contributing factor to pink colour development in cooked turkey breast meat. Ahn and Maurer (1990b) reported that the pink colour of turkey breast could be formed by the binding of denatured ferrocytochrome c with several ligands. The cytochrome c content of poultry meat was closely correlated with the colour of meat and with the content of total haem pigment (Pikul et al., 1987).

Proper slaughtering practices (minimizing stress conditions such as exhaust gases, prolonged transport, high temperature and carcass washing using nitrate-free water) prevent poultry pink colour defect (Ahn and Maurer, 1987; Bowker et al., 2014). The pinking in poultry meat was inhibited by addition of several ingredients such as citric acid, non-fat dry milk and whey protein concentrate (Dobson and Cornforth, 1993; Schwarz et al., 1999; Kieffer et al., 2000). Those chelating additives may have the ability to bind haem iron during heat processing and to inhibit pink pigments formation. Despite the pigments that cause pinking is known, a better understanding of how processing factors affect endogenous conditions is essential to control the pink colour defect in uncured cooked poultry breast products (Holownia et al., 2003).

5 The colour of cured poultry meat

The main objective of curing meat is not only to extend shelf life but also to develop an attractive pink colour. Nitric oxide myoglobin is a natural cured pigment before heating, and its colour is bright-red. During cooking, the nitric oxide pigment is stabilized by the denaturation of globin of myoglobin and is converted to nitrosyl haemochrome. The pigment is bright pink colour characteristic of cured meat. Nitrate or nitrite, which can be added as a curing agent to meat, cannot be directly combined to myoglobin. If nitrate is used in curing, it must be converted to nitrite by nitrate-reducing microorganisms, which is, in turn, reduced to nitric oxide. Nitrite also may be reduced to nitric oxide by natural reducing activity of post-mortem muscle tissues (Judge et al., 1989).

Since nitrite is a very efficient oxidizing agent for myoglobin, the initial reaction is probably the conversion of reduced myoglobin or oxymyoglobin into metmyoglobin. Nitric oxide then combines with the haem portion of metmyoglobin to produce nitric oxide metmyoglobin. Reduction of nitric oxide metmyoglobin must be accomplished either naturally in meat or by reductants included in the cures. Therefore, cured meat pigment formation is also absolutely dependent on reduced conditions of meat (Kim et al., 1988). Usually reducing agents such as ascorbic acid and erythorbic acid are included in curing agents. Sulphydryl groups released during heat processing are also strong reducing compounds. After further heating, denaturation of globin is complete and the other site of haem can be taken up by another nitric oxide. The pigment is called dinitrosyl haemochrome, which is more stable and brightly redder than nitrosyl haemochrome (Lee and Cassens, 1976).

Nitric oxide myoglobin and nitrosyl haemochrome are very susceptible to light fading. During long-term display of cured meat products, the fading of the cured meat colour can be brought about by either lipid oxidation or light dissociation of the nitric oxide from the ferrous porphyrin coordination complex and subsequent formation of not only metmyochromogen but also green pigments. A brownish grey colour develops on exposed meat surface during light fading because faded pigment has its haem group in the ferric state. Therefore, to protect the light fading of cured product colour, the elimination of oxygen from meat surfaces by vacuum packaging is recommended (Min and Ahn, 2012). In the absence of oxygen, the nitric oxide split from haem by light will not be oxidized, and it can then recombine with the haem.

6 The colour of irradiated poultry meat

Ionizing radiation is an excellent method to reduce microbial numbers, but can significantly influence meat quality. Ionizing radiation can affect the rate of lipid oxidation and colour of meat. Irradiated raw chicken and turkey breast muscles had increased redness, and the increased red or pink colour was stable during refrigerated storage (Millar et al., 1995; Nanke et al., 1998). The increases of redness in meat by irradiation vary depending on species, muscle type, irradiation dose and packaging environment (Luchsinger et al., 1996; Ahn et al., 1998; Nanke et al., 1999). The *a** value (redness) of raw poultry breast was increased by irradiation in both aerobically and vacuum-packaging systems (Luchsinger et al., 1996), but the vacuum-packaged meat was significantly redder than aerobically packaged ones during storage (Nanke et al., 1998, 1999). Irradiation also increases the pink colour in precooked poultry meat (Nam et al., 2002). The increased red colour in cooked meat by irradiation may be a problem as discussed above.

Many researchers proposed that the red colour in irradiated light meat is oxymyoglobin, oxymyoglobin (oxyMb)-like pigment (Giddings and Markakis, 1972; Nanke et al., 1998), or carboxyl-myoglobin or nitric oxide myoglobin (Millar et al., 1995). However, the red pigment cannot be an oxymyoglobin because the red colour formed by irradiation can be produced under anoxic conditions, and the carboxyl group is too big to get inside of haem pocket to form the sixth ligand with haem iron. Nam and Ahn (2002a,c) proposed the pigment responsible for the red colour in irradiated light meat as carbon monoxide myoglobin (CO-Mb). Furuta et al. (1992) first reported that carbon monoxide could be produced from organic components such as alcohols, aldehydes, ketones, carboxylic acids, amides and esters by irradiation. Subsequently, Ahn and Lee (2006) found that meat components such as glycine, asparagine, glutamine, pyruvate, glyceraldehydes, α-ketoglutarate and phospholipids were good substrates for CO production by irradiation. It is also well known that irradiation produces hydrated electrons (aqueous e^-), a powerful reducing agent, and decreases the ORP of meat (Thakur and Singh, 1994). The decrease of ORP in meat can play an important role in CO-Mb formation because the CO-Mb complex can only be formed when haem pigment is in reduced form (Nam and Ahn, 2002a,c). The lower ORP in irradiated meat, however, was maintained only under vacuum-packaging conditions (Nam and Ahn, 2002b). Maintaining reducing conditions (low ORP conditions) in meat is important for CO-Mb formation because the CO-Mb complex can only be formed when the haem pigment is in reduced form (Cornforth et al., 1986). Therefore, the decrease of ORP and CO production by irradiation are major factors for CO-Mb ligand formation of irradiated light meat (Nam and Ahn, 2002a,b; Brewer, 2004). In irradiated red meat, however, ORP or the status of haem pigments is more important than CO production or CO-Mb ligand formation. The content of haem pigments in beef is about 10 times greater than that of light meat and the proportion of CO-Mb to total haem pigments in irradiated beef is small (Ahn and Lee, 2004).

Aerobic packaging was more desirable than vacuum packaging in reducing the intensity of irradiated meat colour, as exposing irradiated poultry meat to aerobic conditions decreased the formation of carbon monoxide-myoglobin due to the competition of CO with oxygen and the oxidation of haem iron under aerobic conditions (Du et al., 2002).

7 Objective colour measurement for meat products

Measurement of colour also requires three dimensions of colour (hue, saturation and lightness). Any colour can be matched exactly using a suitable mixture of the three primaries (red, blue and green) and the relative amounts required to match a given colour are known as tristimulus value. In principle, meat colour can be measured by visual appraisal or instrumental colour methodology.

Objective colour measurement is necessary to understand mechanisms leading to colour changes and to devise strategies for elimination of colour defect in poultry meat. Colorimeter is a tricolorimetric system used for measuring hue, saturation and lightness of meat. Usually, the Hunter Lab or CIE Lab values and reflectance at specific wavelengths are used to express colour data (Bianchi et al., 2006). Numerous researchers have used *L*-, *a*-, and *b*-values to document treatment effects on colour. Ratio of *a*/*b*, hue angle or saturation index is also used for discolouration studies (MacDougall, 1982). Tapp et al. (2011) reviewed the instrumental colour measurement information in meat science research using 1068 peer-reviewed journal articles. The paper reported that most researchers used Minolta (60.0%) over Hunter (31.6%), and most of the research was done using illuminant D_{65} (32.3%). Various reflectance values have been used to measure meat colour, to follow colour change and to quantitate myoglobin forms. A major advantage of reflectance is that repeated measurements can be made on individually packaged sample. However, measurement reproducibility may be less as much of the light is scattered on translucent meat surfaces (Guidi and Castigliego, 2010).

Extraction procedures can give sharper peaks and better separation than reflectance measurements. However, it is possible to overestimate oxymyoglobin and metmyoglobin, and underestimate deoxymyoglobin because of changes occurring during extraction and measurement (AMSA, 2015). Each of the haem pigments has a characteristic absorbance spectrum, which can be used to determine the proportion of each pigment at a meat surface (Krzywicki, 1979; Millar et al., 1996). Reduced myoglobin has a diffuse absorption band with a maximum at 555 nm. Oxymyoglobin has two sharp peaks at approximately 540 and 580 nm. Metmyoglobin has an absorption peak at 505 nm and second weaker peak at 630 nm (Bandman, 1987). CO-Mb responsible for the red or pink colour of irradiated poultry meats showed absorption peaks at 541 and 577 nm (Nam and Ahn, 2002a).

Equations to estimate the myoglobin forms were summarized by Hunt (1980). For cured meat colour, a reflectance ratio of wavelengths 650/570 nm (Erdman and Watts, 1957) and the amount of total haem pigment and the percentage conversion total pigment to nitrosohaemochrome are widely used (Hornsey, 1956). The brown colour of cooked meat is difficult to measure instrumentally. Therefore, lack of redness or the percentage of denatured pigments is used in the colour evaluation of cooked meat (Trout and Gutzke, 1996).

A computer vision system (CVS) has been applied to the colour measurement of foods. The CVS has been proved efficient and reliable even when the samples have nonhomogeneous colour, shapes and surfaces (Girolami et al., 2013). Wu and Sun (2013) suggested that further in-depth research is required on the development of a faster, lighter/smaller and less-expensive CVS hardware, which is necessary to increase the image resolution, reduce image acquisition and analysis time, and improve the speed and space of storage for detailed colour measurement.

8 Conclusions

Numerous research works on the colour of fresh, cooked, cured and irradiated poultry meats; the fundamental chemistry of poultry meat pigments and the mechanisms of discolouration have been conducted. However, there are still a few significant basic concepts and treatments that need to be answered. In particular, the chemistry of pink colour regeneration in oven-roasted poultry breast meat during storage, the prevention of pinking in uncured cooked poultry breast meat, the role of genetics in meat colour and the relationships between metmyoglobin reduction and oxygen consumption have been suggested for the future research to solve practical meat colour problems. Developing new methods for colour detection using advanced CVS or biomedical techniques is also necessary to apply the new technologies to practical systems. The effect of additives, packaging materials, processing, oxidative stress of live birds and slaughtering conditions on breast meat colour needs further attention.

9 Where to look for further information

For an overview of the fundamental basis of myoglobin's interactions with biomolecules in postmortem skeletal muscles, factors influencing meat colour and the chemistry of meat colour phenomena; current research in meat colour:

Suman, S. P. and Joseph, P. (2013). Myoglobin chemistry and meat color. *Annual Review in Food Sci. Technol.* 4: 79–99.

Manchini, R. A. and Hunt, M. C. Review: Current research in meat color. *Meat Sci.* 71(1): 100–21.

For meat colour defects and factors affecting the colour of raw and cooked meat and poultry: USDA-FSIS, 2013. The Color of Meat and Poultry. http://www.fsis.usda.gov/wps/portal/fsis/topics/food-safety-education/get-answers/food-safety-fact-sheets/meat-preparation/the-color-of-meat-and-poultry/the-color-of-meat-and-poultry/ct_index.

For irradiation effect on meat colour: Brewer, S. (2004). Irradiation effects on meat color – a review. *Meat Sci.* 68: 1–17.

10 References

Ahn, D. U. and Maurer, A. J. (1987). The concentration of nitrate and nitrite in raw turkey breast meat and the microbial conversion of added nitrate to nitrite in tumbled turkey breast meat. *Poult. Sci.* 66: 1957–60.

Ahn D. U. and Lee, E. J. (2004). Mechanisms and prevention of off-odor production and color changes in irradiated meat. In *Food Irradiation Research and Technology*, edited by C. H. Sommers and X. Fan, 127–42. Blackwell Publishing Professional, Ames, IA.

Ahn, D. U. and Maurer, A. J. (1989a). Effects of added nitrite, sodium chloride, and phosphate on color, nitrosoheme pigment, total pigment and residual nitrite in oven roasted turkey breast meat. *Poult. Sci.* 68: 100–6.

Ahn, D. U. and Maurer, A. J. (1989b). Effects of sodium chloride, phosphate, nitrite, dextrose and conalbumin on the heat stability of purified myoglobin and cytochrome c. *Poult. Sci.* 68: 1218–25.

Ahn, D. U. and Maurer, A. J. (1989c). Effects of added pigments, salt, and phosphate on color, extractable pigment, and oxidation-reduction potential in turkey breast meat. *Poult. Sci.* 68: 1088–99.

Ahn, D. U. and Maurer, A. J. (1990a). Poultry meat color: kinds of heme pigments and concentrations of the ligands. *Poult. Sci.* 69: 157–65.

Ahn, D. U. and Maurer, A. J. (1990b). Poultry meat color: heme-complex forming ligands and color of cooked turkey breast meat. *Poul. Sci.* 69: 1769–74.

Ahn, D. U. and Maurer, A. J. (1990c). Poultry meat color: pH and the heme-complex forming reaction. *Poult. Sci.* 69: 1218–25.

Ahn, D. U., Olson, D. G., Jo, C., Chen, X., Wu, C. and Lee, J. I. (1998). Effect of muscle type, packaging, and irradiation on lipid oxidation, volatile production, and color in raw pork patties. *Meat Sci.* 47: 27–39.

Allen, C. D., Fletcher, D. L., Northcutt, J. K. and Russell, S. M. (1998). The relationship of broiler breast color to meat quality and shelf-life. *Poult. Sci.* 77: 361–6.

Allen, C. D., Russell, S. M. and Fletcher, D. L. (1997). The relationship of broiler breast meat color and pH to shelf-life and odor development. *Poult. Sci.* 76: 1042–6.

Alnahhas, N., Berri, C., Boulay, M., Baéza, E., Jégo, Y., Baumard, Y., Chabault, M. and Le Bihan-Duval, E. (2014). Selecting broiler chickens for ultimate pH of breast muscle: analysis of divergent selection experiment and phenotypic consequences on meat quality, growth, and body composition traits. *J. Anim. Sci.* 92: 3816–24.

AMSA. (2015). Guidelines for meat color evaluation. In *Proceedings of the 68th Reciprocal Meat Conference*. American Meat Science Association, Champaign, IL.

Andersen, H. J. and Skibsted, L. H. (1992). Kinetics and mechanism of thermal oxidation and photooxidation of nitrosylmyoglobin in aqueous solution. *J. Agric. Food Chem.* 40: 1741–50.

Arroyo, C., Eslami, S., Brunton, N. P., Arimi, J. M., Noci, F. and Lyng, J. G. (2015). An assessment of the impact of pulsed electric fields processing factors on oxidation, color, texture, and sensory attributes of turkey breast meat. *Poult. Sci.* 94: 1088–95.

Bandman, E. (1987). Chemistry of animal tissues. In *The Science of Meat and Meat Products*, edited by J. F. Price and B. S. Schweigert, 61–101. Food & Nutrition Press, Inc., Westport, CT.

Barbut, S. (1996). Estimates and detection of the PSE problem in young turkey breast meal. *Can. J. Anim. Sci.* 76: 455–7.

Bendall, J. R. and Taylor, A. A. (1972). Consumption of oxygen by the muscles of beef animals and related species. II. Consumption of oxygen by post-rigor muscle. *J. Sci. Food Agric.* 23: 707–19.

Berthelsen, G. and Skibsted, L. H. (1987). Photooxidation of oxymyoglobin – wavelength dependence of quantum yields in relation to light discoloration of meat. *Meat Sci.* 19: 243–51.

Bianchi, M. and Fletcher, D. L. (2002). Effects of broiler breast meat thickness and background on color measurements. *Poult. Sci.* 81: 1766–9.

Bianchi, M., Petracci, M. and Cavani, C. (2006). The influence of genotype, market live weight, transportation, and holding conditions prior to slaughter on broiler breast meat color. *Poult. Sci.* 85: 123–8.

Bowker, B. C., Zhuang, H. and Buhr, R. J. (2014). Impact of carcass scalding and chilling on muscle proteins and meat quality of broiler breast fillets. *LWT – Food Sci. Technol.* 59: 156–62.

Chan, J. T. Y., Omana, D. A. and Betti, M. (2011). Effect of ultimate pH and freezing on the biochemical properties of proteins in turkey breast meat. *Food Chem.* 127: 109–17.

Chouliara, E., Karatapanis, A., Savvaidis I. N. and Kontominas, M. G. (2007). Combined effect of oregano essential oil and modified atmosphere packaging on shelf-life extension of fresh chicken breast meat, stored at 4°C. *Food Microbiol.* 24: 607–17.

Cornforth, D. P., Rabovitser, J. K., Ahuja, S., Wagner, J. C., Hanson, R., Cummings, B. and Chudnovsky, Y. (1998). Carbon monoxide, nitric oxide, and nitrogen dioxide levels in gas ovens related to surface pinking of cooked beef and turkey. *J. Agric. Food Chem.* 46: 255–61.

Cornforth, D. P., Vahabzadeh, F., Carpenter, C. E. and Bartholomew, D. T. (1986). Role of reduced hemochrome in pink color defect of cooked turkey rolls. *J. Food Sci.* 51: 1132–5.

Cornish, D. G. and Froning, G. W. (1974). Isolation and purification of turkey heme proteins. *Poult. Sci.* 53: 365–77.

Dobson, B. N. and Cornforth, D. P. (1993). Nonfat dry milk inhibits pink discoloration in turkey rolls. *Poult. Sci.* 71: 1943–6.

Du, M., Hur, S. J. and Ahn, D. U. (2002). Raw-meat packaging and storage affect the color and odor of irradiated broiler breast fillets after cooking. *Meat Sci.* 61: 49–54.

Erdman, A. M. and Watts, B. M. (1957). Spectrophotometric determination of color change in cured meat. *J. Agric. Food Chem.* 5: 453–8.

Faustman, C. and Cassens, R. G. (1990). The biochemical basis for discoloration in fresh meat: a review. *J. Muscle Foods* 1: 217–43.

Faustman, C., Cassens, R. G. and Greaser, M. L. (1988). Reduction of metmyoglobin of extracts of bovine liver and cardiac muscle. *J. Food Sci.* 53: 1065–7.

Faustman, C., Chan, W. K. M., Lynch, M. P. and Joo, S. T. (1996). Strategies for increasing oxidative stability of fresh meat color. In *Proceedings of the 49th Reciprocal Meat Conference*. 49: 73–8.
Fletcher, D. L., Qiao, M. and Smith, D. P. (2000). The relationship of raw broiler breast meat color and pH to cooked meat color and pH. *Poult. Sci.* 79: 784–8.
Fletcher, D. L. (1999). Broiler breast meat color variation, pH, and texture. *Poult. Sci.* 78: 1323–7.
Fox, J. B. (1987). The pigments of meat. In *The Science of Meat and Meat Products*, edited by J. F. Price and B. S. Schweigert, 193–216. Food & Nutrition Press, Inc., Westport, CT.
Froning, G. W., Daddario, J. and Hartung, T. E. (1968). Color and myoglobin concentration in turkey meat as affected by age, sex, and strain. *Poult. Sci.* 1827–35.
Froning, G. W., Daddario, J., Hartung, T. E., Sullivan, T. W. and Hill, R. M. (1969). Color of poultry meat as influenced by dietary nitrates and nitrites. *Poult. Sci.* 48: 668–74.
Furuta, M., Dohmaru, T., Katayama, T., Toratoni, H. and Takeda, A. (1992). Detection of irradiated frozen meat and poultry using carbon monoxide gas as a probe. *J. Agric. Food Chem.* 40: 1099–100.
Gatellier, P., Anton, M. and Renerre, M. (1993). A rapid method of oxymyoglobin purification. *Meat Sci.* 33: 401–7.
Ghorpade, V. M. and Cornforth, D. P. (1993). Spectra of pigments responsible pink color in pork roasts cooked to 65 or 82°C. *J. Food Sci.* 58: 51–2, 89.
Giddings, G. G. (1977). Symposium: The basis of quality in muscle foods, The basis of color in muscle foods. *J. Food Sci.* 42: 288–94.
Girolami, A., Napolitano, F., Faraone, D. and Braghieri, A. (2013). Measurement of meat color using a computer vision system. *Meat Sci.* 93: 111–18.
Guardia, S., Lessire, M., Corniaux, A., Métayer-Coustard, S., Mercerand, F., Tesseraud, S., Bouvarel, I. and Berri, C. (2014). Short-term nutritional strategies before slaughter are effective in modulating the final pH and color of broiler breast meat. *Poult. Sci.* 93: 1764–73.
Hagler, L., Coppes, R. I. and Herman, R. H. (1979). Metmyoglobin reductase. Identification and purification of a reduced nicotinamide adenine dinucleotides dependent enzyme from bovine heart, which reduces metmyoglobin. *J. Biol. Chem.* 254: 6505–14.
Heaton, K. M., Cornforth, D. P., Moiseeve, I. V., Egbert, W. R. and Carpenter, C. E. (2000). Minimum sodium nitrite levels for pinking of various cooked meats as related to use of direct or indirect-dried soy isolates in poultry rolls. *Meat Sci.* 55: 321–9.
Hoen, T. and Lankhaar, J. (1999). Controlled atmosphere stunning of poultry. *Poult. Sci.* 78: 287–9.
Holownia, K., Chinnan, M. S. and Reynolds, A. E. (2003). Pink color defect in poultry white meat as affected by endogenous conditions. *J. Food Sci.* 68: 742–7.
Hornsey, H. C. (1956). The color of cooked cured pork. I. Estimation of the nitric oxide-heme pigments. *J. Sci. Food Agric.* 7: 534–40.
Howe, J. L., Gullett, E. A. and Usborne, W. R. (1982). Development of pink color in cooked pork. *Can. Inst. Food Sci. Technol. J.* 15: 19–23.
Hunt, M. C. (1980). Meat color measurements. In *Proceedings of the Reciprocal Meat Conference*, 41–6. National Live Stock and Meat Board, Chicago.
Judge, M. D., Aberle, E. D., Forrest, J. C., Hendrick, H. B. and Merkel, R. A. (1989). Properties of fresh meat. In *Principles of Meat Science*, 2nd edited by M. D. Judge,178–85. Kendall/Hunt Publishing Company, Dubuque, IA.
Kanner, J. (1994). Oxidative processed in meat and meat products: quality implications. *Meat Sci.* 36: 169–89.
Kieffer, K. J., Claus, J. R. and Wang, H. (2000). Inhibition of pink color development in cooked, uncured ground turkey by the addition of citric acid. *J. Muscle Foods* 11: 235–43.
Kim, C., Carpenter, C. E., Cornforth, D. P., Mettanant, O. and Mahoney, A. W. (1988). Effect of iron form, temperature, and inoculation with *Clostridirum botulinum* spores on residual nitrite in meat and model system. *J. Food Sci.* 52: 1464–70.
Kropf, D. H. (1980). Effects of retail display conditions on meat color. In *Proceedings of the 33rd Reciprocal Meat Conference*, 15–32. National Live Stock and Meat Board, Chicago.

Krzywicki, K. (1979). Assessment of relative content of myoglobin, oxymyoglobin and metmyoglobin at the surface of the beef. *Meat Sci.* 3: 1–5.

Kuttappan, V. A., Lee, Y. S., Erf, G. F., Meullenet, J. F. C., McKee, S. R. and Owens, C. M. (2012). Consumer acceptance of visual appearance of broiler breast meat with varying degrees of white striping. *Poult. Sci.* 91: 1240–7.

Kuttappan, V. A., Shivaprasad, H. L., Shaw, D. P., Valentine, B. A., Hargis, B. M., Clark, F. D., McKee, S. R. and Owens, C. M. (2013). Pathological changes associated with white striping in broiler breast muscles. *Poult. Sci.* 92: 331–8.

Lawrie, R. A. (1983). Review. Aspects of the biochemistry of meat. *Int. J. Biochem.* 15: 233–42.

Ledward, D. A. (1970). Metmyoglobin formation in beef stored in carbon dioxide enriched and oxygen depleted atmospheres. *J. Food Sci.* 35: 33–6.

Ledward, D. A. (1984). Haemoproteins in meat and meat products. In *Developments in Food Proteins-3*, edited by B. J. F. Hudson, 197–8. Elservier Appl. Sci. Publ. Co., New York.

Lee, B. J., Hendricks, D. G. and Cornforth, D. P. (1999). A comparison of carnosine and ascorbic acid on color and lipid stability in a grou(nd beef pattie model system. *Meat Sci.* 51: 245–53.

Lee, S. H. and Cassens, R. G. (1976). Nitrite binding sites on myoglobin. *J. Food Sci.* 41: 969–70.

Lehninger, A. L. (1982). Globular proteins: the structure and function of hemoglobin. In *Principles of Biochemistry*, 169–206. Worth Publishers, Inc., New York.

Li, K., Zhao, Y. Y., Kang, Z. L., Wang, P., Han, M. Y., Xu, X. L. and Zhou, G. H. (2015). Reduced functionality of PSE-like chicken breast meat batter resulting from alterations in protein conformation. *Poult. Sci.* 94: 111–22.

Liu, Q., Lanari, M. C. and Schaefer, D. M. (1995). A review of dietary vitamin E supplementation for improvement of beef quality. *J. Anim. Sci.* 73: 3131–40.

Livingston, D. J. and Brown, W. D. (1981). The chemistry of myoglobin and its reactions. *Food Technol.* 35: 244–52.

Luchsinger, S. E., Kropf, D. H., Garcia Zepeda, C. M., Hunt, M. C., Marsden, J. L., Rubio Canas, E. J., Kastner, C. L., Kuecher, W. G. and Mata, T. (1996). Color and oxidative rancidity of gamma and electron beam-irradiated boneless pork chops. *J. Food Sci.* 61: 1000–5.

Lynch, M. P., Faustman, C., Chan, W. K. M., Kerry, J. P. and Buckley, D. J. (1998). A potential mechanism by which α-tocopherol maintains oxymyoglobin pigment through cytochrome b_5 mediated reaction. *Meat Sci.* 50: 333–42.

Lyon, B. G., Davis, C. E. and Lyon, C. E. (1986). Detection of blood (hemoglobin) contamination in poultry muscle tissue. *Poult. Sci.* 65: 1919–24.

MacDougall, D. B. (1982). Changes in the color and opacity of meat. *Food Chem.* 9: 75–81.

Mancini, R. A. and Hunt, M. C. (2005). Current research in meat color. *Meat Sci.* 71: 100–21.

Millar, S. J., Moss, B. W. and Stevenson, M. H. (1996). Some observations on the absorption spectra of various myoglobin derivatives found in meat. *Meat Sci.* 42: 277–88.

Millar, S. J., Moss, B. W., MacDougall, D. B. and Stevenson, M. H. (1995. The effect of ionizing radiation on the CIELAB color co-ordinates of chicken breast meat as measured by different instruments. *Int. J. Food Sci. Technol.* 30: 663–74.

Min, B. and Ahn, D. U. (2012). Sensory properties of packaged fresh and processed poultry meat, in the book. In *Advances in Meat, Poultry and Seafood Packaging*, edited by J. P. Kerry, 112–53. Woodhead Publishing, Cambridge.

Nam, K. C. and Ahn, D. U. (2002a). Carbon monoxide-heme pigment is responsible for the pink color in irradiated raw turkey breast meat. *Meat Sci.* 60(1): 25–33.

Nam, K. C. and Ahn, D. U. (2002b). Mechanisms of pink color formation in irradiated precooked turkey breast meat. *J. Food Sci.* 67: 600–7.

Nam, K. C. and Ahn, D. U. (2003). Effects of irradiation on meat color. *Food Sci. Biotechnol.* 12: 198–205.

Nam, K. C., Kim Y. H., Du, M. and Ahn, D. U. (2002). Off-odor volatiles and pink color development in precooked, irradiated turkey breast during frozen storage. *Poult. Sci.* 81: 269–75.

Nanke, K. E., Sebranek, J. G. and Olson, D. G. (1998). Color characteristics of irradiated vacuum-packaged pork, beef, and turkey. *J. Food Sci.* 63: 1001–6.
Ngoka, D. A. and Froning, G. W. (1982). Effect of free struggle and preslaughter excitation on color of turkey breast muscles. *Poult. Sci.* 61: 2291–3.
Nocito, J. S., Bayne, B. H., Penfield, M. P. and Meyer, B. H. (1973). Myoglobin content and color of raw pork loin roasts as affected by freezing at two rates. *J. Anim. Sci.* 37: 1339–43.
Pearson, A. M. and Young, R. B. (1989). Skeletal muscle fiber types. In *Muscle and Meat Biochemistry*, edited by A. M. Pearson, 235–65. Academic Press, Inc., New York.
Pedrão, M. R., Kato, T., Soares, A. L., Ida, E. I., Coró, F. A. G., Grespan, M., Paião F. and Shimokomaki, M. (2015). Influence of cooling on the glycolysis rate and development of PSE (pale, soft, exudative) meat. *Brazilian Arch. Biol. Technol.* 58(2): 272–7.
Petracci, M., Betti, M., Bianchi, M. and Cavani, C. (2004). Color variation and characterization of broiler breast meat during processing in Italy. *Poult. Sci.* 83: 2086–2.
Petrou, S., Tsiraki, M., Giatrakou, V. and Savvaidis, I. N. (2012). Chitosan dipping or oregano oil treatments, singly or combined on modified atmosphere packaged chicken breast meat. *Int. J. Food Microbiol.* 156: 264–71.
Pikul, J., Niewwiarowicz, A. and Kupijaj, H. (1986). The cytochrome c content of various poultry meats. *J. Sci. Food Agric.* 37: 1236–40.
Pikul, J., Niewwiarowicz, A. and Kupijaj, H. (1987). Cytochrome c levels in poultry. *Broiler Ind.* 4: 48.
Popp, J., Krischek, C., Janisch, S., Wicke, M. and Klein, G. (2013). Physico-chemical and microbiological properties of raw fermented sausages are not influenced by color differences of turkey breast meat. *Poult. Sci.* 92: 1366–75.
Potthast, K. (1987). Meat colour, colour stability and reddening. *Fleischwirtsch.* 67: 70.
Qiao, M., Fletcher, D. L., Smith, D. P. and Northcutt, J. K. (2001). The effect of broiler breast meat color on pH, moisture, water-holding capacity, and emulsification capacity. *Poult. Sci.* 80: 676–80.
Qiao, M., Fletcher, D. L., Northcutt, J. K. and Smith, D. P. (2002b). The relationship between raw broiler breast meat color and composition. *Poult. Sci.* 81: 422–7.
Reddy, I. M. and Carpenter, C. E. (1991). Determination of metmyoglobin reductase activity in bovine skeletal muscles. *J. Food Sci.* 56: 1161–4.
Renerre, M. (1999). Biochemical basis of fresh meat colour. In *Proceedings of the 45th International Congress of Meat Science and Technology. Tokyo, Japan*, 344–53.
Renerre, M. and Labadie, J. (1993). Fresh meat packaging and meat quality. In *Proceedings of the 39th Reciprocal Meat Conference*. National Live Stock and Meat Board, Chicago.
Rey, A. I., Segura, J., Olivares, A., Cerisuelo, A., Piñeiro, C. and López-Bote, C. J. (2015). Effect of micellized natural (D-α-tocopherol) vs. synthetic (DL-α-tocopheryl acetate) vitamin E supplementation given to turkeys on oxidative status and breast meat quality characteristics. *Poult. Sci.* 94: 1259–69.
Rickansrud, D. A. and Henrickson, R. L. (1967). Total pigments and myoglobin concentration in four bovine muscles. *J. Food Sci.* 32: 57–61.
Romans, J. R., Costello, W. J., Carlson, C. W., Greaser, M. L. and Jones, K. W. (1994). Meat curing and smoking. In *The Meat We Eat*, edited by J. D. Turton, 727–72. Interstate Publishers, Inc., Danville, IL.
Saffle, R. L. (1973). Quantitative determination of combined hemoglobin and myoglobin in various poultry meats. *J. Food Sci.* 38: 968–70.
Satterlee, L. D. and Zachariah, N. Y. (1972). Porcine and ovine myoglobin: isolation, purification, characterization and stability. *J. Food Sci.* 37: 909–12.
Schilling, M. W., Radhakrishnan, V., Vizzier-Thaxton, Y., Christensen, K., Williams, J. B. and Joseph, P. (2015). Sensory quality of broiler breast meat influenced by low atmospheric pressure stunning, deboning time and cooking methods. *Poult. Sci.* 94: 1379–88.
Schneider, B. L., Renema, R. A., Betti, M., Carney, V. L. and Zuidhof, M. J. (2012). Effect of holding temperature, shackling, sex, and age on broiler breast meat quality. *Poult. Sci.* 91: 468–77.

Schwarz, S. J., Claus, J. R., Wang, H., Marriott, N. G., Graham, P. P. and Fernandes, C. F. (1999). Inhibition of pink color development in cooked, uncured turkey breast through ingredient incorporation. *Poult. Sci.* 78: 255–66.

Seideman, S. C., Cross, H. R., Smith, G. C. and Durland, P. R. (1984). Factors associated with fresh meat color: a review. *J. Food Qaul.* 6: 211–37.

Sheldon, B. W., Curtis, P. A., Dawson, P. L. and Ferket, P. R. (1997). Effect of dietary vitamin E on the oxidative stability, flavor, color, and volatile profiles of refrigerated and frozen turkey breast meat. *Poult. Sci.* 76: 634–41.

Smith, D. P., Northcutt, J. K., Sigmon C. S. and Parisi, M. A. (2015). Effect of sex, bird size and marination on duck breast meat quality. *Int. J. Poult. Sci.* 14(4): 191–5.

Suman, S. P. and Joseph, P. (2014). Color and pigment. *Encycl. Meat Sci.* 3: 244–51.

Tapp, W. N., Yancey, J. W. S. and Apple, J. K. (2011). How is the instrumental color of meat measured? *Meat Sci.* 89: 1–5.

Tarladgis, B. G. and Ehtashan-Ud-Din, A. M. F. (1965). Structure of the pigments of precooked irradiated meats. *Nature* 207: 489–91.

Taylor, A. A. and MacDougall, D. B. (1973). Fresh beef packed in mixtures of oxygen and carbon dioxide. *J. Food Technol.* 8: 453–61.

Thakur, B. R. and Singh, R. K. (1994). Food irradiation – chemistry and applications. *Food Rev. Int.* 10: 437–73.

Trout, G. R. and Gutzke, D. A. (1996). A simple rapid preparative method for isolating and purifying oxymyoglobin. *Meat Sci.* 43: 1–13.

Van Laack, R. L. J. M., Liu, C. H., Smith, M. O. and Loveday, H. D. (2000). Characteristics of pale, soft, exudative broiler breast meat. *Poult. Sci.* 79: 1057–61.

Walter, C. L. and Taylor, A. M. (1965). The reduction of nitrite by skeletal muscle mitochondria. *Biochem. Biophys. Acta.* 96: 522–4.

Wu, D. and Sun, D. W. (2013). Colour measurements by computer vision for food quality control – a review. *Trends Food Sci. Technol.* 29: 5–20.

Yin, M. C., Faustman, C., Fiesen, J. W. and Williams, S. N. (1973). The effect of α-tocopherol and ascorbate upon oxymyoglobin and phospholipid oxidation. *J. Food Sci.* 58: 1273–6.

Zhu, X. H., Xu, X. L., Min, H. H. and Zhou, G. H. (2012). Occurrence and characterization of pale, soft, exudative-like broiler muscle commercially produced in China. *J. Integr. Agric.* 11(8): 1384–90.

Zhu, X., Ruusunen, M., Gusella, M., Zhou, G. H. and Puolanne, E. (2011). High post-mortem temperature combined with rapid glycolysis induces phosphorylase denaturation and produces pale and exudative characteristics in broiler Pectoralis major muscles. *Meat Sci.* 89: 181–8.

Zhu, X., Ruusunen, M., Gusella, M., Ylä-Ajos, M., Xu, X., Zhou, G. H. and Puolanne, E. (2013). High early post-mortem temperature induces activation of AMP-activated protein kinase and development of pale, soft and exudative characteristics in turkey muscles. *Meat Sci.* 93: 600–6.

Chapter 15

Enhancing texture and tenderness in poultry meat

Iksoon Kang, California Polytechnic State University, USA; and Yuan H. Brad Kim, Purdue University, USA

1 Introduction

Poultry has the highest per-capita consumption compared with other types of meats in the United States, and potentially the highest consumption in the world over the next five to ten years (USDA, Economic Research Service, 2015; OECD-FAO, 2015). Compared with other meats, poultry has many advantages such as low production cost, high nutritional value, rapid growth rate and a considerable number of processed products (Barbut, 2002). In addition, unlike some other types of meat, poultry meat is not affected by cultural and religious barriers. Poultry products meet the current demand for low-fat, less-saturated fatty acids, and low sodium and low cholesterol levels. For poultry growers and processors, birds can be raised in a short period of time and processed easily to produce a quality product. Out of many meat quality attributes, product taste and textural tenderness are primary factors in meeting consumers' expectations of quality. These characteristics are significantly affected by how birds are handled before, during and after slaughter. In this

http://dx.doi.org/10.19103/AS.2016.0010.18

chapter, a number of processing factors are discussed for improvement of poultry product quality attributes with a focus on tenderness in particular.

2 Post-mortem pH decline and rigor development

In a living animal, more than 600 muscles exist that are highly structured and specialized for various kinds of activities such as movement, digestion, chemical-to-mechanical energy conversion and heart beat (Fig. 1a). After an animal dies, the properties of muscle structure and their activities are closely related with meat tenderness and final product quality (Fig. 1b). Immediately after slaughter, muscle cells struggle to live, respire and respond to stimuli. Having no oxygen supply, the cells switch from aerobic metabolism

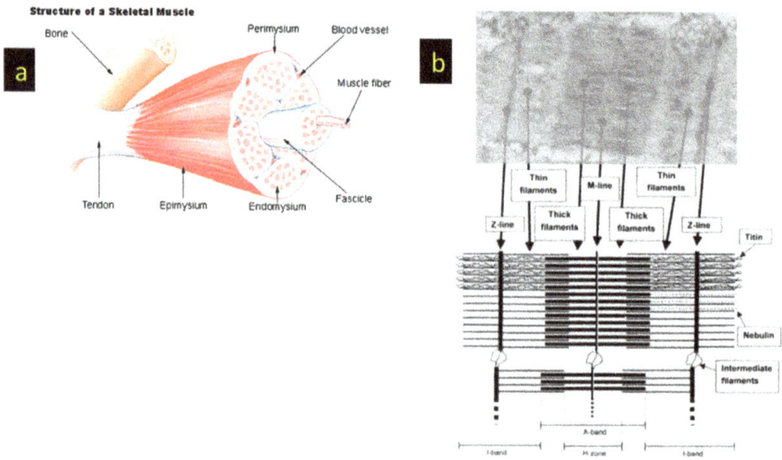

Figure 1 Schematic representation of the muscle structure: (a) structure of a skeletal muscle (adapted from SEER Training Modules at National Cancer Institute); (b) ultramicrography of a longitudinal section of skeletal muscle.

Source: Adapted from Schreurs, 2000.

Table 1 Delay time before the onset of rigor mortis

Species	Hours
Beef	6–12
Lamb	6–12
Pork	¼–3
Turkey	<1
Chicken	<½
Fish	<1

Source: Adapted from Aberle et al., 2001.

to anaerobic metabolism, resulting in a large amount of lactate accumulation. As a result, the muscle pH declines from 7.0 to 5.3–5.7 at 24-h post-mortem (PM) while muscle is converted to meat with stiffness.

Development of muscle stiffness (or rigor mortis) occurs due to insufficient adenosine triphosphate (ATP) as it is required for contraction and relaxation cycle of muscle fibre (DeFremery, 1966). When about 80% of muscle ATP is depleted, actomoysin bonds are formed between actin and myosin filaments as a permanent cross-bridge. The times for onset and completion of rigor mortis vary from animal to animal (Table 1). In broilers, the entire process takes at least 4 h, which is a limiting factor for the production of boneless poultry meat.

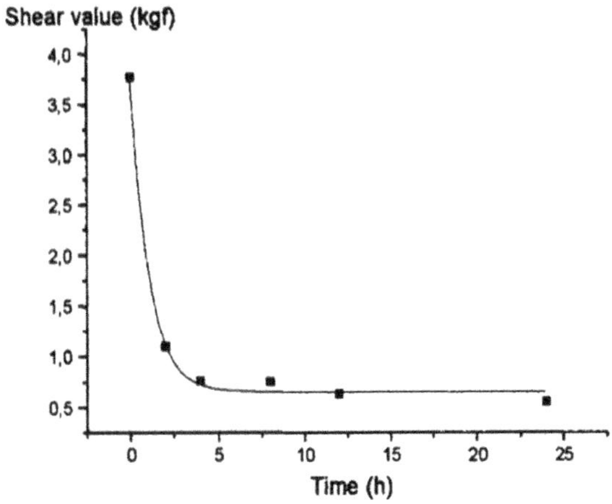

Figure 2 Shear values of chicken carcasses during ageing.
Source: Adapted from Santos et al., 2004.

Table 2 Variations of sensorial tenderness of chicken breast fillets with carcass ageing

Carcass ageing Time (h)	Sensorial tenderness scores for chicken breast fillets (Carcasses with 1.0 ± 0.1 kg)
0	1.75[d]
2	2.65[c]
4	3.06[bc]
8	3.69[ab]
12	4.00[a]
24	4.13[a]

[a–c] Means within a column with unlike superscripts are different ($P < 0.05$).
Source: Adapted from Santos et al., 2004.

To investigate the effect of PM toughness and tenderness on broiler breasts, Santos et al. (2004) stored broiler carcasses at 3°C for 0, 2, 4, 8, 12 and 24 h. During that time, the shear force of breast fillets sharply decreased in the first 2 h, with additional reduction observed for the rest of storage (Fig. 2). In correspondence to the shear force reduction, sensory panel scores increased for tenderness, resulting in the highest score at 8-h PM ageing or more (Table 2).

3 Carcass chilling and muscle tenderness

The quality of poultry carcasses, including tenderness, is largely influenced by chilling type, chilling temperature and chilling time (Carroll and Alvarado, 2008; Alvarado and Sams, 2002; Thompson et al., 1987). Comparing three chilling methods of water immersion chilling (WIC), evaporative air chilling (EAC) and air chilling (AC), Jeong et al. (2011b) reported that broiler chilling required 57, 93 and 125 min for WIC, EAC and AC, respectively (Fig. 3). At the end of chilling, carcasses in WIC showed higher brightness (L*) than those in AC or EAC, whereas AC carcasses exhibited more redness (a*) and yellowness (b*) than the others (Jeong et al., 2011a,b). In a consumer sensory evaluation, AC breasts had a higher juiciness score than did WIC and EAC breasts, with no significant difference observed for flavour, texture and overall acceptability.

During chilling from 37 to 4°C or less, poultry muscles are subjected to hot shortening (or rigor shortening) and cold shortening (Papinaho and Fletcher, 1996). De Fremery and Pool (1960) found that chicken breast fillets showed higher, intermediate and lower shear force when they were cooked after hot boning and placing in different temperatures at 40°C, 0°C and 10–20°C, respectively. In red mammalian muscle, it has been known that hot shortening occurs when PM muscles are exposed to high temperatures (≥40°C) with pH 6.3–6.0 during rigor onset (Honikel et al., 1980). In contrast, cold shortening occurs

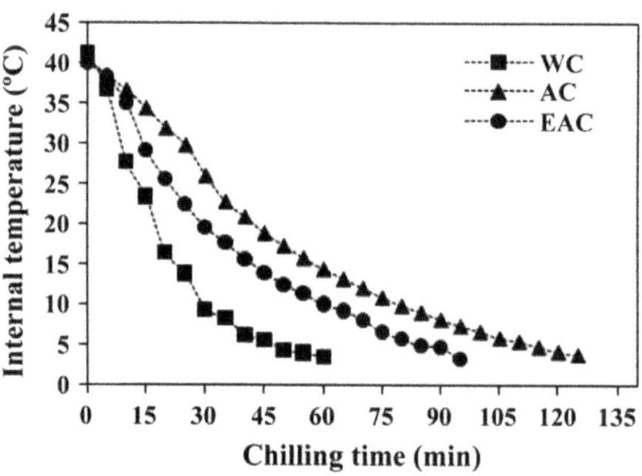

Figure 3 Temperature change profiles of broiler breast fillets during water chilling (WC; ■), air chilling (AC; ▲) and evaporative air chilling (EAC; ●).

Source: Adapted from Jeong et al., 2011b.

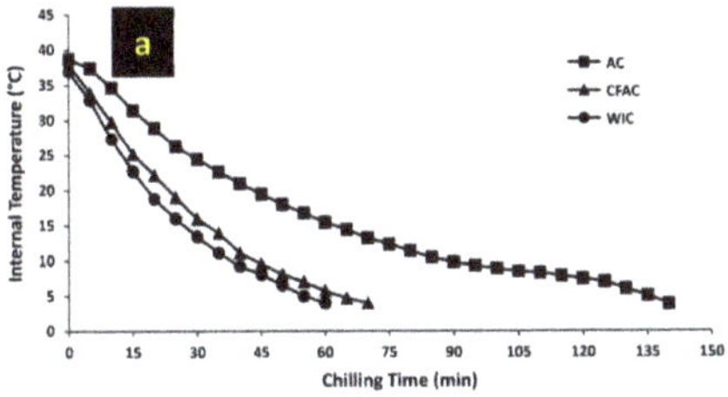

b

Parameter/ Chilling	pH	R-value	Sarcomere (μm)	Shear force (N)
AC	5.78	1.38	1.97^{ab}	11.6
CFAC	5.79	1.39	2.09^{a}	12.9
WIC	5.73	1.37	1.88^{b}	11.4

Figure 4 Carcass temperature change during chilling (a) and breast fillet properties after chilling (b) of broiler carcass using air chilling (AC), CFAC and WIC.

Source: Adapted from Sansawat et al., 2014.

when the PM muscles are subjected to low temperatures (≤15°C) at the early stage of PM glycolysis with high pH 6.7–7.0 (Locker and Hagyard, 1963; Honikel et al., 1983).

In 1993, Dunn et al. (1993) eliminated broiler breast fillets immediately after slaughter and incubated at 0, 5, 10, 20, 30 and 40°C up to 24 h. During the incubation, sarcomere lengths decreased from 1.50 to 1.41 μm at 0°C within 90 min, whereas a minimum sarcomere length (0.86 μm) was observed at 40°C after 380 min. In the evaluation of meat tenderness, the shear force increased from 2.94 at 0°C/30 min to 4.34 kg/cm^2 at 0°C/90 min storage, indicating an early occurrence of muscle shortening at that temperature. For the fillets at remaining temperatures (5–40°C), the shear force also increased from 2.6–3.6 to 5.2–7.2 kg/cm^2 after 500–600 min, which significantly decreased thereafter.

Using two AC temperatures of 0°C (AC control) and −12°C (simulating the chilling rate of broiler in WIC at 0°C), Dunn et al. (1995) found that the fillets of AC at −12°C improved tenderness with almost no hot shortening for the muscle at pH < 6.70 at 15-min PM. However, both chilling temperatures caused cold shortening and muscle toughness when the fillet's pH was ≥6.7 at 15-min PM. Generally, broiler carcasses chilled in air at −12°C had longer sarcomeres, lower cooking losses and lower shear force values than those chilled in air at 0°C (Dunn et al., 2000).

Using similar AC temperatures, Sansawat et al. (2014) reported that the breast temperature was reduced from 40 to 4°C in an average chilling time of 62, 68 and 140 min for WIC at 0°C, crust-freeze air-chilling (CFAC) at −12°C/1.5 m/s and AC at 1°C/1.5 m/s, respectively (Fig. 4a). After chilling, the fillets of CFAC had greater sarcomere length than that of WIC, with an intermediate value seen for AC. However, no significant differences were found for pH, *R*-value and shear force, regardless of the chilling method (Fig. 4b). The potential advantages of CFAC are further discussed in the subject of muscle tensioning for improved tenderness.

4 Pale, soft and exudative meat

The characteristics of pale, soft and exudative (PSE) meat, observed in heavily muscled commercial turkeys, are known to be similar to those of PSE meat found in the pork industry (Briskey, 1964; Ma and Addis, 1973). Several studies indicated that 22–30% broiler breast and 40% turkey breast showed PSE in commercial slaughter houses (Barbut, 1997a,b; Soares et al., 2003; Owens et al., 2000).

Table 3 The effect of seasonal heat stress on drip loss, cooking loss and expressible moisture (%) in turkey *pectoralis*

	Environmental treatments	
Parameters	Heat-stressed	Control
Drip loss	0.44[a]	0.06[b]
Cooking loss	24.6[a]	19.1[b]
Expressible moisture	22.0	25.2

[a–b] Means within a row with unlike superscripts are different ($P < 0.05$).
Source: Adapted from McKee and Sams, 1997.

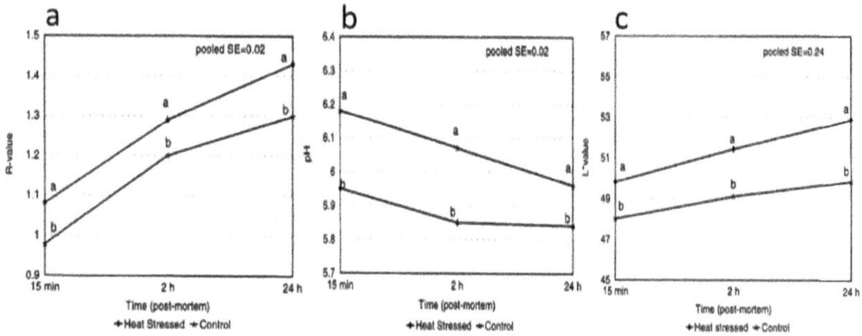

Figure 5 The effect of seasonal heat stress on *R*-value (a), pH (b) and *L*-value (c) in turkey *pectoralis*. Means ($n = 61$ per mean) within the level of PM time with different superscripts are significantly different ($P < 0.05$).

Source: Adapted from McKee and Sams, 1997.

For PSE to occur, both genetic and environmental factors play important roles, including a rapid growth production cycle, pre-slaughter handling conditions and post-slaughter chilling factors (Chiang et al., 2008; McKee and Sams, 1998). While comparing two genetic lines, turkey breasts that were obtained from a genetically selected commercial line showed higher lightness value and lower marinade absorption than those from the control of non-genetically selected turkeys (Chiang et al., 2008). In the comparison of normal and PSE turkey breast meats, the PSE meat showed a significant delay in the upregulation of ryanodine (RYR) isoforms (the sarcoplasmic reticulum calcium channel protein), αRYR and βRYR, and the regulation of calsequestrin, a high-capacity, low-affinity Ca^{2+}-binding protein located in the lumen of the sarcoplasmic reticulum (Sporer et al., 2012). Recently, the breast of PSE turkey has been observed to have both disorganized muscle fibres and altered oxidative metabolism compared with those of normal turkey breast (Malila et al., 2013).

To investigate the meat characteristics of turkeys in sustained seasonal heat stress, McKee and Sams (1997) exposed live turkeys to controlled temperatures (16°C at night/24°C during day) or heat-stressed temperatures (32°C at night/38°C during day). At 21 wk of age, including the last 4 wk of the temperature environments, turkeys were slaughtered and breast samples were taken at 15-min (pre-chill), 2-h (post-chill) and 24-h PM. The breasts from heat-stressed turkeys showed higher *R*-value, faster pH decline and paler colour as early as 15 min PM, with the trend continued throughout 24-h PM compared with the non-stressed counterpart (Fig. 5). The pale-coloured meat was suggested as one of the indicators of PSE meat, resulting from protein denaturation (Lawrie, 1991; Warris and Brown, 1987). In accordance with the lower pH and higher *L**-value, the fillets from heat-stressed turkey showed significantly higher drip loss and cooking loss after cooking, but no difference was found for expressible moisture loss compared with the non-stressed counterpart (Table 3). The reason for no expressible moisture differences was expected from the delayed time (24 PM) of sample analysis, during which much of the expressible moisture had been lost as drip.

Following the study of a sustained seasonal heat stress to live turkeys, McKee and Sams (1998) evaluated the effect of carcass chilling temperatures on muscle pH, colour, water holding capacity and meat tenderness by subjecting turkey carcasses to 0, 20 and 40°C water. During 4-h PM, the carcass temperature was reduced to <10°C at 0°C water and approximately 25°C at 20°C water, whereas the initial temperature (40°C) was maintained at 40°C water (Fig. 6a).

The levels of muscle glycogen indicated that the rate of PM metabolism linearly decreased as the PM temperature was decreased by showing the higher glycogen level at

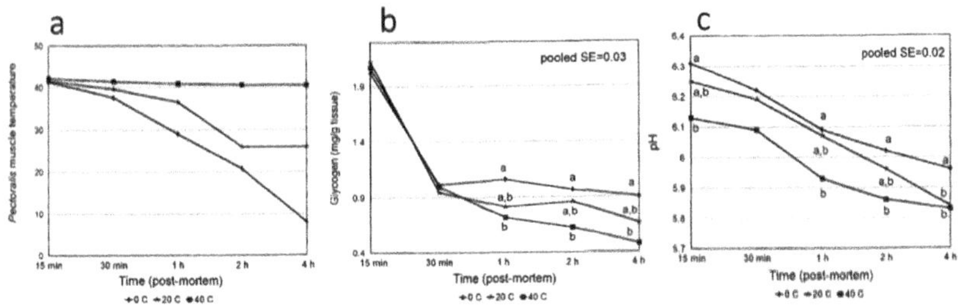

Figure 6 Temperatures and pH of *pectoralis* muscle from turkey carcasses held at 0, 20 or 40°C during PM ageing.
Source: Adapted from McKee and Sams, 1998.

Table 4 Physical parameters of *Pectoralis* muscles from turkey carcasses held at 0, 20 and 40°C for 4-h PM

	PM temperature treatments		
Parameters	0°C	20°C	40°C
Drip loss (%)	0.01^{b}	0.05^{b}	1.88^{a}
Cooking loss (%)	24.l^{b}	27.0^{ab}	28.9^{a}
Shear value (kg/g)	7.3l^{b}	7.43^{b}	9.69^{a}
Sarcomere length (μm)	1.89^{a}	1.87^{a}	1.79^{b}

$^{a–b}$Means within a row with unlike superscripts are different ($P < 0.05$).
Source: Adapted from McKee and Sams, 1998.

0°C water than at 40°C water, with an intermediate level at the 20°C treatment (Fig. 6b). In accordance with the glycogen level, the fillets at 40°C treatment showed a lower pH than at 0°C, with an intermediate pH at 20°C as early as 15 min (Fig. 6c). It is known that muscle protein is denatured when high carcass temperature is combined with a rapid pH decline at early PM (Warris and Brown, 1987).

On evaluation of physical properties, the fillets at 40°C water showed shorter sarcomere length and more drip loss during 24-h PM storage compared with the fillets at 0 and 20°C. As a result, more cooking loss and higher shear value were seen for the fillets of 40°C water treatment (Table 4). Both higher drip loss and cooked loss are attributed to the result of denatured protein and shorter sarcomere length (Fernandez et al., 1994), whereas the tougher meat and shorter sarcomere lengths were presumed from 'heat rigor' developed at the elevated temperature (Lawrie, 1991).

5 The crust-freeze air-chilling technique

As early as 1964, Briskey postulated that the combination of low pH and high carcass temperatures caused extensive protein denaturation in the muscle, which is a primary factor associated with the development of PSE-like meat characteristics (Fernandez et al., 1994; Santos et al., 1994). Poor or delayed chilling process may lead to the condition of accelerated pH decline and high temperature in muscle, causing damage to protein integrity.

In an attempt to expedite carcass chilling process, Medellin-Lopez et al. (2014) hot-boned (HB) turkey breast after evisceration and placed them, with/without quarter sectioning, to an air-freezing room (1.0 m/s, −12°C) for CFAC. For a control, whole carcasses were conventionally chilled in ice slurry water.

The initial temperature (41.3°C) of eviscerated turkey carcasses continuously decreased to 4°C, with the average time of 5.5 h in traditional water immersion chilling (Fig. 7A′), whereas the HB fillets in CFAC (HB-CFAC; Fig. 7B′) took 1.5 h, which was further reduced to 1 h with quarter sectioning (Fig. 7C′). After chilling, HB quarter-sectioned CFAC (HB-¼CFAC) fillets had higher pH and lower *R*-value than those of CB fillets (Table 5), indicating that

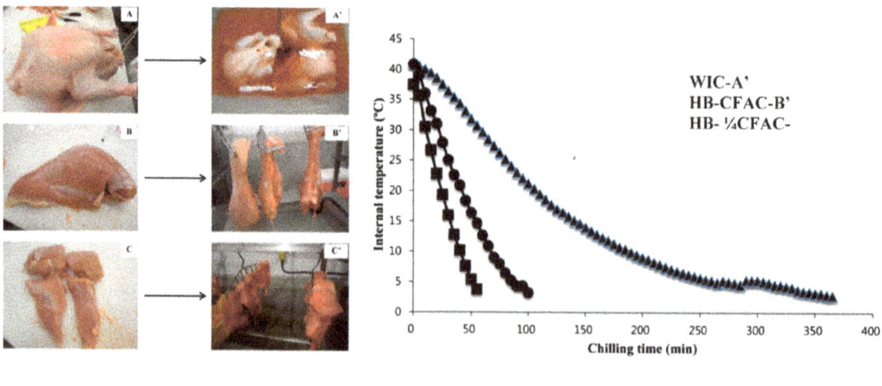

Figure 7 Temperature change profiles of turkey breast fillets during WIC, HB-CFAC and HB-¼CFAC.

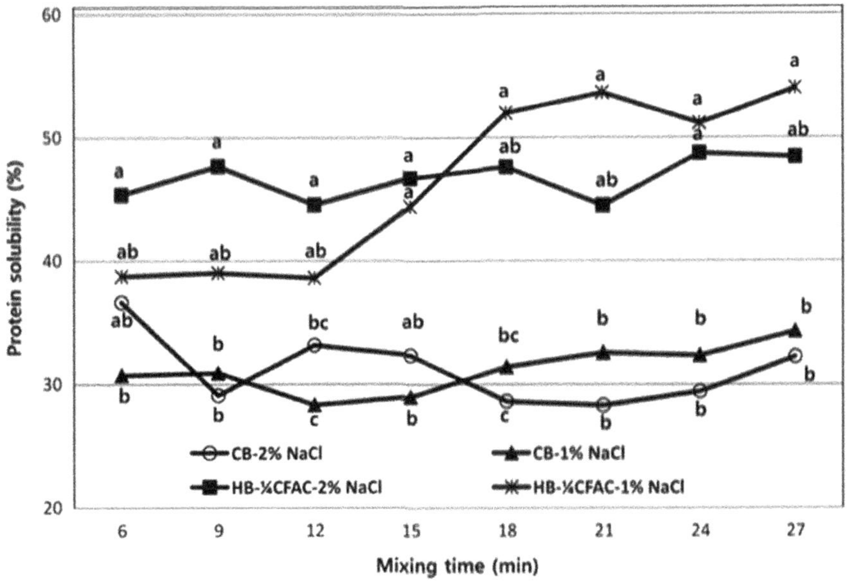

Figure 8 Protein solubility (%) of meat batters during 27-min mincing of turkey fillets that were CB or HB-¼CFAC. Means (n = 8) with no common letters (a–c) within the same mixing time differ ($P < 0.05$). Source: Adapted from Lee et al. (2014).

Table 5 pH and R-value (±SEM) of turkey breasts before and after being CB or HB-¼CFAC

Parameter/chilling	CB	HB-¼CFAC
pH before chill	6.04 ± 0.18^{a}	6.14 ± 0.21^{a}
pH after chill	5.73 ± 0.09^{b}	5.94 ± 0.10^{a}
R-value before chill	1.01 ± 0.09^{a}	0.98 ± 0.08^{a}
R-value after chill	1.32 ± 0.05^{a}	1.19 ± 0.13^{b}

$^{a-b}$Means within a row with unlike superscripts are different ($P < 0.05$).
Source: Adapted from Lee et al., 2014.

less glucose and ATP have been hydrolysed in the HB-¼CFAC fillets presumably due to a rapid chilling.

When turkey breasts were minced, proteins were solubilized for 40–55% in HB-¼CFAC fillets, whereas those of chill-boned (CB) controls did not exceed 37% (Fig. 8). It has been known that pre-rigor muscle proteins are extracted more easily than those of post-rigor muscle (Bernthal et al., 1989; Claus and Sorheim, 2006).

In scanning electron micrographs of meat batters, stabilized matrix complex was clearly detected with more extracted protein, entrapped fat and connective tissues in 2% salt

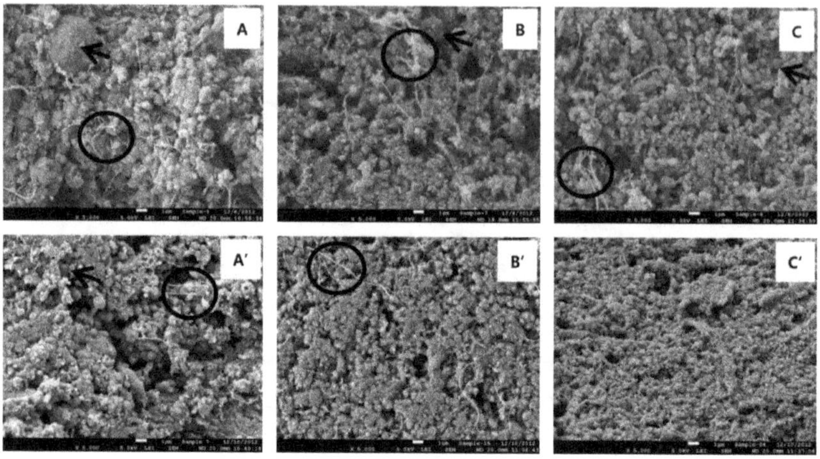

Figure 9 Scanning electron microscopy image of meat batters (2% salt) at 6-, 12- and 24-min mincing of turkey fillets that were CB or HB-¼CFAC. Arrow: emulsified fat globule; circle: connective tissue; bar = 1 μm. (A) Meat batter of HB-¼CFAC fillet minced for 6 min and (A′) meat batter of CB fillet minced for 6 min. (B) Meat batter of HB-¼CFAC fillet minced for 12 min and (B′) meat batter of CB fillet minced for 12 min. (C) Meat batter of HB-¼CFAC fillet minced for 24 min and (C′) meat batter of CB fillet minced for 24 min.

Source: Adapted from Lee et al., 2014.

Table 6 Chilling time and chilling time difference of turkey carcasses and fillets for COMM and RB

	Carcasses		Fillets	
Treatments[1]	WIC	TA-WIC	HB-¼CFAC	TA-HB-¼CFAC
Chilling time (h)				
COMM	4.5	5.75	1.25 (28%)	2.25 (50%)
RB	2.25	2.75	0.75 (33%)	1.5 (67%)
Chilling time difference (h)				
COMM-RB	2.25	3.0	0.5	0.75

[1]Treatments: strains (COMM, RB), TA and chilling methods (WIC, HB-¼CFAC). Parentheses show the percentage (%) of the chilling time compared with that of WIC
Source: Adapted from Lee et al., 2016.

HB-¼CFAC batter (Fig. 9A, B, C) than those of 2% salt CB batter (Fig. 9A′, B′, C′) at 6, 12 and 24 min. With the combination of HB-¼CFAC and cold batter mincing techniques, high quality of raw turkey meats was obtained and excellent protein gels were generated with 50% reduced salt (from 2 to 1% NaCl) (Lee et al., 2014).

Recently, Lee et al. (2016) have applied the HB-¼CFAC technique to chill the carcasses of commercial (COMM) and random-bred (RB) turkeys that were reared to 14.5 kg (an average live weight) in 15–17 wk and 11.0 kg in 20–21 wk, respectively (Fig. 10). When

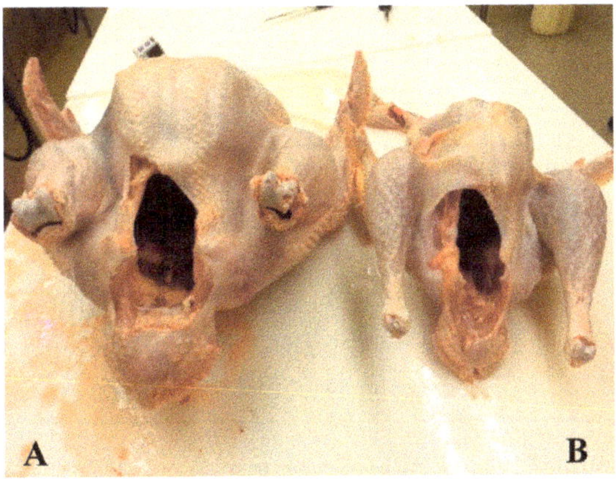

Figure 10 Carcass size of commercial (COMM, A) and random-bred (RB, B) turkeys.

Table 7 Pooled means[1] (±SEM) of cooked turkey gels prepared from two turkey strains after WIC or HB-¼CFAC with/without TA

	Strains		TA		Chilling methods	
Parameter/factors	COMM[1]	RB[1]	NTA[1]	TA[1]	WIC[1]	HB-¼CFAC[1]
Functional properties						
Cook yields (%)	86.9[a]	87.0[a]	88.6[a]	85.2[b]	85.8[b]	88.0[a]
Expressible moisture (%)	30.4[a]	26.8[b]	26.5[b]	30.7[a]	29.6[a]	27.6[b]
Torsion tests						
Shear stress (kPa)	37.6[b]	51.7[a]	49.9[a]	39.4[b]	40.8[b]	48.5[a]
Shear strain	1.04[b]	1.22[a]	1.25[a]	1.01[b]	1.07[a]	1.18[a]
Texture profile analysis						
Hardness (N)	19.3[b]	24.8[a]	24.5[a]	19.6[b]	20.5[b]	23.6[a]
Springiness	0.72[b]	0.78[a]	0.76[a]	0.75[a]	0.75[a]	0.75[a]
Gumminess	9.2[b]	11.2[a]	11.8[a]	8.6[b]	9.1[b]	11.3[a]
Chewiness	6.6[b]	8.7[a]	8.9[a]	6.4[b]	6.8[b]	8.5[a]

[a,b]Means with no common superscript within each row for each factor differ ($P < 0.05$).
Source: Adapted from Lee et al. (2016).

those birds were slaughtered and chilled with/without temperature abuse (TA) at 45°C for 30 min, the average chilling time without TA was 4.5 h for COMM carcasses in WIC and 1.25 h for HB fillets in HB-¼CFAC, which were delayed to 5.75 and 2.25 h with TA, respectively (Table 6).

Similarly, the chilling time of RB carcasses and RB fillets without TA was 2.25 and 0.75 h, respectively, which was extended to 2.75 and 1.5 h with TA (Table 6). These results

Table 8 pH, *R*-value, sarcomere length and shear force (±SEM) of broiler fillets at 3- and 24-h PM following WIC, AC, or CFAC with or without MT

Chilling/parameter	pH[1]	*R*-value[1]	Sarcomeres[1] (μm)	Shear force[1] (N)
WIC[2] 3 h PM[3]	6.14 ± 0.03[ab]	0.99 ± 0.03[d]	1.64 ± 0.04[c]	14.6 ± 0.65[a]
24 h PM[4]	5.73 ± 0.06[d]	1.37 ± 0.01[a]	1.88 ± 0.04[b]	11.4 ± 0.44[def]
AC[5] 3 h PM	6.16 ± 0.04[ab]	1.00 ± 0.02[cd]	1.65 ± 0.07[c]	14.4 ± 0.6l[ab]
24 h PM	5.7B ± 0.08[d]	1.38 ± 0.01[a]	1.97 ± 0.01[ab]	11.6 ± 0.27[cde]
CFAC[6] 3 h PM	5.99 ± 0.04[c]	1.19 ± 0.03[b]	1.89 ± 0.06[b]	13.2 ± 0.50[abc]
24 h PM	5.79 ± 0.05[d]	1.39 ± 0.01[a]	2.09 ± 0.06[a]	12.9 ± 0.65[bcd]

[a–f] Means within a column with unlike superscripts are different ($P < 0.05$).
[1] The number of observations in each chilling, n = l0–15.
[2] WIC: breast fillets deboned after 62 min WIC at 0.2°C and 118 min ageing at 4°C.
[3] 3 h PM: breast fillets were cooked at 3 h PM.
[4] 24 h PM: breast fillets were cooked at 24 h PM.
[5] AC: breast fillets deboned after 140 min AC at 1.5°C and 40 min ageing at 4°C.
[6] CFAC: breast fillets deboned after 68 min CFAC at −12°C and 112 min ageing at 4°C.
Source: Adapted from Sansawat et al., 2014.

indicated that the required chilling time for HB fillets of COMM carcasses in HB-¼CFAC and TA-HB-¼CFAC was 28 and 50% of whole carcasses in WIC, respectively. The required chilling time for HB fillets of RB carcasses in HB-¼CFAC and in TA-HB-¼CFAC was 33 and 67% of whole carcasses in WIC, respectively.

When the breast meats were minced and cooked, the cooked gels from no TA/ HB-¼CFAC/RB fillets had higher cooking yield, lower expressible moisture, higher stress/ strain values and better textural property than those from TA/WIC/COM fillets (Table 7). Both higher cooking yield and lower expressible moisture content indicated that the rapidly chilled fillets had more moisture during heating and centrifugal pressure potentially due to higher protein functionality. It is known that gel firmness, hardness and sensory panel firmness have a positive correlation with higher protein context (or extraction) and quality (Simon et al., 1965; Baker et al., 1969; Bloukas and Paneras,1993).

Three chilling techniques of CFAC, AC and WIC were applied to broiler carcasses to compare the properties of breast fillets after chilling. At 3-h PM, breast pH (5.99–6.16) was significantly higher than the pH (5.73–5.79) at 24-h PM, regardless of chilling method, whereas no significant differences were seen between chilling methods at 24-h PM.

Sarcomere length was significantly longer at 24-h PM than at 3-h PM, regardless of the chilling method, with an intermediate length observed for the CFAC fillets at 3-h PM. In accordance with the results of sarcomere length, lower shear force was observed at 24-h PM than at 3-h PM, with an intermediate value seen for the CFAC fillets at 3-h PM (Table 8).

6 Meat tenderness via enzymatic activity

At the turn of the twentieth century, scientists reported that the increase of meat tenderness during ageing was due to the results of endogenous enzymatic activity remaining in the

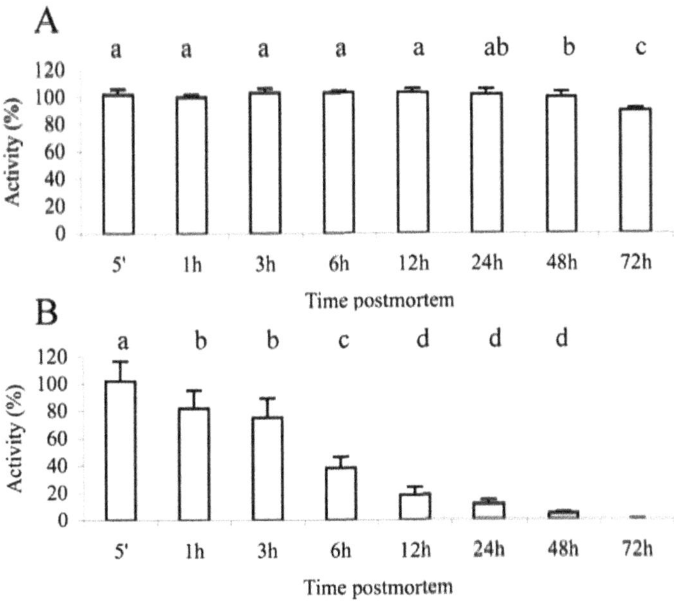

Figure 11 Caseinolytic activity of (A) μ/m-calpain and (B) μ-calpain in *pectoralis* muscle at different PM times. Activities are expressed as percentages of the reference muscle activity, which is taken as 100%.

Source: Adapted from Lee et al., 2008.

muscle (Lehmann, 1907; Hoagland et al., 1917). After an animal dies, numerous enzymes in muscle cells are believed to be active during extended PM times. In the role of PM tenderization, degradation of key muscle proteins by endogenous proteases plays an important role. These kinds of muscle proteins were suggested to be inter-myofibrillar protein (desmin), intra-myofibrillar proteins (titin and nebulin), and linking proteins to sarcolemma and cell membrane (dystrophin and vinculin) (Koohmaraie, 1996). Following considerable research over the past decades, three essential groups of enzymes were reported for PM tenderization as follows: (1) calpains/calpastatin system, (2) cathepsin peptidases and (3) proteasomes – multicatalytic proteinase complexes (Huff Lonergan et al., 1996; Koohmaraie and Geesink, 2006).

Cathepsins comprise a family of proteinases that exist in the lysosomes and sarcoplasmic reticulum of muscle cells, with the involvement of intracellular protein degradation in living tissues (Decker et al., 1980; Kirschke and Barrett, 1987). Out of 15 lysosomal cathepsin enzymes, 8 (cathepsins B, L, H, S, F, K, D and E) are found in skeletal muscle fibre and four (cathepsins B, D, H and L) are observed to degrade some proteins during PM ageing (Sentandreu et al., 2002). Proteasomes are very large protein complexes, and their proteolysis requires the attachment of at least four ubiquitin molecules to lysine residue on the target substrate (Goll et al., 2003).

Although both cathepsins and proteasomes are involved in meat tenderization to a certain degree, calpain enzymes are believed to have a primary role in meat tenderization by weakening intra- and inter-myofibrillar bonds (Huff Lonergan et al., 1996; Koohmaraie,

1994, 1996; Koohmaraie and Geesink, 2006; Taylor et al., 1995). In chicken, Okumura et al. (2005) identified polymorphisms in four calpain genes which are μ-calpain gene, m-calpain gene, p94 gene and μ/m-calpain gene (unique to birds compared with mammals). While comparing the activity of μ-calpain and μ/m-calpain genes in two genetic lines of broiler, gene expression was higher in the fast-growing line than in the slow-growing line. Subsequently, lower shear force was observed in the fast-growth broilers compared with the slow-growth broilers (Piorkowska et al., 2015).

In 2008, Lee et al. investigated the role of calpains in PM proteolysis in chicken muscle. During the chilling of carcass at 4°C, the activity of μ/m-calpain was very stable with no significant reduction until 48-h PM due to the enzyme stability and requirement of more calcium concentration (Fig. 11A). However, the activity of μ-calpain had a step-wise reduction with a remarkable drop until 12-h PM, with almost no detectable activity at 72-h PM (Fig. 11B).

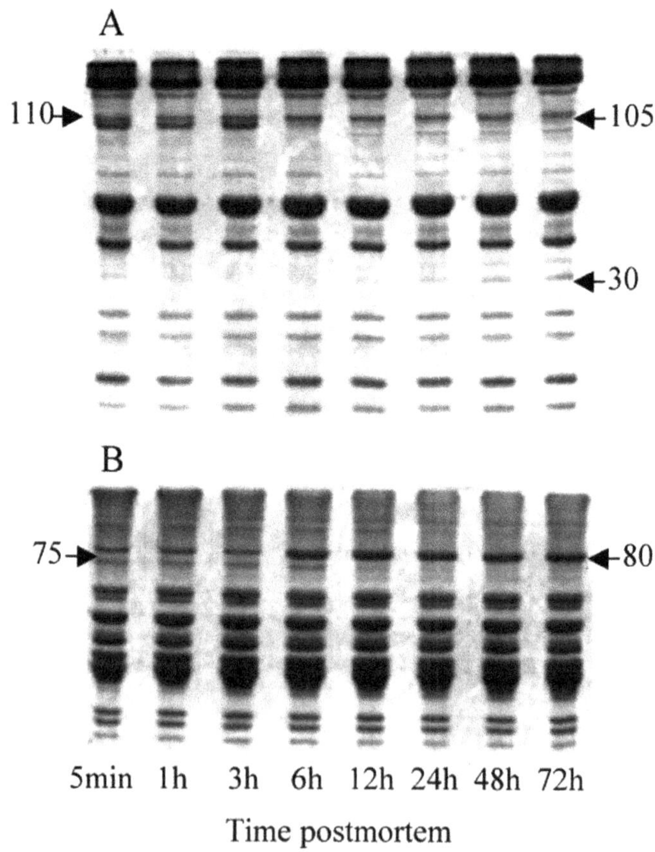

Figure 12 Sodium dodecyl sulfate gel electrophoresis of (A) myofibrillar and (B) sarcoplasmic proteins of chicken *pectoralis* muscle at different times PM.

Source: Adapted from Lee et al., 2008.

In comparison with gel electrophoresis patterns for myofibrillar and sarcoplasmic proteins of chicken, one protein band at 110 kDa (presumed α-actinin) disappeared between 3- and 6-h PM with two new protein bands appearing at 105 and 30 kDa. The 30 kDa has been reported to be a good marker of PM ageing as a partial hydrolysis of troponin T in beef and chicken (Fig. 12A) (Huff Lonergan et al., 1996; Negishi et al., 1996; Hay et al., 1973). A similar pattern of gel electrophoresis was found in sarcoplasmic proteins after 6-h PM, reducing one band size at 75 kDa and increasing the band size at 80 kDa (Fig. 12B).

7 Mechanical tenderization

Mechanical tenderization can be achieved by physically disrupting both connective tissue and myofibrillar protein. Numerous reports indicated that mechanical tenderization reduced shear force, improved tenderness and minimized cooking time with improved flavour (Shackelford et al., 1989; Glover et al., 1975; Smith et al., 1979). Muscle structure

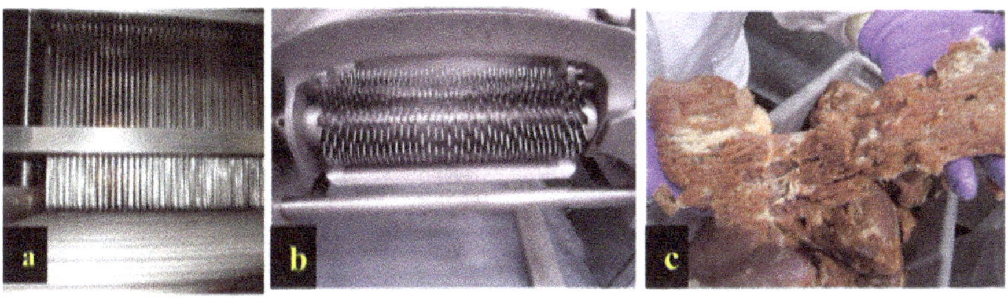

Figure 13 (a) Needles or blades, (b) macerator and (c) tenderized turkey breast fillet.

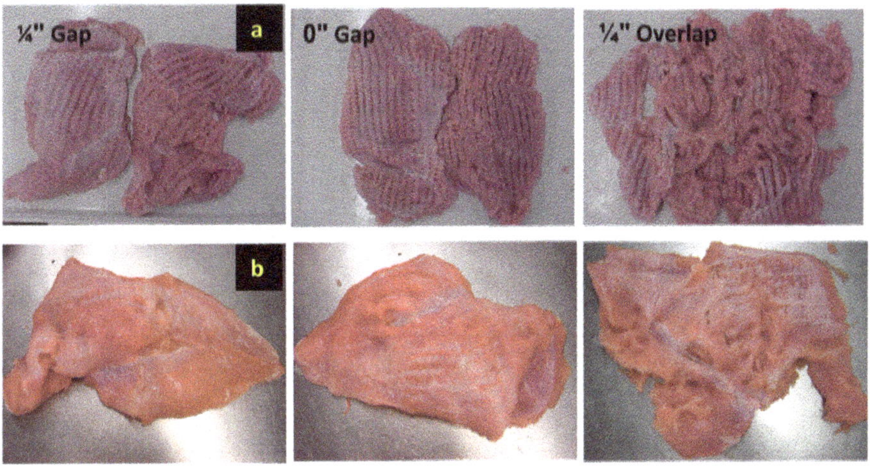

Figure 14 Degree of maceration of CB (a) and HB (b) turkey breast fillets using ¼" Gap, 0" Gap and ¼" Overlap.

can be destroyed with blades that are repeatedly moved up and down or rolled in knife rollers (Fig. 13a,b), while meats are moved on a conveyer belt. The degree of muscle tenderization can be controlled using the speed of the blade and the conveyer belt (Fig. 13c). When comparing the degrees of maceration, HB fillets were less destroyed than CB and aged fillets, presumably due to the integrity of HB muscle with less degradation by proteolytic enzymes (Fig. 14a,b).

8 Muscle tensioning for improved tenderness

Carcasses are chilled immediately after slaughter primarily to prevent microbial growth and weight loss. However, the muscle chilled rapidly before the completion of rigor mortis becomes tough due to muscle contraction known as cold shortening. In the beef industry, physical stretch of muscle has been applied to minimize muscle contraction as a result of gravity during chilling by modification of carcass suspension and is named tenderstretch (Tenderstretch, 2006; Macfarlane et al., 1974).

Papa and Fletcher (1988) investigated the possibility of tenderness improvement by restraining broiler carcasses wings with four alternative positions (Fig. 15). Results indicated that both elbow clasp and wing stretch did not reduce the shear values of cooked breast muscles compared with the no restrained control, whereas the complete wing tie resulted in tougher meat. Birkhold and Sams (1993) applied muscle tensioning (MT) by binding the broiler wings behind the back at the elbow with plastic cable ties. Compared with the control at 1-h PM, the breast fillets with MT lowered shear forces from 12.6 to 10.7 kg/g and increased the sarcomere length from 1.64 to 1.76 μm. In 2002, Cason et al. used physical stretch by clamping HB fillets between rigid aluminium plates. The Warner-Bratzler shear value of the HB, CB and HB/clamped fillets was 13.0, 9.2 and 5.1 kg, respectively, indicating that the combination of hot boning and physical clamping significantly reduced the shear value.

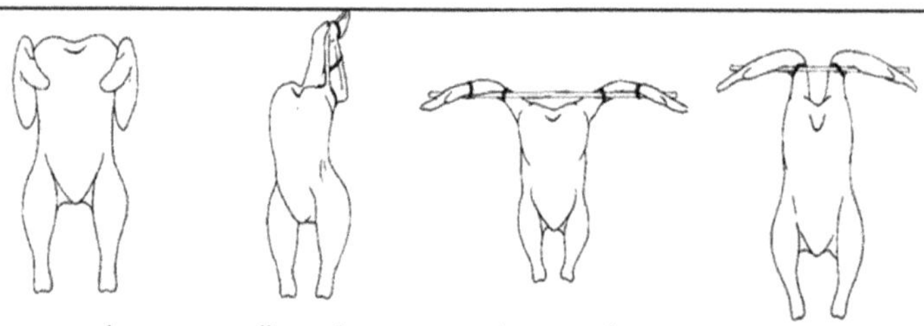

Types of wing restraints	control	elbow clasp	wing stretch	complete wing tie
Allo-Kramer shear value (Cooked sample)	3.9^b	3.8^b	3.7^b	4.3^a (kg/g)

(adapted from Papa and Fletcher, 1988)

Figure 15 Effect of PM wing restraints on the textural quality of broiler *pectoralis*.

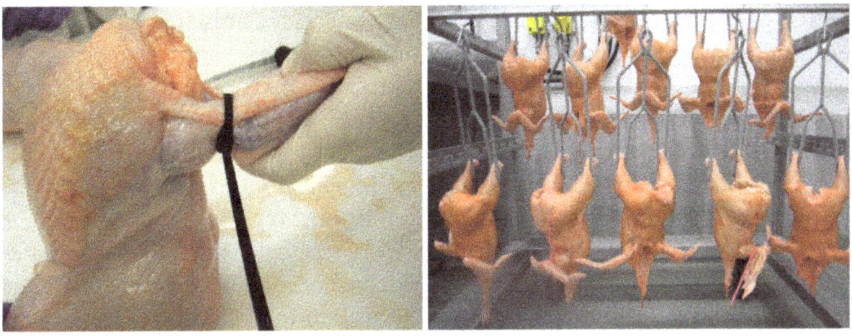

Figure 16 CFAC of broiler carcasses with MT.

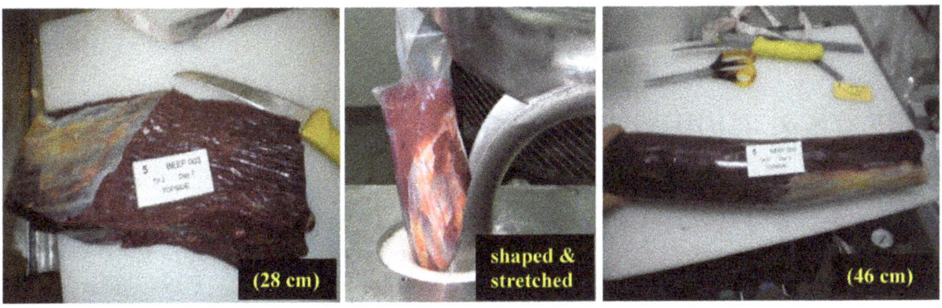

Figure 17 Beef topside muscle of before (28 cm) and after (46 cm) SmartStretch™
Source: Adapted from Taylor et al., 2011a.

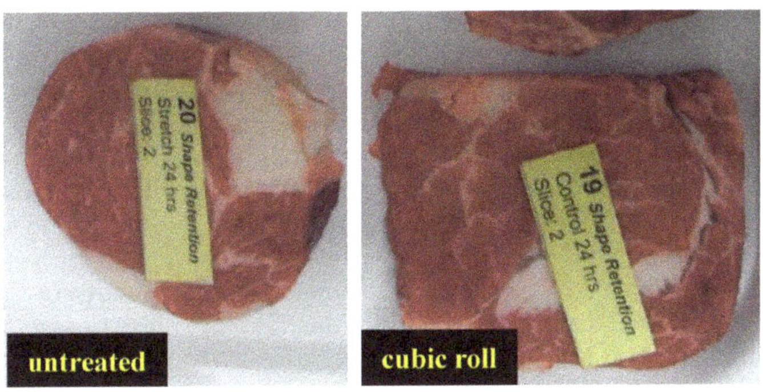

Figure 18 Steaks from SmartShape™-treated cube roll (left) and untreated equivalent cube roll (right) (Adapted from Taylor et al., 2011b).

Recently, the effect of muscle tension on broiler carcasses has been compared after subjecting carcasses to WIC, AC and CFAC with/without MT (Sansawat et al., 2014). The breast fillets with MT significantly improved meat tenderness than the fillets without MT at 24-h PM in AC and CFAC, but no difference was seen in WIC (Fig. 16). Based on the results, the adaptation of MT appears to be more applicable and feasible to AC than to WIC due to the tenderness improvement and less interruption with unshackling/reshackling for chilling. More interestingly, the combination of MT and CFAC could provide three advantages in poultry processing: (1) high efficacy of chilling rate, (2) improvement of muscle tenderness and (3) reduction of water consumption. However, an ideal method for automatic MT on carcasses needs to be designed to prevent any delaying on processing in current poultry plants. Following the technology of MT on whole beef carcasses, SmartStretch™ has been developed to stretch excised muscles after hot boning. Taylor et al. (2011a) stretched beef topside muscle from 28 to 46 cm (64%) using the technology (Fig. 17), and the tenderness of the meat was improved by stretching up to 34% of its initial length (Toohey et al., 2010).

SmartShape™ is an additional technology to change muscles to a uniform shape/size after rigor mortis, bringing several advantages such as portion control, reduced trim, consistent cooking times and lower storage space (Fig. 18) (Taylor et al., 2011b). Both techniques can be adapted to various poultry products, including HB and CB muscles.

9 Electrical stimulation of poultry carcasses

The primary purpose of PM electrical stimulation (ES) on poultry carcasses is to accelerate boneless meat harvest, whereas ES stimulation on red meat carcasses has been conducted to tenderize meats (Sams, 2002; Savell, 1985). The recommended time for deboning of broiler carcasses is 4–8-h PM, including WIC time (1.5-h PM) (Sams, 1999; Savenije, 2002).

Deboning of breast fillets earlier than 4 h possesses many advantages such as higher processing yield, less cooling space, lower labour and lower maintenance fee (Hirschler and Sams, 1998). However, the early boned muscle becomes tougher due to the contraction of the muscle or development of rigor mortis. Immediately after slaughter, ES induces electrical pass through the carcass, accelerates ATP depletion and develops rigor mortis at an early stage so that the resulting meats become tender with less sarcomere shortening (Hamm, 1982; Sams, 2002; Skarovsky and Sams, 1999). Sams (1999) summarized the effects of ES on birds in two systems: (1) low amperages (0–200 mA per bird) that induce muscle contractions and rigor acceleration, allowing some reduction of ageing period although the entire period of ageing cannot be eliminated (Sams, 1990; Lyon and Dickens, 1993) and (2) high amperages (350–500 mA per bird) that induce strong muscle exercises to tear and accelerate ATP depletion to prevent toughening after deboning at 1.5–2-h PM (Birkhold and Sams, 1993; Hirschler and Sams, 1997).

Using an electrical stimulator installed in a commercial broiler processing plant, Hirschler and Sams (1998) reported that the shear force of the ES-treated fillets was reduced by 50% over the control fillets by accelerating PM metabolism, with sensory scores marked for at least 'slightly tender' or 'moderately to very tender'. The breast fillets deboned at 2-h PM instead of 11 h PM (9 h of ageing) increased meat yield by 3.4% (Hirschler and Sams, 1998). With the results, the saved labour cost at the plant (processing 1.3 million broilers per week) could be approximately $210 000 annually due to the elimination of ageing time, with additional revenue ($5 200 000) per year at $1.00/lb due to additional meat yields for 100 000 lb per week.

Similarly, pulsed ES was introduced to broiler birds during bleeding under commercial conditions. Results indicated that the shear value of breast fillets deboned from ES carcasses at 1.5 h was equal to the control fillets deboned from no ES carcasses at 3.5 h,

Table 9 *R*-values (absorbance 250 nm/absorbance 260 nm), pH, evaporative moisture loss and cooked meat shear values for breast fillets deboned at 2-, 4- or 8-h PM from with or without ES carcasses

Treatment	PM deboning time (h)	*R*-Value	pH	Evaporative moisture loss[1] %	Shear value (kg/g)
ES[2]	2	0.93[b]	6.30[b]	1.95[bc]	7.92[a]
Control	2	0.90[b]	6.58[a]	1.88[c]	6.90[a]
ES	4	1.25[a]	5.97[c]	2.88[a]	4.67[b]
Control	4	0.97[b]	6.35[b]	2.25[ab]	7.59[a]
Control	8	1.17[a]	5.98[c]	4.49[ab]	4.13[b]
Pooled SEM		0.03	0.07	0.12	0.50

[a,b,c] Means within a column with no common superscript differ significantly ($P < 0.05$).
[1] Evaporative moisture loss = [(prechil II carcass weight ± postchi II carcass weight)/ Prechil I carcass weight] x 100.
[2] Electrical stimulation.
Source: Adapted from Skarovsky and Sams, 1999.

including 1.5-h chilling (Lyon et al., 2002). However, the stimulated carcasses showed 22.8% of cooking loss compared with 22.0% in non-stimulated carcasses (Lyon et al., 2002). In accordance with the finding of higher cooking loss, trained sensory panelists detected the differences in textural properties due to the stimulation: less moisture release during chewing and higher force to chew for the stimulated fillets over the unstimulated controls (Lyon and Lyon, 2000).

Skarovsky and Sams (1999) applied ES to the broiler carcasses that were air chilled. The fillets of ES broilers had a lower shear value than the control fillets at 4-h PM but not at 2-h PM. Compared to control, the ES reduced the ageing time up to 50% with no difference for evaporative moisture loss (Table 9).

10 Marination for improved tenderness

To improve meat tenderness, several processing technologies have been tested, including ES (Sams et al., 1989), MT (Birkhold and Sams, 1993) and high-temperature conditioning (Sams, 1990), but none of these methods has been widely used in the poultry industry. However, marination is a commonly used processing method to improve product tenderness, yield, flavour and shelf life in the poultry industry (Smith and Acton, 2010; Fletcher, 2004).

There are four margination techniques: soaking, blending, tumbling and injection. Soaking is defined as 'the act of submerging or keeping in a liquid for thorough wetting or softening for hydrotherapy' (Guralink, 1974). It is the simplest and oldest method that is the least efficient and most time consuming. It involves placing meat in marinade solutions so that marinade ingredients can penetrate into the meat through diffusion. This method is commonly used at home, individual restaurants or small processing plants, particularly for boneless meats in pieces. It requires submerging meats in solution for at least 24 h at refrigeration temperatures. Advantages of this method are simplicity, low cost, good adhesion and implementation to specialty products in small batches, whereas

Table 10 Cooked product yield of chicken fillets tumble-marinated in phosphate and salt solutions for various time periods[1]

Tumble Time (min)	No NaCl							8% NaCl						
		PP		TPP		HMP			PP		TPP		HMP	
	Control	Low	High	Low	High	Low	High	Control	Low	High	Low	High	Low	High
5	$80.2^{ef,x}$	$85.9^{bcd,z}$	$89.6^{b,z}$	$85.1^{cd,z}$	$89.3^{bc,z}$	$85.0^{d,y}$	$78.2^{f,y}$	$86.6^{bed,y}$	$78.7^{f,z}$	$83.6^{de,z}$	$85.6^{bcd,z}$	$86.9^{bcd,z}$	$87.7^{bcd,y}$	$96.1^{a,x}$
15	$78.9^{i,x}$	$95.2^{bcde,y}$	$104.3^{a,y}$	$92.3^{efg,y}$	$97.1^{bc,y}$	$88.5^{h,x}$	$89.5^{gh,x}$	$93.3^{def,x}$	$96.0^{bcd,y}$	$98.1^{b,y}$	$96.4^{bcd,y}$	$94.5^{cde,y}$	$85.0^{i,y}$	$90.4^{fgh,y}$
30	$77.3^{j,y}$	$106.0^{de,x}$	$112.0^{c,x}$	$103.9^{e,x}$	$100.4^{f,x}$	$88.7^{i,x}$	$90.0^{i,x}$	$93.5^{h,x}$	$114.9^{b,x}$	$119.4^{a,x}$	$111.9^{c,x}$	$108.3^{d,x}$	$95.3^{gh,x}$	$97.5^{g,x}$

[a–i]Means in the same row with no common superscript differ significantly ($P < 0.05$).
[x–z]Means for each parameter group in the same column with no common superscript differ significantly ($P < 0.05$).
[1]PP = sodium pyrophosphate; TPP = sodium tripolyphosphate; HMP = sodium hexametaphosphate.
Source: Adapted from Xiong and Kuppski, 1999a.

disadvantages include inconsistent marinade uptake, extra labour, extended marination time, contamination issue and large cooler space (Smith and Acton, 2010).

Blending or massage is described as rubbing or kneading of body parts to make the muscle pliable and flexible, whereas tumbling is described as an act of tossing around to disrupt the intact muscle structure (Guralink, 1974). Blending of meat is conducted through meat-to-meat contact and abrasion against the walls of the vessel (Romans et al., 2001). A typical blender contains multiple smooth paddles aligned horizontally on a vertical shaft that agitates slowly to minimize tearing. Blending is generally used to chop and form products, whereas tumbling is designed for a variety of products, including skinless, skin-on, bone-in, whole-muscle and chopped pieces (Smith and Acton, 2010).

Tumbling of muscle is conducted by dropping muscles from the top to the bottom of the drum as it rotates. As a result, muscle pieces are rubbed against each other and proteins are extracted for binding together (Romans et al., 2001). Most tumblers are designed as a cement mixer for a simple and efficient mixing, with a helix-shaped baffle or paddles inside the drum (Toldrá, 2007). Both methods of blending and tumbling are used for commercial marination in large sizes with rapid marinade application to improve functionality, yield and quality (Young et al., 1992; Yang and Chen, 1993).

When skinless chicken breast fillets were ground, marinated and made into 2.54-cm-thick patties, cooking loss of the patties decreased by 1.2 and 11.6%, respectively, as each unit of ionic strength and sodium tripolyphosphate (STPP) was increased after the blend-marinade. In a similar matter, chewiness decreased by 0.31 and 4.39 kg-cm, respectively, as pH and ionic strength units were increased (Young et al., 1992). Skinless chicken breast fillets were tumble marinated in 1.6 and 3.2% of pyrophosphate (PP), tripolyphosphate (TPP) and hexametaphosphate (HMP) with/without 8% NaCl for 0, 5, 15 and 30 min. Using PP or TPP, the marinade absorption rate was the greatest in the first 5 min of tumbling, which was then markedly increased (Xiong and Kupski, 1999b). The cooking yield was 77.3% for unsalted control fillets, 93.5% for salted fillets and 119% for salted and PP fillets (Table 10). Xiong and Kupski (1999a) reported that marinade penetration was the

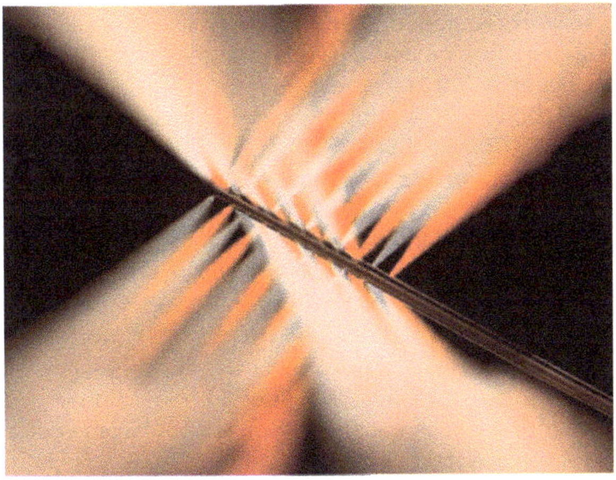

Figure 19 Spray plus system.
Source: Adapted from Xargayó, 2013.

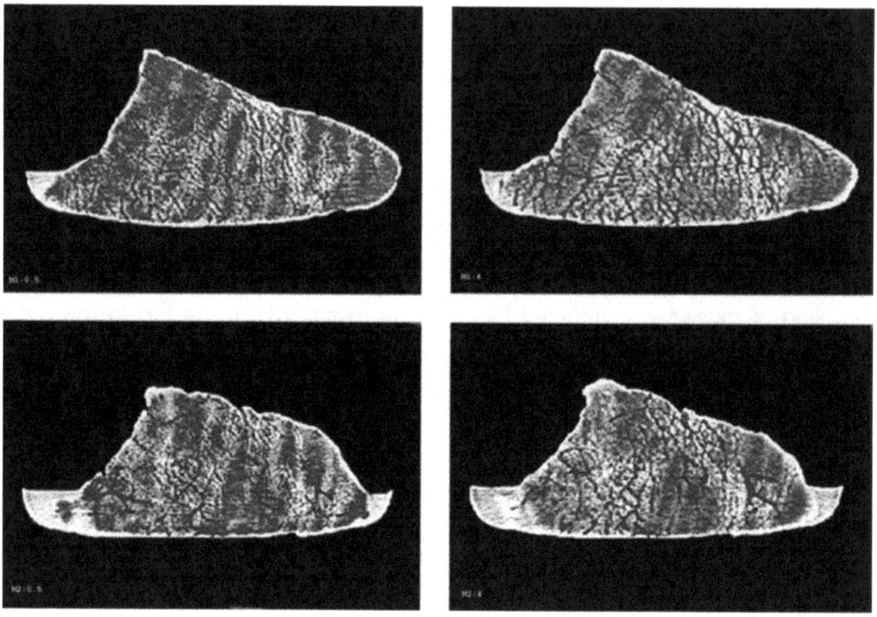

Figure 20 CT scans of cross-sections of brine-injected *semimembranosus* muscles. Scans were taken 0.5 (left) and 4 (right) h after injection. Light area in the muscles indicate distribution of brine with NaCl. Source: Adapted from Claus and Sorheim, 2006.

greatest in the initial 5 min, with further enhancement observed for 196, 171 and 138%, respectively, by adding PP, TPP and HMP.

Mechanical injection is the most widely used method in the meat industry using high-speed complex machines, equipped with 100–250 stainless steel needles and multiple orifices per needle. This machine can inject several thousand kilograms of product per hour, regardless of bone-in or boneless meats. Recently, a spray injection system has been developed for injecting brine or cure solutions into meats using high-velocity microdroplets (Fig. 19) (Meatingplace, 2010). The injection system is usually conducted with tumbling or massaging to extract protein for improvement of water holding capacity and gel-forming ability.

Considering the importance of ingredient delivery into the muscle, Claus and Sørheim (2006) visualized salt distribution in muscle after injection, using X-ray densities in computed tomography (CT) scans. The bovine *semimembranosus* muscles, injected for 28% absorption of brine solution (10.18% sodium chloride), indicated that the range of salt distribution at the cross-sections of the muscle was approximately 0–7% NaCl at 4–5.5 h of injection, which changed to 1–6% NaCl after 24-h injection (Fig. 20). These results indicated that the salt concentration was still uneven although some diffusion occurred 24 h after injection. Previously, Froystein et al. (1989) reported that 1% NaCl corresponded to approximately 11 Housefield units (HU) in the X-ray densities in CT images.

Perumalla et al. (2011) evaluated marination properties and sensory attributes of breast fillets from immersion-chilled and air-chilled broiler carcasses. Results showed that

Table 11 Tenderness, sarcomere length and cooking loss measurements from early PM margination of broiler breast fillets

Parameter measured	Control 3-h PM[1]	Inject 3-h PM	Tumble 3-h PM	Inject + tumble 3-h PM	Control 6-h PM
Shear value (kg/g)	14.09[a]	7.81[b]	14.49[a]	9.05[b]	5.34[c]
Sarcomere length (μm)	1.60[b]	1.61[b]	1.56[b]	1.63[b]	1.80[a]
Cooking loss (%)	23.58[ab]	22.58[ab]	26.30[a]	26.78[a]	19.56[b]

[a–c] Means within a row differ ($P < 0.05$).
[1] PM deboning time
Source: Adapted from Alvarado and Sams, 2004.

marination significantly improved cooking yield and sensory attributes (texture, flavour and juiciness), regardless of the chilling method.

When marinating three groups of broiler breast fillets categorized by pale, normal and dark, the resulting fillets showed significantly lower cooking loss than the non-marinated controls, regardless of the fillet colour. Among the marinated fillets, the least cooking loss was observed in dark fillets, followed by normal and pale fillets (Zhuang et al., 2014). Alvarado and Sams (2004) applied injection, tumble and injection/tumble to broiler breast fillets after deboning at 3-h PM (Table 11). The injected/tumbled fillets were significantly more tender than the tumbled or control fillets, presumably due to more brine penetration and better physical disruption by needle penetration and fluid pressure during injection. No tender effect from the tumbled fillets was expected due to the physical stimulation of the muscle deboned at 3-h PM instead of 4-h PM, the least time required for the development of rigor mortis.

11 Conclusions

The world's meat consumption is on rise, with an annual growth being approximately 1.9% for the next decade. In the consumption of red meat and poultry, poultry is assumed to be the world's most consumed meat in the next five years as pork consumption, currently possessing the greatest share in the world meat production, is slowing. Part of the reasons for the increase of poultry meat is to meet the consumer's needs such as tender meat, tasteful product, convenience and health foods.

In the United States, approximately 8.9 billion broilers are slaughtered annually, and the majority (92.5%) of the birds are further processed, whereas the remaining (7.5%) are sold as whole carcasses. During primary poultry processing, carcasses need to be aged for 4–6 h to prevent meat toughness. As a result, post-chill carcasses cannot be further processed until the completion of rigor mortis through ageing. The ageing processing is expensive due to the chilling space, energy cost, labour and facility maintenance.

Processing technology has rapidly been developed to meet consumer expectation and processor needs. ES has been studied to reduce meat toughness by accelerating rigor development. MT has been attempted to improve muscle tenderness by preventing rigor shortening or muscle contraction. CFAC has been recently researched to expedite the chilling process with no quality loss.

Day by day, there are new developments in processing technique, handling equipment and scientific knowledge. The trends of consumer consumption habits have also changed rapidly from whole carcasses to cut-up meats, to ready-to-eat products and to healthy and specialty products. The processing techniques and skills described in this chapter can be used directly or with modification to improve processing efficiency, product shelf life and meat quality for consumer and processor satisfactions.

12 Where to look for further information

For further information, see the following sources:

Periodical Journals: Poultry Science, British Poultry Science, Journal of Applied Poultry Research, World's Poultry Science Journal, Meat Science, Journal of Food Science.

Books: Poultry Meat Processing, Poultry Meat Processing and Quality, Poultry Processing and Quality, Meat Science,

Trade Journal: The National Provisioner, Meat and Poultry Magazine, Meat Marketing and Technology, Watt Poultry USA, Poultry International.

13 References

Aberle, E. D., Forrest, J. C., Gerrard, D. E., Mills, E. W., Hedrick, H. B., Judge, M. D. and Merkel, R. A. (2001). Conversion of muscle to meat and development of meat quality. In *Principles of Meat Science*, edited by E. D. Aberle, J. C. Forrest, D. E. Gerrard and E. W. Mills, 4th ed, Kendell/Hunt Publishing, Dubuque, IA, pp. 117–54.

Alvarado, C. Z. and Sams, A. R. (2002). 'The role of carcass chilling rate in the development of pale, exudative turkey pectoralis'. *Poult. Sci.* 81: 1365–70.

Alvarado, C. Z. and Sams, A. R. (2004). 'Early postmortem injection and tumble margination effects on broiler breast meat tenderness'. *Poult. Sci.* 83: 1035–8.

Baker, R. C., Darfler, J. and Vadehra, D. V. (1969). 'Type and level of fat and amount of protein and their effect on the quality of chicken frankfurters'. *Food Technol.* 23: 100–3.

Barbut, S. (1997a). Occurrence of pale soft exudative meat in mature turkey hens. *Br. Poult. Sci.* 38: 74–7.

Barbut, S. (1997b). Problem of pale soft exudative meat in broiler chickens. *Br. Poult. Sci.* 38: 355–8.

Barbut, S. (2002). Primary processing of poultry. an industry guide. In *Poultry Products Processing*, edited by S. Barbut, CRC Press LLC, Boca Raton, FL, pp. 1–107.

Bernthal, P. H., Booren, A. M. and Gray, J. I. (1989). 'Effect of sodium chloride concentration on pH, water-holding capacity and extractable protein of prerigor and postrigor ground beef'. *Meat Sci.* 25: 143–54.

Birkhold, S. G. and Sams, A. R. (1993). 'Fragmentation, tenderness, and post-mortem metabolism of early harvested broiler breast fillets from carcasses treated with electrical stimulation and muscle tensioning'. *Poult. Sci.* 72: 577–82.

Bloukas, J. G. and Paneras, E. D. (1993). 'Substituting olive oil for pork for pork back-fat affects quality of low-fat frankfurters'. *J. Food Sci.* 58: 705–9.

Briskey, E. J. (1964). 'Etiological status and associated studies of pale, soft, exudative porcine musculature'. *Adv. Food Res.* 13: 89–105.

Carroll, C. D. and Alvarado, C. Z. (2008). 'Comparison of air immersion chilling on meat quality and shelf life of marinated broiler breast fillets'. *Poult. Sci.* 87: 368–72.

Carson, J. A., Lyon, C. E. and Dickens, J. A. (2002). 'Tenderization of hot-boned broiler breast meat by clamping during chilling'. *Poult. Sci.* 81: 121–5.

Chiang, W., Booren, A. and Strasburg, G. (2008). 'The effect of heat stress on thyroid hormone response and meat quality in turkeys of two genetic lines'. *Meat Science,* 80: 615–22.

Claus, J. R. and Sørheim, O. (2006). 'Preserving pre-rigor meat functionality for beef patty production'. *Meat Sci.* 73: 287–94.

Decker, R. S., Poole, A. R. and Wildenthal, K. (1980). 'Distribution of lysosomal cathepsin D in normal, ischemic, and starved rabbit cardiac myocytes'. *Circ Res.* 46: 485–94.

De Fremery, D. (1966). Some aspects of post mortem changes in poultry muscle. In *The Physiology and Biochemistry of Muscle as a Food*, Volume 1, edited by E. J. Briskey, R. G. Cassens and J. C. Trautmann, University of Wisconsin Press, Madison, pp. 205–12.

De Fremery, D. and Pool, M. F. (1960). 'Biochemistry of chicken muscle as related to rigor mortis and tenderization'. *J. Food Sci.* 25: 73–87.

Dunn, A. A., Kilpatrick, D. J. and Gault, N. F. S. (1993). 'Effect of postmortem temperature on chicken *M. pectoralis* major: muscle shortening and cooked meat tenderness'. *Br. Poult. Sci.* 34: 689–97.

Dunn, A. A., Kilpatrick, D. J. and Gault, N. F. S. (1995). 'Contribution of rigor shortening and cold shortening to variability in the texture of pectoralis major muscle from commercially processed broilers'. *Br. Poult. Sci.* 36: 401–13.

Dunn, A. A., Tolland, E. L. C., Kilpatrick, D. J. and Gault, N. F. S. (2000). 'Relationship between early post-mortem muscle pH and shortening-induced toughness in the *Pectoralis major* muscle of processed broilers air-chilled at 0°C and −12°C'. *Br. Poult. Sci.* 41: 53–60.

Fernandez, X., Forslid, A. and Tornberg, E. (1994). 'The effect of high post-mortem temperature on the development of pale, soft and exudative pork: interaction with ultimate pH'. *Meat Sci.* 37: 133–47.

Fletcher, D. L. (2004). Further processing of poultry. In *Poultry Meat Processing and Quality*, edited by G. C. Mead, CRC Press LLC, Boca Raton, FL, pp. 110–34.

Frøystein, T., Sørheim, O., Berg, S. A. and Dalen, K. (1989). 'Salt distribution in cured hams studied by computer X-ray tomography'. *Fleischwirtschaft* 69: 220–2.

Glover, E., Forrest, J. C., Johnson, H. R., Bramblett, V. D. and Judge, M. D. (1975). 'Palatability and cooking characteristics of mechanically tenderized beef'. *J. Anim. Sci.* (Abstr.), 41: 292.

Goll, D. E., Thompson, V. F., Li, H., Wei, W. and Cong, J. (2003). 'The calpain system'. *Physiol. Rev.* 83: 731–801.

Guralnik, D. B. (1974). *New World Dictionary of the American Language*, 2nd ed, William Collins + World Publishing Co., Inc, New York.

Hamm, R. (1982). 'Post mortem changes in muscle with regard to processing of hot-boned beef'. *Food Technol.* 36(11): 105–15.

Hay, J. D., Currie, R. W. and Wolfe, F. H. (1973). 'Polyacrylamide disc gel electrophoresis of fresh and aged chicken muscle proteins in sodium dodecylsulfate'. *J. Food Sci.* 38: 987–90.

Hirschler, E. M. and Sams, A. R. (1997). 'Commercial-scale electrical stimulation of poultry: the effects on tenderness, breast meat yield, and production costs'. *J. Appl. Poult. Res.* 7: 99–103.

Hoagland, R., McBryde, C. N. and Powick, W. C. (1917). 'Changes in fresh beef during cold storage above freezing'. *U.S. Agr. Bull.* 433: 1–100 [Online], Available: http://books.google.com/books?id=eVFSZgbYZmsC&printsec=frontcover&source=gbs_ge_summary_r&cad=0#v=onepage&q&f=false (November 2015).

Honikel, K. O., Fischer, C. and Hamm, R. (1980). 'Einfluss der largertemperature von schlachtfrischem rinder-muskel auf das wasserbindungsvermogen von fleisch und braten'. *Fleischwirschaft* 60: 1577–86.

Honikel, K. O., Roncales, P. and Hamm, R. (1983). 'The influence of temperature on shortening and rigor onset in beef muscle'. *Meat Sci.* 8: 221–41.

Huff-Lonergan, E., Mitsuhashi, T., Beekman, D. D., Parrish, Jr, F. C., Olson, D. G. and Robson, R. M. (1996). 'Proteolysis of specific muscle structural proteins by μ-calpain at low pH and temperature is similar to degradation in postmortem bovine muscle'. *J. Anim. Sci.* 74: 993–1008.

Jeong, J. Y., Janardhanan, K. K., Booren, A. M., Harte, J. B. and Kang, I. (2011b). 'Breast meat quality and consumer sensory properties of broiler carcasses chilled by water, air, or evaporative air'. *Poult. Sci.* 90: 694–700.

Jeong, J. Y., Janardhanan, K. K., Booren, A. M., Karcher, D. M. and Kang, I. (2011a). 'Moisture content, processing yield, and surface color of broiler carcasses chilled by water, air, or evaporative air'. *Poult. Sci.* 90: 687–93.

Kirschke, H. and Barret, A. J. (1987). Chemistry of lysosomal proteases. In *Lysosomes: Their Role in Protein Breakdown*, edited by H. Glaumann and F. J. Ballard, Academic press, London, pp. 193–238.

Koohmaraie, M. (1994). 'Muscle proteinases and meat aging'. *Meat Sci.* 36: 93–104.

Koohmaraie, M. (1996). 'Biochemical factors regulating the toughening and tenderization processes of meat'. *Meat Sci.* 43: S193–201.

Koohmaraie, M., Doumit, E. and Wheeler, T. L. (1996). 'Meat toughening does not occur when rigor shortening is pre-vented'. *J. Anim. Sci.* 74: 2935–42.

Koohmaraie, M. and Geesink, G. H. (2006). 'Contribution of postmortem muscle biochemistry to the delivery of consistent meat quality with particular focus on the calpain system'. *Meat Sci.* 74: 34–43.

Lawrie, R. A. (1991). *Meat Science*, 5th ed, Pergamon Press, New York, NY, pp. 56–60, 188 and 206.

Lee, H. C., Medellin-Lopez, M., Singh, P., Sansawat, T., Chin, K. B. and Kang, I. (2014). 'Cold-batter mincing of hot-boned and crust-frozen air-chilled turkey breast allows for reduced sodium content in protein gels'. *Poult. Sci.* 93: 2327–36.

Lee, H. C., Erasmus, M. A., Swanson, J. C., Hong, H. G. and Kang, I. (2016). Improvement of turkey breast meat quality and cooked gel functionality using hot-boning, quarter sectioning, crust-freeze-air-chilling and cold-batter-mincing technologies'. *Poult. Sci.* 95: 138–43.

Lee, H. L., Santé-Lhoutellier, V., Vigouroux, S., Briand, Y. and Briand, M. (2008). 'Role of calpains in postmortem proteolysis in chicken muscle'. *Poult. Sci.* 87: 2126–32.

Lehmann, K. B. (1907). 'Studies of the causes for the toughness of meats'. *Arch. F. Hyg.* 63: 134.

Locker, R. H. and Hagyard, C. J. (1963). 'A cold shortening effect in beef muscle'. *J. Sci. Food Agr.* 14: 787–93.

Lyon, C. E. and Lyon, B. G. (2000). 'Broiler breast meat due to electrical stimulation, deboning time, and margination'. *J. Appl. Poult. Res.* 9: 234–41.

Lyon, C. E. and Dickens, J. A. (1993). 'Effects of electric treatments and wind restraints on the rate of post-mortem biochemical changes and objective texture of broiler *Pectoralis major* muscles deboned after chilling'. *Poult. Sci.* 72: 1577–83.

Lyon, C. E., Dickens, J. A. and Lyon, B. G. (2002). 'Effect of electrical stimulation and postchill deboning time on texture and cook loss of broiler breasts processed under commercial conditions'. *J. Appl. Poult. Res.* 11: 217–22.

Ma, R. T. I. and Addis, P. B. (1973). 'The association of struggle during exsanguinations to glycolysis, protein solubility and shear in turkey pectoralis muscle'. *J. Food Sci.* 38: 995–7.

Macfarlane, J. J., Harris, P. V. and Shorthose, W. R. (1974). Manipulation of meat quality, particularly tenderness, by the processor. Paper presented at the 10th Australian Society of Animal Production, pp. 219–26 [Online], Available: http://livestocklibrary.com.au/handle/1234/6732 [November, 2015].

Malila, Y., Tempelman, R. J., Sporer, K. R. R., Ernst, C. W., Velleman, S. G., Reed, K. M. and Strasburg, G. M. (2013). 'Differential gene expression between normal and pale, soft, and exudative turkey meat'. *Poult. Sci.* 92: 1621–33.

McGuire, A. R. (2003). 'Improving carcass quality'. *Poultry* 10(1): 25–6.

McKee, S. R. and Sams, A. R. (1997). 'The effect of seasonal heat stress on rigor development and the incidence of pale, exudative turkey meat'. *Poult. Sci.* 76: 1616–20.

McKee, S. R. and Sams, A. R. (1998). 'Rigor mortis development at elevated temperatures induces pale exudative turkey meat characteristics'. *Poult. Sci.* 77: 169–74.

Meatingplace (2010). *Spray Injection System Features Improved Brine Distribution* [Internet]. Meatingplace, Chicago, IL, http://meatingplace.com.

Medellin-Lopez, M., Sansawat, T., Strasburg, G., Marks, B. P. and Kang, I. (2014). 'Cold-batter mincing of hot-boned and crust-freezing air-chilled turkey breast improved meat turnover time and product quality'. *Poult. Sci.* 93: 711–18.

Negishi, H., Yamamoto, E. and Kuwata, T. (1996). 'The origin of the 30 kDa component appearing during post-mortem ageing of bovine muscle'. *Meat Sci.* 42: 289–303.

OECD/FAO (2015). 'Meat'. In *OECD-FAO Agricultural Outlook 2015*, OECD Publishing, Paris. Accessed March 2016. http://www.oecd-ilibrary.org/agriculture-and-food/oecd-fao-agricultural-outlook-2015_agr_outlook-2015-en.

Okumura, F., Shimogiri, T., Shinbo, Y., Yoshizawa, K., Kawabe, K., Mannen, H., Okamoto, S., Cheng, H. H. and Maeda, Y. (2005). 'Linkage mapping of four chicken calpain genes'. *Anim. Sci. J.* 76: 121–7.

Owens, C. M., Hirschler, E. M., McKee, S. R., Martinez-Dawson, R. and Sams, A. R. (2000). 'The characterization and incidence of pale, soft, exudative turkey meat in a commercial plant'. *Poult. Sci.* 79: 553–8.

Papa, C. M. and Fletcher, D. L. (1988). 'Effect of wing restraint on postmortem muscle shortening and the textural quality of broiler breast meat'. *Poult. Sci.* 67: 275–9.

Papinaho, P. A. and Fletcher, D. L. (1996). 'The influence of temperature on broiler breast muscle shortening and extensibility'. *Poult. Sci.* 75: 797–802.

Perumalla, A. V. S., Saha, A., Lee, Y., Meullenet, J. F. and Owens, C. M. (2011). 'Marination properties and sensory evaluation of breast fillets from air-chilled and immersion-chilled broiler carcasses'. *Poult. Sci.* 90: 671–9.

Piorkowska, K., Nowak, J. and Poltowicz, K. (2015). 'The normalisation of CAPN gene in *M. pectoralis superficialis* in broiler lines differing in growth rate and their relationship to breast muscle tenderness'. *Br. Poult. Sci.* 56: 452–8.

Romans, J. R., Costello, W. J., Carlson, C. W., Greaser, M. L. and Jones, K. W. (2001). Meat curing and soaking. In *The Meat We Eat*, edited by J. R. Romans, W. J. Costello, C. W. Carlson, M. L. Greaser and K. W. Jones, 14th ed, Interstate, Chicago, IL, pp. 731–78.

Sams, A. R. (1990). 'Electrical stimulation and high temperature conditioning of broiler carcasses'. *Poult. Sci.* 69: 1781–6.

Sams, A. R. (1999). 'Commercial implementation of postmortem electrical stimulation'. *Poult. Sci.* 78: 290–4.

Sams, A. R. (2002). 'Post-mortem electrical stimulation of broilers'. *Word's Poult. Sci. J.* 58: 147–57.

Sams, A. R., Janky, D. M. and Woodward, S. A. (1989). 'Tenderness and R-value changes in early harvested broiler breast tissue following postmortem electrical stimulation'. *Poult. Sci.* 68: 1232–5.

Sansawat, T., Lee, H. C., Singh, P., Kim, H., Chin, K. B. and Kang, I. (2014). 'Combination of muscle tension and crust-freeze air-chilling improved efficacy of air chilling and quality of broiler fillets'. *Poult. Sci.* 93: 2314–19.

Santos, C., Roserio, L. C., Goncalves, H. and Melo, R. S. (1994). 'Incidence of different pork quality categories in a Portuguese slaughterhouse: a survey'. *Meat Sci.* 38: 279–87.

Santos, H. C., Brandelli, A. and Ayub, M. A. Z. (2004). 'Influence of post-mortem aging in tenderness of chicken breast fillets'. *Cienc. Rural* 34: 905–10.

Savell, J. W. (1985). Industrial applications of electrical stimulation. In *Advances in Meat Research: Volume 1 – Electrical Stimulation*, edited by A. M. Pearson and T. R. Dutson, AVI Publishing Company, Inc., Westport, CT, pp. 219–36.

Savenije, B., Lambooij, E., Gerritzen, M. A. and Korf, J. (2002). 'Development of brain damage as measured by brain impedance recordings, and changes in heart rate, and blood pressure induced by different stunning and killing methods'. *Poult. Sci.* 81: 572–8.

Schreurs, F. J. G. (2000). 'Postmortem changes in chicken muscle'. *World's Poult. Sci. J.* 56: 319–46.

SEER Training Modules, U.S. National Institute of Health [Online], Available: http://training.seer.cancer.gov/anatomy/muscular/structure.html (March 2016).

Sentandreu, M. A., Goulis, G. and Quail, A. (2002). 'Role of muscle endopeptidases and their inhibitors in meat tenderness'. *Trends Food Sci. Technol.* 13: 398–419.

Shackelford, S. D., Reagan, J. O., Mann, T. F., Lyon, C. E. and Miller, M. F. (1989). 'Effects of blade tenderization, vacuum massage time, and salt level on chemical, textural and sensory characteristics of precooked chuck roasts'. *J. Food Sci.* 54: 843–45, 905.

Simon, S., Field, J. C., Kramlich, W. E. and Tauber, F. W. (1965). 'Factors affecting frankfurter texture and a method of measurement'. *Food Technol.* 19: 410–16.

Skarovsky, C. J. and Sams, A. R. (1999). 'Tenderness, moisture loss and post-mortem metabolism of broiler pectoralis muscle from electrically stimulated and air chilled carcasses'. *Br. Poult. Sci.* 40: 622–5.

Smith, G. C., Seideman, S. C. and Carpenter, Z. L. (1979). 'Blade tenderization effects on cooking and palatability characteristics of steaks from bullock and cow carcasses'. *J. Food Sci.* 42: 563–6.

Smith, D. P. and Acton, J. C. (2010). Marination, cooking, and curing of poultry products. In *Poultry Meat Processing*, edited by C. M. Owens, C. Z. Alvarado and A. R. Sams, CRC Press LLC, Boca Raton, FL, pp. 311–36.

Soares, A. L., Lara, J. A. F., Ida, E. I., Guarnieri, P. D., Olivo, R. and Shimokomaki, M. (2003). Influence of Preslaughter Handling Practices on Broiler Meat Color in a Commercial Plant. 76F-13, IFT annual Meeting Book of Abstracts, 12–16 July, Chicago, IL.

Sporer, K. R. B., Zhou, H. R., Linz, J. E., Booren, A. M. and Strasburg, G. M. (2012). 'Differential expression of calcium-regulating genes in heat-stressed turkey breast muscle is associated with meat quality'. *Poult. Sci.* 91(6): 1418–24.

Taylor, J., Toohey, E. and Hopkins, D. (2011a). Final report – Accelerated processing of sheep/beef meat SmartStretchTM/SmartShapeTM. Meat & Livestock Austrailia Limited.

Taylor, J., Toohey, E. and Hopkins, D. (2011b). 'Outcomes of the SmartshapeTM/SmartStretchTM meat science project'. *Aust. Farm Bus. Manag. J.* 8: 69–72.

Taylor, R. G., Geesink, G. H., Thompson, V. F., Koohmaraie, M. and Goll, D. E. (1995). Is Z-disk degradation responsible for postmortem tenderization?' *J. Anim. Sci.* 73: 1351–67.

Tenderstretch. (2006, reprinted). Meat technology update. 98/2 – July 1998.

Thompson, L. D., Janky, D. M. and Woodward, S. A. (1987). 'Tenderness and physical characteristics of broiler breast fillets harvested at various times from post-mortem electrically stimulated carcasses'. *Poult. Sci.* 66: 1158–67.

Toldrá, F. (2007). Ham. In *Handbook of Food Products Manufacturing: Health, Meat, Milk, Poultry, Seafood, and Vegetables*, edited by Y. H. Hui, John Wiley & Sons, New York, pp. 233–49.

Toohey E. S., Kerr, M. J., van de Ven, R. and Hopkins, D. L. (2010). The effect of SmartStretch technology on the tenderness of beef topsides and cube rolls. In *Proceedings of the 56th International Congress of Meat Science and Technology*, 8–13 August 2010, Session E072, Jeju, Korea, pp. 1–4.

USDA, Economic Research Service (2015). Food availability: Spread sheets: Red meat, poultry, and fish (Red meat and poultry – Per capita availability, carcass weight). http://www.ers.usda.gov/data-products/chart-gallery/detail.aspx?chartId=40060&ref=collection&embed=TrueAccessed February 2015. .

Warris, P. D. and Brown, S. N. (1987). 'The relationship between initial pH, reflectance and exudation in pig muscle'. *Meat Sci.* 20: 65–72.

Xargayó, M., Lagares, J., Fernandez, E., Gumà, J., Sanz, D. and Brugué, J. M. (2013). *Spray Plus System®*, evolution of injection of cooked meat products. Tedhnologic book part 1, pp. 1–33. http://en.metalquimia.com/upload/document/article-sprayplussystem-eng.pdf (accessed March 2016).

Xiong, Y. L. and Kupski, D. R. (1999a). 'Monitoring phosphate marinade penetration in tumbled chicken filets using a thin-slicing, dye-tracing method'. *Poult. Sci.* 78: 1048–52.

Xiong, Y. L. and Kupski, D. R. (1999b). 'Time-dependent marinade absorption and retention, cooking yield, and palatability of chicken filets marinated in various phosphate solutions'. *Poult. Sci.* 78: 1053–9.

Yang, C. C. and Chen, T. C. (1993). 'Effects of refrigerated storage, pH adjustment, and marinade on color of raw and microwave cooked chicken meat'. *Poult. Sci.* 72: 355.

Young, L. L., Papa, C. M., Lyon, C. E. and Wilson, R. L. (1992). 'Moisture retention and textural properties of ground chicken meat as affected by sodium tripolyphosphate, ionic strength and pH'. *J. Food Sci.* 57: 1291.

Zhuang, H., Bowker, B. and Samuel, D. (2014). 'Effect of postmortem aging on margination performance of broiler breast pectoralis major categorized by color lightness'. *Poult. Sci.* 93: 3123–9.

Chapter 16

Preventing spoilage of poultry meat

Arthur Hinton Jr., U. S. National Poultry Center – USDA-ARS, USA

1 Introduction

There was a 51% worldwide increase in poultry consumption between 2000 and 2012 (The Poultry Site, 2012). The total farm value of poultry production in the United States is estimated at over $20 billion, which makes it the world's largest poultry-producing country (USDA, ERS, 2012). In the United States, approximately 80% of processed poultry meat is sold as fresh, unfrozen products (Russell et al., 1996; Russell, 1997). Due to the nutritional content, favourable pH and moisture content of fresh poultry, poultry meat is a highly perishable commodity (Bruckner et al., 2012).

Spoilage is defined as the changes in the meat that render the products unacceptable for human consumption (Gram et al., 2002). These changes may begin post harvest and continue throughout the processes of distribution and storage until the meat is ready to be prepared for consumption (Roller, 1999). It is estimated that between 3.7 and 4.2% of chicken meat and between 3.4 and 3.5% of turkey meat may be lost due to spoilage, moisture loss and other factors (Buzby et al., 2009). Spoilage of the meat may be due to prolonged distribution or storage time, improper storage temperatures, contamination, or high pH of the processed meat. Food spoilage produces physical, chemical or biological changes that produce off-flavours, off-colours, off-odours or microbial growth in the foods. Microbial spoilage, the most common form of spoilage, also results in the production of slime and bacterial colonies on foods.

The shelf-life of fresh poultry meat is determined by the level of contamination of processed meat by spoilage microorganisms, storage temperature and storage

http://dx.doi.org/10.19103/AS.2016.0010.19

atmosphere (Cerveny, 2009; Russell, 2009). The shelf-life of fresh poultry is approximately 4–10 days when stored at refrigeration temperatures (approximately 4°C) under normal atmospheres (Marenzi, 1986). The feet, feathers, skin and alimentary tract of broilers entering commercial processing facilities are heavily contaminated by a diverse indigenous microbiome (Barnes, 1972, 1979; Kotula and Pandya, 1995), and freshly processed poultry meat may be contaminated by microorganisms carried into the processing facility by the live birds and by microorganisms that contaminate the carcass during processing. These microorganisms vary in the range of temperatures that they can grow, and optimal growth temperature for these microorganisms will determine which microorganisms will play a role in the spoilage of the poultry meat (Barnes, 1976). Bacteria that are responsible for spoilage are a small portion of the native microflora of freshly processed poultry; however, these are the bacteria that can multiply at refrigerated storage temperatures to produce malodours, slime and discoloration associated with spoilage (Russell, 2009). Higher initial levels of contamination of the freshly processed chicken meat will decrease the shelf-life of the product (Geornaras et al., 1994). Methods for extending the shelf-life of fresh poultry are based on reducing or eliminating the growth of these spoilage microorganisms until it is time to prepare the product for consumption.

2 Spoilage microorganisms

The microorganisms of primary concern to commercial processors include those related to food safety and those that can cause undesirable organoleptic changes in fresh poultry meat stored at refrigeration temperatures. Most human food-borne pathogens in the microbiota of poultry require optimal temperatures of 35–45°C for growth and have minimal growth temperatures between 5°C and 10°C. These microorganisms are defined as mesophiles because of the optimal temperatures required for growth (Ingram, 1951). The alimentary tracts of healthy poultry are the primary reservoirs for mesophiles, such as *Salmonella* and *Campylobacter* (Painter et al., 2013), and these pathogens may be spread to carcasses during processing. Since mesophiles cannot grow at refrigerated temperatures, these bacteria play no role in the spoilage of fresh poultry (Doyle, 1990; Upmann et al., 2000); however, mesophiles may survive refrigeration and freezing temperatures and may still pose food safety risks if contaminated meat is not properly handled or prepared before consumption.

The microorganisms responsible for spoilage of fresh, refrigerated poultry are cold-loving bacteria and fungi that can grow at temperatures from −3°C to 34°C (Barnes and Impey, 1968; Barnes, 1976). Unlike mesophilic bacteria that may survive only at refrigerated temperatures, psychrotrophic and psychrophilic microorganisms can grow at refrigeration temperatures and produce undesirable end products that produce colours, odours and physical changes that make foods undesirable for consumption (Morita, 1975). It is the growth of these cold-loving microorganisms that must be controlled to extend the shelf-life of fresh poultry. Microorganisms that can grow at temperatures around 0°C but grow best at temperatures greater than 20°C are defined as psychrotrophs. The optimum temperature for growth of psychrotrophs is between 25°C and 30°C. Further classification of Gram-negative psychrotrophic bacteria has suggested the term 'stenopsychrotrophs' for those bacteria such as *Pseudomonas fragi* that can grow rapidly at 7°C, but not at

40°C, and the term 'eurypsychrotrophs' for those bacteria such as *Yersinia enterocolitica* that can grow above 40°C and slowly at 7°C (Jay, 1987). These cold-loving microorganisms that can grow at temperatures around 0°C but cannot grow at temperatures greater than 20°C are defined as psychrophiles (Gounot, 1986). Psychrotrophs and psychrophiles possess metabolic enzymes, such as lipases and proteases, which allow these bacteria to continue to grow and proliferate at refrigerated temperatures (Russell, 2009). Additionally, psychrotrophs and psychrophiles generally have higher levels of unsaturated fatty acids in their cellular membranes than bacteria that are unable to grow at these lower temperatures.

Though psychrotrophs and psychrophiles are both defined as 'cold-loving' microorganisms, the growth range for psychrophiles is more limited than the growth range for psychrotrophs (Eddy, 1960). Therefore, psychrophiles are usually found only in cold environments where cold temperatures are constantly maintained. In these cold environments, psychrophiles play a role in the degradation of organic waste materials to help maintain the ecological balance. Since psychrotrophs have a wider range of growth than those of psychrophiles, psychrotrophic microorganisms are more ubiquitous and can be isolated from water, soil, air and foods; thus, psychrotrophs play a major role in the spoilage of poultry and other refrigerated foods (Gounot, 1986).

The shelf-life of poultry meat is determined by the numbers and types of spoilage microorganisms present on the carcass after processing, the storage temperature and the type of muscle meat that has been contaminated (Barnes, 1976). Most psychrotrophs are carried on the feathers, skin and feet of the live bird instead of the alimentary tract of the animals where most food-borne pathogens may be found in the normal microbiota of the birds (Russell, 2009). Commercial poultry processing includes operations such as stunning, bleeding, scalding, picking, evisceration and chilling. As the carcasses pass through the various processing operations, microorganisms carried into the facility by the live bird may be spread to the carcass meat. Additionally, poultry meat may be contaminated by microorganisms found in the processing water, pickers, chill tanks and ice in the facility. Nonetheless, processing generally decreases carcass contamination by psychrotrophs and other microorganisms; therefore, the number of spoilage microorganisms on processed carcasses is significantly less than the number of these bacteria recovered from broilers entering the processing facility (Hinton et al., 2004).

During chilling operations, poultry carcasses are rapidly cooled to reduce temperature of the carcasses in order to inhibit the growth of mesophilic food-borne pathogens and other microorganisms found on the carcass (Farrell and Barnes, 1964). Psychrotrophic bacteria isolated from ice water used in immersion chilling of carcasses in poultry processing facilities can grow at these colder temperatures (Farrell and Barnes, 1964). *Acinetobacter* and *Aeromonas* sp. were the primary bacterial isolates recovered from carcasses immediately after processing (Hinton et al., 2004) and both of these bacteria are considered major spoilage bacteria in poultry (Barnes, 1976). The psychrotrophs *Pseudomonas* and *Achromobacter* have also been isolated from the chill water of poultry processing facilities (Farrell and Barnes, 1964). Those psychrotrophic bacteria and fungi that remain on the carcass after processing can multiply on the carcasses during refrigerated storage (Hinton et al., 2004). Pseudomonads comprise a relatively small portion of the microbiome of freshly processed chicken when compared to *Acinetobacter* and *Aeromonas* spp.

As few as 20% of the bacteria found on freshly processed poultry meat are responsible for spoilage of these foods during refrigerated storage. When the chicken meat is refrigerated in normal atmospheres, *Pseudomonas* spp. rapidly become the predominant species (Ayres et al., 1950; Russell et al., 1994; Hinton et al., 2004), and fresh poultry meat that is contaminated with a higher level of pseudomonads will spoil faster than meat that is contaminated with lower levels of these spoilage bacteria (Arnaut-Rollier et al., 1999).

Pseudomonas spp. are considered as major food spoilage bacteria because of their ability to grow at refrigerated temperatures, to degrade proteins in foods and to produce undesirable flavours, odours and pigments in spoilt foods (Russell et al.,1996). In a research that was conducted, 250 bacterial isolates were recovered from chicken breast muscle, isolates that had been stored at 2°C for up to 16 days. Most of the isolates were identified as *Pseudomonas* groups I, II, IV or as enteric *Pseudomonas* sp. (McMeekin, 1975). Primary spoilage bacteria associated with refrigerated poultry are pigmented and non-pigmented pseudomonads, such as *Pseudomonas fluorescens*, *Pseudomonas putida*, *P. fragi* and *Shewanella* (formerly *Pseudomonas*) *putrefaciens* (Russell, 2009).

Isolates from *Pseudomonas* group I produce a fluorescent pigment, while isolates from *Pseudomonas* group II do not produce a pigment (Shewan et al., 1960). Group II includes *Pseudomonas fragi*, *Pseudomonas lundensis* and *P. fluorescens* biovars A, B, C and F (Arnaut-Rollier, 1999). Research has demonstrated that early spoilage of chicken breasts is primarily due to *Pseudomonas* group I bacteria, whereas later spoilage is attributed to *Pseudomonas* group II bacteria. Only small numbers of *Pseudomonas* group IV and enteric pseudomonads are recovered from the chicken breasts after 8 or 12 days of refrigerated storage. In chicken breast meat that was inoculated with *Pseudomonas* bacteria and stored at 2°C for up to 16 days, approximately 73% of *Pseudomonas* type II produced spoilage odours, while approximately 22% of *Pseudomonas* type I produced off-odours (McMeekin, 1975). Similar research indicated that when fresh, minced poultry breast and leg muscle meat were inoculated with *Pseudomonas* I and II, *Pseudomonas putrefaciens* and *Acinetobacter* groups B and C and then stored for 4 days at 1°C, no *P. putrefaciens* or *Acinetobacter* were recovered from breast meat; however, these isolates were recovered from inoculated leg meat samples. Furthermore, the non-pigmented *Pseudomonas* type II multiplied during storage of breast samples to become the predominant spoilage flora. Differences in the growth of the *P. putrefaciens* or *Acinetobacter* in breast meat and in leg meat were determined to be related to lower pH of the breast meat (5.7–5.9) when compared to the higher pH of the leg meat (6.4–6.7) (Barnes, 1976).

Although *Pseudomonas* are not considered pathogens, the production of amines by these bacteria can theoretically cause food-borne intoxications due to the production of biogenic amines during spoilage. Over 90% of *Pseudomonas* strains isolated from poultry were able to produce biogenic amines *in vitro*, while 100% of *Salmonella* isolates were able to produce these compounds (Geornaras et al., 1995). No strains of *Listeria* that were isolated from poultry produced biogenic amines *in vitro* (Geornaras et al., 1995).

Along with *Pseudomonas*, *Acinetobacter* and *Aeromonas* (Beuchat, 1990) many other psychrotrophs responsible for spoilage of poultry are Gram-negative rods, such as *Alcaligenes*, *Chromobacterium*, *Corynebacteria*, *Flavobacterium* and *Moraxella* (Ayres, 1960). Gram-positive spoilage bacteria include *Lactobacillus viridescens*, *Brochothrix thermosphacta*, *Listeria monocytogenes* and *Clostridium botulinum* type E. *L. monocytogenes* and *C. botulinum* type E cause major concern because these bacteria are also pathogenic and can cause human food-borne illnesses (Schofield, 1992).

Yeasts also form a major portion of psychrotrophic microorganisms found on fresh poultry stored at 4, 7 or 10°C (Al-Mohizea et al., 1994). Although yeasts are outnumbered by psychrotrophic bacteria, during refrigerated storage of fresh poultry, the number of yeasts on fresh broiler carcasses also increases (Hinton et al., 2002). In broiler carcasses taken from a commercial processing facility and stored at 4°C for 14 days, a significant increase in the number of yeasts was noticed on the carcasses. The same strain of yeast was isolated from carcasses taken from the same processing facility on different days. Yeasts isolated from the stored carcasses included *Candida* spp., *Cryptococcus* spp., *Yarrowia lipolytica*, *Trichosporon beigelii* and *Zygosaccharomyces bailli*. These microorganisms produce lipases and proteases that can also cause undesirable organoleptic changes in refrigerated poultry.

When populations of spoilage bacteria on carcasses reach approximately 10^5 cfu/cm^2, undesirable odours, flavours and appearances associated with spoilage become apparent (Russell, 2009). The numbers of psychrophilic bacteria on spoilt meat can reach as high as 10^7–10^8/cm^2 of meat, and slime becomes apparent when values of 10^7–10^9 are reached. There is still no consensus among food microbiologists for a single plating medium, incubation temperature or length of incubation for enumerating psychrotrophs on poultry and other food products (Cousin et al., 2001). Incubation at 10°C for 7 days on plate count agar is currently considered to be the standard method for enumerating psychrotrophs (Jay, 2002). Since microorganisms that cause spoilage of refrigerated poultry grow at low temperatures, using the total plate count procedure that requires incubation of plates at 35°C is of little value since many of the microorganisms that are enumerated during this procedure are mesophiles (Huss et al., 1974). In the 26 studies that conducted psychrotrophic counts of meat and poultry products from 1985 through 2002, incubation temperatures from 4°C to 25°C were used. Over 15 different plating media and 29 time–temperature incubation protocols were reported. Additionally, the use of rapid techniques such as a selective medium for rapid enumeration of *Pseudomonas fluorescens* (Russell, 1997b) and the use of capacitance (Russell, 1997a) to determine the level of contamination of poultry meat by *P. fluorescens* have been examined (Russell, 1997a,b).

3 Extending the shelf-life of poultry meat: vacuum and modified atmosphere packaging (MAP)

Several techniques are used to extend the shelf-life of fresh poultry. All of these preservation techniques are designed to reduce or inhibit the growth of those psychrotrophs that can grow on poultry meat during refrigerated storage. This section looks at packaging techniques.

3.1 Vacuum packaging

Vacuum packaging involves the complete removal of air from the fresh poultry meat placed in packages (Sawaya, 1993; Rao and Sachindra, 2002). Pseudomonads are strict aerobes that require oxygen for growth, and the reduced oxygen content in the packaging limits the growth of these aerobic psychrotrophs. Vacuum packaging does not inhibit all bacterial growth on these products, but it does produce changes in the composition of the microflora of the chicken meat (Arafa and Chen, 1975). In refrigerated poultry stored under normal atmospheres, *Pseudomonas* spp. predominates because these aerobic psychrotrophs have shorter generation times than other bacteria that can grow on chicken meat stored at refrigeration temperatures (Geornaras et al., 1994). However, the lower concentrations of oxygen produced by creating vacuums in packaged products allow facultative or anaerobic bacteria to become the predominant species in the stored meat.

Lactobacillus and *Microbacterium*, instead of *Pseudomonas*, are the major microbiota of refrigerated meats stored under a vacuum. Growth by lactic acid-producing bacteria may cause souring of the poultry, but this spoilage is not as pronounced as spoilage from aerobic psychrotrophs (Thomas et al., 1984). Since organoleptic changes produced in chicken meat by growth of the lactic acid bacteria are not as objectionable as changes produced by the proteolytic pseudomonads, it has been reported that the chilled meats that are stored under vacuum may have a shelf-life of up to three months. Facultative Enterobacteriaceae are another group of bacteria that are able to outgrow pseudomonads when meat is vacuum packaged (Arafa and Chen, 1975). Enterobacteriaceae have been found to comprise up to 96% of the microbiota of spoilt vacuum-packaged poultry in a study in which broiler chicken carcasses were packaged under a vacuum, compared to carcasses packaged under carbon dioxide and then stored at −1.5 or 3.0°C (Gill et al., 1990). The microorganisms of the vacuum-packaged carcasses were predominately enterobacteria, and putrid odours developed after 2 weeks at 3°C or 3 weeks at −1.5°C. However, growth of Enterobacteriaceae was inhibited on carcasses stored under carbon dioxide, and lactobacilli developed as the primary microbiota. Strict anaerobes, such as the food-borne pathogen *Clostridium perfringens*, may also be able to grow in vacuum-packaged products, but the growth of this pathogen can be controlled by cold storage since it will not grow at temperatures less than 15°C (ICMSF, 1998).

Combining vacuum packaging with other treatments has also been examined as a method for extending the shelf-life of fresh poultry. For example, studies have been conducted on dipping broiler carcasses in 20 ppm of chlorine in chilled water before vacuum packaging. Results showed that significantly fewer bacteria were recovered from meat packaged under vacuum than meat packaged under normal atmospheres and stored at 5°C for up to 10 days. Vacuum packaging also increased the shelf-life of eviscerated broiler carcasses by up to 8, 6 and 4 days when carcasses were stored at 4, 7 or 9°C, respectively. Washing the chickens in chlorine had no effect on extending the shelf-life of the poultry, but the shelf-life of meat packaged under a high barrier film was longer than the shelf-life of meat packaged under a low barrier film.

Other studies have examined the shelf-life of chicken breasts dipped in a phosphate-spice marinade, packaged under a vacuum and stored at 2°C for up to 28 days (Buses and Thompson, 2003). Findings indicated that the growth of coliforms was inhibited on the chicken breast and that the shelf-life of these products was extended for up to 14 days longer than the shelf-life of untreated, control. The marinade treatment also improved the chicken breasts by adding flavour to the product, thus increasing the convenience of product preparation.

3.2 Modified atmosphere packaging

Another method of reducing the growth of aerobic spoilage bacteria on fresh poultry is to package the processed meat under modified atmospheres with increased carbon dioxide and reduced oxygen concentrations. MAP involves changing the composition of gases inside the package by flushing these gases into impermeable packages (Warburton et al., 1990). MAP normally contains carbon dioxide to inhibit the growth of bacteria and moulds, nitrogen to avoid oxidation of fats and to provide positive pressure to prevent collapse of the packaging and oxygen to prevent the growth of strict anaerobes (Farber et al., 1990; Rao and Sachindra, 2002). Concentrations of reduced oxygen and increased carbon dioxide in food packages can increase shelf-life of foods by inhibiting the growth of aerobic spoilage microbiota, such as *Pseudomonas* (Marshall et al., 1991). MAP packaging of poultry allows the shelf-life of fresh poultry to be extended without any noticeable changes in the organoleptic changes of these products (Narasimha and Sachindra, 2002). Meats retain their colour better when packaged under MAP than under a vacuum, even though the colour may change when meats are removed from vacuum and exposed to air. When the odour, slime and bacterial growth of poultry packaged under normal atmospheres, vacuums or carbon dioxide and stored for 17 days at 5°C are examined, the results showed that poultry flushed with carbon dioxide had the longest shelf-life (Thomas et al., 1984). Although the shelf-life of vacuum-packed poultry was less than the shelf-life of poultry flushed with carbon dioxide, the shelf-life of these products was longer than meat packed with air.

The use of higher concentrations of CO_2 in MAP has also been studied. Findings indicate that increasing CO_2 concentrations increased the lag phase of growth for *Listeria monocytogenes* and *Pseudomonas fluorescens* (Hendricks and Hotchkiss, 1997). A concentration of 20% CO_2 decreased the growth rate of *P. fluorescens*, while a concentration of 80% CO_2 was required to decrease the growth rate of *L. monocytogenes* in buffered nutrient broth. The shelf-life of chicken stored under oxygen impermeable film (vinylidene chloride–vinyl chloride copolymer) was 16 days, while the shelf-life of chicken stored under oxygen permeable polyethylene film was only 11 days. The concentration of carbon dioxide in the impermeable plastic increased 9–10%, and the development of spoilage microbiota was significantly slower. Pseudomonads were the main spoilage organisms on carcasses wrapped in permeable film, while *Alteromonas putrefaciens* were the predominant microorganisms on carcasses wrapped in impermeable film. Growth of pseudomonads was inhibited by the accumulation of carbon dioxide in the impermeable films (Barnes, 1976). A concentration of 10–20% CO_2 delayed the growth pseudomonads and some moulds at temperatures between 0°C and 4°C; however, carbon dioxide concentrations of 10–20% CO_2 can cause undesirable organoleptic changes in the poultry meat.

The shelf-life of fresh chicken breasts package under modified atmospheres of 30% carbon dioxide and 70% nitrogen or atmospheres of 70% carbon dioxide and 30% nitrogen was extended when the meat was stored at 4°C (Jimenez et al., 1997). The meat was examined after 3, 7, 14 or 21 days, and findings indicated that the modified atmospheres inhibited the growth of pseudomonads; however, the growth of lactobacilli, Enterobacteriaceae and *Brochothrix thermosphacta* was not inhibited. The growth of pseudomonads was not inhibited on chicken breasts stored under air or vacuum (Jimenez et al., 1997). The shelf-life of the breasts stored under modified atmospheres was increased to 21 days, while the shelf-life of the breasts stored under air was only 5 days.

Packaging films with an oxygen permeability of 100 cm^3/m^2/atm/day at 25°C are generally used for MAP and vacuum packaging (Rao and Sachindra, 2002). Research has been conducted comparing the shelf-life of freshly processed broiler meat placed in tray packages and wrapped in stretch wrap under normal atmospheres or wrapped in low or high barrier plastic and packaged under a vacuum (Kraft et al., 1982).

Research has also shown that packaging damage may produce perforations that will allow gases of the modified atmosphere to escape and decrease the shelf-life of the stored products (Church et al., 2014). Leaks in packages of chicken fillets that had been dipped in a suspension of 1.0% chitosan then packaged under a modified atmosphere of 70% CO_2 and 30% N_2 allowed changes in the atmosphere of the product that reduced the shelf-life of the poultry meat (Rossaint et al., 2014). Perforations of 0.2 mm were made in packaging material of broilers stored at 4°C for 20 days with the atmosphere of high oxygen (70% oxygen and 30% CO_2) or high nitrogen (70% N_2 and 30% CO_2). Perforations in the packaging led to increases in the population of pseudomonads, *Brochothrix thermosphacta* and Enterobacteriaceae, and the shelf-life of the poultry was decreased by 26% due to the holes in the packaging materials (Church, 1994).

MAP can increase the shelf-life of poultry products by 50 to 400% (Rao and Sachindra, 2002). Disadvantages of this technology include high costs, the need for special equipment and training, the need to maintain refrigerated temperature and standardization of different gas mixtures for each type (Rao and Sachindra, 2002).

4 Extending the shelf-life of poultry meat: bacteriocins, herbs and spices

4.1 Bacteriocins

Bacteriocins are antimicrobial peptides that are produced by other bacteria. These peptides have the ability to kill or prevent the growth of other bacteria by inhibiting cell wall synthesis, permeabilizing cellular membranes or inhibiting RNase or DNase activity in bacteria that are attacked by the compounds (Cleveland et al., 2001). Bacteriocins differ from antibiotics because bacteriocins are synthesized in bacterial ribosomes, while antibiotics are produced as secondary metabolites by the bacteria, and bacteriocins are utilized primarily as food preservatives, while antibiotics have clinical applications (Hurst, 1981). Although bacteriocins may be produced by Gram-negative and Gram-positive bacteria, most research has been conducted on the bacteriocins produced by the Gram-positive lactic acid bacteria for use as preservatives in meat products. Additionally, these antimicrobials are naturally produced in some foods in which lactic acid bacteria compose a significant portion of the microbiota, and bacteriocins have been shown to be safe when used as food preservatives (Cleveland et al., 2001).

Nisin is an anti-bacterial polypeptide produced by the lactic acid bacterium, *Lactococcus lactis* (Delves-Broughton, 1990). This bacteriocin has been used as a food preservative for over 60 years. The use of nisin as a preservative has been approved in over 40 countries, and it is the only bacteriocin approved for use as a preservative in the United States. Research has been conducted on the use of nisin and other bacteriocins as preservatives in meats. When applied to meats as food preservatives, bacteriocins are used as part of a multiple hurdle approach (Zhou et al., 2010) that includes one or more other preservatives (Cleveland et al., 2001). Nisin has been coupled with nitrates to prevent the growth of *Clostridium* in meats, although the efficacy of the preservatives has been questioned (Rayman et al., 1981).

Another bacteriocin produced by *Pediococcus acidilactici*, pediocin AcH, has been studied for use as a privative to inhibit the growth of *L. monocytogenes* in raw chicken (Coventry et al., 1995). As compared to controls with no bacteriocin, pediocin produced approximately a 6 log reduction in the growth of the pathogen after 28 days of storage at 5°C.

4.2 Herbs and spices

Some of the dried aromatic plant products that are used to flavour foods may possess anti-bacterial activity. Most spices possess more anti-bacterial activity against Gram-positive bacteria than against Gram-negative bacteria (Shelef, 1983). Garlic is commonly used as a condiment in poultry preparation, but it also possesses anti-bacterial activity due to the concentration of allicin in the plant (Oliveira et al., 2005). Concentrations of 10 and 15% garlic extract inhibited bacterial growth on poultry carcasses stored at 4°C for 3 days. Also, synergistic activity of a concentration of 4% aqueous garlic extract and cultures of *Lactobacillus acidophilus* extended the shelf-life of minced chicken meat from 4 to 6 days (Yadav and Singh, 2004).

5 Extending the shelf-life of poultry meat: high hydrostatic pressure (HPP) and electrolyzed water

5.1 High hydrostatic pressure

HHP processing is a food preservation method that can be used to increase the shelf-life of foods with nutritional, sensory and functional characteristics that are sensitive to high temperatures (Carlez et al., 1994). High pressure has been used to extend the shelf-life of food products such as ready-to-eat meats, ground beef, fruit juices and guacamole. HPP in ranges of 100–800 megapascals (MPa) can be used to inactivate microorganisms and some enzymes found in food (Tuboly et al., 2003). High pressure can kill microorganisms that cause human food-borne illnesses and microorganisms that cause food spoilage of foods (Carlez et al., 1993). Cultures of *Pseudomonas fluorescens* were reduced over 5 logs to completely inactivate the cultures in fresh minced beef muscle inoculated with the bacterium and treated with 200 MPa for 20 min at 20°C, while treatments of 400 MPa were required to inactivate cultures of *Listeria innocua* in the beef (Carlez et al., 1993).

Bacteria vary in their sensitivity to HPP, and the medium in which the bacteria are suspended may play a role in the sensitivity of microorganisms to high-pressure treatments (Styles et al., 1991). A suspension of 10^6 cfu/ml of the Gram-negative bacterium, *Vibrio parahaemolyticus*, that was suspended in commercially sterile heat-processed clam juice was inactivated by exposure to 172 MPa of hydrostatic pressure after 10 min at 23°C, while 345 MPa of hydrostatic pressure for 80 min was required to inactivate 10^6 cfu/ml of *L. monocytogenes* suspended in ultra-high temperature-processed milk. Additionally, research has indicated that cultures of *Salmonella* Senftenberg and *Salmonella* Typhimurium were more sensitive to HHP when suspended in a phosphate buffer solution (63 mM, pH 7.0) than when suspended in a chicken-based medium (Metrick et al., 1989). Furthermore, *Salmonella* Senftenberg cultures were inactivated by lower pressures than cultures of *Salmonella* Typhimurium. Research also indicates that psychrotrophic bacteria are more sensitive to HHP than are mesophilic bacteria (Yuste et al., 1998). Injury to psychrotrophs by HHP can cause the bacteria to lose the ability to grow at low temperatures; therefore, the treatment may extend the shelf-life of poultry stored at refrigerated temperatures.

HPP can cause death or injury to psychrotrophic spoilage bacteria (Linton et al., 2004). No pseudomonads were recovered from minced chicken meat immediately after the chicken was treated with high pressure, but these bacteria were recovered from the chicken samples that were placed under cold storage. These findings may indicate that some pseudomonads that had been injured by the pressure treatments were able to repair the injuries and grow during cold storage of the meat. Further experiments indicated that portions of the *Pseudomonas* population were able to survive in pressure-treated minced meat and that these bacteria could be recovered after the treatment and resume growth in the meat after a repair phase of 3–9 days in meat stored at 3°C (Carlez et al., 1994). HHP also produced significant reductions in the number of bacteria recovered from mechanically deboned turkey meat, which resulted in an increase in the shelf-life of the product (Tuboly et al., 2003). Treatment of fresh poultry meat with HHP produces changes in the microbiota of the meat (Linton et al., 2004). Gram-negative bacteria, such as *Pseudomonas*, appear to be more susceptible to high pressures than are Gram-positive bacteria, such as *Staphylococcus*. Additionally, HHP may cause injury to some bacteria that may cause spoilage of poultry (Metrick et al., 1989). Those bacteria that are injured by these treatments have slower growth rates than bacteria that have not been exposed to HHP.

Combinations of pressure and temperatures (40–60°C) may produce synergistic effects that can reduce contamination of poultry by spoilage bacteria. Pressures of 400 mPa at 40°C reduced populations of *Brochothrix thermosphacta*, *Carnobacterium divergens* and *Pseudomonas fluorescens* to levels below the detection limits in minced poultry meat that had been inoculated with the bacteria (Liu et al., 2012).

Combinations of HHP and vacuum packing also extend the shelf-life of poultry. There was no significant ($p < 0.05$) increase in the number of aerobic, psychrotrophic and anaerobic bacteria recovered from vacuum-packaged, minced chicken that was pressure-treated at 500 MPa for 15 min at 40°C and then stored for 182 days at 3°C. However, the number of bacteria recovered from untreated control meat samples increased to approximately 10^7 cfu/g after only 8 days of storage at 3°C (Linton et al., 2004). Spoilage of minced chicken occurs faster than spoilage of fresh chicken because preparation of the product spreads bacteria throughout the meat. Minced chicken meat has been shown to spoil as rapidly as 4 days of storage at 5°C (Froning, 1987). *Bacillus* and *Staphylococcus* were the primary bacteria isolated from minced chicken meat that was pressure-treated and vacuum packaged for 31 days (Linton et al., 2004). After 182 days of storage, although *Staphylococcus* remained a major component of the microbiota, other bacteria such as *Pseudomonas*, coryneforms and *Micrococcus* were present in high numbers. The combination of high pressure, vacuum packaging and cold storage successfully extended the shelf-life of minced chicken, but the appearance and texture of the chicken was changed by these preservation methods (Linton et al., 2004).

When high-pressure treatments were combined with nisin to treat mechanically recovered poultry meat, the population of psychrotrophs in the meat was reduced by approximately 7.5 log units and no psychrotrophs were detected in the meat (Yuste et al., 1998). Findings indicate that psychrotrophs appear to be more sensitive to the combined treatments because the combination of treatments may cause these bacteria to lose the ability to grow at low temperatures (Yuste et al., 1998).

Other studies have shown that HHP might have an undesirable effect on the quality and sensory characteristics of poultry meat. Treating chicken breasts fillets with 450 and 600 MPa eliminated *Salmonella* Typhimurium, *Escherichia coli* and *L. monocytogenes* on the meat (Kruk et al., 2011). Treatments of 600 MPa reduced bacterial contamination of the meat by 6–8 logs during 7–14 days of storage. However, the pressure treatments produced changes in the flavour, aroma and juiciness of the meat. These treatments also increased cooking loss, hardness, gumminess and chewiness of the meat.

5.2 Electrolyzed water

Electrolyzed water is a sanitizer that is produced by the electrolysis of dilute solutions of sodium chloride or other salts. Acidic electrolyzed water can kill microorganisms because of the acidic pH, high oxidation reduction potential and high concentration of hypochlorous acid produced during electrolysis of the salt molecules (Park et al., 2004). Research has demonstrated that significantly fewer psychrotrophic bacteria and yeasts were recovered from freshly processed broiler carcasses immediately after the carcasses were sprayed for 5 seconds with electrolyzed water (Hinton et al., 2007). Additionally, significantly fewer psychrotrophs and yeasts were recovered from carcasses treated with electrolyzed water than from carcasses treated with tap water or with 50 ppm chlorine then stored at 4°C for up to 14 days. *Pseudomonas* spp. were the primary psychrotrophic bacteria, and *Candida* spp. were the primary yeasts isolated from stored carcasses that had been sprayed with electrolyzed water.

6 Irradiation, non-thermal plasma and oxygen absorbers

Extending the shelf-life of fresh poultry will reduce product losses and increase the time that these products can be stored by consumers. Research on increasing the efficacy of current technology and developing new technologies to reduce or inhibit the growth of spoilage microorganisms on fresh poultry will ensure that new methods will be available to provide consumers with wholesome, fresh poultry products.

Irradiation of poultry can also kill spoilage microorganisms on poultry and extend the shelf-life of these products, and use of this technology has been approved by the United States Department of Agriculture (FSIS, 1999). Low-dose irradiation can extend the shelf-life of fresh chicken refrigerated under normal atmospheres (Balamatsia et al., 2006). Radiation treatments of 0.5, 1.0 and 2.0 kilogreys (kGy) reduced the number of total bacteria, *Brochothrix thermosphacta*, and lactic acid bacteria recovered from treated breast meat stored at 4°C for 21 days. No pseudomonads, yeasts, moulds and Enterobacteriaceae were recovered from breasts treated with only 0.5 kGy and stored for 21 days. Doses of 0.5 and 1.0 kGy increased the shelf-life of the fillets by 4–5 days while treatments with 2.0 kGy increased the shelf-life of these products by up to 15 days. Research has indicated that electron beam irradiation of minced chicken meat under MAP produced a synergistic effect resulting in increased shelf-life (Grandison and Jennings, 1993). No undesirable organoleptic changes are produced in fresh poultry irradiated with up to 2.86 kGy (Hashim et al., 1995). Consumer acceptance of these products has limited the widespread adoption of these products (Roller, 1999), and currently there are no facilities in the United States for producing irradiated poultry meat. This technology is being used to extend the shelf-life of ground beef, oysters, fruits and vegetables (Savage, 2014).

Research has been conducted on the use of non-thermal dielectric barrier discharge plasma systems to reduce microbial contamination and extend the shelf-life of broiler breast fillets that had been packaged in polyolefin plastic bags (Kronn et al., 2015). Non-thermal plasma can be generated by the breakdown of gases by dielectric fields (Gaunt et al., 2006). Findings of research demonstrated that the shelf-life of breasts treated in packaging with a gaseous atmosphere of 65% oxygen, 30% carbon dioxide and 5% nitrogen was greater than the shelf-life of breasts in packaging with breathing air. The increase in shelf-life was due to the increased production of bactericidal forms of oxygen (reactive oxygen species) in the modified atmospheres by the plasma systems. Ozone was one of the main forms of reactive oxygen produced during treatment with the non-thermal plasma system. Breasts treated with the plasma system produced a 1.53 log cfu/ml after 24 h of storage when compared to the controls, and after 14 days of storage, the treated breast meat had 1.49 log cfu/ml fewer bacteria than untreated samples. Reduction of bacterial contamination on the surface of the chicken breasts extended the shelf-life of the meat.

These products have received widespread use in Japan since 1997 and are currently available in North America for use as an alternative to nitrogen and vacuum packaging for food preservation and other processes that require low oxygen concentrations (Mexis et al., 2012). The products are reported to be non-toxic and residue free. Ageless® is currently the most widely used oxygen absorber that is available, and the ability of the product to absorb oxygen is based on iron oxidation (Grattan and Gilbert, 1994). By lowering the concentration of oxygen available in packages, oxygen absorbers can reduce the growth of aerobic bacteria and moulds. The products are available in small sachets that contain the oxygen absorbent in gas-permeable, water-impermeable flam. The Ageless produce is reported to be able to reduce oxygen concentration to less than 0.0001% in sealed containers.

7 Where to look for further information

- The National Provisioner is a publication designed for meat processing professionals. The publication provides the latest information on meat processing, packaging, food safety, new products and industry trends. An online version of the publication is also available at www.provisioneronline.com.
- Watt Publishing Co., Inc. publishes magazines for agribusiness and food-processing professionals. The company's publications are available online at the Poultry Information Network.
- The Poultry Site magazine provides news on the poultry industry and poultry farming. The Poultry Site-Digital version of the magazine allows readers to view this information online.

8 References

Al-Mohizea, I., Mashhadi, A. S., Fawwal, A. and Al-Shalhat, A. (1994), Microbiological and shelf life assessment of chilled eviscerated whole chicken broilers in Saudi Arabia, *Br. Poult. Sci.*, 35, 519–26.

Arafa, A. S. and Chen, T. C. (1975), Effect of vacuum packaging on microorganisms on cut-up chickens and in chicken products, *J. Food Sci.*, 40, 50–2.

Arnaut-Rollier, I., de Zutter, L. and van Hoof, J. (1999), Identities of the *Pseudomonas* spp. in flora from chilled chicken, *Int. J. Food Microbiol.*, 48, 87–96.

Ayres, J. C. (1960), The relationships of organisms of the genus *Pseudomonas* to the spoilage of meat, poultry and eggs, *J. Appl. Bacteriol.*, 23, 471–86.

Ayres, J. C., Ogilvy, W. S. and Stewart, G. F. (1950), Postmortem changes in stored meat. 1. Microorganisms associated with development of slime on eviscerated cut-up poultry, *Food Technol.*, 4, 199–205.

Balamatsia, C. C., Rogga, K., Badeka, A., Kontominas, M. G. and Savvaidis, I. N. (2006), Effect of low-dose radiation on microbiological, chemical, and sensory characteristics of chicken meat stored aerobically at 4°C, *J. Food Prot.*, 69, 1126–33.

Barnes, E. M. (1972), Food poisoning and spoilage bacteria in poultry processing, *Vet. Rec.*, 90, 720–2.

Barnes, E. M. (1976), Microbiological problems of poultry at refrigerator temperatures-A review, *J. Sci. Fd. Agric.*, 27, 777–82.

Barnes, E. M. (1979), Intestinal microflora of poultry and game birds during life and after storage, *J. Appl. Bact.*, 46, 407–19.

Barnes, E. M. and Impey, C. S. (1968), Psychrophilic spoilage bacteria of poultry, *J. Appl. Bacteriol.*, 31, 97–107.

Beuchat, L. R. (1990), Behavior of *Aeromonas* species at refrigeration temperatures, *Int. J. Food Microbiol.*, 13, 217–24.

Bruckner, S., Albrecht, A., Petersen, B. and Kreyenschmidt, J. (2012), Characterization and comparison of spoilage processes in fresh pork and poultry, *J. Food Qual.*, 35, 372–82.

Buses, H. and Thompson, L. (2003), Dip application of phosphates and marinade mix on shelf life of vacuum-packaged chicken breast fillets, *J. Food Prot.*, 66, 1701–3.

Buzby, J. C., Wells, H. F., Axtman, B. and Mickey, J. (2009), Supermarket loss estimates for fresh fruit, vegetables, meat, poultry, and seafood and their use in the ERS loss-adjusted food availability data, Unites States Department of Agriculture, Economic Research Service, Economic Information Bulletin, No. 44, March 2009.

Carlez, A. Rosec, J.-P., Richard, N. and Cheftel, J.-C. (1993), High pressure inactivation of *Citrobacter freundii, Pseudomonas fluorescens* and *Listeria innocua* in inoculated minced beef muscle, *Lebensm.-Wiss. Technol.*, 26, 357–63.

Carlez, A. Rosec, J.-P., Richard, N., and Cheftel, J.-C. (1994), Bacterial growth during chilled storage of pressure-treated minced meat, *Lebensm.-Wiss. Technol.*, 27, 48–54.

Cerveny, J., Meyer, J. D. and Hall, P. A. (2009), Microbiological spoilage of meat and poultry products, in W. H. Sperber and M. P. Doyle (eds), *Compendium of the Microbiological Spoilage of Foods and Beverages*, New York, Springer.

Church, N. (1994), Developments in modified-atmosphere packaging and related technologies, *Trends Food Sci. Technol.*, 5 (11), 345–52.

Cleveland, J., Montville, T. J., New, I. G. and Chikindas, M. L. (2001), Bacteriocins: safe, natural antimicrobials for food preservation, *Int. J. Food Microbiol.*, 71, 1–20.

Cousin, M. A., Jay, J. M. and Vasavada, P. C. (2001), Psychrotrophic microorganisms, in F. P. Downes and K. Ito (eds), *Compendium of Methods for the Microbiological Examination of Foods*, pp. 159–66, 4th ed. American Public Health Association, Washington, DC.

Coventry, M. J., Muirhead, K. and Hickey, M. W. (1995), Partial characterization of pediocin PO2 and comparison with nisin for biopreservation of meat products, *Int. J. Food Microbiol.*, 26, 133–45.

Delves-Broughton, J. (1990), Nisin and its application as a food preservative, *J. Soc. Dairy Technol.*, 43,164–7.

Doyle, M. P. (1990), Campylobacter jejuni, in D. O. Cliver (ed.), *Foodborne Diseases*, pp. 217–22, San Diego, California, USA, Academic Press Inc.

Eddy, B. P. (1960), The use and meaning of the term 'psychrophilic', *J. Appl. Bact.*, 23 (2),189–90.

Farber, J. M., Warburton, D. W, Gour L. and Milling, M. (1990). Microbiological quality of foods packaged under modified atmospheres. *Food Microbiol.* 7, 327–34.

Farrell, A. J. and Barnes, E. M. (1964), The bacteriology of chilling procedures used in poultry processing plants. *Brit. Poult. Sci.*, 5 (1), 89–95.

Food Safety and Inspection Service (1999), USDA issues final rule on meat and poultry irradiation, United States Department of Agriculture, Washington, DC.

Froning, G. W. (1987), Further processed poultry meat products, in F. E. Cunningham and N. A. Cox (eds), *The Microbiology of Poultry Meat Products*, pp. 317–30, London, UK, Academic Press Inc.

Gaunt, L. F., Beggs, C. B. and Georgihiou, G. E. (2006), Bactericidal action of the reactive species produced by gas-discharge nonthermal plasma at atmospheric pressure: a review, *IEEE Trans. Plasma Sci.*, 34 (4), 1257–69.

Geornaras, I., Dykes, G. A. and von Holy, A. (1994), Microbial populations associated with refrigerated poultry, *Suid-Afrikaanse Tydskrif vir Wetenskap.*, 90, 579–82.

Geornaras, I., Dykes, G. A. and von Holy, A. (1995), Biogenic amine formation by poultry-associated spoilage and pathogenic bacteria, *Lett. Appl. Microbiol.*, 21, 164–6.

Gill, C. O., Harrison, J. C. L. and Penney, N. (1990), The storage life of chicken carcasses packaged under carbon dioxide, *Int. J. Food Microbiol.*, 11, 151–8.

Gounot, A. M. (1986), Psychrophilic and psychrotrophic microorganisms, *Experientia*, 42, 1192–7.

Gram, L., Ravn, L., Rasch, M., Bruhn, J. B., Christensen, A. B. and Givskov, M. (2002), Food spoilage-interactions between food spoilage bacteria, *Int. J. Food Microbiol.*, 78, 79–97.

Grandison, A. S. and Jennings, A. (1993), Extension of the shelf-life of fresh minced chicken meat by electron beam irradiation combined with modified atmosphere packaging, *Food Control.*, 4 (2), 83–8.

Grattan, D. W. and Gilbert, M. (1994), Ageless oxygen absorber: chemical and physical properties, *Stud. Conserv.*, 39 (3), 210–14.

Hashim, I. B., Resurreccion, A. V. A. and McWatters, K. H. (1995), Descriptive sensory analysis of irradiated frozen or refrigerated chicken, *J. Food Sci.*, 60, 664–6.

Hendricks, M. T. and Hotchkiss, J. H. (1997), Effect of carbon dioxide on the growth of *Pseudomonas fluorescens* and *Listeria monocytogenes* in aerobic atmospheres, *J. Food Prot.*, 60, 1548–52.

Hinton, A., Jr., Cason, J. A. and Ingram, K. D. (2002), Enumeration and identification of yeasts associated with commercial poultry processing and spoilage of refrigerated broiler carcasses, *J. Food Prot.*, 65, 993–8.

Hinton, A., Jr., Cason, J. A. and Ingram, K. D. (2004), Tracking spoilage bacteria in commercial poultry processing and refrigerated storage of poultry carcasses, *Int. J. Food Microbiol.*, 91, 155–65.

Hinton, A., Jr., Northcutt, J. K, Smith, D. P., Musgrove, M. T. and Ingram, K. D. (2007), Spoilage microflora of broiler carcasses washed with electrolyzed oxidizing or chlorinated water using an inside-outside bird washer, *Poult. Sci.*, 86, 123–7.

Hurst, A. (1981), Nisin, *Adv. Appl. Microbiol.*, 27, 85–125.

Huss, H. H., Dalsgard, D., Hansen, L., Ladefoged, H., Pedersen, A. and Zittan, L. (1974), The influence of hygiene in catch handling on the storage life of cod and plaice, *J. Food Technol.*, 9, 213–21.

ICMSF (1998), Poultry and poultry products, in T. A. Roberts, J. I. Pitt, J. Farkas and F. H. Grau (eds), *Microorganisms in Foods 6. Microbial Ecology of Food Commodities*, pp. 75–129, London, UK, Blackie Academic and Professional.

Ingram, M. (1951), The effect of cold on micro-organisms in relation to food, *Proc. Soc. Appl. Bacteriol.*, 14, 243–60.

Jay, J. M. (1987), The tentative recognition of psychrotrophic Gram-negative bacteria in 48 h by their surface growth at 10°C, *Int. J. Food Microbiol.*, 4, 25–32.

Jay, J. M. (2002), A review of aerobic and psychrotrophic plate count procedures for fresh meat and poultry products, *J. Food Prot.*, 65, 1200–6.

Jimenez, S. M., Salsi, M. S., Tiburzi, M. C., Rafaghelli, R. C., Tessi, M. A. and Coutaz, V. R. (1997), Spoilage microflora in fresh chicken breast stored at 4°C: influence of packaging methods, *J. Appl. Microbiol.*, 83, 613–18.

Kotula, K. And Yoga, P. (1995), Bacterial contamination of broiler chickens before scalding. *J. Food Prot.* 12, 1326–9.

Kraft, A. A., Reddy, K. V., Hasiak, R. J., Lind, K. D. and Galloway, D. E. (1982), Microbiological quality of vacuum packaged poultry with or without chlorine treatment, *J. Food Sci.*, 47, 380–5.

Kronn, T. G., Lawrence, K. C., Zhuang, H., Hiett, K. L., Rothrock, M. J., Huang, Y. W., Kenner, K. M. and Abdo, Z. (2015), Nonthermal plasma system for extending shelf life of raw broiler breast fillets, *Trans. ASABE.*, 58 (2), 493–500.

Kruk, Z. A., Yun, H., Rutley, D. L., Lee, E. J., Kim, Y. J. and Jo, Cheorun (2011), The effect of high pressure on microbial population, meat quality and sensory characteristics of chicken breast fillet, *Food Control.*, 22, 6–12.

Linton, M., McClements, M. M. J. and Patterson, M. F. (2004), Changes in the microbiological quality of vacuum-packaged, minced chicken treated with high hydrostatic pressure, *Innov. Food Sci. Emerg. Technol.*, 5, 151–9.

Liu, Y., Betti, M. and Ganzle, G. (2012), High pressure inactivation of *Escherichia coli*, *Campylobacter jejuni*, and spoilage microbiota on poultry meat, *J. Food Prot.*, 75, 497–503.

Marenzi, C. (1986), Proper meat storage prevents spoilage, *Poultry*, 6, 12–15.

Marshall, D. L., Wiese-Lehigh, P. L., Wells, J. H. and Farr, A. J. (1991), Comparative growth of *Listeria monocytogenes* and *Pseudomonas fluorescens* on precooked chicken nuggets stored under modified atmospheres, *J. Food Prot.*, 54, 841–3.

McMeekin, T. A. (1975), Spoilage association of chicken breast muscle, *Appl. Microbiol.*, 29, 44–7.

Metrick, C., Hoover, D. G. and Farkas, D. F. (1989), Effects of high hydrostatic pressure on heat-resistant and heat-sensitive strains of *Salmonella*, *J. Food Sci.*, 54, 157–60.

Mexis, S. F., Chouliara, E. and Kontominas, M. G. (2012), Shelf life extension of ground chicken meat using an oxygen absorber and a citrus extract, *LWT-Food Sci. Technol.*, 49, 21–7.

Morita, R. Y. (1975), Psychrophilic bacteria, *Bacteriol. Rev.*, 39, 144–67.

Narasimha, R. D. and Sachindra, N. M. (2002), Modified atmosphere and vacuum packaging of meat and poultry products, *Food Rev. Int.*, 18, 263–93.

Oliveira, K. A. D.-M., Santos-Mendonca, R. C., Gomide, L. A. D.-M., and Vanetti, M. C. D. (2005) Aqueous garlic extract and microbiological quality of refrigerated poultry meat. *J. Food Process. Preserv.*, 29, 98–108.

Painter, J. A., Hoekstra, R. M., Ayers, T., Tauxe, R. V., Braden, C. R., Angula, F. J. and Griffin, P. M. (2013), Attribution of foodborne illnesses, hospitalizations, and deaths to food commodities by using outbreak data, United States, 1998–2008, *Emerg. Infect. Dis* [Internet], March 2013 [5 November 2015].

Park, H., Hung, Y.-C. and Chung, D. (2004), Effects of chlorine and pH on efficacy of electrolyzed water for inactivating *Escherichia coli* O157:H7 and *Listeria monocytogenes*, *Int. J. Food Microbiol.*, 91, 13–18.

Rao, D. N. and Sachindra, N. M. (2002), Modified atmosphere and vacuum packaging of meats and poultry products, *Food Rev. Int.*, 18 (4), 263–93.

Rayman, M. K. Aris, B. and Hurst, A. (1981), Nisin: a possible alternative or adjunct to nitrite in the preservation of meats, *Appl. Environ. Microbiol.*, 41, 375–80.

Roller, S. (1999), Physiology of food spoilage organisms, *Int. J. Food Microbiol.*, 50, 151–3.

Rossaint, S., Klausmann, S., Herbert, U. and Kreyenschmidt, J. (2014), Effect of package perforation on the spoilage process of poultry stored under different modified atmospheres, *Food Pack Shelf Life.*, 1 (1), 68–76.

Russell, S. M. (1997), A rapid method for predicting the potential shelf life of fresh broiler chicken carcasses, *J. Food Prot.*, 60, 148–52.

Russell, S. M. (1997), A rapid microbiological method for enumeration of *Pseudomonas fluorescens* from broiler chicken carcasses, *J. Food Prot.*, 60, 385–90.

Russell, S. M. (2009), Understanding poultry products spoilage, WattAgNet.com. Retrieved from www.wattagnet.com/Understanding_poultry_sproducts_spoilage.html.

Russell, S. M., Fletcher, D. L. and Cox, N. A. (1996), Spoilage bacteria of fresh broiler chicken carcasses, *Poult. Sci.*, 75, 2041–7.

Savage, B. (2014), Irradiation: a work in progress, *Meat and Poultry News*, http://www.meatpoultry.com/Writers/Bryan%20Salvage/Irradiation%20A%20work%20in%20progress.aspx.

Sawaya, W. N., Abu-Ruwaida, S. S., Hussain, A. J., Khalafawi, M. S. and Dashti, B. H. (1993), Shelf-life of vacuum packaged eviscerated broiler carcasses under simulated market storage conditions, *J. Food Saf.*, 13, 305–21.

Schofield, G. M. (1992), A review: Emerging food-borne pathogens and their significance in chilled foods, *J. Appl. Bacteriol.*, 72, 267–73.

Shelef, L. A. (1983), Antimicrobial effects of spices, *J. Food Saf.*, 6, 29–44.

Shewan, J. M., Hobbs, G. and Hodgkiss, W. (1960), A determinative scheme for the identification of certain genera of bacteria with special reference to the Pseudomonadacea, *J. Appl. Bacteriol.*, 23, 379–390.

Styles, M. F., Hoover, D. G. and Farkas, D. F. (1991), Response of *Listeria monocytogenes* and *Vibrio parahaemolyticus* to high hydrostatic pressure, *J. Food Sci.*, 56, 1404–7.

The Poultry Site (2012), Global poultry trends 2012-Chicken consumption steady in Oceania, rising in Africa, Available online at http://www.thepoultrysite.com/articles/2640/global-poultry-trends-2012-chicken-consumption.

Thomas, Y. O., Kraft, A. A., Rust, R. E. and Hotchkiss, K. K. (1984), Effect of carbon dioxide flushing and packaging methods on the microbiology of packaged chicken, *J. Food Sci.*, 49, 1367–71.

Tuboly, E., Levovic V. K., Gaal, O., Meszaros, L. and Farkas, J. (2003), Microbiological and lipid oxidation studies on mechanically deboned turkey meat treated by high hydrostatic pressure, *J. Food Eng.*, 56, 241–4.

United States Department of Agriculture, Economic Research Service (2012), Poultry and Eggs. http://www.ers.usda.gov/topics/animal-products/poultry-eggs.aspx.

Upmann, M., Paulsen, P., James, S. and Smulders, F. J. M. (2000), The microbiology of refrigerated meat. *Fleischwirtschaft Int.*, 3, 38–45.

Warburton, D. W., Gour, L. and Milling, M. (1990), Microbiological quality of foods packaged under modified atmospheres, *Food Microbiol.*, 7, 327–34.

Yadav, A. S. and Singh, R. P. (2004), Natural preservatives in poultry meat products, *Nat. Prod. Radiance.*, 3 (4), 300–3.

Yuste, J., Mor-Mur, M., Capellas, M., Guamis, B. and Pla, R. (1998), Microbiological quality of mechanically recovered poultry treated with high hydrostatic pressure and nisin, *Food Microbiol.*, 15, 407–14.

Zhou, G. H., Xu, X. L. and Liu, Y. (2010), Preservation technologies for fresh meat-A review, *Meat Sci.*, 86, 119–28.

Part 3

Sustainability

Chapter 17

Life cycle assessment (LCA) of intensive poultry production systems

Ilkka Leinonen, Newcastle University, UK

1 Introduction

Among agricultural production systems, livestock production has been considered to have serious environmental impacts, including extensive resource use, nutrient leaching and especially a significant contribution to global warming in the form of emission of greenhouse gases (Steinfeld et al. 2006). These emissions are difficult to handle and arise from various sources. For example, ruminant production produces high amounts of methane (from enteric fermentation) and requires large areas of grazing land, and therefore, some parts of the world are associated with land use changes (LUC), such as deforestation, which in turn contributes to global greenhouse gas emissions (e.g. Opio et al. 2013).

In non-ruminant production systems, such as poultry and pig production, some of the above-mentioned problems related to greenhouse gas emissions can be avoided, and when compared to other livestock systems, poultry production has been found to be relatively environmentally friendly (Williams et al. 2006; ADAS 2007; de Vries and de Boer 2010). However, problems in terms of environmental sustainability exist in non-ruminant production systems (Eriksson et al. 2005; van der Werf et al. 2005; Pelletier 2008). For example, these systems are very much dependent on external feed production and require feed ingredients, especially protein sources such as soya, the production of which is limited to certain areas of the world and has been associated with various environmental impact issues. Furthermore, poultry systems in particular have some other features that require special attention in terms of their environmental impacts. These include nitrogen emissions, for example, in the forms

http://dx.doi.org/10.19103/AS.2016.0010.20

of ammonia emissions to air, nitrous oxide emissions that contribute to global warming and nitrate leaching contributing to eutrophication (e.g. Wathes et al. 1997).

An essential requirement for improving the environmental sustainability of poultry production is to have a systematic tool that can quantify holistically the level of environmental impacts arising from production and then identify the potential target areas for improvement. Currently, there is a general agreement that a method called life cycle assessment (LCA) is the most suitable for quantifying such impacts for agricultural production systems (e.g. MacLeod et al. 2013). In this chapter, recent studies are reviewed, studies that apply the LCA modelling framework to quantify the contribution of different subsystems of poultry production (production of feed, housing emissions, manure management, etc.) to their overall environmental impacts and to suggest measures which could make intensive poultry production more environmentally friendly.

2 Life cycle assessment (LCA) of poultry production: general principles

Improving environmental sustainability of agricultural production has often been linked to modifications in direct farm activities; for example, reduction of energy use on farms or reduction of direct emissions to the environment, including ammonia and methane emitted from animal houses or fields (Leinonen et al. 2014b). However, a large part of the overall environmental impacts related to livestock production actually arises from indirect sources. For example, the production of feed for animals requires growing of feed crops (which consumes energy and generates emissions to the environment), processing and mixing the feed ingredients and transporting them sometimes over very long distances. Furthermore, the environmental impacts originating from direct and indirect sources can result in complicated interactions, and in theory, improving one aspect of production can actually have harmful indirect effects on some of the other links in the production chain. For example, changing the composition of the livestock feed simultaneously affects the nutrient emissions from manure and the environmental impacts related to feed production. In addition, some activities in livestock production may have both environmental benefits and burdens at the same time. For example, using livestock manure as a fertilizer may reduce the environmental impacts associated with the production of manufactured fertilizers, but it may also increase nutrient leaching and emissions to the atmosphere (Leinonen et al. 2014b).

To handle such complex interactions and trade-offs, a systematic, quantitative and holistic approach is needed to evaluate the environmental impacts in agricultural systems. In recent years, systems-modelling-based LCA methods have been developed to be specifically used for agricultural production (Williams et al. 2006, 2007, 2010; Leinonen et al. 2012a,b, 2016a). Like LCA approaches in general, these methods account for all environmental burdens occurring during the production cycle, starting from raw material extraction through to the end products. However, a unique feature of this systems-based approach is that it combines structural models of the agricultural industries with process-based crop and livestock simulation models in order to handle all major interactions between separate subsystems. In this approach, a structural model for poultry production system calculates all the inputs required to produce a specific 'functional unit' (e.g. 1000 kg of broiler or turkey meat, or 1000 kg eggs), together with other outputs coming from the systems, including useful products and unwanted, necessary outcomes, such as wastes, mortalities and emissions to the environment. In the systems model, the quantities of any inputs or outputs are functionally connected to other subsystems. For example, the production of a certain amount of eggs requires a certain number of laying hens, which requires a certain number of pullets, which in turn requires a certain number of parent birds and so on, and all these need to be quantified by the model simulation.

As an output, an LCA model generally quantifies the emissions to the environment (per functional unit) and these can be then aggregated into environmentally functional groups (or 'potentials'), generally including the following (e.g. Guinée et al. 2002; Leinonen et al. 2012a).

Global warming potential is a measure of the greenhouse gas emissions to the atmosphere. The main sources of these emissions in poultry production are carbon dioxide (CO_2) from fossil fuel and LUC, nitrous oxide (N_2O) and methane (CH_4). The sum of global warming potential per functional unit is also known as the 'carbon footprint'.

Eutrophication potential is used to assess the over-supply (or unnatural fertilization) of nutrients as a result of them reaching water systems by leaching, run-off or atmospheric deposition. The main sources are nitrate (NO_3^-) and phosphate (PO_4^{3-}) leaching to water and ammonia (NH_3) emissions to air.

Acidification potential is mainly an indicator of potential reduction of soil pH. The main source is ammonia emissions, together with sulphur dioxide (SO_2) from fossil fuel combustion.

In addition to emissions to environment, LCA can quantify the consumption of various resources associated with production of the functional unit in question. These include, for example, primary energy use, which is quantified in terms of the cumulative use of different 'energy carriers' (including the energy needed for their extraction and supply), such as gas, oil, coal, nuclear and renewable. Other possible indicators of resource use are land occupation, which describes the area of the land required to produce a unit of the product (in the case of poultry production, this mainly consists of the arable land for producing crops for feed), abiotic resource use, which describes the use of non-renewable raw materials, such as fossil fuels and minerals, and water use.

3 The main environmental impacts of intensive poultry production

The overall environmental impacts arising from non-ruminant livestock production systems can be roughly divided to three categories, on the basis of their sources, namely (1) feed production, (2) direct farm energy use and (3) emissions from housing and manure management. The relative contribution of these sources was recently quantified for intensive and alternative broiler and egg production systems in the UK (Leinonen et al. 2012a,b), in order to identify opportunities where greatest environmental improvements would be possible. This was done by applying the systems-based LCA model described above, together with detailed production data provided by the poultry industry.

The results of Leinonen et al. (2012a,b) showed that the first category, that is, feed production, including growing the feed crops and processing and transporting the ingredients, was the main component of several environmental impact categories, most notably the global warming potential, both in broiler and egg production systems

(Leinonen et al. 2012a,b), with its contribution being over 70% in the intensive systems. This result is consistent with LCA studies from other countries, including Finland (Katajajuuri et al. 2006; Katajajuuri 2008), USA (Pelletier 2008) and the Netherlands (Mollenhorst et al. 2006). Sources of the greenhouse gas emissions contributing to the global warming potential of the poultry feed production include the use of fossil energy for manufacturing fertilizers, together with emissions from field operations and transport (Leinonen et al. 2012a,b). Furthermore, the gaseous emissions of N_2O as a result of growing the feed crops significantly contribute to global warming potential, and leaching of nutrients also affects the eutrophication potential (Leinonen et al. 2012a,b).

The high contribution of feed production to global warming potential is also partly a result of LUC related to the production of certain feed crops (Leinonen et al. 2012a,b, 2013). Among ingredients generally used in broiler diets, especially soya and palm oil are known to be largely produced on land that has been only recently converted from natural vegetation to agricultural use in South America and South Asia. In the LCA model, the partial losses of ecosystem carbon storage as a result of such conversion need to be added to the CO_2 emissions and consequently to the global warming potential arising from the system (IPCC 2006; BSI 2011). Although it is generally agreed that such LUC emissions should be taken into account when quantifying the environmental impacts of livestock production, unfortunately there is currently no full agreement on the exact method how this should be done, especially when the exact origin (e.g. whether it comes from recently converted or 'mature' agricultural land) of a certain crop is not known, which is quite often the case in agricultural production. For example, Leinonen et al. (2012a,b) applied the guidelines of the carbon foot printing method PAS2050 (BSI 2011) when calculating the LUC effect on global warming potential. Following this 'best estimate' approach (Leinonen et al. 2012a,b, 2013), in cases where it is not possible to know the exact origin of a certain agricultural commodity, the following procedure is applied. First, the overall LUC emissions from the production of a certain crop within a bigger area, for example, a country or a region are specified, and second, these are allocated evenly to that crop, produced at any location within that area. However, other commonly used LCA approaches may use completely different methods to account for LUC, and this has potentially a considerable effect on the estimate of the environment impacts of feed and livestock production (Leinonen et al. 2013, 2015, see also discussion below).

The second source of the impacts, direct farm energy use, includes the electricity, gas and oil consumed during broiler production, egg production and at breeder farms and hatcheries. The relative contribution of the emissions arising from these activities to the overall impacts has been found to vary between production systems. Leinonen et al. (2012a) found that in the intensive UK indoor broiler system, its proportion was less than 15% of the total global warming potential of the whole system, with the main sources being the liquid propane gas (LPG) used for heating the broiler houses and electricity used mainly for ventilation. In the case of eutrophication potential and acidification potential, the contribution of farm energy use was found to be minimal. These results demonstrate the fact that, although reducing farm energy use is often considered as the primary target when aiming to improve the environmental friendliness of the system, the effects of such reductions on the overall environmental impacts are actually rather limited (Leinonen et al. 2012a,b).

The third source of the impacts, namely emissions from housing and manure management, significantly contributes to both eutrophication potential and acidification potential in both the broiler and egg production systems (Leinonen et al. 2012a,b; Mollenhorst et al. 2006). This is mainly a result of ammonia emissions, which contribute to both eutrophication potential and acidification potential, together with nitrate leaching after land application, which affects eutrophication potential. Housing and manure management also contribute to global warming potential, mainly through nitrous oxide emissions. However, despite these considerable emissions, poultry manure also has environmental benefits. In agricultural crop production, manure can replace some of the manufactured fertilizers and this effect will partially counterbalance the greenhouse gas emissions arising from poultry production (Leinonen et al. 2012a,b).

Leinonen et al. (2012a,b) also compared the environmental impacts of different broiler and egg production systems, including intensive indoor production, free range and organic production. The result of those studies showed that in both broiler and egg production the differences in the impacts were largely related to the efficiency of resource use of the system. In broiler production, the intensive indoor production system that had a shorter production cycle (average of 39 days in the data included in that study) compared to the alternative systems (58 days in free range production and 73 days in organic production) was found to be the most efficient in terms of feed conversion and therefore also had the lowest feed consumption and manure production per kilogram of carcass produced. However, this effect was partially counterbalanced by the slightly higher final weight of birds in the alternative systems where fewer birds were, therefore, needed to produce the functional unit (in this case, 1000 kg broiler carcass). Also in egg production, the alternative systems (barn, free range, organic) were generally less efficient than the conventional cage system. For example, on an average, the birds in the alternative systems produced fewer eggs, while having higher feed consumption and higher mortality than birds in the cage system. As a result, a general trend was found that less intensive poultry systems had higher environmental impacts than more intensive systems, although there was also a large variation between farms within each system (Leinonen et al. 2012a,b).

The lower efficiency especially of the organic systems could be found both in the production of organic feed crops and in the animal production itself. However, it should also be noted that the relatively low productivity of organic and other extensive systems can be compensated at least in part by the low input of resources to these systems, including the inputs for production of organic feed crops. Although, for example, Leinonen et al. (2012 a,b) found that organic poultry systems have higher acidification potential and eutrophication potential than other systems considered, opposite results have also been observed, showing that extensive forms of livestock production can reduce the use of fossil fuels, fertilizers and other inputs (Haas et al. 2001; Basset-Mens et al. 2009; Boggia et al. 2010) or have lower emissions from housing Dekker et al. (2011). Other potential reduction of environmental impacts associated with organic production is that in those systems the origin of the feed must be certified and the use of crops originating from 'unsustainable' sources can be avoided. This will allow, for example, removing greenhouse gas emissions related to direct LUC (and thus reducing global warming potential), as discussed in more detail below.

Although most of the studies found in literature on environmental impacts of poultry production concentrate on broiler and chicken egg production, LCA studies have also been conducted for other poultry products, such as turkey meat (Williams et al. 2006) and *foie gras* (Arroyo et al. 2013). However, only recently a national-level study based on detailed farm data provided by large-scale producers was conducted in order to quantify the environmental impacts of the main UK turkey production systems (Leinonen et al. 2016a).

The results of that study demonstrate that in terms of the sources of the environmental impacts, one of the major differences in turkey production when compared to the broiler systems is a considerably longer production cycle. As a result of this, the feed intake and consequently the emissions from housing and manure are higher per mass unit of turkey meat compared to broiler meat. The same effect can be found when comparing two separate turkey systems, namely the male and female turkey production. Leinonen et al. (2016a) found that systems producing male turkeys had higher feed consumption per unit of output than systems producing females, again as a result of a longer production cycle. In contrast, this effect was found to be partially compensated by the fact that male birds require less inputs such as heating fuel and farm electricity per unit of output, due to higher meat output per bird. Furthermore, high variation in production figures was found between farms, so in general no clear statistically significant trends in the environmental impacts between different intensive UK turkey systems were found (Leinonen et al. 2016a).

4 Reducing environmental impacts: feed consumption and composition

The results discussed in the above section demonstrate that the amount of the feed the birds consume per unit of output (meat or eggs) has a direct effect on the overall environmental impacts in different poultry systems (Leinonen et al. 2012a,b, 2016a; Williams et al. 2006). Therefore, it is clear that improving feed efficiency can have a major effect on the reduction of greenhouse gas emissions related to poultry production, as a majority of these emissions are caused by growing, processing and transporting the feed. In the UK poultry systems (similarly as in many parts of Europe), soya bean meal is the most important protein source in the diets, due to its favourable amino acid composition. This brings in problems other than the 'standard' environmental impacts related to feed production. For example, currently a large part of the soya used as animal feed in the UK is produced in South America, where recent large-scale LUC have occurred, and as stated above, these must be taken into account when quantifying the overall emissions of greenhouse gases to the atmosphere arising from the crop production and from the subsequent use of that crop as animal feed.

In addition to the global warming potential, the quantity and composition of feed strongly affects the eutrophication and acidification potentials of poultry production, both directly and indirectly. The production of feed crops is associated with various direct emissions, including leaching of nitrate to water and emissions of ammonia to the atmosphere. The indirect effects include the emissions from poultry housing and manure management, which similarly contribute to the eutrophication and acidification potential related to poultry production. The magnitude of these emissions depends directly on the nitrogen content of the manure, that is, the amount of nitrogen excreted by the birds. This in turn is dependent on the conversion efficiency of the birds, that is, the amount of nitrogen retained in the bird body or eggs (in a form of protein) divided by the amount of nitrogen obtained from the feed (again in a form of protein). A large part of this surplus nitrogen will be released to the environment as various emissions, although some of it can also be utilized as crop nutrient if the manure is spread on fields as a fertilizer.

One way to reduce the feed-related environmental impacts in poultry production is to select feed ingredients that have lower environmental impacts during the production stage, or have an optimally balanced nutrient content, meaning that the birds can actually utilize a large proportion of the dietary nutrient intake and thus reduce the excretion of nitrogen and phosphorous. One suggested step towards more 'environmentally friendly' livestock diets in Europe is to replace at least part of soya with other, preferably local, protein sources. In the modelling study by Leinonen et al. (2013), the environmental impacts of UK poultry production were quantified for alternative scenarios with different protein crops used as a basis of diet formulations for broilers and laying hens. In the alternative diets, protein crops grown in Europe, namely beans, peas, rapeseed and sunflower meal, replaced some or most of the soya bean meal, which was used as the main protein source in the baseline diets. In the alternative diets used in the model simulations, the inclusion rates of the ingredients were modified so that the metabolizable energy and nutrient content, including essential amino acids, of the diets remained unchanged. Furthermore, the essential amino acid content of the diets was balanced when needed by using pure (synthetic) amino acids. The use of pure amino acid varied between the diets. For example, the bean and pea-based diets required a higher additional level of pure amino acids than the standard soya diets, which also (unintentionally) reduced the level of crude protein in those diets.

The results by Leinonen et al. (2013) showed that when relatively high inclusion rates of beans or peas (up to 300 g/kg) were used in the diets, the global warming potential of broiler and egg production could be reduced by more than 10% when compared to the soya-based baseline diet. However, the inclusion of such high levels would not normally be used by the poultry industry, especially as these ingredients are known to have anti-nutritional effects which may reduce the bird performance (Wiseman et al. 2003; McNeill et al. 2004). Instead, when lower, 'realistic' inclusion rates were used in the model scenarios, the reductions in global warming potential were much more modest, and when all uncertainties in the calculations were taken into account, it was found that these effects were not statistically significant. These results highlight the complexity of the issue of replacing the soya in poultry feed. For example, in this modelling scenario, there was an unwanted consequence that when adding large amounts of alternative ingredients with relatively low energy content to the diets, additive energy needed to be provided, for example, in the form of vegetable oils, which have large environmental impacts themselves. Furthermore, in order to maintain the nutrient balance of the alternative diets, higher amounts of pure amino acids had to be added to alternative diets compared to the original soya diets. Although the amount of these ingredients still remained relatively low, their greenhouse gas emissions per mass unit were high and, as a result, they partly counteracted the favourable effect of soya reduction.

The study by Leinonen et al. (2013) also demonstrated the significant role of the selected method of LUC accounting when quantifying the global warming potential of agricultural production. The default method in that study was the 'best estimate' LUC accounting discussed above, that is, taking a weighted average of 'new' (associated with greenhouse gas emissions from LUC) and 'mature' (no LUC emissions) agricultural land used for production of a certain feed crop in a certain country. One of the alternative approaches to LUC accounting also considered in the study by Leinonen et al. (2013) was the so-called 'top down' method (Audsley et al. 2009). The principle behind this method is that all agricultural production, no matter which crop in which country, has an effect on the global pressure of land use, and therefore has indirect consequences on LUC. Following this principle, the global LUC emissions should be evenly allocated to all agricultural land, so equal emissions per unit of land area will be applied for all crops used in poultry feed. With this alternative approach, only minimal differences between global warming potentials of the different diets were found. So, the overall conclusion of the study by Leinonen et al. (2013) was that there might be a potential to reduce the global warming potential of Northern European poultry production by using locally grown protein sources instead of soya in the feed, but there are limitations in the magnitude of reduction that can be reached through this strategy, and methodological issues (i.e. the LUC accounting method) may also make such benefits questionable. Similar conclusions can also be found in other livestock environmental impact assessment studies. For example, Baumgartner et al. (2008) suggested that there is a rather limited potential of European legumes to reduce the environmental impacts when used in livestock feed.

Among other environmental impact categories, Leinonen et al. (2013) found that the use of beans and peas instead of soya in poultry diets had only a minor effect on the eutrophication potential. The reason for this was that the relatively high level of nitrate leaching is associated with the growing of beans and peas due to the surplus nitrogen these crops fix directly from the atmosphere. However, this effect was partly counterbalanced by the crude protein content of feed which was 'accidentally' reduced in the diets with European protein crops with high inclusion of pure amino acids as mentioned above. A relatively great improvement occurred in acidification potential when diets with high levels of inclusion of peas and beans were applied. However, as in the case of eutrophication potential, this effect occurred only because the alternative protein sources were combined with pure amino acids. This result is consistent with an earlier modelling study comparing alternative diets in pig production in Sweden (Eriksson et al. 2005), which also showed that the acidification potential can be reduced if pure amino acids are applied together with alternative protein crops, due to reduction of the crude protein content of the feed. With a high inclusion of peas without additional pure amino acids, the acidification potential actually increased in their study compared to the baseline soya diet.

Another suggested strategy to replace part of the South American soya in poultry feed and thus reduce some of their environmental impacts is the use of agricultural co-products. Such products include wheat-based Dried Distillers Grains with Solubles (DDGS) from bioethanol production and processed animal protein (PAP) from animal by-products (ABP) (i.e. body parts not intended for human consumption), although the use of the latter in livestock feed is currently forbidden by European legislation. The potential environmental friendliness of such products is based on the principle of 'economic allocation' of the environmental burdens in LCA modelling. The starting point of this approach is that all environmental impacts are distributed to various co-products originating from a certain process in the proportions similar to the relative economic values of these products. As a result, less burdens should be allocated, for example, to DDGS (low economic value) than to bioethanol (high value). Similarly, ABP can be considered as a waste (no economic value), and therefore, following the same principle, no environmental burdens originating from animal production should be included in the LCA of PAP production (Ramirez et al. 2012). The principle of economic allocation is commonly utilized in agricultural LCA studies due to its practicality and suggested causality (economic value can be seen as an ultimate driver of producing any commodity) and most of the LCA studies reviewed in this chapter apply this principle. For example, in soya production, only a part of the overall impacts associated with growing this crop is allocated to soya meal which is used in animal feed (the other product is soya oil), and this is done on the basis of the economic value. However, alternative allocation methods have also been suggested. Some of these methods are based on physical properties of the products such as mass or energy content, because such properties can be considered to be more robust than the economic value. Furthermore, the international standards for LCA, such as ISO 14040, actually advise the avoidance of any allocation in the first place (BSI 2006) and suggest distribution of the burdens among the co-product using physical separation of their production processes (which is very difficult in the case of agricultural products originating from biological processes). Therefore, the environmental friendliness of agricultural co-products depends very much on the subjectively selected method for accounting their impacts. Processes related to bioethanol production and pig industry have relatively high environmental impacts themselves, so the co-products originating from these can be seen as environmentally friendly alternatives only if it is accepted that only a small proportion of the environmental impacts should be allocated to them due to their low economic value.

An important aspect to consider when using alternative 'environmentally friendly' ingredients in poultry diets is the potential effect on the bird performance, for example, the growth rate and feed use efficiency. The results presented above are an outcome of theoretical calculations assuming that the performance would remain unchanged when the alternative diets are used. However, if any effects on the bird performance occur as a result of changing diets, the conclusions on the environmental benefits could be totally different. For example, poorer feed conversion efficiency would automatically increase the environmental impacts per unit of the final product due to higher resource requirement (i.e. a larger amount of feed would need to be produced to produce a certain amount of meat or eggs). Increases in mortality would also have similar effects, because in that case the amount of output would be reduced. Changes in bird performance would also have effects arising from manure management. Worsening feed conversion efficiency would automatically mean that a higher proportion of the nitrogen obtained from the feed would be excreted, which in turn would directly affect the emissions to the environment.

While negative effects on feed efficiency will result in negative environmental consequences, improving the feed utilization will have opposite, favourable effects. It is apparently clear that reducing the amount of feed needed for a certain body weight gain or egg production would reduce the environmental impacts of poultry production arising from both feed production and manure management. One way to achieve this is to improve feed utilization by feed processing techniques or by using additives such as enzymes to improve digestion. This is especially important in the case of protein digestion, since excess use of protein has multiple environmental consequences, including greenhouse gas emissions from the production of certain protein crops and nitrogen emissions arising from manure management. One possible method to reduce the required amount of dietary protein is adding a specific enzyme, protease, to the feed in order to improve the protein utilization (Oxenboll et al. 2011). In a recent study by Leinonen and Williams (2015), the environmental consequences of using a protease product RONOZYME® ProAct in broiler diets was investigated by applying LCA modelling, together with bird performance data, from several feeding experiments where broiler diets with and without protease were applied. By adding the protease to diet, the required digestible protein intake could be achieved with a smaller proportion of soya in the feed, which would result in the overall reduction of the global warming potential arising from the feed production (by up to 12% per mass unit of feed). Furthermore, when the whole broiler production chain was considered, it was found that in most feeding experiments, the largest relative improvements occurred in two environmental impact categories: eutrophication potential and acidification potential. The reason for this was that when protease was used in the diets and the crude protein content of the feed could be reduced, the amount of nitrogen excreted by the birds was also lower. The direct consequences of this were the reductions in the emission of ammonia (affecting both the acidification and eutrophication potentials) and leaching of nitrate, (affecting the eutrophication potential). In most of the experiments analysed by Leinonen and Williams (2015), the low protein diet had no negative effect on bird performance, and therefore, the environmental benefits could be fully achieved. However, there was an exception in one experiment, where the birds fed with a low protein diet had a slightly higher feed consumption per unit live weight and higher mortality than the birds fed with a standard diet. This immediately resulted in the loss of any environmental benefits in terms of reduction in the amount of dietary soya and demonstrates the importance of maintaining the bird performance discussed above.

5 Reducing environmental impacts: breeding, housing and manure management

As demonstrated above, there are potential environmental benefits of producing more output (poultry meat or eggs) with less inputs (e.g. poultry feed). One solution to achieve this, namely, changing the feeding strategies, was already discussed, but as an alternative or complementary option, genetic selection aiming to improve the feed efficiency has been suggested, and it is believed to have a significant role in reducing the environmental impacts of livestock production (Neeteson-van Nieuwenhoven et al. 2013). In broiler production, the most notable genetic trend during the last decades has been an increase in growth rate and therefore reduction of the time required to reach the slaughter weight (Laughlin 2007). An automatic outcome of the shortening of the production cycle is that an increasingly higher proportion of the metabolizable energy obtained by the animal from feed is retained in the body (as protein and lipid) and a smaller proportion is released as heat. In practical farming, this trend has been observed as a smaller feed consumption by the birds (de Beer et al. 2011; Laughlin 2007; Mussini 2012).

It has been suggested that the genetic changes in recent decades, especially the reduced feed consumption, have considerably reduced the greenhouse gas, ammonia and nitrate emissions arising from poultry production (Defra 2008; Pelletier 2010). However, it should be noted that many studies considering the effects of genetic changes on environmental impacts have been based on rather simplified assumptions, for example, expecting that the observed or predicted reduction in the feed intake per unit of the product would be directly proportional to the changes in the environmental impacts arising from feed production. For more accurate predictions, however, it might be necessary to apply a more mechanistic modelling approach in order to understand the overall consequences of continuing genetic changes. An attempt towards this direction was made by Leinonen et al. (2016b). In their study, a system approach-based LCA framework, combined with a simple mechanistic model of bird energy balance, was used to predict the potential effects of 15 years of prospective broiler breeding on the environmental impacts of the standard UK broiler production system. The results showed that predicted changes in biological performance (mainly increased growth rate) due to selective breeding could potentially lead to reduced environmental impacts of the broiler production chain, most notably in the eutrophication potential, acidification potential, abiotic resource use and global warming potential. Such reductions are a direct result of the reduced maintenance energy requirement and thus lower feed intake, resulting from the shorter production cycle. However, some of these environmental benefits might be limited by the required changes in feed composition (e.g. increased inclusion of soy meal and vegetable oil) as a result of the changes in bird nutrient requirements. An expected outcome of shifting the energy use of the bird from metabolic heat production towards growth (especially protein growth) is an increase in the use of protein crops in the diets, with potential environmental problems discussed earlier in this chapter.

As discussed above, the conventional, usually intensive, poultry systems have been generally considered to be environmentally friendly due to their efficiency, that is, low resource use and low emissions per unit of product (Leinonen et al. 2012a,b). However, in modern poultry production, efficiency is not the only aim in the process of improving the systems, but measures to enhance the bird welfare are getting a lot of emphasis. In a study by Leinonen et al. (2014a), the environmental consequences of new, 'animal welfare enhancing' production systems that have recently been introduced to broiler and egg production in the EU were quantified by using the systems-based LCA model and data collected from the UK poultry industry. In broiler production, a new, low-density system was evaluated where the maximum live weight per square meter was reduced from the standard practice of the industry. In some of the low-density units, an additional feature was a heat exchanger, which was used to circulate the heat otherwise lost in ventilation, in order to compensate the expected increase of heating requirement, which was a result of the fact that a smaller number of birds in a house produce less heat. In egg production, the traditional battery cage system was completely replaced by a new, lower-density 'colony cage' system, following the changes in EU legislation, and the consequences of this change were assessed by using the LCA model.

The results of Leinonen et al. (2014a) showed that the low-density broiler system increased the primary energy use and global warming potential compared to the standard system, mainly due to increased liquefied petroleum gas consumption during housing as a result of increased heating requirement. However, the increase in farm energy use was partly compensated by a better feed efficiency, which was a result of a shorter production cycle typically used in this system. As a combined effect of these changes, the overall increase of global warming potential was only about 2%, when the low-density system was compared to the standard UK production system. When the heat exchanger was applied in the low-density system, the overall primary energy use in applying the heat exchanger was similar to the baseline system and the global warming potential was actually reduced by 3%. Furthermore, the low-density systems resulted in reduction in the eutrophication potential and acidification potential, mainly as a result of better feed efficiency compared to the standard system (Leinonen et al. 2014a).

The comparison of the old and new egg production systems showed that the colony cage system had 8% lower primary energy use and 3% lower global warming potential than the earlier battery cage system, due to lower energy use in housing and slightly improved productivity. There were only minor differences in the eutrophication and acidification potentials between the systems, as there were no significant changes in feed consumption or nutrient excretion per unit of product. In general, the results by Leinonen et al. (2014a) suggest that measures taken to improve bird welfare in poultry production do not necessarily have a major effect on the efficiency of the system, and therefore, such measures can be put into action without risking the environmental sustainability of the production.

As discussed above, a large proportion of the eutrophication and especially acidification arising from poultry systems originates from housing and manure management. These emissions can be reduced by improving feed efficiency, but it can be expected that direct measures aiming to improve the management of manure both at the farm and in connection of its end use could also reduce those emissions. Currently, a routine practice in the UK is to spread the manure directly to the field as a crop fertilizer. Although this has some environmental benefits, that is, it can reduce the demand of production of synthetic fertilizers and also increase the soil carbon storage (thus removing CO_2 from the atmosphere), the problem in this practice is considerable emissions of ammonia and nitrate to the environment. Therefore, alternative strategies for manure management have been considered (e.g. Lynch et al. 2013; Huang et al. 2015). One of the most promising new practices is the use of poultry litter (i.e. manure and used bedding) as a fuel to generate electricity at a power station. Only recently, a study was published where the LCA framework was used in a systematic way to compare the alternative uses of turkey litter produced in the UK (Williams et al. 2016). The results of that study show substantial environmental benefits when the litter was used as a fuel compared with direct land application as fertilizer. Among all environmental impact categories considered in that study, the greatest relative effects occurred in eutrophication and acidification potentials, which were reduced by about 55% and 70%, respectively, when compared to the traditional use of manure as a fertilizer. These reductions were mainly the result of eliminating ammonia emissions during storage and land spreading. The fuel use of litter also resulted in a reduction in primary energy use (by about 14%) and a reduction of a similar magnitude was also found in abiotic resource use. However, the effect of the litter fuel use on reducing global warming potential was relatively small (about 3%). This outcome was mainly a consequence of two processes with opposite effects. First, there was a large reduction of greenhouse gas emissions as a result of replacing fossil fuels used to generate electricity with a renewable energy source. This is because, according to international standards, poultry litter is considered a biogenic fuel with no (long-term) greenhouse gas emissions. However, a contrasting effect occurred when the changes in the soil carbon storage were taken into account (which is required, e.g. by the IPCC [2006] standards). Using the litter as a fuel resulted in loss of carbon to the atmosphere, while part of this carbon would have been stored in the soil for a relatively long period following the direct fertilizer use of the litter. This resulted in a gradual reduction in soil carbon content, which was taken into account in the calculations. Other negative effects of the fuel use of the litter included increased emissions of NO_x and microparticles. However, in general, the authors considered the overall environmental effect of litter fuel use as beneficial (Williams et al. 2016).

6 Summary

Environmental sustainability of agricultural production, especially of livestock systems, is a complex issue, and any attempts to make improvements in this area require understanding of a network of multiple interactions, ranging from the resources used for feed crop production to the end use of manure. Systems modelling-based LCA provides a useful tool to systematically handle such interactions and consequently identify where significant improvements can be made to the sustainability of livestock systems. Although poultry production has been found to be relatively environmentally friendly compared to the production of other livestock commodities, it still contributes to environmental impacts such as global warming, eutrophication and acidification; so there are further opportunities to reduce these impacts, for example, through a combination of changing feeding strategies, feed processing, genetic selection and improvements in housing and manure management.

Among the different subprocesses in the poultry production chain, feed production and transport is the largest contributor to the global warming potential of poultry systems (e.g. in the standard UK indoor broiler production, its contribution was found to be about 70%), whereas manure management substantially contributes to their eutrophication and acidification potentials. All these impacts can be reduced by improving the feed efficiency, for example, by making the feed more digestible by using additives such as enzymes. Furthermore, genetic selection has the potential to reduce the resources needed for broiler production, especially feed consumption. However, possible changes in bird performance may change the need of certain feed ingredients (most notably protein sources), which may limit the benefits of this strategy. The use of alternative feed ingredients, such as locally grown protein crops and agricultural by-products, as a replacement for South American grown soya can also potentially lead to improvements in several environmental impact categories, as long as such feeding strategies have no negative effects on bird performance. Other management options, such as improving poultry housing and formulating new strategies for manure management, also have a considerable potential in further improving the environmental sustainability of the poultry industry.

7 Future trends

Various dimensions of sustainability will continue to be an essential part of any improvements in livestock systems, including poultry production. Environmental sustainability especially is an important aspect of performance of any agricultural system, partly because environmental impacts (e.g. greenhouse gas and ammonia emissions) of production are being increasingly regulated by national and international legislation.

Among the methods quantifying the environmental sustainability of production systems, LCA has become a more or less the standard approach. It has various benefits compared to other assessment methods, including its ability to take into account the whole production cycle and to handle different environmental impact categories. Recently, commercial software and the ever-expanding Life Cycle Inventory databases have become publicly available, which makes the use of LCAs easier. However, in complex production chains, including agricultural production, approaches based on systems modelling provide clear benefits compared to simple 'carbon calculators', and research and model development in this area is expected to continue in the future.

Among the environmental 'hotspots' of poultry production, feed efficiency and feed composition continue to be the main research areas. Reducing the dependency of European livestock systems on South American soya is a continuous target of research. In addition to locally grown protein crops, alternative protein sources for poultry production have been suggested, including ABP (currently banned in Europe) and completely new solutions, such as yeast products. As the examples presented in this chapter demonstrate, assessing the environmental sustainability of such solutions is an essential step when evaluating their applicability in poultry production.

Another environmental hotspot of poultry production, namely environmental issues related to manure management, also remains a subject of future research. Recent studies have quantified the environmental impacts (and environmental benefits) of the centralized use of poultry litter as a fuel in power stations. In addition to that, the current UK legislation also allows the fuel use of litter at local, farm-based heating and combined heat and power units. The potential environmental benefits of such new technologies, similar to other measures aiming to improve the energy efficiency of poultry farms, are among the subjects of further research.

8 Where to look for further information

The following publications present detailed analysis of the environmental impacts of the main broiler, chicken egg and turkey meat production systems in the UK:

- Leinonen, I., Williams, A. G., Wiseman, J., Guy, J. and Kyriazakis, I. 2012. Predicting the environmental impacts of chicken systems in the UK through a Life Cycle Assessment: broiler production systems. *Poult. Sci.* 91: 8–25.
- Leinonen, I., Williams, A. G., Wiseman, J., Guy, J. and Kyriazakis, I. 2012. Predicting the environmental impacts of chicken systems in the UK through a Life Cycle Assessment: egg production systems. *Poult. Sci.* 91: 26–40.
- Leinonen, I., Williams, A. G. and Kyriazakis, I. 2016. Comparing the environmental impacts of UK turkey production systems using analytical error propagation in uncertainty analysis. *J. Clean. Prod.* 112: 141–8.

The following quantify the effects of some strategies for reducing the environmental impacts of poultry production, including changes in feeding and manure management:

- Leinonen, I., Williams, A. G., Waller, A. H. and Kyriazakis, I. 2013. Comparing the environmental impacts of alternative protein crops in poultry diets: the consequences of uncertainty. *Agric. Sys.* 121: 33–42.
- Leinonen, I. and Williams, A. G. 2015. Effects of dietary protease on nitrogen emissions from broiler production: a holistic comparison using Life Cycle Assessment. *J. Sci. Food Agric.* 95: 3041–6.
- Williams, A. G., Leinonen, I. and Kyriazakis, I. 2016. Environmental benefits of using turkey litter as a fuel instead of a fertiliser. *J. Clean. Prod.* (in press).

General assessments of greenhouse gas emissions of agricultural production in the UK and poultry production globally are presented in the following reports:

- Audsley, E., Brander, M., Chatterton, J., Murphy-Bokern, D., Webster, C. and Williams, A. 2009. How low can we go? An assessment of greenhouse gas emissions from the UK food system and the scope to reduce them by 2050. WWF-UK.
- MacLeod, M., Gerber, P., Mottet, A., Tempio, G., Falcucci, A., Opio, C., Vellinga, T., Henderson, B. and Steinfeld, H. 2013. Greenhouse gas emissions from pig and chicken supply chains – A global life cycle assessment. Food and Agriculture Organization of the United Nations (FAO), Rome.

Detailed instruction of performing LCA following international standards can be found in this book:

- Guinée, J. B., Gorrée, M., Heijungs, R., Huppes, G., Kleijn, R., de Koning, A., van Oers, L., Wegener Sleeswijk, A., Suh, S., Udo de Haes, H. A., de Bruijn, H., van Duin, R. and Huijbregts, M. A. J. 2002. Handbook on life cycle assessment. Operational guide to the ISO standards. I: LCA in perspective. IIa: Guide. IIb: Operational annex. III: Scientific background. Kluwer Academic Publishers, ISBN 1-4020-0228-9, Dordrecht, p. 692.

9 References

ADAS. 2007. The Environmental Impact of Livestock Production. Report for Defra FFG.

Arroyo, J., Fortun-Lamothe, L., Auvergne, A., Dubois, J. P., Lavigne, F., Bijja, M. and Aubin, J. 2013. Environmental influence of maize substitution by sorghum and diet presentation on goose foie gras production. *J. Clean. Prod.* 59: 51–62.

Audsley, E., Brander, M., Chatterton, J., Murphy-Bokern, D., Webster, C. and Williams, A. 2009. How low can we go? An assessment of greenhouse gas emissions from the UK food system and the scope to reduce them by 2050. WWF-UK.

Basset-Mens, C., Ledgard, S. and Boyes, B. 2009. Eco-efficiency of intensification scenarios for milk production in New Zealand. *Ecol. Econom.* 68: 1615–25.

Baumgartner, D. U., de Baan, L. and Nemecek, T. 2008. European Grain Legumes – Environment-Friendly Animal Feed? Life Cycle Assessement of Pork, Chicken Meat, Egg, and Milk Production. Grain Legumes Integrated Project. Final Report WP2.2, Environmental Analysis of the Feed Chain. Agroscope Reckenholz-Tänikon Research Station ART, Zürich.

Boggia, A., Paolotti, L. and Castellini, C. 2010. Environmental impact evaluation of conventional, organic and organic-plus poultry production systems using life cycle assessment. *World's Poult. Sci. J.* 66: 95–114.

BSI. 2006. Environmental management: Life cycle assessment – Principles and framework. http://www.iso.org/iso/iso_catalogue/catalogue_tc/catalogue_detail.htm?csnumber=37456.

BSI. 2011. Specification for the assessment of the life cycle greenhouse gas emissions of goods and services. PAS 2050:2011. http://www.bsigroup.com/Standards-and-Publications/How-we-can-help-you/Professional-Standards-Service/PAS-2050/.

de Beer, M., Elfick, D. and Emmerson, D. A. 2011. Is a feed conversion ratio 1:1 realistic and appropriate goal for broiler chicken in the next 10 years? *Recent Adv. Anim. Nutr. – Aust.* 18: 9–14.

Defra. 2008. A study of the scope for the application of research in animal genomics and breeding to reduce nitrogen and methane emissions from livestock based food chains. Defra project AC0204. http://randd.defra.gov.uk/Default.aspx?Menu=Menu&Module=More&Location=None&Completed=0&ProjectID=14662.

Dekker, S. E. M., de Boer, I. J. M., Vermeij, I., Aarnink, A. and Groot Koerkamp, P. W. G. 2011. Ecological and economic evaluation of Dutch egg production systems. *Livest. Sci.* 139: 109–21.

de Vries, M. and de Boer, I. J. M. 2010. Comparing environmental impacts for livestock products: a review of life cycle assessments. *Livest. Sci.* 28: 1–11.

Eriksson, I. S., Elmquist, H., Stern, S. and Nybrant, T. 2005. Environmental systems analysis of pig production – The impact of feed choice. *Int. J. Life Cycle Ass.* 10: 143–54.

Guinée, J. B., Gorrée, M., Heijungs, R., Huppes, G., Kleijn, R., Koning, A. de, Oers, L. van, Wegener Sleeswijk, A., Suh, S., Udo de Haes, H. A., Bruijn, H. de, Duin, R. van and Huijbregts, M. A. J. 2002. Handbook on life cycle assessment. Operational guide to the ISO standards. I: LCA in perspective. IIa: Guide. IIb: Operational annex. III: Scientific background. Kluwer Academic Publishers, ISBN 1-4020-0228-9, Dordrecht, 2002, p. 692.

Haas, G., Wetterich, F. and Köpke, U. 2001. Comparing intensive, extensified and organic grassland farming in southern Germany by process life cycle assessment. *Agric. Ecosyst. Environ.* 83: 43–53.

Huang, Y., Anderson, M., McIlveen-Wright, D., Lyon, G. A., McRoberts, W. C., Wang, Y. D., Roskilly, A. P. and Hewitt, N. J. 2015. Biochar and renewable energy generation from poultry litter waste: a technical and economic analysis based on computational simulations. *Appl. Energ.* 160: 656–63.

IPCC. 2006. IPCC Guidelines for National Greenhouse Gas Inventories. (2006 Guidelines) http://www.ipcc-nggip.iges.or.jp/.

Katajajuuri, J. M. 2008. Experiences and improvement possiblities – LCA case study of broiler chicken production. *Proc. 6th Int. Conf. Life Cycle Assessment in the Agri-Food Sector*. Zurich, Switzerland.

Katajajuuri, J. M., Grönroos, J., Usva, K., Virtanen, Y., Sipilä, I., Venäläinen, E., Kurppa, S. Tanskanen, R., Mattila, T. and Virtanen, H. 2006. Broilerin fileesuikaleiden tuotannon ympäristövaikutukset ja kehittämismahdollisuudet. *Maa- ja elintarviketalous* 90: 1–118.

Laughlin, K. 2007. The evolution of genetics, breeding and production. Temperton Fellowship Report No.15. Harper Adams University College.

Leinonen, I., Williams, A. G., Wiseman, J., Guy, J. and Kyriazakis, I. 2012a. Predicting the environmental impacts of chicken systems in the UK through a Life Cycle Assessment: broiler production systems. *Poult. Sci.* 91: 8–25.

Leinonen, I., Williams, A. G., Wiseman, J., Guy, J. and Kyriazakis, I. 2012b. Predicting the environmental impacts of chicken systems in the UK through a Life Cycle Assessment: egg production systems. *Poult. Sci.* 91: 26–40.

Leinonen, I., Williams, A. G., Waller, A. H. and Kyriazakis, I. 2013. Comparing the environmental impacts of alternative protein crops in poultry diets: the consequences of uncertainty. *Agric. Syst.* 121: 33–42.

Leinonen, I., Williams, A. G. and Kyriazakis, I. 2014a. The effects of welfare-enhancing system changes on the environmental impacts of broiler and egg production. *Poult. Sci.* 93: 256–66.

Leinonen, I., Williams, A. G., Wiseman, J. and Kyriazakis, I. 2014b. The potential to mitigate the environmental impacts of broiler production systems through changes in their diets. In P. C. Garnsworthy and J. Wiseman (eds), *Recent Adv. Anim. Nutr. – 2013*. Context Products Ltd., pp. 123–44.

Leinonen, I. and Williams, A. G. 2015. Effects of dietary protease on nitrogen emissions from broiler production: a holistic comparison using Life Cycle Assessment. *J. Sci. Food Agric.* 95: 3041–6.

Leinonen, I., Williams, A. G. and Kyriazakis, I. 2015. Evaluating methods to account for the greenhouse gas emissions from Land Use Changes in agricultural LCA. In R. Schenck and D. Huizenga (eds), *Proc. 9th Int. Conf. on Life Cycle Assessment in the Agri-Food Sector* (LCA Food 2014), ACLCA, Vashon, WA, USA, pp. 711–17.

Leinonen, I., Williams, A. G. and Kyriazakis, I. 2016a. Comparing the environmental impacts of UK turkey production systems using analytical error propagation in uncertainty analysis. *J. Clean. Prod.* 112: 141–8.

Leinonen, I., Williams, A. G. and Kyriazakis, I. 2016b. Potential environmental benefits of prospective genetic changes in broiler traits. *Poult. Sci.* (in press).

Lynch, D., Henihan, A. M., Bowen, B., Lynch, D., McDonnell, K., Kwapinski, W. and Leahy, J. J. 2013. Utilisation of poultry litter as an energy feedstock. *Biomass Bioenergy* 49: 197–204.

MacLeod, M., Gerber, P., Mottet, A., Tempio, G., Falcucci, A., Opio, C., Vellinga, T., Henderson, B. and Steinfeld, H. 2013. Greenhouse gas emissions from pig and chicken supply chains – A global life cycle assessment. Food and Agriculture Organization of the United Nations (FAO), Rome.

McNeill, L., Bernard, K. and MacLeod, M. G. 2004. Food intake, growth rate, food conversion and food choice in broilers fed on diets high in rapeseed meal and pea meal, with observations on sensory evaluation of the resulting poultry meat. *Br. Poult. Sci.* 45: 519–23.

Mollenhorst, H., Berentsen, P. B. M. and De Boer, I. J. M. 2006. On-farm quantification of sustainability indicators: an application to egg production systems. *Br. Poult. Sci.* 47: 405–17.

Mussini, F. J. 2012. Comparative response of different broiler genotypes to dietary nutrient levels. PhD thesis, University of Arkansas, Fayetteville, AR.

Neeteson-van Nieuwenhoven, A. M., Knap, P. and Avendaño, S. 2013. The role of sustainable commercial pig and poultry breeding for food security. *Animal Frontiers* 3/2013: 52–7.

Opio, C., Gerber, P., Mottet, A., Falcucci, A., Tempio, G., MacLeod, M., Vellinga, T., Henderson, B. and Steinfeld, H. 2013. Greenhouse gas emissions from ruminant supply chains – A global life cycle assessment. Food and Agriculture Organization of the United Nations (FAO), Rome.

Oxenboll, K. M., Pontoppidan, K. and Fru-Nji, F. 2011. Use of a protease in poultry feed offers promising environmental benefits. *Int. J. Poult. Sci.* 10: 842–8.

Pelletier, N. 2008. Environmental performance in the US broiler poultry sector: life cycle energy use and greenhouse gas, ozone depleting, acidifying and eutrophying emissions. *Agric. Syst.* 98: 67–73.

Pelletier, N. 2010. Breeding poultry for environmental performance: a life cycle-based supply chain perspective. *Proc. 9th World Congr. Genet. Appl. Livest. Prod. Paper 75*. Leipzig, Germany.

Ramirez, A. D., Humphries, A. C., Woodgate, S. L. and Wilkinson, R. G. 2012. Greenhouse gas life cycle assessment of products arising from the rendering of mammalian animal byproducts in the UK. *Environ. Sci. Technol.* 46, 447–53.

Steinfeld, H., Gerber, P., Wassenaar, T., Castel, V., Rosales, M. and de Haan, C. 2006. Livestock's long shadow: environmental issues and options. FAO, Rome, Italy.

van der Werf, H. M. G., Petit, J. and Sanders, J. 2005. The environmental impacts of the production of concentrated feed: the case of pig feed in Bretagne. *Agric. Syst.* 83: 153–77.

Wathes, C. M., Holden, M. R., Sneath, R. W., White, R. P. and Phillips, V. R. 1997. Concentrations and emission rates of aerial ammonia, nitrous oxide, methane, carbon dioxide, dust and endotoxin in UK broiler and layer houses. *Br. Poult. Sci.* 38: 14–28.

Williams, A. G., Audsley, E. and Sandars, D. L. 2006. Determining the environmental burdens and resource use in the production of agricultural and horticultural commodities. Main Report. Defra Research Project IS0205. www.agrilca.org.

Williams, A. G., Audsley, E. and Sandars, D. L. 2007. Environmental burdens of livestock production systems derived from life cycle assessment (LCA). In P. C. Garnsworthy and J. Wiseman (eds), *Recent Adv. Anim. Nutr.* Nottingham University Press, Nottingham, pp. 171–200.

Williams, A. G., Audsley, E. and Sandars, D. L. 2010. Environmental burdens of producing bread wheat, oilseed rape and potatoes in England and Wales using simulation and system modeling. *Int. J. Life Cycle Assessment* 15: 855–68.

Williams, A. G., Leinonen, I. and Kyriazakis, I. 2016. Environmental benefits of using turkey litter as a fuel instead of a fertiliser. *J. Clean. Prod.* (in press).

Wiseman, J., Al-Mazooqi, W., Welham, T. and Domoney, C. 2003. The apparent ileal digestibility, determined with young broilers, of amino acids in near-isoenergentic lines of peas (*Pisum sativum* L) differing in trypsin inhibitor activity. *J. Sci. Food Agric.*, 83: 644–51.

Chapter 18

Minimizing the environmental impact of poultry production through improved feed formulation

Hector E. Leyva-Jimenez and Christopher A. Bailey, Texas A&M University, USA

1 Introduction

The poultry industry began as a backyard activity in the 1800s, when families raised small flocks of double-purpose chickens to supply eggs and meat. Years later, the development of new technologies allowed farmers to have excess chickens and eggs that were sold to their neighbours and friends. This rapidly became a very profitable economic activity due to the increasing demand for eggs and meat (NCC, 2015). After the development of a specialized meat-type chicken known as a broiler around the year 1920 (Muir and Aggrey, 2003), the egg and meat production began to differentiate in two independent segments of the global poultry industry. Today, the poultry industry has become a fully vertically integrated global business, an industry that arose from small independent backyard flocks and evolved into well-organized multinational companies (NCC, 2015).

http://dx.doi.org/10.19103/AS.2016.0010.21

According to the summary of the United States Department of Agriculture (USDA) Poultry-Production Value 2014, the combined value of production from broilers, eggs, turkeys and chickens (spent fowl from breeder flock and table-egg layers) in the United States was $48.3 billion. From the combined total value of production, 8.54 billion broilers and 238 million turkeys were raised, which represented a value of $32.7 and $5.30 billion, respectively; the egg production value was $10.2 billion. A total of 99.8 billion eggs were estimated to be produced in the United States in 2014 (USDA-NASS, 2015).

The accelerated production rate of the poultry industry over the last 30 years has led to multiple negative environmental concerns. Generation of waste related to intensive poultry production has become a major issue due to the generation of dust, nauseous odours, greenhouse gases and alteration and deterioration of ecosystems (Steinfeld et al., 2006). Therefore, the implementation of sustainable production strategies to reduce the environmental impact of the poultry industry is necessary to ensure the conservation of our ecosystems.

Soil accumulation, surface run-off and leaching to groundwater systems of excess nutrients such as nitrogen (N) and phosphorus (P), metal elements, antibiotics and coccidiostats found in poultry manure cause soil toxicity and eutrophication of water systems and represent a human health hazard. Diet modification to reduce the output of excess nutrients could be accomplished by precisely meeting animal requirements, formulating diets with the most digestible ingredients available and using feed additives such as enzymes to increase digestibility of feedstuffs, ultimately reducing the adverse environmental impact of poultry farms and achieving sustainable production of poultry meat and further processed products (Jongbloed and Lenis, 1992).

The purpose of this chapter is to provide a good understanding of the potential pollutant sources derived from intensive poultry production, as well as discuss different nutritional strategies to reduce the negative environmental impact of the poultry industry through precise feed formulation and manufacture. Finally, it is to recognize the importance of a sustainable use of all available natural resources to ensure the supply of animal protein products for future generations.

2 Environmental impact of livestock and poultry production

Commercial poultry production concentrates huge flocks of birds, and consequently, the waste generated from live production operations and processing plants is one of the biggest environmental concerns. Livestock and poultry waste contributes to the deterioration of the environment, principally by contaminating water, soil and air. The odour of manure attracts flies and rodents that carry diseases, leaching of excess nutrients, pathogens and heavy metals pollute water systems. Environmental contamination is generally caused by poor manure management and disposal systems (Gerber et al., 2007).

Nitrogen (N) and phosphorus (P) are the most deleterious components of animal manure. Surface run-off and leaching of these elements to the surface and groundwater systems can cause eutrophication, which causes dense growth of plant life and death of animals due to the lack of oxygen (Sharpley, 1999). Several strategies have been described to reduce the environmental impact of animal operations, such as precise feed formulation, and the addition of feed supplements, such as probiotics, that improve nutrient efficiency and can result in significant decreases in N, P, odour and dry matter weight of manure (Nahm, 2002).

The demand of livestock and poultry products is increasing rapidly with the growing human population. Industries must have a good understanding of the causes of contamination and the effects of livestock and poultry waste on the environment. This allows the design of effective strategies to reduce adverse environmental impact and achieve sustainable production of meat and further processed products. In the following section of this chapter, we will discuss some of the environmental impacts of livestock and poultry waste in water, soil and air in more detail.

2.1 Environmental impact of livestock and poultry waste in soil and water

Excess nutrients fed to animals that are not absorbed in their intestine will be excreted. Barker (1996) reported estimated values for manure production of a hen, broiler, a heavy turkey tom and a duck to be 117, 142, 348 and 148 g/day, respectively. Chastain and Camberato (2004) reported a production of fresh manure in dairy cows to be about 1.61 ft^3 (12.0 gal) per 1000 lb average live weight per day. Chastain et al. (1999) reported swine manure production of a finisher hog to be around 84 lb per 1000 lb average live weight per day. These examples give us an idea of the manure volume that must be managed every year. Contamination of soil, surface and groundwater systems is due to accumulation, run-off and leaching of these excess nutrients that are present in livestock and poultry manure (Gerber et al., 2007). Proper management and storage of animal waste is fundamental to reduce the negative impact to the environment.

Nitrogen leaching from manure produced in intensive production operations, often used as fertilizer in agriculture, is the main source of excessive N losses to the environment. Nitrogen compounds contained in manure are very dynamic and mobile in soil, and can be transported to surface and groundwater easily (Williams, 2013). Animal manure is usually composed of ammonium, ammonia and organic-N. Ammonium (NH_4^+) is readily available to plants, but organic-N must undergo a process of mineralization before it can be utilized by the plants. Hence, excess nitrogen in manure applied to crops at a time when the plant cannot utilize it will result in greater leaching losses (Hooda et al., 2000). Waste lagoons, spray fields and litter fields are also possible sources of N losses to groundwater resources (Mallin and Cahoon, 2003).

Excessive accumulation of P in agricultural soils due to fertilizers and animal manure result in high P losses into the groundwater resources which negatively impact the aquatic ecosystems (Heckrath et al., 1995). However, contrary to N, phosphorus is not very mobile in soil and in order to affect the environment it must to be transported to surface waters first. This P transport is generally via eroding soil as sediment-bound P or in surface run-off as soluble inorganic or organic P (Williams et al., 1999). The enrichment of water bodies with chemical nutrients such as P and N is known as eutrophication. The negative consequences of eutrophication include excessive production of noxious phytoplankton that reduces water clarity and harms water quality, affecting the growth of surface plants and benthic organisms (Sims and Wolf, 1994). Benthic organisms live in and on the bottom sediments of a body of water. The dense growth of plant life can result in anoxia conditions in water ecosystems, causing the death of animals due to the lack of oxygen (Chislock et al., 2013). High rates of photosynthesis can deplete dissolved inorganic carbon and raise pH to extreme levels, affecting animals that rely on their chemosensory abilities to catch preys (Turner and Chislock, 2010).

Other nutrients that enter water systems and cause negative effects in plants and animals are toxic metals like zinc, copper or arsenic. In excess, these elements pollute the water through surface run-off and leaching (Gupta and Charles, 1999). Animal diets contain heavy metal elements due to contaminated feedstuffs or as dietary nutrients used to meet the animal requirements (Gerber et al., 2007). However, overformulation of trace elements is a common practice in modern livestock and poultry production to ensure optimal performance; very often these elements are added in amounts that exceed the animal requirements (Williams et al., 1999). Additionally, leaching of pesticides, antibiotics and coccidiostats used in animal feeds are also a potential risk of pollution to the water systems (Sims and Wolf, 1994).

2.2 Environmental impact of livestock and poultry waste in air quality

The emission of odours and gases to the atmosphere such as carbon dioxide (CO_2), methane (CH_4) and ammonia (NH_3) generated in livestock and poultry facilities can affect air quality (Steinfeld et al., 2006). Microbial decomposition of manure leads to the production of unpleasant odours (Zhu, 2000). When manure is stored, transported or discarded, the emitted odours affect not only the employees but also residential and recreational areas that are located near the production operations (Burnett, 1969).

Emissions of CO_2 and NH_3, which can be transformed into nitrous oxide (N_2O) (Denmead et al., 2008), are known to be important greenhouse gases that concentrate in the atmosphere and can potentially affect ecosystems, biodiversity and the lives of people worldwide by slowly increasing the temperature of the earth. Carbon dioxide is one of the main air pollutants from intensive husbandry due to CO_2 expelled during animal respiration, microbial fermentation of manure and open flame heaters that elevate CO_2 concentration in the atmosphere (Harry, 1978). Methane, another major air pollutant, is produced from bacterial decomposition of fresh or stored manure (Zeeman, 1994), accumulates in the atmosphere and contributes to the greenhouse effect as well.

Ammonia is the major air pollutant from poultry facilities. This colourless gas is able to solubilize in water and react with other compounds to form ammonia-containing compounds that are able to acidify soil and water (Gerber et al., 2007). Also, it is a source of N that causes eutrophication of water systems and increases emissions of the greenhouse gases such as N_2O (Lee et al., 1978; Gerber et al., 2007; Denmead et al., 2008). Dietary N from poultry feeds that is not converted into eggs, meat or other products will be excreted as organic-N, which is rapidly converted into ammonia and then released to the atmosphere with the housing ventilation system or after the manure is stored and exposed to the environment (Williams, 2013).

Dust is another potential hazard as it causes irritation of the respiratory tract when it is inhaled and lowers the resistance to respiratory diseases (Gross et al., 1959). Additionally, it plays a key role as a carrier of pathogenic microorganisms excreted by livestock and poultry (Harry, 1978). Improper processing, storage, disposal and transportation contributes to the release of dust, especially during dry hot weather (Mukhtar et al., 2004).

3 Strategies to reduce the environmental impact of poultry production through improved feed formulation and manufacturing

The poultry industry is one of the most versatile and dynamic industries in the world and has experienced an accelerated growth over the last 30 years. Although it is true that the poultry industry supplies meat and eggs at a relative low cost to consumers, thanks to the intensive genetic selection of modern poultry strains, improvements in poultry nutrition and welfare, modern poultry facilities, new processing technologies and better biosecurity procedures (Havenstein et al., 2003; Applegate and Angel, 2014; NCC, 2015), multiple negative environmental impacts have arose from the intensive growth of broilers, turkeys, layers and other species as previously described.

The overexploitation of natural resources to supply the feedstuffs required in animal feeds, along with the excess nutrients fed to animals that are not bioavailable and thus are excreted, represents the biggest challenge to reduce the environmental impact of intensive poultry production through precise feed formulation and manufacturing.

Commercial poultry feed processing consists of several steps that include the grinding, mixing and conditioning the ingredients to create a pellet (Svihus, 2006). The main objective of precise feed formulation and manufacturing is to achieve a nutritionally uniform diet that precisely meets the animal requirements for maintenance, growth and production. Additionally, it involves the reduction of the risk of water, soil and air pollution by limiting excess nutrient intake and improving nutrient bioavailability through the use of supplements, genetically modified feed stuffs and highly digestible synthetic amino acids and trace minerals (Behnke and Beyer, 2004; Nahm, 2007; Gerber et al., 2007). Some of the strategies to achieve an environmental friendly poultry feed will be described in the following sections of this chapter.

4 Nutrient uniformity, particle size and pelleting

Uniformity is defined in the Oxford dictionary as the state or condition in which everything remains the same in all cases and at all times, homogeneous, or unvarying. In feed manufacturing, uniformity is achieved through a good ingredient selection to formulate a nutritionally complete diet that meets the animal requirements for maintenance, growth and production. Additionally, the mixing process is important for combining the ingredients previously selected (Behnke and Beyer, 2004; Nahm, 2007). Reduced uniformity of the diet because of an improper mixing process results in poor animal performance and increased nutrient excretion into the environment (Nahm, 2002). Particle size is recognized as a factor that influences bird's performance, as the selection of certain particles will disrupt the nutrient balance (Amerah et al., 2007). It has been shown that chickens have a natural preference for larger particle sizes and this preference increases with the age of the bird (Schiffman, 1968; Portella et al., 1988). Other factors such as trajectory, angle of repose and percolation are sources of particle segregation and are known to decrease diet uniformity during the feed-manufacturing process (Axe, 1995).

Samples of the final feed are collected to measure uniformity after mixing. Usually, the chloride ion of salt is used as a marker to measure uniformity. This method is widely used because it is rapid, accurate and inexpensive. Also, coloured iron particles can be used to measure mixer performance (Axe, 1995; Behnke and Beyer, 2004). The coefficient of variation (CV) obtained after the analysis of the feed samples is used as the indicator of quality or uniformity. A maximum-accepted CV in commercial feed manufacturing is 10% (Wicker and Poole, 1991). McCoy et al. (1994) conducted two experiments to study the effects of inadequate mixing time on broilers' performance. Experiment 1 (20, 40 and 80 mixer revolutions) and experiment 2 (5, 20 and 80 mixer revolutions) with three different mixing times were selected to represent poor, intermediate and adequate diet uniformities. For both experiments, they found that diet variability significantly decreased as the mixing time increased from poor to intermediate uniformity and a small change was found as the mixing time changed from intermediate to adequate uniformity. In the second experiment, daily weight gain, average daily feed intake and gain-to-feed ratio were improved as the mixing time increased from poor to intermediate uniformity, and thus it can be concluded that diet uniformity greatly influences broiler performance. Johnston and Southern (2000) evaluated the effect of varying mix uniformity of phytase on growth performance, mineral retention and bone mineralization in broiler chickens fed with diets containing microbial phytase. They reported that broiler P excretion numerically increased ($p < 0.07$) as phytase units CV increased from 0 to 103, and bone breaking and bone ash significantly decreased in 18-day-old broilers.

4.1 Effect of particle size and pelleting in performance and nutrient utilization

The importance of particle size reduction of commonly used feedstuffs such as corn, sorghum, wheat, soybean and cottonseed is to break the outer seed coat and increase the surface area of exposure to digestive enzymes (Goodband et al., 2002). Also, particle size reduction is used to decrease the segregation of ingredients during mixing and improve the process of pelleting or extrusion (Nir, 1996). The hammer mill and the roller mill are the two most common methods of ingredient particle size processing. The hammer mill will reduce the particle size by using a set of hammers that move at a certain speed in a grinding chamber until the grains are able to pass through a screen with a specific size (Koch, 1996; Amerah et al., 2007). The efficiency of the hammer mill is correlated to the type of grain, number of hammers, mill design, power of the motor and screen size (Martin, 1985). On the other hand, the roller mill uses one or more pairs of horizontal rollers located in a supporting frame; the grains are reduced by a constant compression force as they pass between the rotating rollers (Koch, 1996; Amerah et al., 2007). The roller mill produces uniform particles that are more spherical in shape than the ones obtained from a hammer mill and requires less processing energy (Reece et al., 1985). However, the hammer mill is more commonly used than the roller mill for grinding grain in commercial feed manufacturing because it is easier to use and to maintain (Amerah et al., 2007).

The effect of grinding type in white layer hens was studied by Deaton et al. (1989) with corn processed by using a hammer mill or a roller mill and fed to commercial egg layers. Particle size ranged from 1343 to 1501 μm and 814 to 873 μm for the roller mill and hammer mill, respectively. They found that the different particle sizes obtained from the two mills did not significantly influence the performance of commercial laying hens as measured by body weight, percentage of hen-day egg production, egg weight, average daily feed consumption, feed efficiency, egg shell-breaking strength and mortality. Nir et al. (1990) also compared the effect of the grinding type on the performance of broiler chickens, where sorghum was ground by using a hammer mill and a roller mill into three different particle sizes: fine (≈0.5 mm) medium (≈0.7 mm) and coarse (≈0.9 mm) and fed to broiler chickens. No significant differences in feed intake, body weight gain and feed conversion were found between grinding methods. Their results also indicated that the broiler chickens preferred the coarse grinding rather than the fine particle size in mash diets. The Nir et al. (1990) results were in agreement with Schiffman (1968), Portella et al. (1988) who recognized the ability of chickens to naturally prefer larger particle sizes, which is more evident as the birds age.

The positive effect of grinding on nutrient utilization was studied by Kilburn and Edwards (2001) by using two corn particle sizes, coarse (2897 μm) and fine (869 μm), that were fed to broiler chickens. The true metabolized energy was significantly greater in mash diets that contained the fine corn particles, although, no difference was found after the diets were pelleted. In agreement with Kilburn and Edwards, Peron et al. (2005) found that fine wheat (2 mm) improved the digestibility of starch and nitrogen-corrected apparent metabolized energy in broiler chickens. Parsons et al. (2005) found that broiler performance and energy metabolism were decreased when corn particle size exceeded 1042 μm. Also, Jacobs et al. (2010) compared the performance, gizzard weight and cecal microbial populations in young broiler chickens that were fed different corn particle sizes (557 858, 1210, or 1387 μm). They found a significant linear decrease of nitrogen-corrected metabolized energy values as particle size increased, although amino acid digestibility was not affected by particle size. Additionally, gizzard weight and *Lactobacillus* population in the ceca increased significantly with bigger particle size.

In contrast to the previous studies, coarse particles seem to have a positive effect in increasing mineral bioavailability (Amerah et al., 2007). Calcium and phytate phosphorus retention were greatest with the coarse (2897 μm) maize diets fed to broiler chicks (Kilburn and Edwards, 2001). Feeding coarse (1239 μm) soybean meal to broiler chickens improved bone ash and plasma P levels (Kilburn and Edwards, 2004). Bone mineralization was enhanced in hens that received a mixture of Ca particles sizes in a 67 and 33 % proportion for ground Ca and coarse Ca supplemented as oyster shell, respectively (Saunders-Blades et al., 2009).

The pelleting process is defined as the agglomeration of small particles into larger particles through a mechanical process that usually involves treating ground feed with steam, heat and pressure (Falk, 1985; Leeson and Summers, 2005). The benefits or advantages of pelleting animal feeds include improved animal performance, decreased feed wastage, reduced selective feeding, improved bulk density, destruction of deleterious organisms and customer acceptability (Behnke, 1994; Behnke and Beyer, 2004). The quality of the pellet is related to the chemical composition of the ingredients and to the particular machinery properties (Garcia-Maraver and Carpio, 2015). Pellet quality depends on a number of factors, including feed formulation (40%), particle size (20%), conditioning (20%), die specifications (15%) and cooling and drying (5%) (Reimer, 1992). Therefore, 60% of the factors affecting pellet quality are determined before any conditioning process occurs (Behnke, 1994). Pellet durability index (PDI) and pellet hardness are two physical parameters used to measure pellet quality (Amerah et al., 2007). Parsons et al. (2006) found that feeding hard pellets (1856 g of pellet breaking force) to broiler chickens improved nutrient retention, true metabolized energy and performance when compared with broilers fed with soft pellets (1662 g of pellet breaking force). Briggs et al. (1999) concluded that the ingredients in the diet strongly influence pellet quality, and pellet durability could be improved with high protein content. In contrast, the inclusion of fat above 2% in corn-soy diets will decrease pellet durability (Richardson and Day, 1976). If high-fibre ingredients are used to formulate a diet such as bran, alfalfa and bagasse (pulpy residue left after the extraction of juice from sugarcane or sorghum stalks), a large amount of energy will be used to compress the mash to the density of the subsequent pellet (Behnke, 1994). Pelleting at too low temperatures or with too little steam results in a 'shiny pellet' due to the increased friction on the pellet going through the die (Leeson and Summers, 2005). The addition of 10–15% wheat or wheat middlings will improve the pellet hardness and the water applied to the mash (2%) will help produce a better-quality pellet, but the drying process should be extended to prevent the formation of mold (Leeson and Summers, 2005).

The positive effect of pelleting on nutrient utilization and performance has been widely studied. Hussar and Robblee (1962) compared the performance of broiler chickens fed with mash and pelleted diets. Chickens that consumed the pelleted feed gained 25% more weight, increased feed intake by 15%, and improved feed efficiency by 10% with respect to chickens fed with the mash feed. Mitchell et al. (1972) studied the effect of pelleting and particle size of roasted soybeans in broiler chickens. They demonstrated that pelleting diets containing the roasted soybeans significantly improved 7–24-day gains and feed-conversion ratios as compared to the chickens fed with the mash diet. Additionally, diets containing roasted full fat beans can be used in broiler diets without negative effects in performance if the diets are pelleted. Jensen and Falen (1973) reported that the inclusion of fat up to 3% in pelleted turkey diets significantly improved feed efficiency when compared to all mash diets. Proudfoot and Hulan (1982) concluded that feed conversion and body weight was significantly higher on turkeys fed with pelleted feed than with those fed with mash feed. Douglas et al. (1990) reported a significant improvement in weight gain and feed conversion of broiler chickens fed with pelleted diets containing yellow corn, low tannin or high tannin sorghum when compared to those fed with mash diets. Svihus et al. (2004) found that pelleting was able to even-out the differences in particle size obtained from grinding wheat in a hammer mill and a roller mill, as well as increase performance, and increase the apparent metabolizable energy of the diets from 11.6 to 11.8 MJ/kg. Jia and Slomiski (2010) observed that pelleting whole flaxseed-containing diets showed a positive effect on performance when compared to finely ground mash diets fed to broiler chickens.

5 Optimizing nutrient density and multiphase feeding

As previously mentioned, the main objective of precise feed formulation is to achieve a nutritionally uniform diet that precisely meets the animal requirements for maintenance, growth and production. Additionally, precisely, feed formulation reduces the amount of nutrients excreted into the environment (Behnke and Beyer, 2004; Nahm, 2007; Gerber et al., 2007). However, it is well known that modern livestock and poultry production very often add nutrients in amounts that exceed the animal requirements to ensure maximum performance, especially trace minerals (Williams et al., 1999), P and N (Nahm, 2002), which represents a major environmental concern as discussed in the previous section of this chapter.

Nutrient excretion due to overfeeding can be reduced by alternating the nutrient density of feeds. The reduction of dietary crude protein (CP) and the use of highly digestible synthetic amino acids have been described as feasible strategies to reduce the N emissions (Ferket et al., 2002). Balir et al. (1999) reported a 10–27% reduction in N excretion in broiler chickens during the six-week rearing period after a dietary CP reduction from 25 to 21% at 3200 kcal/kg ME for the starter phase and 21 to 18% at 3200 kcal/kg ME for the grower phase, respectively. Performance was not severely affected by the CP reduction and use of synthetic amino acids, although body fat was increased. They also reported a reduction of 30 to 35% N excretion reduction in daily output in layer hens after a dietary CP reduction from 17 to 13% at 2850 kcal/kg ME. No significant negative effects in egg production were observed after eight weeks. Aletor et al. (2000) investigated the effect of low-protein diets supplemented with synthetic amino acids in broiler chickens. Diets containing 22.5 (control) 21, 19, 17.2 and 15.3% CP were formulated, and synthetic amino acids were used to meet the NRC (1994) recommendations for each individual essential amino acid. Also, in a second experiment, non-essential amino acids were supplemented to the lower CP diets (21 to 15.3%) so that they became isonitrogenous with the 22.5% CP control diet. Neither lowering of the CP nor supplementation with non-essential amino acids diets had a significant effect on weight gain or relative weights of various carcass cuts. However, performance was significantly affected when the chickens were fed with the lowest CP diets, and fat deposition increased linearly. Dietary energy, protein retention efficiency and protein efficiency ratio were significantly improved in the lower-protein diets. The N excretion was linearly decreased after the dietary CP was reduced from 22.5 to 15.3%. Brickett et al. (2007) reported that low-energy/protein-reduced density diets (starter: 3100 kcal/kg 22% CP to 2800 kcal 20.7% CP; grower: 3150 kcal/kg 20.7% CP to 2850 kcal/kg 19.6% CP; finisher: 3200 kcal/kg 18.6% CP to 2900 kcal/kg 18% CP) fed to broiler chickens had no significant effect on body weight when the diets were offered as pellet when compared to diets offered as mash. Feed intake decreased as nutrient density increased and was lower for chickens fed with mash diets. Mortality was not affected by nutrient density but was significantly reduced when the diets were mash (3.8%) compared with feeding pellets (5.6%).

The concept of ideal protein is defined as a balance of essential amino acids that exactly meets the animal's requirements, with no deficiencies and no excesses (Baker and Chung, 1992; Baker and Han, 1994; Firman and Boling, 1998). Lysine is used as a reference amino acid, with the requirements of all other essential amino acids expressed as a percentage of lysine (Table 1). Lysine is preferred as the reference amino acid to set the ideal ratios because it is the second-limiting amino acid in typical corn-soy diets. Lysine analysis is simple and accurate, dietary lysine is used for protein accretion and maintenance, and lysine requirements for different varieties of diet compositions, environmental conditions and different stages of growth are widely available (Emmert and Baker, 1997).

Ideal protein is a powerful tool to improve the estimation of amino acid requirements and therefore reduce excretion of excess N into the environment (Klopfenstein et al., 2002), particularly when combined with true or standardized amino acid digestibility data. In a study conducted by Wijtten et al. (2004) to evaluate the response of male broilers of different age categories to different dietary ideal protein concentrations, apparent faecal digestible lysine/kg values (9.0 to 14.4 g) were used to formulate different ideal protein-level diets. The ratios between digestible lysine and essential amino acids were as follows: methionine (53%), methionine + cystine (75%), threonine (63%), tryptophan (18%), arginine (111%), valine (76%), isoleucine (0.67%), leucine (132%) and histidine (44%). All diets had a constant metabolizable energy content of 3078 kcal/kg. Results showed that the weight gain and feed conversion of male broilers positively respond to higher dietary ideal protein ratios than would be expected from individual amino acid requirements. The ideal protein concept allows nutritionists to formulate low-protein diets on the basis

Table 1 Estimated ideal protein amino acid ratios expressed as percentages of the lysine requirement for growing chickens (0–21-day old)

Amino acid	NRC 1994[1]	Illinois ideal chick protein (IICP)[2]
Lysine	100	100
Arginine	114	105
Methionine	46	36
Methionine + Cystine	82	72
Tyrosine	56	50
Threonine	73	67
Tryptophan	18	16
Leucine	109	111
Isoleucine	73	67
Valine	82	77
Phenylalanine	66	55
Histidine	32	37

[1] NRC (1994) Ratios based on total amino acid requirements.

[2] IICP Baker and Chung (1992), Baker and Han (1994) Ratios based on digestible amino acid requirements.

of amino acid digestibility values to precisely meet the birds' requirements and avoid overfeeding (Firman and Boling, 1998), especially when non-starch polysaccharides (NSP) are present in the diet ingredients (Nahm, 2007).

Digestible lysine requirement values that could be used to formulate poultry diets by using the ideal protein concept have been reported in the literature for different phases of broiler chickens and turkey growth, environmental conditions and diet compositions (Han and Baker, 1991; Firman and Boling, 1998; Boling and Firman, 1998; Garcia et al., 2006; Dozier et al., 2008, 2009; Mejia et al., 2012). However, mathematical models like the Illinois ideal chick protein (IICP) model could be used to estimate the digestible lysine requirement; such models allow the development of equations that predict digestible amino acid requirements as a function of age (Baker and Han, 1994; Baker, 1997; Emmert and Baker, 1997).

It is probable that modern poultry strains have altered their nutritional requirements due to intensive genetic selection, better biosecurity procedures, technological improvements in poultry facilities and animal welfare (Havenstein et al., 2003; NCC, 2015). The requirement definition has certainly evolved over the past 30 years, from being a requirement to prevent a nutrient deficiency to a requirement to optimize growth or egg production response per unit of nutrient intake (Applegate and Angel, 2014). Therefore, it is necessary to re-evaluate the ability of modern poultry to respond to modern feeding strategies.

5.1 Multiphase feeding

One effective strategy to precisely balance nutrients in poultry feeds is the use of multiphase feeding. In order to enhance nutrient utilization and consequently reduce the excretion of excess nutrients like N or P, it is essential to accurately estimate nutritional requirements at various stages of life (Nahm, 2007). Estimating nutrient requirements for any animal population can be difficult, as the response to changes in nutrient composition is variable and can be influenced by many factors such as genetics, gender, environment (Pomar et al., 2010) and age (Nahm, 2007). The NRC (1994) provides phase-feeding references for growing meat-type chickens and turkeys divided in fixed phases. However, the NRC (1994) recommendations have several limitations for nutritionists to accurately estimate nutritional requirements for modern-industry meat-type poultry strains, which are usually grown for longer periods to maximize meat yield after processing (Warren and Emmert, 2000). The NRC (1994) recommendations are based on total amino acids requirements and do not take into account amino acid digestibility values. Also, they encompass male and female for the same recommendations (Pope and Emmert, 2001).

Multiphase feeding programmes should be designed to precisely estimate nutritional requirements more closely at specific points in the broilers' life cycle (Gutierrez et al., 2008). For this purpose, growth equation models are used to predict requirements at any point during the broilers' development. Mathematical modelling can describe relationships among dietary nutrients' inputs and outputs and tissue development as a function of time or environment (Oviedo-Rondón and Waldroup, 2002). More accurate equations to predict birds' requirements should consider different environmental conditions and information about the bird and feed composition (Emmans and Fisher, 1986). This will result in an equation able to estimate requirements under multiple environmental and economic conditions (Oviedo-Rondón and Waldroup, 2002).

As an example, the Gompertz (1825) equation has been described as a useful growth curve (Winsor, 1932) and is currently used to model broilers' growth in the following form:

$$W_t = W_o e(L \div K)(1 - e(-K_t))$$

where:

W_t = Weight at time t
W_o = Hatch weight
L = Initial growth rate
K = Rate of exponential decay of L

In this form, the mature weight is described as $A = W_o e^{(L/K)}$ (Hurby et al., 1994; Oviedo-Rondón and Waldroup, 2002). This empirical model describes a mathematical relationship between a dependent variable and an independent variable on the basis of theoretical assumptions concerning growth without giving any explanation of the biological processes involved (Oviedo-Rondón and Waldroup, 2002).

Emmert and Baker (1997) used the IICP model (IICP; Baker and Han, 1994; Baker, 1997) to develop regression equations that predict amino acid requirements in a multiphase feeding programme for broilers on the basis of digestible amino acid requirements as a function of age. By using these equations, it is possible to predict amino acid requirements at any period of time, including periods beyond eight weeks of age (Warren and Emmert, 2000). Other requirement prediction models include broken-line regression analysis with linear or quadratic ascending portions, a saturation kinetics model, a logistic model and a compartmental model (Pesti et al., 2009), all useful to estimate nutrient requirements using nutrient dose response data. The advantages of broken-line regression were described by Robbins et al. (2006) as relatively simple by using standard nonlinear regression software that provides a functional description of a response to different nutrient dose levels and a break-point estimate and standard error that can be interpreted as the optimal nutrient requirement above which there is no significant change in the dependent variable. The saturation kinetics model, compartmental model and logarithmic model are useful to predict requirements using an economic approach known as the law of diminishing marginal productivity to determine the 'most economical feeding level' (Pesti et al., 2009).

Computerized software is also commercially available. Some of them include OmniPro® II. This program is based on the model proposed by Hurwitz et al. (1978) and is used to estimate amino acid requirements for maintenance, carcass growth and feather growth, taking into account growth curves of the genetic line according to sex and environmental constraints (temperature, flock density, feed restriction, etc.) (Rondón and Waldroup, 2002). The EFG Software® of Natal, South Africa, can also be used to estimate the optimum concentration of amino acids and energy to determine the economical optimum feeding programme for broilers and other species (Gous and Fisher, 2001; Gous, 2001). Research done at Edinburg University (*Edinburg, UK*) (Emmans, 1981, 1987; Emmans and Fisher, 1986) was taken as a base to develop the program that relates the Gompertz equation, diet composition, feeding schedule, feed restriction, temperature and stock density as parameters used to estimate broiler amino acids and energy requirements (Rondón and Waldroup, 2002).

The effect of multiphase feeding on broilers' performance was studied by Pope et al. (2002) who evaluated a phase-feeding programme with lysine, sulfur amino acids (SAA) and threonine reductions every other day from 42 to 63 days of age by using Emmert and Baker (1997) prediction equations and comparing the broilers' performance with the reference NRC (1994) regimen. Additionally, modifications to the linear regression equations were done to decrease 10% or increase 15% the amino acid requirements previously predicted. They concluded that the phase-feeding regimen with lysine, SAA and threonine reductions every other day was able to support maximum growth performance during the experimental period. No significant differences were found in percentage breast, wing, leg or abdominal fat between treatment diets even when amino acid requirements were lowered by 10% or increased by 15%.

In another experiment, Pope et al. (2004) evaluated the beneficial effects of multiphase feeding in reducing the N excretion and performance of broilers from 21 to 63 days of age. Two multiphase feeding treatments were obtained by using Emmert and Baker (1997) amino acid prediction equations. One treatment (PF) consisted of a lysine, SAA, and threonine reduction every other day according to the predicted values obtained from the linear regression equations, and the second treatment (PF-10%) was modified to obtain a 10% decrease relative to the normal calculated values of the previous treatment PF. Both treatments were compared to the reference NRC (1994) feeding regimen. Their results showed a significant improvement in weight gain and feed efficiency in broilers fed with the PF-10% when compared to the NRC reference diet. Also, CP intake was reduced in PF-10% relative to the NRC reference diet. Finally, both the PF and PF-10% feeding programmes reduced N excretion by 3.9 and 8.6% respectively from 43 to 63 days of age.

Cooley and Classen (2003) evaluated the effects of essential (EAA) acid formulation and delivery on manure nutrient content in broiler chickens from 7 to 41 days of age. Iso-energetic diets of various EAA levels were fed in a three-phase feeding programme (starter, grower, finisher) or by blending diets daily to gradually reduce EAA content. The EAA level and feeding programme significantly reduced N and P excretion per bird. The daily blending feeding programme significantly reduced total N excretion by 2 g per bird.

Multiphase feeding has been shown to be a powerful tool to precisely meet animal requirements. However, some factors should be considered before applying an 'aggressive multiphase' feeding programme. In many cases feeding more than five or six diets may be economically impractical as constant formulation and feed manufacturing could represent an additional cost and could be time-consuming (Warren and Emmert, 2000) A cost–benefit analysis is recommended to evaluate the efficacy of multiphase feeding programmes under commercial conditions.

6 Exogenous enzymes as feed additives

Exogenous enzymes such as xylanase, β-glucanase, α-amylase, phytase and β-mannanase are broadly used in modern poultry feed formulations as additives. The main purpose of using exogenous enzymes in poultry feeds is to increase the nutrient bioavailability of feedstuffs, to reduce the nutrient variation of ingredients and to reduce the incidence of wet litter (Bedford, 2000). Corn is preferred in feed formulations to other cereal grains because of its good digestibility and grains such as wheat, barley, rye and oats have been related with negative nutritional characteristics due to their fibre content and various NSP (Friesen et al., 1992; Nahm, 2002).

The effect of NSP contained in grains such as wheat, barley, rye and oats increases the intestinal viscosity, which slows the rate of digestion and also impedes access to their contents by digestive enzymes (Bedford, 2000). Excess N and P excretion can also be reduced with the addition of dietary exogenous enzymes that improve protein digestibility or destroy the anti-nutritional factors found in feedstuffs (Nahm, 2007). Therefore, exogenous enzymes are useful in reducing the negative effect of viscous grains, allowing proper digestion processes and increasing nutrient bioavailability.

The beneficial nutritional effect of exogenous enzymes has been extensively studied. However, the specific modes of action of exogenous enzymes are still not well understood. The nature of dietary feedstuffs, feed processing, appropriate use of enzymes for specific substrates contained in the diet (Acamovic, 2001), age of the animal and interaction with other dietary ingredients are shown to influence the effectiveness of enzymes in poultry feeds (Bedford and Schulze, 1998).

Although a great variety of exogenous enzymes are commercially available, xylanase, phytase and β-glucanase are probably the most widely used enzymes in commercial poultry feed formulation. The nutritional significance of these enzymes is that they are essentially absent in the intestinal tract of poultry and their target substrates are abundant in commonly used ingredients for poultry feeds (Bedford and Schulze, 1998).

Xylanases are hydrolytic enzymes that randomly cleave the β 1, 4 backbone of the complex plant cell wall polysaccharide xylan (Collins et al., 2005), which constitutes the major component of hemicellulose along with xyloglucan, glucomannan, galactoglucomannan and arabinogalactan (Shallom and Shoham, 2003).

B-glucanase is another enzyme that is usually added to barley-based diets to hydrolyze the β-glucans, NSPs that increase gut-viscosity (Pluske and Lindemann, 1998), also found in the plant cell wall (Theander et al., 1993). These hydrolytic enzymes break down polysaccharides and provide shorter-chain 18–24 carbon oligosaccharides or simple sugars to the same organisms that naturally produced the enzymes as part of their metabolism (Bedford and Schulze, 1998).

Pan et al. (1998) evaluated the effect of enzyme (pentosanase/xylanase) supplementation and inorganic P (Biophos; 18% Ca and 21% P) on wheat or rye-based diets on laying hen performance, energy and P availability. Enzyme supplementation, regardless of the cereal grain used, significantly improved the nitrogen-corrected apparent metabolized energy and feed efficiency by 6.2 and 3%, respectively. Egg production had a numerical increase from 87.6 to 90.1% when compared with non-enzyme diets. Wang et al. (2005) used wheat-based diets to study the effect of enzyme supplementation on performance, nutrient digestibility, gastrointestinal morphology and volatile fatty acid profile in the hindgut of broilers. Treatment diets were supplemented with 0, 200, 400, 600, 800 or 1000 mg/kg enzyme (xylanase/β-glucanase) and fed to broilers from 7 to 42 days of age. Daily weight gain and feed conversion was significantly improved with increasing concentrations of enzyme. The digestive organs and gastrointestinal tract was reduced in size with enzyme inclusion. There was a linear increase in total volatile fatty acid content in ileum on day 21 and in the cecum on day 21 and 42 of age. Finally, apparent CP digestibility increased linearly, but no difference in apparent metabolized energy was found. Fowler et al. (2014) reported that the inclusion of two commercially available enzyme formulas containing xylanase, amylase, cellulase and β-glucanase activities supplemented to 45-day feeding programme corn-based broiler diets with a 148–157 kcal/kg energy reduction had a significant positive effect on performance. Enzyme supplementation was able to overcome a loss in performance due to the energy reduction. Zhu et al. (2014) conducted a study by using an enzyme cocktail consisting of xylanase (1800 IU/g), β-glucanase (500 IU/g) and α-amylase (800 U/g) activities to supplement corn-soy broiler diets at 0.1%. The enzyme cocktail had no significant effect on general performance. However, enzyme supplementation resulted in an increase in pancreatic amylase, trypsin, lipase, pepsin and maltase activity. Also, increased jejunum and ileum villus height and villus surface area was observed. Munyaka et al. (2015) studied the effect of combined β-glucanase and xylanase enzymes (2500 U of xylanase and 250 U of β-glucanase per kg of complete feed) on growth performance, nutrient utilization and gut microbiota in broiler chickens fed with corn or wheat-based diets for 21 days. They found a significant improvement in body weight (4%) and feed conversion ratio (7%) in birds fed with the supplemental enzyme mix. Birds fed with wheat diets had higher neutral digestible fibre (46.5%) and less CP digestibility (− 5.4%) than those fed with corn-based diets. Jejunal digesta viscosity was reduced more effectively in wheat diets (− 31%) than in corn-based diets (− 10%). Enzyme supplementation increased starch digestibility (3.5%) when compared to diets without the enzyme mix.

Vegetable oil and/or animal fats are usually added to high caloric poultry diets. However, it has been shown that poultry do not utilize fats efficiently when they are young probably due to low lipase activity (Krogdahl and Sell, 1989). Sell et al. (1986) found that fat retention improved with age in young turkey poults, regardless of the supplementation of saturated or unsaturated fats to the diet. Krogdahl and Sell (1989) reported similar results; dietary tallow and a blended animal–vegetable fat were not efficiently utilized by young turkey poults until lipase activity reached maximum levels at around 42 to 56 days of age after hatch. Noy and Sklan (1995) observed that duodenal secretions of amylase, trypsin and lipase in young broiler chickens increased by 20, 50 and 100 fold, respectively, at 21 days after hatch. Also, they found that body weight and feed intake increased more rapidly after 10 days hatch.

The beneficial effect of lipase enzyme supplementation on poultry diets, especially in the starter phases (0–3 wk), to enhance fat digestibility has been studied. Polin et al. (1980) reported that the supplementation of 0.04% cholic acid and/or 0.10% lipase improved the absorption of tallow by 8 and 4% in young leghorn chicks of one and three weeks of age, respectively. Al-Marzooqi and Leeson (1999) found that the supplementation of two commercially available lipase enzymes sources (0.714%) to corn-soy broiler starter diets containing 4 and 8% of animal–vegetable blend fat had a significant increase in diet-metabolized energy and apparent fat digestibility when enzymes were used. However, both enzymes caused lower feed intake and lower body weight gain. Reduced performance was suspected to be caused by association with contaminants such as cholecystokinin hormone. Brenes et al. (2008) evaluated the performance of broiler chickens fed with diets containing 250 g/kg or 150 g/kg high oleic acid sunflower seeds and supplemented with 1 g/kg or 0.5 g/kg lipase, 1 g/kg or 0.5 g/kg phospholipase, and a combination of these obtained from a selected strain of *Aspergillus oryzae*. Enzyme addition to the sunflower seeds at 1 g/kg increased weight gain, feed consumption, relative pancreas and liver weights, fat digestibility, amylase and lipase activities, plasma uric acid, calcium, serum lactate dehydrogenase and creatine phosphokinase, and total protein concentration. Feed conversion, relative spleen weight, relative duodenum, jejunum, ileum and ceca lengths, plasma cholesterol and glucose were reduced when compared with the unsupplemented enzyme diet.

Beneficial nutritional effects of other exogenous enzymes have been studied as well. Friesen et al. (1992) reported that broiler diets containing up to 70% wheat, barley, oats or rye supplemented with cellulase (0.04 g/kg of diet) increased the value of nitrogen-corrected metabolized energy 4, 7, 42, 33 and 14%, respectively. Also, weight gain and feed conversion was improved with the enzyme supplementation in growing broiler chickens. Lee et al. (2003) studied the effect of three levels of β-mannanase (0, 1×, and 4×; 1× = 1.09×10^5 units/kg) on broiler chickens fed with guar germ and hull fractions. Enzyme addition was found to significantly reduce intestinal viscosity, increase body weight and reduce feed-conversion ratio in diets containing guar hull fractions. The use of enzymes is not limited to viscous grains. Jackson et al. (2004) reported that an inclusion of 80 MU Hemicell®/ton ($1 MU = 10^6$ enzyme activity units β-mannanase) significantly improved weight gain and feed efficiency in broilers fed with a corn-soy diet.

The use of exogenous enzymes as feed additives in poultry feeds is a good strategy to increase nutrient value of feedstuffs and reduce the negative effects of anti-nutritional compounds that ultimately allow a reduction of nutrients excreted to the environment. However, their use must be carefully studied since the target enzyme substrates found in cell wall structures and starch/protein complexes are considerably heterogeneous (Bedford, 2000).

Commercial exogenous enzymes products are available in different forms such as powder, granule and liquid. The different methodologies to supplement exogenous enzymes in poultry diets were reviewed by Acamovic (2001), Anjum and Chaudhry (2010) and Amerah et al. (2011) and are as follows:

- Enzyme powders are added to pre-mixes before their mixing with other ingredients as coarse mixtures.
- Enzyme powders or granules are mixed directly with a coarse feed.
- Liquid enzymes are either mixed with other ingredients or directly in the drinking water.
- Liquid enzymes could also be added after the pelleting process to avoid problems associated with enzyme denaturation due to pelleting at high temperatures. However, the enzyme is coated to the surface of the pellet and not by direct contact with the internal components of the pellet which may reduce the activity of the enzyme, reducing its beneficial effect.

- Mixing exogenous enzymes with acidifiers, probiotics, growth promoters or symbiotic amino acids is feasible.
- Combination of enzymes with a glucose solution has also been suggested.

Due to the considerable variety of commercial enzyme products in the market, it can be difficult to decide which enzyme and in what form would have the best impact for our precision diets. Moreover, some commercial products contain not only one enzyme but a mixture of enzyme activities (Ferket, 1993), making it more difficult for the consumer to decide on one or the other product. However, an ideal feed-enzyme supplement should have a high specific activity per unit of protein, good thermostability during feed processing, be highly active in the pH range of the animal gut, be resistant to gastric proteases and have a long shelf life (Selle and Ravindran, 2007). These characteristics could be used as a guideline when deciding on an enzyme product.

7 Use of phytase to reduce phosphorus pollution

Phytate is the plant storage form of P in seeds and plant tissues (Ravindran, 1997), and it is known to have a considerable anti-nutritive effect for most animals (Bedford, 2000), due to a potent binding activity on Fe, Zn, Ca, Mg, Mn, Cu, Se and Ni (Harland and Oberleas, 1997; Gibson et al., 2010), proteins and/or starch (Kies, 1997). Phytate is not a good source of P either. Young chickens have limited ability to use phytate P because they lack the phytase enzyme to hydrolyze the phytate and release the P (Ravindran et al., 1995); this leads nutritionists to formulate diets with inorganic P dietary supplements to meet the bird's requirement. To ensure an adequate performance and to avoid P deficiency, feed formulations usually contain total phosphates in amounts that exceed the animal requirements. Therefore, the excess P is excreted into the environment (Bedford, 2000; Bougouin et al., 2014), polluting water and soil resources (Heckrath et al., 1995).

If the P contained in the phytate molecule, which is around 28.2% (Ravindran, 1997), is released, it can serve as a bioavailable nutrient for growing animals and fish (Ketola and Harland, 1993). Phytase is a widely used phosphatase exogenous enzyme additive supplemented in poultry diets to hydrolyze the phytate molecule and minimize mineral binding (Harland and Oberleas, 1997; Kies, 1997). Consequently, increases in P bioavailability subsequently reduce inorganic P supplementation and P excretion into the environment (Harland and Oberleas, 1997; Bougouin et al., 2014).

Several phytase sources have been described in the literature, such as plant, fungi, microbial and yeast (Harland and Oberleas, 1997). The study of phytase effects on performance and nutrient availability has been a topic of great interest for a number of years now, since plan-derived ingredients are used to formulate poultry diets (Ravindran, 1997). Moreover, it is a viable strategy to reduce the excretion of excess P into the environment and is actually required in poultry feeds in some states such as Delaware. Waldroup et al. (2000) reported that the addition of 800 units/kg of a commercially available source of phytase obtained from the non-pathogenic strains of *Aspergillus niger* in large-scale fermentation processes, to highly available P corn (HAPC; 0.27% total P and 0.17% nonphytate P) or yellow dent corn (YDC; 0.23% total P and 0.03% nonphytate P), liberated approximately 50% of the phytate-bound P in diets containing the HAPC or the YDC and supported body weight gain, feed conversion and livability in three-week-old broiler chickens. Additionally, P excretion was successfully reduced by 28% and 46% in faeces from broilers fed with HAPC+phytase and YDC+phytase, respectively, when compared to the control group which contained 0.45% nonphytate P at the NRC (1994) recommendation. Pillai et al. (2006) conducted a series of experiments to compare the efficacy of microbial *Escherichia coli* (*E. coli*) phytase with two commercially available fungal phytases. Treatment diets were supplemented with 250, 500, 1000, 2000 or 4000 phytase units/kg of *E. coli* phytase or one of the two commercial fungal phytases in adequate or P-deficient corn-soy broiler diets. Their results indicate that *E. coli* phytase supplemented to P-deficient diets improves growth, bone and carcass performance and is more effective at releasing phytate-bound P than the other phytase products that were tested. Powell et al. (2008) detected a significant reduction in total P (22%), soluble P (11.7%) and reactive soluble P (14.3%) in the litter of broilers fed with corn-soy diets supplemented with 600 phytase units/kg from a commercially available phytase source obtained from the non-pathogenic strains of *Aspergillus niger* in large-scale fermentation processes. Walk et al. (2012) conducted an experiment to determine the influence of two levels of dietary Ca and three levels of phytase on broiler performance, bone ash, gastrointestinal pH and the apparent ileal digestibility (AID) of Ca, P and amino acids. Corn-soy diets were supplemented with 0, 500 or 5000 FTU/kg of a commercially available *E. coli*-derived phytase source. Feed-conversion ratio was improved as dietary phytase increased. Tibia ash per cent was reduced from 41.4 to 40.0% as dietary Ca decreased but increased with phytase addition. Phytase at 5000 FTU/kg increased pH in the gizzard, duodenum, jejunum and ileum. Phytase supplementation improved AID of CP in broilers fed with 1.0% Ca. Their results suggested that phytase supplementation was beneficial in maintaining broiler performance, bone ash and amino acid digestibility in low Ca diets.

8 Prebiotics and probiotics as feed additives

Other commonly used additives in modern poultry industry feed formulations are probiotics and prebiotics. Probiotics are specific cultures of a single bacteria strain or of a mixture of bacteria stains that are fed to animals to improve health (Griggs and Jacob, 2005) by improving its intestinal balance (Fuller, 1989). On the other hand, prebiotics are non-digestible carbohydrates (Griggs and Jacob, 2005) that beneficially affect the host by selectively stimulating the growth and activity of one or a limited number of bacteria in the colon and, thus, improving host health (Gibson and Roberfroid, 1995). Probiotics and prebiotics are thought to decrease disturbances in the intestinal microbiota by competing for substrates, producing toxic compounds that inhibit pathogens and competing for attachment sites (Ghadban, 2002; Patterson and Burkholder, 2003). Moreover, prebiotics are thought to function as immunostimulants, providing protection against pathogens by upregulatng phagocytosis, increasing bacterial killing and intensifying oxidative bursts (Ganguly, 2013).

Several microbial species have been used as probiotics, such as *Bacillus, Bifidobacterium, Enterococcus, E. coli, Lactobacillus, Lactococcus, Streptococcus, Saccharomyces*, a variety of yeast species and undefined mixed cultures (Simon et al., 2001; Patterson and Burkholder, 2003). Commonly used prebiotics include fructooligosaccharide products (FOS, oligofructose, inulin), trans-galactooligosaccharides, glucooligosaccharides, glycooligosaccharides, lactulose, lactitol, maltooligosaccharides, xylo-oligosaccharides, stachyose, raffinose, sucrose and mannan oligosaccharides (MOS) (Patterson et al., 1997; Patterson and Burkholder, 2003; Jiang et al., 2006; Hajati and Rezaei, 2010; Hashim et al., 2013). MOS is usually obtained from *Saccharomyces* spp. yeast outer cell walls (Griggs and Jacob, 2005; Ganguly, 2013) and is thought to bind and remove pathogens from the intestinal tract and stimulation of the immune system instead of selectively enriching the gut microbiota (Spring et al., 2000). Much of the early work with so-called MOS was actually yeast cell wall (YCW) and not pure preparations of mannan oligosaccharides. YCW is obviously more heterogeneous and contains significant quantities of beta-glucans, various glycoproteins and other components of the cell wall structure.

Ideally, probiotics should be of host origin, non-pathogenic, resistant to gastric acid and bile, be able to adhere to epithelium or mucus, persist in the intestinal tract, modulate immune responses and alter microbial activities (Simmering and Blaut, 2001). On the other hand, ideal prebiotics should be neither hydrolyzed nor absorbed in the upper part of the gastrointestinal tract, but should serve as a substrate for one or a limited number of non-pathogenic bacteria and stimulate bacterial growth (Sinovec and Markovic, 2005) to beneficially alter the intestinal microbiota and their activities (Simmering and Blaut, 2001).

Scientists have reported the beneficial effect of probiotics and prebiotics on poultry performance and immunity for the last several decades. Kim et al. (1988) studied the effect of supplementation of commercial probiotics on broiler diets containing moldy corn. Supplementation of probiotics increased cumulative feed intake and weight gain at two weeks of age. After six weeks supplementation of probiotics increased weight gain by 5.5% in moldy corn diet and 3.3% in the positive control corn diet. A decrease in coliform bacteria in the lower small intestine was also detected. Bailey et al. (1991) studied the effect of fructooligosaccharide (FOS) on the ability of *Salmonella typhimurium* to grow and colonize the gut of broiler chickens. They found that FOS fed at 0.75% resulted in a 12% decrease in positive birds colonized with *Salmonella* when compared with the non-fed FOS control birds. When chicks received a partially protective competitive exclusion (CE) culture, a 42% decrease in positive *Salmonella* chickens was detected when they were supplemented with 0.75% FOS after being challenged with 10^9 *Salmonella* cells at seven years of age. When chickens were subjected to stress by feed and water deprivation on day 13 and then challenged with 10^9 *Salmonella* cells on day 14, 92% of the chickens fed with the control diet were colonized compared with 25% chickens fed with the 0.75% FOS diet.

Previous research done in our laboratory has shown the beneficial effect of pre and probiotics in poultry feeds to improve performance and stimulate the immunological system. Hashim et al. (2103) investigated the influence of YCW supplementation at two concentrations (250 and 500 ppm) on early laying hen performance (21–36 wk). They reported that YCW supplementation at 250 ppm resulted in significantly higher egg weight from 21 to 28-week-old and higher than 500 ppm from 29 to 36 weeks. The average cumulative liquid egg yield over four production periods for 250 ppm YCW was 5.47 kg/bird versus 5.29 and 5.13 kg/bird for the control and 500 ppm YCW, respectively. Additionally, the supplementation at 500 ppm YCW improved specific gravity, egg shell thickness, egg shell weight and per cent shell weight when compared with 250 ppm YCW and control groups. Fowler et al. (2015a) evaluated the effect of supplemental YCW, live yeast culture (LY) or their combination in broiler chickens subjected to weekly feed ingredient variation. After the chickens were fed a corn-soy diet or the variable-ingredient diets for six weeks with or without prebiotic supplementation, they concluded that the supplementation of YCW and LY had a consistent ameliorative effect in both groups containing the prebiotic, especially during the starter and grower phase. In a different study, Fowler et al. (2015b) evaluated the growth-promoting effects of prebiotic YCW products in broiler chickens under an immune stress and *Clostridium perfringens* challenge. A significant improvement in growth rate compared to control birds fed with no YCW was detected. The results show that prebiotics may be considered as alternative growth promoters for broilers in their first three weeks of age.

9 Antibiotics and coccidiostat growth promoters in poultry feeds

Due to consumer perception about the use of antibiotics and coccidiostats in poultry production, 'antibiotic free' products have started to become very popular in global markets. Antibiotic-free products can be defined as 'no use of antibiotics (including ionophore anticoccidials) (Cervantes, 2015) and non-ionophore anticoccidials (NCC, 2014) at the farm'. It is believed that the presence of antibiotics in poultry and other animal feeds has contributed to the increasing ability of pathogen resistance to human antibiotics (Barton, 2000). Additionally, the negative impact of antibiotics on the environment has also been addressed (Kümmerer, 2003). Although the negative environmental effects of antibiotics depend on their photostability, binding and adsorption to soil solids, biodegradation and water-solubility properties (Kumar et al., 2005), antibiotics released into the environment can form chelate structures that bind metals, some of them even at low concentrations are toxic and affect plant growth. They can alter the natural biota of the soil, causing potential terrestrial microbial resistance, and are able to be transported to groundwater systems when excess antibiotics are excreted (Nygaard et al., 1992; Kumar et al., 2005).

The use of medicated products as feed additives in animal production in the United States is regulated by the Association of American Feed Control Officials (AAFCO) in accordance with the Food and Drug Administration (FDA) which states the rules of operation and approved drugs for specific animal species to ensure their correct usage (Table 2). Moreover, governmental agencies such as the USDA routinely test tissues and poultry products to ensure that drug residues are below the maximum tolerance limit (USDA, 2012). Specific withdrawal periods for each drug are clearly specified to ensure that residues are below the maximum tolerance limit or are not present at all in meat and eggs (Cervantes, 2015) to be safe for human consumption.

In 2006, the European Union banned the use of antimicrobials to promote growth in animals after finding evidence of increased bacterial resistance to antibiotics (Cogliani et al., 2011). Some years later, in 2013, the FDA issued the final industry guidance implementing voluntary plans to phase out the use of antibiotics in the United States as growth promoters in livestock production (Sneeringer, 2015). Additionally, the poultry industry supports the FDA's proposed Veterinary Feed Directive (VFD) that ensures that all antibiotics for food-producing animals are administered only under the supervision and prescription of licensed veterinarians (NCC, 2014). New laws are currently being developed in the United States as a consequence of the increasing concern regarding the effects of inappropriate use of antibiotics in animal production. In fact, California Governor Jerry Brown signed in October 2015 Senate Bill No. 27, the strictest law regulating the use of antibiotics in livestock. This law prohibits the administration of antimicrobial drugs to livestock unless it is under the supervision and prescription of licensed veterinarians and prohibits the use of antibiotics for the purpose of promoting weight gain or improving feed efficiency. The new law takes effect on January, 2018 (Amelinckx, 2015).

Table 2 Commonly authorized antibiotics and coccidiostats used in poultry production

Type	Drug	Indications for use*
Coccidiostat	Amprolium	As an aid in prevention and treatment of coccidicois
Coccidiostat	Decoquinate	For the prevention of coccidiosis
Coccidiostat	Lasalocid	For the prevention of coccidiosis
Coccidiostat	Monensin	As an aid in the prevention of coccidiosis
Coccidiostat	Narasin	For the prevention of coccidiosis
Coccidiostat	Narasin/Nicarbazin	For the prevention of coccidiosis
Growth Promoter	Bacitracin Methylene Disalicylate	For increased rate of weight gain and improved feed efficiency
Growth Promoter	Bacitracin Zinc	For increased rate of weight gain and improved feed efficiency
Growth Promoter	Bambermycins	For increased rate of weight gain and improved feed efficiency
Growth Promoter	Tylosin	For increased rate of weight gain and improved feed efficiency
Growth Promoter	Lincomycin	For increased rate of weight gain and improved feed efficiency
Growth Promoter	Virginiamycin	For increased rate of weight gain and improved feed efficiency

*Specific doses for prevention or treatment apply.

Source: Feed Additive Compendium, 2014.

Usually one or a combination of drugs is supplemented in poultry diets to prevent or treat specific diseases that affect poultry during their growth. In order to avoid overfeeding growth promoters and consequently increasing the risk of negative environmental impacts, specific dose levels are calculated on the basis of active ingredient concentration as follows:

We will use a commercial coccidiostat product Monteban 45™ as an example.

Monteban 45™
Package: 50-lb. bag; Concentration: Each pound contains 45 g Narasin;

Narasin
Chemical Name: Narasin **Use level 54-72 g/ton for Chicken (broilers)**
For the **prevention** of coccidiosis caused by *Eimeria tenella, Eimeria acervulina, Eimeria brunetti, Eimeria mivati* and *Eimeria maxima.*

Min
Monteban 45™
453.59237 g Source → 45 g Narasin
X → 54 g

X = 544.311 g of Monteban 45™ or 1.2 lb per ton

Max
Monteban 45™
453.59237 g Source → 45 g Narasin
X → 72 g

X = 725.748 g of Monteban 45™ or 1.6 lb per ton

where:
ton = 2000 lbs
1 lb = 453.59 g

(*Feed Additive Compendium, 2014*)

The demand of 'antibiotic free' products from consumers is increasing. Industries will face multiple challenges in live production, management, disease control and animal welfare in their effort to adopt antibiotic-free systems (Cervantes, 2005). If growth promoters are to be used in poultry formulations, critical control measures are required to ensure the delivery of safe and healthy animal products to consumers (Barton, 2000).

10 Summary and future trends

The efforts to reduce the negative environmental impact of poultry and livestock have increased considerably over the last years. Surface run-off and leaching of excess nutrients like N and P into the groundwater systems associated with the high concentrations in manure has received special attention as major pollutants of soil and water. Moreover, intensive poultry and livestock production facilities have been associated with the emission of greenhouse gases like CO_2, N_2O, and CH_4 that concentrate in the atmosphere and affect ecosystems, biodiversity, and potentially the lives of billions of people worldwide by perhaps increasing the temperature of the earth's atmosphere.

Several nutritional strategies to reduce the environmental impact of poultry farms have been implemented to improve the efficiency of nutrient utilization and decrease the excretion of excess nutrients into the environment. The use of highly digestible ingredients and proper grinding, mixing and pelleting processes results in a uniform and nutritionally complete diet that meets the animal requirements. Formulating diets with low dietary CP and supplementing with highly digestible synthetic amino acids are a feasible strategy to reduce the excretion of excess N. Additionally, formulating diets on the basis of amino acid digestible values by using the concept of ideal protein results is a powerful tool to precisely meet the animal amino acid requirement and consequently reduce the N excretion as well. Multiphase feeding programmes used to meet nutritional requirements at specific points of the lifecycle of poultry reduce the risk of overfeeding nutrients that will be excreted into the environment. The addition of exogenous enzymes such as xylanase, β-glucanase, α-amylase, phytase or β-mannanase and/or probiotics, and prebiotics have been shown to increase nutrient bioavailability, reduce nutrient variation of ingredients, enable the use of potentially less expensive ingredients like barley, wheat or oats, reduce the excretion of excess N and P, decrease disturbances in the intestinal morphology and microbiota, and stimulate the immune system against pathogens.

Growth promoters have shown to improve poultry performance by providing additional aid in preventing or treating diseases. However, their use has drastically decreased over the last years due to consumer perception about the use and effects of antibiotics in animal production. In 2013, the FDA issued the final industry guidance which increased the regulation of antimicrobial drugs used in livestock production and the implementation of voluntary plans to phase out the use of antibiotics as growth promoters. Moreover, stringent laws are currently being developed to prevent the inappropriate use of antibiotics. As an example, the State of California approved in October 2015 Senate Bill No. 27, which banned the use of antibiotics for the purpose of promoting weight gain or improving feed efficiency. The new law takes effect beginning January 2018.

The potential to reduce the adverse environmental impact of intensive poultry production through precise feed formulation and manufacturing is extensive, as new additives and supplements are developed to increase nutrient retention and decrease nutrient excretion, enhance performance and prevent diseases. The poultry industry must implement strategies for the sustainable use and preservation of natural resources to ensure the supply of animal protein for future generations.

With the development of new technologies and the increasing demand for antibiotic-free, free range and organic products, the study of new feed additives such as exogenous enzymes, probiotics and prebiotics has received considerable attention in the last few years. The potential of feed additives in poultry production as a means to increase nutrient bioavailability and to substitute antibiotics as growth promoters has been widely studied. However, the mechanism of action of these new additives is still not well understood as the response of poultry to these products is highly variable and is dependent on the substrate, additive source, concentration and chemical relationship with other feed components. Therefore, further research is necessary to fully understand the beneficial effect of feed additives in intensive poultry production.

On the other hand, the waste generated from livestock and poultry facilities has become a major environmental concern due to the generation of dust, nauseous odours and greenhouse gases, excretion of excess nutrients such as N and P, and deterioration of ecosystems. The development of sustainable agriculture strategies is required to preserve our natural resources for future generations. Specialists need to increase research done in alternative sustainable poultry production systems and precise feed formulation systems to reduce the environmental impact of poultry farms and to achieve a sustainable production of poultry meat and further processed products.

11 Where to look for further information

Poultry production information and statistics

Anon. About the U.S. Egg Industry. American Egg Board (AEB). Park Ridge, IL. http://www.aeb.org/farmers-and-marketers/industry-overview (Accessed 12 October 2015).

Anon. U.S. chicken industry history. National Chicken Council (NCC). Washington, DC. http://www.nationalchickencouncil.org/about-the-industry/history/ (Accessed 12 October 2015).

Anon. 2015. Statistics. National Chicken Council (NCC). Washington, DC. http://www.nationalchickencouncil.org/about-the-industry/statistics/u-s-broiler-performance/ (Accessed 12 October 2015).

Anon. 2013. Statistics. National Turkey Federation (NTF). Washington, DC. http://www.eatturkey.com/why-turkey/stats (Accessed 24 Octber 2015).

Important scientific journals

Poultry Science Journal online access: http://ps.oxfordjournals.org/
Journal of Applied Poultry Research online access: http://japr.oxfordjournals.org/
The Journal of World´s Poultry Research online access: http://jwpr.science-line.com/

Environmental impact of poultry farms and nutritional strategies to reduce the environmental impact

Flanders, F., and Gillespie, J. (2015). *Modern Livestock and Poultry Production*, 9th edn. Boston: Cengage Learning.
Kebreab, E., ed. (2013). *Sustainable Animal Agriculture*. Boston, MA: CABI.

12 References

Acamovic, T. (2001). Commercial application of enzyme technology for poultry production. *World's Poultry Science Journal*, 57(3): 225–42.
Amerah, A. M., Ravindran, V., Lentle, R. G., and Thomas, D. G. (2007). Feed particle size: Implications on the digestion and performance of poultry. *World's Poultry Science Journal*, 63(3): 439–55.
Amelinckx A. (2015). California passes the country´s strongest regulations for antibiotic use in livestock. *Modern Farmer News*. http://modernfarmer.com/2015/10/california-antibiotic-livestock-regulations/ (Accessed 4 March 2016).
Amerah, A. M., Gilbert, C., Simmins, P. H., and Ravindran, V. (2011). Influence of feed processing on the efficacy of exogenous enzymes in broiler diets. *World's Poultry Science Journal*, 67(1): 29–46.
Anjum, M. S., and Chaudhry, A. S. (2010). Using enzymes and organic acids in broiler diets. *The Journal of Poultry Science*, 47(2): 97–105.
Anon. (2012). United States Department of Agriculture, Food Safety and Inspection Service, Office of Public Health Science. National residue program for meat, poultry and egg products, 2010 residue sample results.
Anon. Feed Additive Compendium. (2014). Published by The Miller Publishing Company, Minnetonka, Minn., and The Animal Health Institute NW, Washington, DC.
Anon. U.S. chicken industry history. National Chicken Council (NCC). Washington, DC. http://www.nationalchickencouncil.org/about-the-industry/history/ (Accessed 12 October 2015).
Anon. (2014). Questions and answers about antibiotics in chicken production. National Chicken Council (NCC). Washington, DC. http://www.nationalchickencouncil.org/questions-answers-antibiotics-chicken-production/ (Accessed 4 March 2016).
Anon. (2015). Statistics. National Chicken Council (NCC). Washington, DC. http://www.nationalchickencouncil.org/about-the-industry/statistics/u-s-broiler-performance/ (Accessed 12 October 2015).
Aletor, V. A., Hamid, I. I., Niess, E. and Pfeffer, E. (2000). Low protein amino acid supplemented diets in broiler chickens: effects on performance, carcass characteristics, whole body composition and efficiencies of nutrient utilisation. *The Journal of the Science of Food and Agriculture*, 80(5): 547–54.
Al-Marzooqi, W. and Leeson, S. (1999). Evaluation of dietary supplements of lipase, detergent, and crude porcine pancreas on fat utilization by young broiler chicks. *Poultry Science*, 78(11): 1561–6.
Applegate, T. J. and Angel, R. (2014). Nutrient requirements of poultry publication: history and need for an update. *The Journal of Applied Poultry Research*, 23: 567–75.

Axe, D. E. (1995). Factors affecting uniformity of a mix. *Animal Feed Science and Technology*, 53(2): 211–20.

Bailey, J. S., Blankenship, L. C. and Cox, N. A. (1991). Effect of fructooligosaccharide on *Salmonella* colonization of the chicken intestine. *Poultry Science*, 70(12): 2433–8.

Baker, D. H. and Chung, T. K. (1992). *Ideal Protein for Swine and Poultry*, 1–17. St. Louis, MO: Biokyowa Publishing Co.

Baker, D. H. and Han, Y. (1994). Ideal amino acid profile for chicks during the first three weeks posthatching. *Poultry Science*, 73(9): 1441–7.

Baker, D. H. (1997). Ideal amino acid profiles for swine and poultry and their application in feed formulation. In Biokyowa Technical Review No. 9, 1– 24. St. Louis, MO: Biokyowa Press.

Barker, J. C. (1996). Swine and poultry manure production and characteristics. In *Phytase in Animal Nutrition and Waste Management*, edited by M. B. Coelho and E. T. Kornegay, 1–15. Mt Olive, NJ: BASF Corp.

Barton, M. D. (2000). Antibiotic use in animal feed and its impact on human health. *Nutrition Research Reviews*, 13(02): 279–99.

Bedford, M. R. and Schulze, H. (1998). Exogenous enzymes for pigs and poultry. *Nutrition Research Reviews*, 11(01): 91–114.

Bedford, M. R. (2000). Exogenous enzymes in monogastric nutrition – their current value and future benefits. *Animal Feed Science and Technology*, 86(1): 1–13.

Behnke, K. C. (1994). Factors affecting pellet quality. In Maryland Nutrition Conference. Department of Poultry Science and Animal Science, College of Agriculture, University of Maryland, College Park.

Behnke, K. C. and Beyer, R. S. (2004). Effect of feed processing on broiler performance. PhD Dissertation, Kansas State University, Manhattan.

Blair, R., Jacob, J. P., Ibrahim, S. and Wang, P. (1999). A quantitative assessment of reduced protein diets and supplements to improve nitrogen utilization. *The Journal of Applied Poultry Research*, 8(1): 25–47.

Boling, S. D. and Firman, J. D. (1998). Digestible lysine requirement of female turkeys during the starter period. *Poultry Science*, 77(4): 547–51.

Bougouin, A., Appuhamy, J. A. D. R. N., Kebreab, E., Dijkstra, J., Kwakkel, R. P. and France, J. (2014). Effects of phytase supplementation on phosphorus retention in broilers and layers: a meta-analysis. *Poultry Science*, 93(8): 1981–92.

Brenes, A., Centeno, C., Viveros, A. and Arija, I. (2008). Effect of enzyme addition on the nutritive value of high oleic acid sunflower seeds in chicken diets. *Poultry Science*, 87(11): 2300–10.

Brickett, K. E., Dahiya, J. P., Classen, H. L. and Gomis, S. (2007). Influence of dietary nutrient density, feed form, and lighting on growth and meat yield of broiler chickens. *Poultry Science*, 86(10): 2172–81.

Briggs, J. L., Maier, D. E., Watkins, B. A. and Behnke, K. C. (1999). Effect of ingredients and processing parameters on pellet quality. *Poultry Science*, 78(10): 1464–71.

Burnett, W. E. (1969). Air pollution from animal wastes. Determination of malodors by gas chromatographic and organoleptic techniques. *Environmental Science & Technology*, 3(8): 744–9.

Cervantes, H. M. (2015). Antibiotic-free poultry production: is it sustainable? *The Journal of Applied Poultry Research*, 24(1): 91–7.

Chastain, J. P. and Camberato, J. J. (2004). Dairy manure production and nutrient content. Confined Animal Manure Manager Certification Program Manual Dairy Version, 1–16.

Chastain, J. P., Camberato, J. J., Albrecht, J. E. and Adams, J. (1999). Swine manure production and nutrient content. In *South Carolina Confined Animal Manure Managers Certification Program*, edited by J. R. Gentry and J. E. Albrecht, 1–17. Clemson, SC: Clemson University.

Chislock, M. F., Doster, E., Zitomer, R. A. and Wilson, A. E. (2013). Eutrophication: causes, consequences, and controls in aquatic ecosystems. *Nature Education Knowledge*, 4(4): 10.

Cogliani, C., Goossens, H. and Greko, C. (2011). Restricting antimicrobial use in food animals: lessons from Europe. *Microbe*, 6(6): 274–9.

Collins, T., Gerday, C. and Feller, G. (2005). Xylanases, xylanase families and extremophilic xylanases. *FEMS Microbiology Reviews*, 29(1): 3–23.

Cooley, L. R. and Classen H. L. (2003). Effects of diet essential amino acid formulation and delivery on manure nutrient content in broiler chickens. *Poultry Science*, 82 (Suppl. 1): 13 (Abstr.)
Deaton, J. W., Lott, B. D. and Simmons, J. D. (1989). Hammer mill versus roller mill grinding of corn for commercial egg layers. *Poultry Science*, 68(10): 1342–4.
Denmead, O. T., Chen, D., Griffith, D. W. T., Loh, Z. M., Bai, M. and Naylor, T. (2008). Emissions of the indirect greenhouse gases NH3 and NOx from Australian beef cattle feedlots. *Animal Production Science*, 48(2): 213–18.
Douglas, J. H., Sullivan, T. W., Bond, P. L., Struwe, F. J., Baier, J. G. and Robeson, L. G. (1990). Influence of grinding, rolling, and pelleting on the nutritional value of grain sorghums and yellow corn for broilers. *Poultry Science*, 69(12): 2150–6.
Dozier, W. A., Corzo, A., Kidd, M. T. and Schilling, M. W. (2008). Dietary digestible lysine requirements of male and female broilers from forty-nine to sixty-three days of age. *Poultry Science*, 87(7): 1385–91.
Dozier, W. A., Corzo, A., Kidd, M. T., Tillman, P. B. and Branton, S. L. (2009). Digestible lysine requirements of male and female broilers from fourteen to twenty-eight days of age. *Poultry Science*, 88(8): 1676–82.
Edens, F. W. (2003). An alternative for antibiotic se in poultry: probiotics. *Revista Brasileira de Ciencia Avicola*, 5(2): 75–97.
Emmans, G. C., (1981). A model of the growth and feed intake of ad libitum fed animals, particularly poultry. In *Computers in Animal Production* (Occasional Publication No. 5, British Society of Animal Production), 103–10. Edinburgh, Scotland.
Emmans, G. C. (1987). Growth, body composition and feed intake. *World's Poultry Science Journal*, 43: 208–27.
Emmans, G. C. and Fisher, C. (1986). Problems in nutritional theory. In *Nutrient Requirements of Poultry and Nutritional Research*, Br. Poultry Science Symposium, edited by C. Fisher and K. N. Boorman, 9–39. London: Butterworths.
Emmert, J. L. and Baker, D. H. (1997). Use of the ideal protein concept for precision formulation of amino acid levels in broiler diets. *The Journal of Applied Poultry Research*, 6(4): 462–70.
Falk, D. (1985). Pelleting cost center. In *Feed Manufacturing Technology*, edited by M. M. McElihiney, 3rd ed, 167–90. Arlington, VA: American Feed Industry Association.
Ferket, P. R. (1993). Practical use of feed enzymes for turkeys and broilers. *The Journal of Applied Poultry Research*, 2(1): 75–81.
Ferket, P. R., Van Heugten, E., Van Kempen, T. A. T. G. and Angel, R. (2002). Nutritional strategies to reduce environmental emissions from nonruminants. *Journal of Animal Science*, 80: E168–82.
Firman, J. D. and Boling, S. D. (1998). Lysine: ideal protein in turkeys. *Poultry Science*, 77(1): 105–10.
Food and Agriculture Organization of the United Nations (FAO). (2006). World agriculture: towards 2030/2050 interim report. Rome.
Fowler, J., Alsadwi, A. M., Hashim, M., Abdaljaleel, R. A. and Bailey, C. A. (2014). Improving broiler performance in reduced-energy diets with exogenous enzymes. *Poultry Science*, 93 (E-Suppl. 1): 127.
Fowler, J., Hashim, M., Haq, A. and Bailey, C. A. (2015a). Yeast cell wall and live yeast products and their combination in broiler diets formulated with weekly ingredient variations. *Journal of Animal Physiology and Animal Nutrition*, 99: 932–7.
Fowler, J., Kakani, R., Haq, A., Byrd, J. A. and Bailey, C. A. (2015b). Growth promoting effects of prebiotic yeast cell wall products in starter broilers under an immune stress and *Clostridium perfringens* challenge. *The Journal of Applied Poultry Research*, 24(1): 66–72.
Friesen, O. D., Guenter, W., Marquardt, R. R. and Rotter, B. A. (1992). The effect of enzyme supplementation on the apparent metabolizable energy and nutrient digestibilities of wheat, barley, oats, and rye for the young broiler chick. *Poultry Science*, 71(10): 1710–21.
Fuller, R. (1989). Probiotics in man and animals. *Journal of Applied Bacteriology*, 66: 365–78.
Ganguly, S. (2013). Supplementation of prebiotics, probiotics and acids on immunity in poultry feed: a brief review. *World's Poultry Science Journal*, 69(03): 639–48.

Garcia-Maraver, A. and Carpio, M. (2015). *Factors Affecting Pellet Quality. Biomass Pelletization: Standards and Production*. Southampton, UK: WIT Press, 21p.

Garcia, A. R., Batal, A. B. and Baker, D. H. (2006). Variations in the digestible lysine requirement of broiler chickens due to sex, performance parameters, rearing environment, and processing yield characteristics. *Poultry Science*, 85(3): 498–504.

Gerber, P., Opio, C. and Steinfeld, H. (2007). Poultry production and the environment- a review. Food and Agriculture Organization of the United Nations. http://www.fao.org/ag/againfo/home/events/bangkok2007/docs/part2/2_2.pdf (Accessed 24 October 2015).

Ghadban, G. S. (2002). Probiotics in broiler production-a review. *Archiv fur Geflugelkunde*, 66(2): 49–58.

Gibson, G. R. and Roberfroid, M. B. (1995). Dietary modulation of the human colonic microbiota: introducing the concept of prebiotics. *Journal of Nutrition*, 125: 1401–12.

Gibson, R. S., Bailey, K. B., Gibbs, M. and Ferguson, E. L. (2010). A review of phytate, iron, zinc, and calcium concentrations in plant-based complementary foods used in low-income countries and implications for bioavailability. *Food and Nutrition Bulletin*, 31(Suppl. 2): 134–46.

Gompertz, B. (1825). On the nature of the function expressive of the law of human mortality, and on a new mode of determining the value of life contingencies. *Philosophical Transactions of the Royal Society of London*, 115: 513–83.

Goodband, R. D., Tokach, M. D. and Nelssen, J. L. (2002). *The Effects of Diet particle Size on Animal Performance* (MF-2050 Feed Manufacturing). Manhattan: Department of Grain Science and Industry and Kansas State University, 6p.

Gous, R. M. (2001). Optimizing the feeding of commercial broilers. In Proceedings do II Simpósio Internacional ACAV-Embrapa sobre Nutrição de Aves. Concórdia, SC. Brazil.

Gous, R. M. and Fisher, C. (2001). Optimising commercial broiler feeding programmes. In Proceedings of 13th European Symposium on Poultry Nutrition and Ascites Workshop, 246–7. Blankenberge, Belgium.

Griggs, J. P. and Jacob, J. P. (2005). Alternatives to antibiotics for organic poultry production. *The Journal of Applied Poultry Research*, 14(4): 750–6.

Gross, P., Westrick, M. L. and McNerney, J. M. (1959). The pulmonary response to certain chronic irritants: an experimental study on rats and guinea pigs. *Archives of Pathology*, 68: 252–61.

Gupta, G. and Charles, S. (1999). Trace elements in soils fertilized with poultry litter. *Poultry Science*, 78(12): 1695–8.

Gutierrez, O., Surbakti, N., Haq, A., Carey, J. B. and Bailey, C. A. (2008). Effect of continuous multiphase feeding schedules on nitrogen excretion and broiler performance. *The Journal of Applied Poultry Research*, 17(4): 463–70.

Hajati, H. and Rezaei, M. (2010). The application of prebiotics in poultry production. *International Journal of Poultry Science*, 9(3): 298–304.

Han, Y. and Baker, D. H. (1991). Lysine requirements of fast-and slow-growing broiler chicks. *Poultry Science*, 70(10): 2108–14.

Harry, E. G. (1978). Air pollution in farm buildings and methods of control: a review. *Avian Pathology*, 7(4): 441–54.

Hashim, M., Fowler, J., Haq, A. and Bailey, C. A. (2013). Effects of yeast cell wall on early production laying hen performance. *The Journal of Applied Poultry Research*, 22(4): 792–7.

Havenstein, G. B., Ferket, P. R. and Qureshi, M. A. (2003). Growth, livability, and feed conversion of 1957 versus 2001 broilers when fed representative 1957 and 2001 broiler diets. *Poultry Science*, 82(10): 1500–8.

Heckrath, G., Brookes, P. C., Poulton, P. R. and Goulding, K. W. T. (1995). Phosphorus leaching from soils containing different phosphorus concentrations in the Broadbalk experiment. *Journal of Environmental Quality*, 24(5): 904–10.

Hooda, P. S., Edwards, A. C., Anderson, H. A. and Miller, A. (2000). A review of water quality concerns in livestock farming areas. *Science of the Total Environment*, 250(1): 143–67.

Hurwitz, S., Sklan, D. and Bartov, I. (1978). New formal approaches to the determination of energy and amino acid requirements of chicks. *Poultry Science*, 57: 197–205.

Hussar, N. and Robblee, A. R. (1962). Effects of pelleting on the utilization of feed by the growing chicken. *Poultry Science*, 41(5):1489–93.
Jacobs, C. M., Utterback, P. L. and Parsons, C. M. (2010). Effects of corn particle size on growth performance and nutrient utilization in young chicks. *Poultry Science*, 89(3): 539–44.
Jackson, M. E., Geronian, K., Knox, A., McNab, J. and McCartney, E. (2004). A dose-response study with the feed enzyme beta-mannanase in broilers provided with corn-soybean meal based diets in the absence of antibiotic growth promoters. *Poultry Science*, 83(12): 1992–6.
Jensen, L. S. and Falen, L. (1973). Effect of pelleting on the 'extra caloric' effect of dietary fat for developing turkeys. *Poultry Science*, 52(6): 2342–4.
Jia, W. and Slominski, B. A. (2010). Means to improve the nutritive value of flaxseed for broiler chickens: the effect of particle size, enzyme addition, and feed pelleting. *Poultry Science*, 89(2): 261–9.
Jiang, H. Q., Gong, L. M., Ma, Y. X., He, Y. H., Li, D. F. and Zhai, H. X. (2006). Effect of stachyose supplementation on growth performance, nutrient digestibility and caecal fermentation characteristics in broilers. *British Poultry Science*, 47(4): 516–22.
Johnston, S. L. and Southern, L. L. (2000). The effect of varying mix uniformity (simulated) of phytase on growth performance, mineral retention, and bone mineralization in chicks. *Poultry Science*, 79(10): 1485–90.
Jongbloed, A. W. and Lenis, N. P. (1992). Alteration of nutrition as a means to reduce environmental pollution by pigs. *Livestock Production Science*, 31(1): 75–94.
Ketola, H. G. and Harland, B. F. (1993). Influence of phosphorus in rainbow trout diets on phosphorus discharges in effluent water. *Transactions of the American Fisheries Society*, 122(6): 1120–6.
Kies, K. (1997). *Phytase-Mode of action. Phytase in Animal Nutrition and Waste Management: A BASF Reference*, 295–12. Mount Olive, NJ: BASF Corp.
Kilburn, J. and Edwards Jr., H. M. (2001). The response of broilers to the feeding of mash or pelleted diets containing maize of varying particle sizes. *British Poultry Science*, 42(4): 484–92.
Kilburn, J. and Edwards, H. M. (2004). The effect of particle size of commercial soybean meal on performance and nutrient utilization of broiler chicks. *Poultry Science*, 83(3): 428–32.
Kim, C. J., Namkung, H., An, M. S. and Paik, I. K. (1988). Supplementation of probiotics to broiler diets containing moldy corn. *Korean Journal of Animal Sciences* (Korea R.), 30: 542–8.
Klopfenstein, T. J., Angel, R., Cromwell, G., Erickson, G. E., Fox, D. G., Parsons, C. and Meyer, D. (2002). Animal diet modification to decrease the potential for nitrogen and phosphorus pollution. *Faculty Papers and Publications in Animal Science*, 518.
Koch, K. (1996). Hammermills and rollermills. Feed Manufacturing (No. MF-2048) Department of Grain Science and Industry Kansas State University, Manhattan, Agricultural Experiment Station and Cooperative Extension Service, 8p.
Krogdahl, A. and Sell, J. L. (1989). Influence of age on lipase, amylase, and protease activities in pancreatic tissue and intestinal contents of young turkeys. *Poultry Science*, 68(11): 1561–8.
Kumar, K., Gupta, S. C., Chander, Y. and Singh, A. K. (2005). Antibiotic use in agriculture and its impact on the terrestrial environment. *Advances in Agronomy*, 87: 1–54.
Kümmerer, K. (2003). Significance of antibiotics in the environment. *Journal of Antimicrobial Chemotherapy*, 52(1): 5–7.
Lee, G. F., Rast, W. and Jones, R. A. (1978). Water report: eutrophication of water bodies: insights for an age old problem. *Environmental Science and Technology*, 12(8): 900–8.
Lee, J. T., Bailey, C. A. and Cartwright, A. L. (2003). Beta-mannanase ameliorates viscosity-associated depression of growth in broiler chickens fed guar germ and hull fractions. *Poultry Science*, 82(12): 1925–31.
Leeson, S. and Summers, J. D. (2005). *Commercial Poultry Nutrition*, 3rd ed. Guelph: University Books.
Mallin, M. A. and Cahoon, L. B. (2003). Industrialized animal production – a major source of nutrient and microbial pollution to aquatic ecosystems. *Population and Environment*, 24(5): 369–85.
Martin, S. (1985) Comparison of hammer mill and roller mill grinding and the effect of grain particle size on mixing and pelleting. Master's Thesis, Kansas State University, Kansas.

Matthews, W. A. and Sumner, D. A. (2014). Effects of housing system on the costs of commercial egg production1. *Poultry Science*, 94: 552–7.

McCoy, R. A., Behnke, K. C., Hancock, J. D. and McEllhiney, R. R. (1994). Effect of mixing uniformity on broiler chick performance. *Poultry Science*, 73(3): 443–51.

Mejia, L., McDaniel, C. D. and Corzo, A. (2012). Dietary influence of digestible lysine concentration on Cobb 500 hen broiler breeder reproductive performance. *Poultry Science*, 91(2): 426–31.

Mitchell, R. J., Waldroup, P. W., Hillard, C. M. and Hazen, K. R. (1972). Effects of pelleting and particle size on utilization of roasted soybeans by broilers. *Poultry Science*, 51(2): 506–10.

Muir, W. M. and Aggrey, S. E., eds. (2003). *Poultry Genetics, Breeding, and Biotechnology*. Cambridge, MA, USA: CABI Publishing.

Mukhtar, S., Ullman, J. L., Carey, J. B. and Lacey, R. E. (2004). A review of literature concerning odors, ammonia, and dust from broiler production facilities: 3. Land application, processing, and storage of broiler litter. *The Journal of Applied Poultry Research*, 13(3): 514–20.

Munyaka, P. M., Nandha, N. K., Kiarie, E., Nyachoti, C. M. and Khafipour, E. (2015). Impact of combined β-glucanase and xylanase enzymes on growth performance, nutrients utilization and gut microbiota in broiler chickens fed corn or wheat-based diets. *Poultry Science*, 95: 528–40.

Nahm, K. H. (2002). Efficient feed nutrient utilization to reduce pollutants in poultry and swine manure. *Critical Reviews in Environmental Science and Technology*, 32(1): 1–16.

Nahm, K. H. (2007). Feed formulations to reduce N excretion and ammonia emission from poultry manure. *Bioresource Technology*, 98(12): 2282–300.

National Research Council. (NRC) (1994). *Nutrient Requirements of Poultry*, 9th ed. Washington, DC, USA: National Academy Press.

Nir, I., Melcion, J. P. and Picard, M. (1990). Effect of particle size of sorghum grains on feed intake and performance of young broilers. *Poultry Science*, 69(12): 2177–84.

Nir, I. (1996). The effects of food particle size and hard – ness on performance: nutritional behavioral and metabolic aspects. In Proceedings of the XX World's Poultry Conference Vol. 2, 173–83.

Noy, Y. and Sklan, D. (1995). Digestion and absorption in the young chick. *Poultry Science*, 74(2): 366–73.

Nygaard, K., Lunestad, B. T., Hektoen, H., Berge, J. A. and Hormazabal, V. (1992). Resistance to oxytetracycline, oxolinic acid and furazolidone in bacteria from marine sediments. *Aquaculture*, 104(1): 31–6.

Oviedo-Rondón, E. O. and Waldroup, P. W. (2002). Models to estimate amino acid requirements for broiler chickens: a review. *International Journal of Poultry Science*, 1(5), 106–13.

Pan, C. F., Igbasan, F. A., Guenter, W. and Marquardt, R. R. (1998). The effects of enzyme and inorganic phosphorus supplements in wheat-and rye-based diets on laying hen performance, energy, and phosphorus availability. *Poultry Science*, 77(1): 83–9.

Parsons, A. S., Buchanan, N. P., Blemings, K. P., Wilson, M. E. and Moritz, J. S. (2006). Effect of corn particle size and pellet texture on broiler performance in the growing phase. *The Journal of Applied Poultry Research*, 15(2): 245–55.

Patterson, J. A. and Burkholder, K. M. (2003). Application of prebiotics and probiotics in poultry production. *Poultry science*, 82(4): 627–31.

Peron, A., Bastianelli, D., Oury, F. X., Gomez, J. and Carre, B. (2005). Effects of food deprivation and particle size of ground wheat on digestibility of food components in broilers fed on a pelleted diet. *British Poultry Science*, 46(2): 223–30.

Pesti, G. M., Vedenov, D., Cason, J. A. and Billard, L. (2009). A comparison of methods to estimate nutritional requirements from experimental data. *British Poultry Science*, 50(1): 16–32.

Pillai, P. B., O'Connor-Dennie, T., Owens, C. M. and Emmert, J. L. (2006). Efficacy of an Escherichia coli phytase in broilers fed adequate or reduced phosphorus diets and its effect on carcass characteristics. *Poultry Science*, 85(10): 1737–45.

Pluske, J. R. and Lindemann, M. D. (1998). Maximizing the response in pig and poultry diets containing vegetable proteins by enzyme supplementation. In *Biotechnology in the Feed Industry*, edited by T. P. Lyons and K. A. Jacques, 375–9. Proceedings of Alltech's 14th Annual Symposium, Nottingham University Press, Nottingham, UK.

Polin, D., Wing, T. L., Ki, P. and Pell, K. E. (1980). The effect of bile acids and lipase on absorption of tallow in young chicks. *Poultry Science*, 59(12): 2738–43.

Pomar, C., Hauschild, L., Zhang, G. H., Pomar, J. and Lovatto, P. A. (2010). Precision feeding can significantly reduce feeding cost and nutrient excretion in growing animals. In *Modelling Nutrient Digestion and Utilisation in Farm Animals*, edited by D. Sauvant, J. van Milgen, P. Faverdin and N. Friggens, 327–34. Wageningen: Wageningen Academic Publishers.

Pope, T. and Emmert, J. L. (2001). Phase-feeding supports maximum growth performance of broiler chicks from forty-three to seventy-one days of age. *Poultry Science*, 80(3): 345–52.

Pope, T., Loupe, L. N., Townsend, J. A. and Emmert, J. L. (2002). Growth performance of broilers using a phase-feeding approach with diets switched every other day from forty-two to sixty-three days of age. *Poultry Science*, 81(4): 466–71.

Pope, T., Loupe, L. N., Pillai, P. B. and Emmert, J. L. (2004). Growth performance and nitrogen excretion of broilers using a phase-feeding approach from twenty-one to sixty-three days of age. *Poultry Science*, 83(4): 676–82.

Portella, F. J., Caston, L. J. and Leeson, S. (1988). Apparent feed particle size preference by broilers. *Canadian Journal of Animal Science*, 68(3): 923–30.

Powell, S., Johnston, S., Gaston, L. and Southern, L. L. (2008). The effect of dietary phosphorus level and phytase supplementation on growth performance, bone-breaking strength, and litter phosphorus concentration in broilers. *Poultry Science*, 87(5): 949–57.

Proudfoot, F. G. and Hulan, H. W. (1982). Feed texture effects on the performance of turkey broilers. *Poultry Science*, 61(2): 327–30.

Ravindran, V., Bryden, W. L. and Kornegay, E. T. (1995). Phytates: occurrence, bioavailability and implications in poultry nutrition. *Poultry and Avian Biology Reviews* (United Kingdom), 6: 125–43.

Ravindran, V. (1997). Occurrence of phytic acid in plant feed ingredients. In *Phytase in Animal Nutrition and Waste Management: A BASF Reference*, edited by M. B. Coelho and E. T. Kornegay, 85–92. Mount Olive, NJ: BASF Corp.

Reece, F. N., Lott, B. D. and Deaton, J. W. (1985). The effects of feed form, grinding method, energy level, and gender on broiler performance in a moderate (21 °C) environment. *Poultry Science*, 64(10): 1834–9.

Reimer, L. (1992). Conditioning. In Proceedings of Northern Crops Institute Feed Mill Management and Feed Manufacturing Technol. Short Course, p 7. California Pellet Mill Co. Crawfordsville, IN.

Richardson, W. and E. J. Day, 1976. Effect of varying levels of added fat in broiler diets on pellet quality. *Feedstuffs*, 48(20): 24.

Robbins, K. R., Saxton, A. M. and Southern, L. L. (2006). Estimation of nutrient requirements using broken-line regression analysis. *Journal of Animal Science*, 84 (13_suppl.): E155–65.

Saunders-Blades, J. L., MacIsaac, J. L., Korver, D. R. and Anderson, D. M. (2009). The effect of calcium source and particle size on the production performance and bone quality of laying hens. *Poultry Science*, 88(2): 338–53.

Schiffman, H. R. (1968). Texture preference in the domestic chick. *Journal of Comparative and Physiological Psychology*, 66(2): 540.

Sell, J. L., Krogdahl, A. and Hanyu, N. (1986). Influence of age on utilization of supplemental fats by young turkeys. *Poultry Science*, 65(3): 546–54.

Selle, P. H. and Ravindran, V. (2007). Microbial phytase in poultry nutrition. *Animal Feed Science and Technology*, 135(1): 1–41.

Shallom, D. and Shoham, Y. (2003). Microbial hemicellulases. *Current Opinion in Microbiology*, 6(3): 219–28.

Sharpley, A. (1999). Symposium: reducing the environmental impact of poultry production: focus on phosphorus. *Poultry Science*, 78: 660–73.

Simmering, R. and Blaut, M. (2001). Pro-and prebiotics–the tasty guardian angels? *Applied Microbiology and Biotechnology*, 55(1): 19–28.

Sims, J. T. and Wolf, D. C. (1994). Poultry waste management: agricultural and environmental issues. *Advances in Agronomy*, 52(1): 2–72.

Simon, O., Jadamus, A. and Vahjen, W. (2001). Probiotic feed additives–effectiveness and expected modes of action. *Journal of Animal and Feed Sciences*, 10 (1 suppl.): 51–67.

Sinovec, Z. and Markovic, R. (2005). Use of pre-biotics in poultry nutrition. *Biotechnology in Animal Husbandry*, 21(5–6): 235–9.

Sneeringer, S. (2015). Restrictions on antibiotic use for production purposes in U.S. livestock industries likely to have small effects on prices and quantities. United States Department of Agriculture (USDA). Economic Research Service. http://www.ers.usda.gov/amber-waves/2015-november/restrictions-on-antibiotic-use-for-production-purposes-in-us-livestock-industries-likely-to-have-small-effects-on-prices-and-quantities.aspx#.VtxUq_nhDIU (Accessed 6 March 2016).

Spring, P., Wenk, C., Dawson, K. A. and Newman, K. E. (2000). Effect of mannan oligosaccharide on different cecal parameters and on cecal concentration on enteric bacteria in challenged broiler chicks. *Poultry Science*, 79: 205–11.

Steinfeld, H., Gerber, P., Wassenaar, T., Castel, V., Rosales, M. and De Haan, C. (2006). *Livestock's Long Shadow: Environmental Issues and Options*. Rome, Italy: FAO.

Svihus, B., Klovstad, K. H., Perez, V., Zimonja, O., Sahlström, S., Schüller, R. B. and Prestlokken, E. (2004). Physical and nutritional effects of pelleting of broiler chicken diets made from wheat ground to different coarsenesses by the use of roller mill and hammer mill. *Animal Feed Science and Technology*, 117(3): 281–93.

Svihus, B. (2006). The role of feed processing on gastrointestinal function and health in poultry. In *Avian Gut Function in Health and Disease*, edited by G. C. Perry, 183–94. Poultry Science Symposium Series, 28. Wallingford, UK: CABI.

Theander, O., Westerlund, E. and Aman, P. (1993). Structure and components of dietary fiber. *Cereal Foods World* (USA), 38: 135–41.

Turner, A. M. and Chislock, M. F. (2010). Blinded by the stink: nutrient enrichment impairs the perception of predation risk by freshwater snails. *Ecological Applications*, 20(8): 2089–95.

USDA-NASS[1]. (2015). Poultry-production value 2014 summary. United States Department of Agriculture-National Agriculture Statistics Service. Washington, DC. http://www.nass.usda.gov/ (Accessed 12 October 2015).

Waldroup, P. W., Kersey, J. H., Saleh, E. A., Fritts, C. A., Yan, F., Stilborn, H. L. and Raboy, V. (2000). Nonphytate phosphorus requirement and phosphorus excretion of broiler chicks fed diets composed of normal or high available phosphate corn with and without microbial phytase. *Poultry Science*, 79(10): 1451–9.

Walk, C. L., Bedford, M. R. and McElroy, A. P. (2012). Influence of limestone and phytase on broiler performance, gastrointestinal pH, and apparent ileal nutrient digestibility. *Poultry Science*, 91(6): 1371–8.

Wang, Z. R., Qiao, S. Y., Lu, W. Q. and Li, D. F. (2005). Effects of enzyme supplementation on performance, nutrient digestibility, gastrointestinal morphology, and volatile fatty acid profiles in the hindgut of broilers fed wheat-based diets. *Poultry Science*, 84(6): 875–81.

Warren, W. A. and Emmert, J. L. (2000). Efficacy of phase-feeding in supporting growth performance of broiler chicks during the starter and finisher phases. *Poultry Science*, 79(5): 764–70.

Wicker, D. L. and Poole, D. R. (1991). How is your mixer performing? *Feed Management*, 42(9): 40.

Wijtten, P. J. A., Prak, R., Lemme, A. and Langhout, D. J. (2004). Effect of different dietary ideal protein concentrations on broiler performance. *British Poultry Science*, 45(4): 504–11.

Winsor, C. P. (1932). The Gompertz curve as a growth curve. *Proceedings of the National Academy of Sciences of the United States of America*, 18(1): 1.

Williams, C. M., Barker, J. C. and Sims, J. T. (1999). Management and utilization of poultry wastes. In *Reviews of Environmental Contamination and Toxicology*, edited by G. W. Ware, 105–57. New York, NY: Springer.

Williams, C. M. (2013). Poultry waste management in developing countries. The role of poultry in human nutrition, Food and Agriculture Organization of the United Nations, 46.

Zeeman, G. (1994). Methane production/emission in storages for animal manure. *Fertilizer Research*, 37(3): 207–11.

Zhu, H. L., Hu, L. L., Hou, Y. Q., Zhang, J. and Ding, B. Y. (2014). The effects of enzyme supplementation on performance and digestive parameters of broilers fed corn-soybean diets. *Poultry Science*, 93: 1704–12.

Zhu, J. (2000). A review of microbiology in swine manure odor control. *Agriculture, Ecosystems and Environment*, 78(2): 93–106.

Chapter 19

Energy and water use in poultry processing

D. Luján-Rhenals, University of Arkansas Fayetteville, USA and Universidad de Córdoba, Colombia; R. Morawicki, University of Arkansas Fayetteville, USA; E. J. Van Loo, Ghent University, Belgium; and S. C. Ricke, University of Arkansas Fayetteville, USA

1 Introduction

As the demands for food production increased in the world, it is imperative to balance this with continued improvement of food safety and food security on a global scale, especially in third-world countries. The implementation of these safety and food security systems may require the use of even more energy and resources. In animal production this can be somewhat complicated, with some aspects more straightforward than others. For example, for ruminants grazing is the major activity of land use; hence livestock producers can have a direct impact on reducing the negative effects on the environment (Costa, 2007). However, in some animal industries such as poultry production, which have become more vertically integrated, efforts in one part of the production chain such as changing the size of the bird to be marketed can impact other elements of the production chain including processing in a more complex manner. This dichotomy is difficult to balance, but the first step is to establish a base line of the environmental impacts to initiate the application of effective strategies to minimize those impacts.

Studies over the past few decades have concentrated on technological developments in the food-processing plant to address consumer market demands and still meet food safety requirements (Ricke, 2010). For example, there has been considerable public health emphasis on minimizing *Salmonella* contamination during poultry processing (Cox

http://dx.doi.org/10.19103/AS.2016.0010.22

et al., 2011; Handley et al., 2015; Jarvis et al., 2016). Hence, the number of cleaning cycles in food production has continued to expand with a higher consumption of energy and hot water due to the tremendous demand for increased levels of hygiene (Dalsgaard and Abbots, 2003). Traditional thermal treatments for food conservation and food safety applications at industrial levels, such as sterilization, pasteurization, drying and cooking, contribute to the energy expenditure with the resulting increase in the food industry footprint (Pereira and Vicente et al., 2010; Jarvis et al., 2016). This will only increase as food production output increases. For instance, worldwide meat production is expected to double from 229 million tons in 1999/2001 to 465 million tons in 2050, with the majority of this increase in China, Brazil and India (FAO, 2007). Likewise, international markets and global trade for commercial egg production will continue to grow and transition as consumer demands change (Windhorst, 2009). Furthermore, the demand for meat and other animal products will continue to increase more than other food products (FAO, 2007).

The other element that will add some variability to these estimations is the growth in the non-conventional food industries, including meat products generated organically, as well as those being marketed as natural and thus raised without antibiotics (Ricke, 2010; Ricke et al., 2012; Seideman et al., 2010). In addition to the food safety challenges related to administration of interventions that would be considered acceptable (O'Bryan et al., 2008; Sirsat et al., 2009; Van Loo et al., 2012a), there are also major nutritional challenges to produce organic crops for feed (Chalova et al., 2012, 2016). While less is known regarding non-conventional and organic processing, it would be anticipated that these would add to the energy and water costs and in some cases such as mobile poultry processing units impact local environments as these are moved from one location to the next (Fanatico, 2010; O'Bryan et al., 2012a,b, 2014). The aim of this chapter is to review and analyse what is known about the energy and water consumption patterns of conventional and non-conventional poultry processing and discuss where limitations in data exist.

2 The potential for life cycle assessment (LCA) of poultry processing systems

While individual improvements can be suggested, a more comprehensive approach will eventually be needed to efficiently transform the entire poultry processing system. It will be necessary to direct future research on identifying the specific effects of the poultry slaughter industry on the environment. Under these circumstances, application of effective methodologies to evaluate the environmental impacts of any industrial activity will need to be comprehensive to account for all inputs and outputs. The most common comprehensive methodologies that have been applied are life cycle assessment (LCA) and carbon footprint analysis. Hence, research must be focused on establishing a detailed LCA and footprint of conventional and non-conventional poultry processing in order to retrieve more precise and effective information on each particular operation applied during the entire poultry slaughter process. According to Finkbeiner et al. (2006), LCA, as it is described in ISO 14040, includes the whole life cycle of a product, from the origin through energy and material production and manufacturing, to use and final disposal. Boggia et al. (2010) describes LCA as a method based on a 'cradle-to-grave' approach to evaluating the environmental impact of products, activities and services. Even though the majority of the LCA studies carried out so far involve livestock production (Pelletier, 2008; de Vries and de Boer, 2010; Castellini et al., 2012), several LCA studies also have included industrial processing in foods (Calderon et al., 2010; Hospido et al., 2010; Lundie and Peters, 2005; Eide, 2002; Andersson et al., 1998; Pelletier, 2008; Castellini et al., 2012).

So far, there are limited studies related to the LCA and footprint of poultry processing, particularly, organic and small processing unit systems. In addition, most studies to date have focused more on the production side with only some inclusion of meat processing elements and only a more recent emphasis on issues such as water use (Wiedemann et al., 2012; Skunca et al., 2015). Therefore, in the future, it will be critical to conduct more in-depth studies that have an integrated description and analysis of the potential effects that this important sector could have on the environment and will be projected to have in the future. The LCA of food-product supply chains in general is a useful, challenging and an ongoing developing tool for defining the carbon footprint of a food product (Johnson, 2010). For instance, the carbon footprint in gas emissions, overall, is related to the sum of all gas emissions, with any impact on climate change linked with any human activity (Dunkley, 2011). Furthermore, in order to participate in the reduction and optimal control of pollutants from this industry, the environment, health and safety (EHS) guidelines in poultry processing projects will need to include water management strategies and energy consumption patterns (International Finance Corporation and World Bank Group, 2007). Such information is crucial to managing and developing new strategies that will provide better analysis and control of the processes in this sector of the food industry as well as its environment impacts. Once collected, this information can be used to develop models for predicting impacts of incorporation of new technologies in commercial poultry processing systems. This helps to prioritize investment in infrastructure, which could have the most impact and avoids extraneous implementation of changes that may not improve water and energy efficiencies sufficiently to justify the cost.

3 The current status of poultry production and processing

Poultry production and processing is certainly expected to continue to be a major part of the future growth of the food industry. According to the USDA (2011), the total number of chickens slaughtered during 2010 was 8.8 billion chickens, 0.24 billion turkeys and 23 000 million ducks. This corresponds, as total live weight, to 50 billion pounds (22.8 billion kg) of chickens, 7.1 billion pounds (3.2 billion kg) of turkeys and 0.16 billion pounds (0.73 billion kg) of ducks. The state of Georgia produced, for the same period, approximately 1.2 billion broilers, nearly 14% of the national total, making it the largest producing state in the United States. Today's modern poultry processing plants make it possible to slaughter more than 200000 birds per day, leaving few older, more basic processing plants in operation (Kiepper, 2011). Alabama and Arkansas were the other two states with the highest number of poultry slaughtered in 2010 with 1.08 billion broilers and 0.97 billion broilers, respectively (USDA, 2011b).

Poultry meat consumption has been increasing rapidly during the past two decades and for the next century its per capita consumption will continue increasing (Uijttenboogaart,

1999; Sams and Alvarado, 2010). In the long term, poultry consumption is expected to continue growing at 2 to 3% per year (Harlan, 2007). The primary reasons for this trend are the marketing of health aspects associated with this product, increasing availability in the market and the relative low price and versatility of poultry meat (Uijttenboogaart, 1999; Greene, 1998). For instance, in Europe, the consumption per year is 20.9 kg/per capita and was expected to increase to 32.9% for total meat consumption in the year 2015 (IDD, 2001).

4 Conventional poultry slaughter processing plants

Poultry processing plants include the integration and implementation of many disciplines of science, including engineering, marketing, microbiology and economics (Sams and Alvarado, 2010). Additionally, these plants not only include the production of human food as the main goal, but also the appropriate management of wastes, non-food uses of poultry and the generation of protein and non-protein ingredients for pet foods or livestock feeds (Morawicki, 2010).

The classic poultry processing plant (slaughtering plant) consists, basically, of five unit operations, namely, slaughtering, scalding and defeathering, eviscerating, washing and chilling and packaging (Jekayinfa, 2001; Owens et al., 2010; Sofos et al., 2013; Handley et al., 2015). Poultry slaughtering plants have markedly changed during the past two decades to improve their processing efficiency. Currently, chickens are typically processed in modern design plants with process capacities of over tens of thousands of birds per day transported from several farms (Papinaho, 1996; Sofos et al., 2013). The details of each of these processes are discussed in the following sections.

4.1 Pre-slaughtering

Before slaughtering, birds are placed in cages or crates and transported via trucks to the processing plant (Sofos et al., 2013). Several considerations must be taking place to avoid any damage in the bird due the environmental temperature, humidity or bruising (Fonkwe et al., 2001). According to Fletcher (1991), there are short-term effects, which affect the quality of poultry meat 24 hours before the slaughtering process. The main short-term effects are harvesting (feed and catching), transportation, plant holding, unloading, shackling, immobilization, stunning and killing.

Once arriving at the loading platform in the processing plant, birds are weighed, removed from the cages or crates and hung upside down on a conveyor rail system where they are immobilized by electrical stunning or exposure to a CO_2 controlled atmosphere (Fonkwe et al., 2001; Sofos et al., 2013). Following USDA humane slaughter regulations, the chickens are electrically stunned prior to killing (Stadelman et al., 1988). Stunning not only immobilizes the birds for efficient killing, but also relaxes the muscles that hold the feathers (Mountney and Parkhurst, 1995). In more recent times, one of the promising alternative technologies is the controlled atmosphere killing (CAK), which uses gas mixtures to render birds unconscious and reduces many of the welfare problems associated with the stunning operation (Shields and Raj, 2010). CAK does not require live bird handling at the slaughter plant and avoids the problem associated with dumping of birds from crates. The birds remain inside the crates and are transported to the gas tunnels where CO_2, inert gases (argon or nitrogen) or a mixture are dispensed sufficiently to kill them instead of stunning (Raj, 1998). The bleed-out operation, as well as carcasses and meat quality, are not affected by CAK (Raj and Gregory, 1991; Raj et al., 1990). There are potentially further refinements that can be made. For example, Vizzier-Thaxton et al. (2010) designed a new humane method of stunning broilers using low atmospheric pressure, which produces a controlled atmosphere with reduced oxygen tension. With this system there are no toxic gases, minimal plant modifications and the ability to view and record the birds throughout the process via digital video, and there is no generation of greenhouse gases, reducing its carbon footprint when compared with other systems.

4.2 Slaughtering

Slaughter is related specifically to the act of killing the animal for food. Wholesomeness and quality of the meat have been the primary factors that are typically considered primary factors during the slaughter processes. Wholesomeness is connected with both the physiological and sociological factors, while quality, conversely, refers basically to aspects such as shelf life, product desirability and market value. Poultry slaughter plants in North America and the European Union (E.U.) have been and continue to be regulated to ensure that the animal's death is under humane treatment and the product is safe (Fletcher, 1999).

After stunning, a conveyor transports the birds to the cutting machine for exsanguinations (Sams and McKee, 2010). The carotid artery and jugular vein of the birds are cut to kill the animal. Birds are bled for no more than 180s (Fonkwe et al., 2001). During the bleed-out period, the bird loses anywhere from 30 to 50% of its body blood in a typical slaughter process (Mountney and Parkhurst, 1995; Sams, 2001; Barbut, 2002; Sams and McKee, 2010). During processing, the blood that drains out of the carcass during bleed-out represents 4% of the yield loss from bird live weight to processed live weight (Dukes et al., 1973; Wilson, 1998; Barbut, 2002; Sams, 2001; Sams and McKee, 2010).

Commercially processed birds are electrically stimulated for 60 s with 200 V AC, 1 s on then 1 s off, during bleeding (Dickens et al., 2002). After this treatment, force to shear values were reduced in fillets deboned immediately after a 2-h chill, which improved the texture, and were comparable to fillets deboned at 22 h post chill. Based on research with this process, Dickens et al. (2002) concluded that the use of electrical stimulation with early fillet deboning would eliminate refrigerated holding areas for carcasses, which can reduce the energy use and required space in the process plant.

4.3 Scalding and defeathering

Scalding is one of the most important processing steps in poultry slaughtering plans and the risk of cross-contamination by *Salmonella* and *Campylobacter* during scalding increases as birds from subsequent flocks are processed during the day (Oosterom et al., 1983; Genigeorgis et al., 1986; Whyte et al., 2004; Handley et al., 2015). The scalder is the first process section where pathogenic bacteria of one bird can be washed free and spread to the surfaces of other birds (Russell, 2008). Due to their attachment in the follicles, feather removal is a difficult task (Sams, 2001). Scalding is done by immersing the birds in hot water controlling time and temperature to facilitate the feather removal (defeathering) (Papinaho, 1996). For fresh birds, typical scalding temperatures are between 50 and 52°C ('soft' or 'semi-scald'), whereas for older birds and frozen meat products, scalding conditions can be between 56 and 63°C ('hard-scald'). Soft scalding does not cause substantial injury to the outer skin layers. Since soft scalding maintains the yellow-pigmented layer of the skin intact, it is the chosen method for producing fresh poultry with yellow skin that is highly desirable in many markets, where very high temperatures can reduce the yield and liquefy the skin fat (Sams, 2001; Sams and McKee, 2010). After the scalding operation, the birds enter into a chamber where rotating rubber fingers located on picking machines release the feathers (Sams and McKee, 2010; Sofos et al., 2013). Following defeathering, birds are washed with pressurized water and transferred to the evisceration section (Fonkwe et al., 2001).

4.4 Evisceration

The evisceration operation removes the viscera by using a series of interconnected mechanical equipment. Currently, conventional modern plants have vacuum systems to extract the viscera faster and to easily facilitate cleanliness and inspection (Barbut, 2002). This step can reduce contamination during processing, which can occur either because of intestinal tract leakage onto the carcass, or through spillage resulting from ruptures of the gastrointestinal tract and viscera during evisceration. Incidence of faecal contamination, assumed to be the primary pathway for pathogenic carcass contamination, can be used as an indicator of process control during slaughter (Russell, 2001; Mead, 2004; Rasekh et al., 2005; Sofos et al., 2013; Handley et al., 2015).

Evisceration is the extraction of edible and inedible viscera of the bird, and can vary depending of the plant and machinery used during this operation (Sofos et al., 2013). Evisceration has three main steps: (1) making an incision through the abdominal wall under the tail of the bird to open the body cavity, (2) extraction of the viscera (gastrointestinal tract and associated organs, reproductive tract, heart and lungs) from the bird and finally (3) separation of the 'giblets' or edible viscera (heart, liver and gizzard) from the extracted viscera (Sams and McKee, 2010). The heart, liver and gizzards are trimmed, washed and saved while the remaining internal organs are discarded with the heads and feet as offal (Fonkwe et al., 2001).

4.5 Washing and chilling

After the birds have passed the inspection phase, they pass through an inside/outside bird washer with spray nozzles directed to the interior and exterior of the carcass to remove additional material adhering to the bird prior to entering the chilling tank. The water in this equipment usually contains chlorine to reduce the contamination for microorganisms (Wei et al., 1985; Sams, 2001; Sams and McKee, 2010). While chlorine has proven quite useful for reducing spoilage and increasing shelf life, there are concerns related to public health hazards due to accumulation of chlorine organic derivatives and corrosion of metals in processing plants (Wei et al., 1985; Waters et al., 2014).

The carcasses are subsequently chilled in the chiller with cold water at a chilling rate that will avoid any damage in the product caused by low temperatures (cold shortening) (Fonkwe et al., 2001). The primary goal of the chilling operation is the reduction of microbial growth to extend the half-life during marketing (James et al., 2006; Grey, 1988; Lillard, 1982; Sams and McKee, 2010). In the United States, processors usually apply water chilling, while in Europe, air chilling is also common (Sams and McKee, 2010; Sofos et al., 2013). The entire process generally includes temperatures of approximately 4°C or less (Allen et al., 2000). Water chilling involves basically two stages, pre-chiller and chiller. In the pre-chiller, the temperature ranges from 7 to 12°C during 10 to 15 minutes with countercurrent water to allow a transition in carcass temperature to avoid a thermal shock. The chiller subsequently ranges from 1 to 4°C during 45 to 110 minutes, with the carcasses absorbing 2–4% of the water in the skin (Sams and McKee, 2010). Air chilling uses less water and the carcasses do not absorb considerable amounts of water. Cold air (−8 to −1°C) is used during 1–3 hours, which is longer than the water chilling process, due to less thermal efficiency associated with the air (Veerkamp, 1989; Sams and McKee, 2010).

4.6 Packaging and storage

The birds are packaged as a whole or cut into pieces before packaging to be stored in freezers or transported under freezing conditions to the final market (Fonkwe et al., 2001). 'Wet shipper' is the oldest packing and distribution system for poultry meat. This method is a wax-coated corrugated box containing ice to preserve the chicken meat. Currently, polymer bags are more commonly used to package whole carcasses. Chicken parts are usually packed into highly permeable polystyrene foam trays with high oxygen-permeable PVC or polymer-based and a stretch-film overwrap. To improve the shelf life of poultry meat vacuum or modified atmosphere, packaging is typically applied (Dawson, 2010).

5 The use of energy in poultry processing plants

The energy demand in a poultry slaughter processing plant depends on several factors such as throughput volume and structure, thermo-physical properties of the raw material, the applied production technology, the degree of mechanization in the production processes, the share of human labour and the rate of utilization in the manufacturing capacity (Marcotte et al., 2008; Corry et al., 2007; Fritzson and Berntsson, 2006; James et al., 2006; Somsen et al., 2004). The per capita energy use for poultry production is 694 thousand BTU in farm and agribusiness activities, corresponding to an increase of 10.4% at an annual rate from 1997 to 2002. In food-processing activities, poultry yielded 585 thousand BTU per capita, corresponding to a 7.9% increase at an annual rate from 1997 to 2002. Furthermore, energy flows for packaging poultry products and beverages categories were the only segments of the supply chain, with increases at an annual rate greater than 6%, whereas most other commodity groups exhibited an increase of 2.5% or even less (USDA, 2010a). The total energy use per ton of carcass in the poultry industry production was reported by Williams et al. (2006) in England and Wales, and it represented 12 Gj/ton carcass, while organic production represented only 0.5% of this data.

Jekayinfa (2001) studied the energy consumption in three different categories of poultry processing plants depending on the mechanization level (least-mechanized, semi-mechanized and high mechanized). The results revealed that the major energy consumption operational steps were scalding and defeathering in all three types of plants, with 44% of the total consumption in the plants. Evisceration consumption was 17.5%; slaughtering, 17%; washing and chilling, 16% and packing, 6%, respectively. However, in contrast to the study carried out by Formentini et al. (2010) and Wojdalski et al. (2009), it was demonstrated by Jekayinfa (2001) that the less mechanized plants consumed the highest amount of energy (50.36 MJ/1000 chickens), while the semi-mechanized consumed

28.04 MJ/1000 chickens and the most mechanized exhibited an energy consumption of 17.83 MJ/1000 chickens.

5.1 Electricity

The International Finance Corporation and World Bank Group (2007) reported that the total electrical energy consumption per manufactured unit in E.U. poultry plants ranged between 0.37 and 0.67 KW-h/chicken. This data can be compared with the global electric power consumption in a small Brazilian poultry slaughter plant, which was estimated as 0.1645 KW-h/1000 chicken units (Formentini et al., 2010). It was demonstrated by Jekayinfa (2001) that, among the mechanized plants, the less mechanized plants consumed the greatest amount of energy (50.36 MJ), while the semi-mechanized consumed 28.04 MJ and the most mechanized had an energy consumption of 17.83 MJ. In contrast, Wojdalski et al. (2009) studied the effectiveness of electrical energy in a small-size poultry processing plant. The average value of electrical energy was 123.6 Wh/1000 poultry units and was much less than the energy expended in a modern plant. This was mainly because there were fewer machines with high electricity consumption in the smaller plant. This particular situation is one of the advantages of organic and small-scale slaughter poultry plants where most of the equipment utilized do not require high-energy consumption rates and in some cases can even be operated manually (O'Bryan et al., 2014).

The U.S. Energy Information Administration, cited by Klingele (2011b), reported that in most industrial facilities light systems comprise approximately 5–10% of global electricity consumption. Furthermore, energy-efficient lighting changes can generate 45% return on investment and can pay for themselves in approximately 2.2 years, according to a 1998 report by the Energy Cost Savings Council cited by Klingele (2011b). This report also demonstrated that processors can make a simple improvement by switching to fluorescent lighting systems or they can install motion sensors into lighting systems, resulting in a savings of up to 50% of energy use. This action will also significantly reduce greenhouse gas emissions. Hence, reducing electrical consumption by optimal management of the lighting system would be one simple and practical way to begin decreasing the footprint of the poultry industry. More recently, another of the best potential options to help save energy in processing plants would be to use solar energy systems.

Another considerable electrical cost in poultry slaughter plants is the stunning process, which is applied prior to slaughtering in order to induce unconsciousness. Overall, the most common method is the electrical water bath stunner setting with constant voltage (Shields and Raj, 2010; Sofos et al., 2013). The electrical parameters of stun baths can vary from plant to plant. Low current application with a high-frequency, pulsed, direct current is the most commonly employed system for most U.S. plants (Craig and Fletcher, 1997; Wilkins et al., 1999). The time of the electrical stun is usually 10–12 s (Sams, 2001) depending of the length of the water bath and the line speed. Usually, there are up to 20 birds at any one time in the water bath (Wilkins et al., 1999). In the United States, 10–28 volts are normally used in the stunners, dispensing 10–45 mA per bird, and the frequency of the current normally is between 350 and 500 Hz (Bilgili, 1999; Gazdziak, 2007). The total current is equivalent to the sum of the current flowing through each particular chicken. This is applied for constant-voltage multiple-bird stunners, which are the most commercial equipment (Sparrey et al., 1993). Two hundred and twenty chickens per minute have been reported as one of the highest processing speeds in a modern slaughter plant (Bilgili, 1999).

5.2 Fossil energy

Fossil fuel uses are the major source of emissions produced by daily human activities (Carlsson-Kanyama and Gonzalez, 2009). Currently, almost every major industrial plant utilizes steam systems as their heat source. Approximately 37% of the fossil fuel burnt in the U.S. industry is used to produce steam and 57% of the fossil fuel consumption is used to produce steam in the food-processing industry (Einstein et al., 2001).

The energy used in the food and beverages sector includes inputs from heating, cooking, drying, cooling, freezing and other unitarian operations. The majority of these energy inputs come from fossil fuel combustion and purchased electricity. Among the overall industrial sectors, the food and beverages sector produces 51 $MMCO_2eq$ of Green House Gas (GHG) emissions represented as CO_2 from the total 106 $MMTCO_2eq$ (EPA, 2008). These values represent CO_2 equivalents over a 100-year time period.

Marculescua and Stana (2011) conducted research focused on combustibility properties quantification for the appropriate poultry waste to energy chain identification. The results of this study highlighted the importance of fuel thermal–physical characteristics and the opportunity of using poultry feathers as renewable fuel for power generation. Wojdalski et al. (2009) studied the effectiveness of thermal energy in a small-size poultry processing plant, where it was estimated that the average value of thermal energy was 485 MJ/1000 poultry units. Fritzson and Berntsson (2006) concluded from studies on non-integrated plants that it would be economically attractive to invest in an increased heat exchanger network or in heat pumps. By adopting these measures, it would be possible to achieve reductions of between 5% and 35% of the total CO_2 emissions in a large non-integrated plant. All this was possible by switching from heavy fuel oil to natural gas, which generates the more cost-effective reduction of CO_2 emissions.

6 The use of water in poultry plants

Water in the poultry-processing industry is crucial not only to maintain and clean process areas but also to enable many other basic operations such as stunning, scalding, chilling, carcass washing and cleaning and disinfection equipment. Historically, before the 1972 U.S. Clean Water Act, the cost of water received or the environmental impact of water discharged from poultry processing plants was a minimum consideration in overall plant management decisions (Merka, 2010). Alvarado (2011) reported that chilling is the main form of water use in the United States and that poultry industry has been increasing the water usage over the past 10–15 years and before, water use was estimated at 22.7–30.3 L/bird, but the current estimate is 37.8–45.4 L/bird due to the food safety and hazard analysis and critical control point (HACCP) regulations.

According to Wojdalski et al. (2009), in a small poultry slaughter plant, the chicken carcass volume influenced the processing plant water consumption per manufactured unit (more than 70%). In addition, the factory electrical energy consumption per manufactured unit was accounted for to a smaller degree (approximately 53% of the slaughtered poultry volume). Overall, there was no significant interdependence between the slaughtered poultry volume and the thermal energy consumption (steam and hot water), which may be the reason why water consumption in smaller slaughter plants is less than in large plants.

6.1 Water uses

Water usage represents one of the most expensive utilities in the food industry, especially for those plants located at long distances from the natural water resources and for those that have to have some type of legal restriction for usage (Klingele, 2011a). Kiepper et al. (2008) reported that the use of water in a commercial broiler processing plant varied widely between 19 and 38 L/2.3 kg of birds processed. Being more specific, Northcutt and Jones (2004) reported an average water use of 20.6 L/bird before the implementation of HACCP in slaughter plants across the United States, while the usage increased to 26.0 L/bird after implementation. Furthermore, Northcutt and Jones (2004) observed that 72% of the plants used city water to process broilers, and 66% reported discharging to city sewers. In other studies, Plumber and Kiepper (2011), as well as Formentini et al. (2010), reported that during 2009, an average of 7 gallons (26 L) of potable water was used per carcass in Georgian and Brazilian poultry slaughter industries, respectively. However, the North Carolina Division of Environment and Natural Resources (2010) reported that typically broiler processing water uses ranged from 13.2 to 37.8 L/bird; for turkeys, 41.6–87 L/bird and for beef slaughtering plants the flow rates were 1325 L/head. Avula et al. (2009) concluded that water use in turkey processing is substantially more efficient when compared to poultry processing, with the average turkey live weight being between 12 and 18 kg and water consumption being approximately between 130 and 150 L/bird (Avula et al., 2009). The International Finance Corporation and World Bank Group (2007) reported that the total water consumption per manufactured unit in E.U. poultry plants ranged between 16.1 and 18.7 L/head.

The North Carolina Division of Environment and Natural Resources (2010) also showed that broiler processing plants require a scalder overflow rate of 0.95 L/bird and a chiller overflow rate of 1.9 L/bird. Eventually, this water is used in the plant for transportation of feathers and offal for discard. Overall, poultry processing accounts for 76% of the water used, approximately 13% is used in clean-up and 12% is used during downtime. Likewise, Huezo et al. (2007) reported that immersion chilling, for example, requires approximately 2.6 L/bird to fill the chill tank at shift start-up and an additional overflow of 1.9 L/bird.

According to Formentini et al. (2010), in a Brazilian poultry slaughter plant, the greatest water-consuming steps are the slaughtering and the cleaning processes (platform trucks waiting, reception, bleeding, scald, evisceration, cooling and cuts room), which represent nearly 71 and 15%, respectively, of the total water consumed. Therefore, evisceration (29%), cooling/chilling (26%), cleaning and washing (15%) and scalding (8%) are the steps in poultry processing with the largest volumes of water consumption. To prevent waste accumulation in equipment and floors, water is used in considerable amounts to wash and clean installations and equipment. At the same time, the evisceration process requires tremendous consumption of water and generation of wastewater containing high concentrations of organic matter, nitrogen and phosphorus. Wojdalski et al. (2009) examined water management in a small-size poultry processing plant, and noted that water use was 7.12 L/poultry, which is approximately 4–5 times lower when compared to large processors.

6.2 Wastewater and reuse

The most significant impact on the poultry processing wastewater stream is due to the operations of bleed-out, scalding and evisceration (Plumber and Kiepper, 2011). The average water consumption reported in the current review is 33.6 L/bird for broiler chicken slaughter plants, which is far less than the water consumption for other food-processing plants such as red meat, swine or even turkey. Nevertheless, it would be even less if the poultry industry developed more initiatives to reduce its water consumption and the global environmental footprint. By making simple daily changes in hygiene practices, wastewater generated by the poultry slaughter industry could be reduced. For example, not reusing the water conveys waste and offal. Water could be replaced by mechanical or air-vacuum equipment-based approaches. As another approach, this water could be reused safely by installing the appropriate filtration systems. By doing that, the flow pressure in the facility could be reduced and it would project less energy consumption. Undoubtedly, the poultry industry needs to face the relevant challenges of high investment costs for water reuse treatment technology or seek less expensive new alternatives. This investment will be recovered in the middle or long term not only economically but environmentally.

Numerous authors have been studying the potential for wastewater reuse in the poultry slaughter industry, specifically the practice of cleaner water production; for instance, Formentini et al. (2010) concluded that the effluent tank desensitization can be reused to wash the cages of birds; also the effluent generated in the cooling tunnel, the storage chambers and cooling towers can be reused to wash the waiting area and unloading platform. Hence, it can reduce approximately 7.5% of the total amount of wastewater generated in a slaughter plant in Brazil. De Sena et al. (2009) reported that the wastewater treatment by dissolved air flotation efficiently reduced the COD, colour, turbidity and total solids contents. Additionally, the advanced oxidation process increased the organic matter removal of the treated samples. Among the more advanced oxidation processes, the photo-Fenton reaction had the best results. Overall, the treatment by dissolved air flotation followed by an advanced oxidation process might be efficient for meat wastewater treatment, whether water reuse is part of the system or not. Avula et al. (2009) pointed out that even though the costs of ultrafiltration are higher, their life-time costs are comparable with conventional treatments. Furthermore, the footprint of ultrafiltration could be 30–50% of conventional systems with less consumption of chemicals. Kist et al. (2009) reported that among the cleaner process actions applied in a Brazilian slaughter plant, there should be focus on the implementation of dry cleaning procedures, control over the amount of water used on a sector-by-sector basis and better collection of blood residue. When incorporating these changes, the plant reached reductions in water consumption up to 13%. In addition, automation of the water system and change in the configuration of the wastewater treatment plant by the use of an anaerobic reactor and a sequential oxic/anoxic system for reduction of the eutrophication influenced the final wastewater product.

Matsumura and Mierzwa (2008) studied the effectiveness of rational water use strategies to identify operational flaws by improvements at operational issues and physical losses identification. Some sectors with high consumption and wastewater generation were identified and the feasibility of water reuse was subsequently evaluated. A reduction of approximately 31% in water consumption was achieved by utilizing these strategies. Amorim et al. (2007) reported that another Brazilian slaughter plant could reduce its hydraulic and organic loads by combining water consumption reduction measurements, recycling water in other processes, controlling excessive contamination of wastewater and substituting the products utilized in its cleaning processes. Reuse of effluents from the desensitization tank, the cooling towers and the defreeze of the cooling tunnel and the storage chambers would achieve up to a 12% reduction in consumed water, 10% of the generated effluent and U.S. $ 6500 per year in the wastewater treatment costs. Other studies with similar measurements and results have been reported (Kupsovic et al., 2007; Sarkar et al., 2006; Manios et al., 2003; Casani et al., 2005; Mohsen and Jaber, 2003).The accumulation of uncollected blood, feathers, eviscerations and cleaning of the live haul area at a poultry slaughter plant corresponds to the poultry processing wastewater (Merka, 2010). In the United States, the total wastewater generated by slaughter plants had been reported to be between 45 and 90 billion gallons (170.34–340.69 L) annually (Kiepper, 2003). The most significant impact on the poultry processing wastewater stream is due to the operations of bleed-out, scalding and evisceration (Plumber and Kiepper, 2011).

In poultry slaughter plants, chiller and scalder water is reused for flushing water to remove offal and feathers (North Carolina Division of Environment and Natural Resources, 2010). Bates (1948) in Plumber and Kiepper (2011) reported a BOD of 100 000 mg/L in the wastewater of a regular poultry slaughter plant, which is close to the 92 000 mg/L value reported by Porges (1950). In addition, Merka (1991) made an extrapolation that 27 billion gal (102.2 billion L) of water is used in the U. S. broiler industry, with 378 million lb (171.82 million kg) of BOD charged annually to process broilers in the United States.

7 Non-conventional poultry processing

At the beginning of the twentieth century, poultry was sold live to consumers and subsequently taken home and processed. Later, during the 1930s, poultry was sold with only the blood and feathers removed and was referred to as 'New York Dressed' (NYD). Barnes and Impey (1975) reported that this type of processing enabled carcasses to remain for longer times under refrigeration conditions, compared to eviscerated carcasses (Mead and Scott, 1997). Fully eviscerated birds average from 7 to 9 days at 4°C, while NYD birds held at the same temperature remained acceptable after 18 days. The carcass when held at low temperatures, dry and intact, limited optimal growth conditions of psychotrophic spoilage bacteria on the surface, which contributed to the longer shelf life (Barnes and Impey, 1975). However, as conventional poultry production became more vertically integrated by increasing the number of birds grown and processed on a much larger scale, the resulting poultry carcasses were further processed to generate much more uniform eviscerated carcasses for retail sales.

More recently, the consumer demand has been changing once again and a non-conventional poultry market is growing in response to more demand for speciality raised poultry (Fanatico, 2010). Today, there are an increasing number of small producers, which raise poultry outdoors on pasture, process the bird on-farm and sell the meat directly to consumers at the farm or at the farmers' market without intermediates (O'Bryan et al., 2012a,b, 2014). Many states allow processing of 1000 birds on a farm per year and selling directly to consumers without inspection (SARE, 2006; Fanatico, 2003; O'Bryan et al., 2012a,b, 2014). Furthermore, the organic poultry market has increased the profitability for these types of producers (USDA, 2006; Ricke et al., 2012). Frequently, the terms 'conventional' and 'non-organic' are used to mean the same; however, non-organic is possibly more descriptive, but 'conventional', implies the aggregation of contemporary non-organic practices (Ricke et al., 2012). Organic is essentially food produced in a manner that complies with organic standards set by national governments as well as international organizations (Pittman et al., 2012; Van Loo et al., 2012b). In the United States, the definition of agricultural products that are to be labelled and marketed as organic require that they comply with the certification and all other applicable requirements to the National Organic Program mandate (Pittman et al., 2012). A livestock and agriculture product is considered organic when, among other requirements, agrochemicals (synthetic pesticides and fertilizers) are avoided during their growing out and processing (Fanatico, 2008; Pittman et al., 2012).

Even though some parts of the organic sector are growing rapidly, currently it still remains a fairly minor fraction of total poultry production. While conventional poultry remains a large proportion of the overall market, organic sources of poultry products continue to increase both domestically and internationally (Sundrum, 2006; Williams et al., 2006; Ricke et al., 2012). The total U.S. certified organic poultry livestock for 2008 was 15.5 million; Nebraska, with 6.5 million, was leading followed by California and Pennsylvania with 2.2 and 1.9 million, respectively (USDA, 2010b). Even though organic production has been applied in the United States for decades, this production system is considerably less developed than conventional poultry processing since the USDA established the National Organic Program in 2002 (Fanatico, 2008; Pittman et al., 2012; Van Loo et al., 2012b). However, because of the increasing markets for organic meats, as well as the development of organic by-products, the demand in food safety regulations for large and even small meat-processing plants will also increase (Van Loo et al., 2012c; O'Bryan et al., 2014). Hence, small plants such as organic processors would be more affected by the implementation of structural changes in terms of food safety, such as the HACCP system, which encourage them to easily exit the industry (Ollinger, 2011).

The organic poultry slaughter processing plants essentially include the same five basic operations also applied in the conventional processing plants, but their differences are mainly in the type of chemical products and additives used in each operation as well as the scale of the operation and the lesser technology associated with the equipment utilized (Fanatico, 2003; O'Bryan et al., 2012a,b, 2014). For instance, conventional processors have numerous antimicrobials available, including chlorine materials, hydrogen peroxide, peracetic acid, phosphoric acid and organic acids, but non-conventional and organic processors have a much more limited range that is considered allowable (Ricke, 2003; Ricke et al., 2005, 2012; Fanatico, 2008; Sirsat et al., 2009; Finstad et al., 2012; O'Bryan et al., 2008, 2014). Other common differences could be in the chill tank water where some organic producers do not use additives and others use ozone in the water or even chilled air (Fanatico, 2008, 2010). The processing must be conducted according to the organic handling standards of the National Organic Program (NOP, Section 205.270–205.272) if meat or eggs are going to be processed on-farm (Baier, 2008).

While the organic and small-scale slaughter poultry plants may be more efficient than the large operations in terms of energy consumption, additional research needs to be conducted on these systems. Further efforts to precisely quantitate energy consumption among these systems will need to take into account the greater variability in operations and infrastructure, compared to the much more uniform conventional processing systems. This is an important advantage for these smaller-scale operations and could be a strong motivating factor to encourage more governmental support for this activity, but market interest will need to be considered before such emphasis can be applied. A critical element in this consideration is the need to maintain optimal hygienic conditions in such processing plants in order to release a safe product for the consumers, as well as further improve their waste management programmes. This will require more research to generate the baseline data of relative effectiveness of current systems.

8 Conclusions and recommendations

Poultry meat has a high overall efficiency of feed conversion in the non-organic sector compared with other meat products such as swine and cattle. The poultry and pig feed conversion efficiency is greater and methane production is less of an issue than ruminants. Nonetheless, it is necessary for poultry processors at the industrial and farm levels to have the knowledge, technology and economic motivation to better control their energy consumption, and manage water use to further reduce the environmental footprint of this important industry.

Electricity cost in a poultry slaughter processing plant could be reduced by switching to other alternative energy sources. For instance, solar energy systems are currently a very sustainable and approachable option with rapid recovery of the initial investment. Likewise, poultry plants might achieve optimal results by switching traditional light systems to other options that use less energy. Furthermore, as the U.S. Department of Energy and the Environmental Protection Agency has recommended, CH_4 generated in processing plants can be recovered to produce heat and electricity, which could significantly reduce the utility costs in a facility. Another economically attractive alternative is switching from heavy fuel oil to natural gas, which generates the most cost-effective reduction of CO_2 emissions. Water consumption in poultry slaughter processing plants has been increasing in the past few years. Currently, the average water consumption is 33.6 L/bird for boiler chicken slaughter plants, which is slightly higher than that reported by most articles (26 L/bird) during the past 10 years.

It is clear that sufficient knowledge, technology and economic motivation to reduce energy consumption and water use is needed in order to decrease the environmental footprint in poultry processing and advance the industry. This can be reached by collecting and effectively applying precise measurements of the corresponding utilities expended during process, especially in those areas with most significant impact such as bleed-out, scalding, defeathering and evisceration, which represent the highest energy and water consumption.

9 Where to look for further information

Boggia, A., L. Paolotti and C. Castellini. 2010. Environmental impact evaluation of conventional, organic and organic-plus poultry production systems using life cycle assessment. *Worlds Poult. Sci. J.* 66(1): 95–114.

Castellini, C., A. Boggia, L. Paolotti, G. J. Thoma and D. Kim. 2012. Chapter 7. Environmental impacts and life cycle analysis of organic meat production and processing. In S. C. Ricke, E. J. Van Loo, M. G. Johnson and C. A. O'Bryan (Eds), *Organic Meat Production and Processing*, pp. 113–36. Wiley Scientific/IFT, New York, NY.

Morawicki, R. O. 2010. Chapter 19. Coproducts and by-products from poultry processing. In C. M. Owens, C. Z. Alvarado and A. R. Sams (Eds), *Poultry Meat Processing*, 2nd ed., pp. 381–96. CRC Press, Boca Raton, FL.

Morawicki, R. O. 2012. *Handbook of Sustainability for the Food Sciences*. Wiley-Blackwell, Hoboken, NJ.

O'Bryan, C. A., P. G. Crandall, M. L. Davis, G. Kostadini, K. E. Gibson, W. Q. Alali, D. Jaroni, S. C. Ricke and J. A. Marcy. 2014. Mobile poultry processing units: A safe and cost-effective poultry processing option for the small-scale farmer in the United States. *World's Poultry Sci. J.* 70: 787–802.

Owens, C. M., C. Z. Alvarado and A. R. Sams (Eds). 2010. *Poultry Meat Processing*. 2nd edition. CRC Press, Boca Raton, FL.

Pelletier, N. 2008. Environmental performance in the US broiler poultry sector: Life cycle energy use and greenhouse gas, ozone depleting, acidifying and eutrophying emissions. *Agric. Syst.* 98(2): 67–73.

Ricke, S. C., E. J. Van Loo, M. G. Johnson and C. A. O'Bryan (Eds). 2012. *Organic Meat Production and Processing*. Wiley Scientific/IFT, New York, NY, p. 444.

Wiedemann, S., E. McGahan and G. Poad. 2012. Using life cycle assessment to quantify the environmental impact of chicken meat production. Australian Government – Rural Industries Research and Development Corporation. RIRDC Publication No. 12/029, p. 73.

10 References

Allen, V., J. Corry, C. Burton, R. Whyte and G. Mead. 2000. Hygiene aspects of modern poultry chilling. *Int. J. Food Microbiol.* 58(1–2): 39–48.

Alvarado, C. 2011. Strategies for maximizing water recycling. http://www.meatingplace.com/MembersOnly/technology/details.aspx?item=19887 (Accessed 22 July 2011).

Amorim, A. K. B., I. R. de Nardi and V. Del Nery. 2007. Water conservation and effluent minimization: Case study of a poultry slaughterhouse. *Resour. Conserv. Recycl.* 51: 93–100.

Andersson, K., T. Ohlsson and P. Olsson.1998. Screening life cycle assessment (LCA) of tomato ketchup: a case study. *J. Clean Prod.* 6: 277–88.

Avula, R. Y., H. M. Nelson and R. K. Singh. 2009. Recycling of poultry process wastewater by ultrafiltration. *Innov. Food Sci. Emerg. Technol.* 10(1): 1–8.

Baier, A. H. 2008. Organic standards for handling (processing). National sustainable agriculture service. https://attra.ncat.org/attra-pub/PDF/nopstandard_handling.pdf (Accessed 25 May 2011).

Barbut, S. 2002. *Poultry Products Processing: An Industry Guide*. CRC Press, Boca Raton, FL, p. 548.

Barnes, E. and C. Impey. 1975. Shelf-life of uneviscerated and eviscerated chicken carcasses stored at 10 degrees C and 4 degrees C. *Br. Poult. Sci.* 16(3): 319–26.

Bates, R. W. 1948. A discussion of packing plant waste disposal. Iowa Sewage Works Digest. In H. S. Plumber and B. H. Kiepper. 2011. Impact of poultry processing by-products onwastewater generation, treatment, and discharges. *Proc. of the 2011 Georgia Water Resour. Conf.* 11–13 April 2011, University of Georgia.

Bilgili, S. F. 1999. Recent advances in electrical stunning. *Poult. Sci.* 78: 282–6.

Boggia, A., L. Paolotti and C. Castellini. 2010. Environmental impact evaluation of conventional, organic and organic-plus poultry production systems using life cycle assessment. *Worlds Poult. Sci. J.* 66(1): 95–114.

Calderon, A. L., L. Iglesias, A. Laca, M. Herrero and M. Diaz. 2010. The utility of life cycle assessment in the ready meal food industry. *Resour. Conserv. Recycl.* 54(12): 1196–207.

Carlsson-Kanyama, A. and A. D. Gonzalez. 2009. Potential contributions of food consumption patterns to climate change. *Am. J. Clin. Nutr.* 89(5): S1704–S1709.

Casani, S., M. Rouhany and S. Knochel. 2005. A discussion paper on challenges and limitations to water reuse and hygiene in the food industry. *Water Res.* 39(6): 1134–46.

Castellini, C., A. Boggia, L. Paolotti, G. J. Thoma and D. Kim. 2012. Chapter 7. Environmental impacts and life cycle analysis of organic meat production and processing. In S. C. Ricke, E. J. Van Loo, M. G. Johnson and C. A. O'Bryan (Eds), *Organic Meat Production and Processing*, pp. 113–36. Wiley Scientific/IFT, New York, NY.

Chalova, V. I., J. H. Kim, P. H. Patterson, S. C. Ricke and W. K. Kim. 2016. Reduction of nitrogen excretion and emission in poultry: A review for organic poultry. *J. Environ. Sci. Health Part B.* 51: 230–5.

Chalova, V. and S. C. Ricke. 2012. Chapter 10. Organic nutrition and feed supplementation. In S. C. Ricke, E. J. Van Loo, M. G. Johnson and C. A. O'Bryan (Eds), *Organic Meat Production and Processing*, pp. 157–75. Wiley Scientific/IFT, New York, NY.

Corry, J. E. L., S. J. James, G. Purnell, C. S. Barbedo-Pinto, Y. Chochois, M. Howell and C. James. 2007. Surface pasteurisation of chicken carcasses using hot water. *J. Food Eng.* 79(3): 913–19.

Costa, N. D. 2007. Reducing the meat and livestock industry's environmental footprint. *Nutr. Diet.* 64: S185–S191.

Cox, N. A., J. A. Cason and L. J. Richardson. 2011. Minimization of *Salmonella* contamination on raw poultry. *Annual Review of Food Science and Technology* 2: 75–95.

Craig, E. W. and D. L. Fletcher. 1997. A comparison of high current and low voltage electrical stunning systems on broiler breast rigor development and meat quality. *Poultr. Sci.* 76: 1178–81.

Dalsgaard, H. and Abbots, A. (2003). Improving energy efficiency. In B. Mattson and U. Sonesson (Eds), *Environmentally-Friendly Food Processing*. CRC, Boca Raton, USA, p. 337.

Dawson, P. L. 2010. Packaging. Chapter 6. In C. M. Owens, C. Z. Alvarado and A. R. Sams (Eds), *Poultry Meat Processing*, 2nd ed., pp. 101–23. CRC Press, Boca Raton, FL.

De Sena R. F., J. L. Tambosi, A. K. Genena, R. de F. P. M. Moreira, H. F. Schröder and H. J. José. 2009. Treatment of meat industry wastewater using dissolved air flotation and advanced oxidation processes monitored by GC–MS and LC–MS. *Chem. Eng. J.* 152: 151–7.

De Vries, M. and I. J. M. de Boer. 2010. Comparing environmental impacts for livestock products: A review of life cycle assessments. *Livest. Sci.* 128(1–3): 1–11.

Dickens, J., C. Lyon and R. Buhr. 2002. The effects of electrical stimulation during bleeding on shear values and cook loss of breast fillets from mature chickens deboned at two or twenty-four hours post-evisceration. *J. Appl. Poult. Res.* 11(1): 111–16.

Dukes, H. H., E. A. Hewitt, G. W. McNutt and S. A. Asdell. 1973. *The Physiology of Domestic Animals*. Comstock Pub. Co., Ithaca, NY., p. 695.

Dunkley, C. S. 2011. Global warming: how does it relate to poultry? University of Georgia. College of Agriculture and Environmental Science. http://www.caes.uga.edu/publications/pubDetail.cfm?pk_id=7939 (Accessed 10 June 2011).

Eide, M. H. 2002. Life cycle assessment (LCA) of industrial milk production. *Int. J. Life Cycle Assess.* 7(2): 115–26.

Einstein, D., Worrell, E. and Khrushch, M. 2001. Steam systems in industry: Energy use and energy efficiency improvement potentials. Lawrence Berkeley National Laboratory. University of California. http://escholarship.org/uc/item/3m1781f1#page-1 (Accessed 25 May 2011).

Environmental Protection Agency (EPA). 2008. Quantifying greenhouse gas emissions from key industrial sectors in the United States. Working draft. Sector strategies. http://www.epa.gov/sectors/pdf/greenhouse-report.pdf (Accessed 12 July 2011).

Fanatico, A. 2010. Chapter 20. Poultry processing under animal welfare and organic standards in the United States. In C. M. Owens, C. Z. Alvarado and A. R. Sams (Eds), *Poultry Meat Processing*, 2nd ed., pp. 397–407. CRC Press, Boca Raton, FL.

Fanatico, A. 2008. Organic poultry production in the United States. National sustainable agriculture service. https://attra.ncat.org/attra-pub/PDF/organicpoultry.pdf (Accessed 25 May 2011).

Fanatico, A. 2003. Small-scale poultry processing. https://attra.ncat.org/attra-pub/PDF/poultryprocess.pdf (Accessed 25 May 2011).

FAO. 2007. Managing Livestock-Environment Interactions. Committee on Agriculture. *Proc., 20th Session*. Rome.

Finkbeiner, M., A. Inaba, R. B. H. Tan, K. Christiansen and H. J. Klüppel. 2006. The new international standards for life cycle assessment: ISO 14040 and ISO 14044. *Int. J. LCA.* 11 (2): 80–5.

Finstad, S., C. A. O'Bryan, J. A. Marcy, P. G. Crandall and S. C. Ricke. 2012. *Salmonella* and broiler production in the United States: Relationship to foodborne salmonellosis. *Food Res. Int.* 45: 789–94.

Fletcher, D. 1999. Slaughter technology. *Poult. Sci.* 78(2): 277–81.

Fletcher, D. L. 1991. Ante mortem factors related to meat quality. *Proc. of the 10th European Symposium on the Quality of Poultry Meat.* Beekbergen, The Netherlands.

Fonkwe, L. G., Singh, R. K. and Lee, J. H. 2001. Utilization of poultry processing wastes. *J. Food Sci. Nutr.* 6(4): 257–62.

Formentini, D. F., R. N. Costanzi, A. M. Hashisuca, G. Paulillo, A. R. P. Neto and J. Morais. 2010. Water conservation and reuse in poultry slaughterhouse of Matelandia-Pr Brazil – a case study. Efficient use of water and slurry management in livestock production systems. http://www.ramiran.net/ramiran2010/docs/Ramiran2010_0210_final.pdf (Accessed 25 July 2011).

Fritzson, A. and T. Berntsson. 2006. Energy efficiency in the slaughter and meat processing industry-opportunities for improvements in future energy markets. *J. Food Eng.* 77 (4): 792–802.

Gazdziak, S. 2007. Kill floor improvements: Automation on the poultry kill and eviscerating lines is increasing efficiency and product quality. The National Provisioner. http://www.provisioneronline.com/articles/kill-floor-improvements-1 (Accessed 22 June 2011).

Genigeorgis, C., M. Hassuneh and P. Collins. 1986. *Campylobacter jejuni* infection on poultry farms and its effect on poultry meat contamination during slaughtering. *Journal of Food Protection* 49: 895–903.

Greene, J. 1998. U.S. Red meat & poultry markets in a global setting. Agricultural Outlook. United State Department of Agriculture (USDA). http://www.ers.usda.gov/publications/agoutlook/jun1998/ao252b.pdf (Accessed 11 July 2011).

Grey, T. C. 1988. The chilling of poultry carcasses and its influence on quality. *Inst. Food Res.* Bristol Lab., Bristol, UK.

Handley, J. A., S. H. Park, Z. Shi, T. M. Dawoud, Y. M. Kwon and S. C. Ricke. 2015. Chapter 6. *Salmonella* and the potential role for microbial process indicators on chicken carcasses. In S. C. Ricke, J. R. Donaldson and C. A. Phillips (Eds), *Food Safety: Emerging Issues, Technologies and Systems*, pp. 81–104. Elsevier, Oxford, UK.

Harlan, D. 2007. Perspectives on the global markets for poultry products. *Proc. Poultry in the 21st Century.* Bangkog. http://www.fao.org/AG/againfo/home/events/bangkok2007/docs/presentation/7.Harlan.pdf (Accessed 23 July 2011).

Hospido, A., J. Davis, J. Berlin and U. Sonesson. 2010. A review of methodological issues affecting LCA of novel food products. *Int. J. Life Cycle Assess* 15(4): 424.

Huezo, R., D. P. Smith, J. K. Northcutt and D. L. Fletcher. 2007. Effect of immersion or dry air chilling on broiler carcass moisture retention and breast fillet functionality. *J. Appl. Poult. Res.* 16 (3): 438–47.

IDD. 2001. Greenhouse gas emissions reduction and material flows. Global change and sustainable development subprogramme 2: to provide scientific support for Belgian politics. 1. http://www.belspo.be/belspo/home/publ/pub_ostc/CG2131/rappCG31_en.pdf (Accessed 8 July 2001).

International Finance Corporation and World Bank Group. 2007. Environmental, health, and safety guidelines for poultry processing. http://www.ifc.org/ifcext/enviro.nsf/AttachmentsByTitle/gui_EHSGuidelines2007_PoultryProcessing/$FILE/Final+-+Poultry+Processing.pdf (Accessed 25 May 2011).

James, C., C. Vincent, T. I. d. A. Lima and S. J. James. 2006. The primary chilling of poultry carcasses – a review. *International Journal of Refrigeration-Revue Internationale Du Froid* 29(6): 847–62.

Jarvis, N. A., C. A. O'Bryan, T. M. Dawoud, S. H. Park, Y. M. Kwon, P. G. Crandall and S. C. Ricke. 2016. An overview of *Salmonella* thermal destruction during food processing and preparation. *Food Control* 68: 280–90.

Jekayinfa, S. O. 2001. Energetic analysis of poultry processing operations. *Leonardo J. Sci.* 6 (10): 77–92.

Johnson, R. 2010. Carbon footprints and food security. http://www.meatpoultry.com/Writers/Guest%20columnists/Carbon%20footprints%20and%20food%20security.aspx (Accessed 26 May 2011).

Kiepper, B. H. 2011. Poultry Processing: Measuring true water use converting your plant from gpb to gpk. University of Georgia, Cooperative Extension Bulletin. http://www.caes.uga.edu/applications/publications/files/pdf/B%201381_3.PDF (Accessed 25 May 2011).

Kiepper, B. H., W. C. Merka and D. L. Fletcher. 2008. Proximate composition of poultry processing wastewater particulate matter from broiler slaughter plants. *Poult. Sci.* 87: 1633–6.

Kiepper, B. H. 2003. Characterization of poultry processing operations, wastewater generation, and wastewater treatment using mail survey and nutrient discharge monitoring methods. M.S. Thesis. University of Georgia, Athens, GA.

Kist, L. T., S. Moutaqi and E. L. Machado. 2009. Cleaner production in the management of water use at a poultry slaughterhouse of Vale do Taquari, Brazil: a case study. *J. Clean. Prod.* 17: 1200–5.

Klingele, L. 2011a. Sustainability 101: Greening up your process. Processing Solutions. http://www.meatingplace.com/MembersOnly/technology/details.aspx?item=19834#1 (Accessed 11 July 2011).

Klingele, L. 2011b. Sustainability 101: Identifying opportunities for energy conservation at the plant. http://www.meatingplace.com/MembersOnly/technology/details.aspx?item=19832#1 (Accessed 4 July 2011).

Kupusovic, T., S. Midzic, I. Silajdzic and J. Bjelavac. 2007. Cleaner production measures in small-scale slaughterhouse industry – case study in Bosnia and herzegovina. *J. Clean. Prod.* 15(4): 378–83.

Lillard, H. S. 1982. Improved chilling systems for poultry. *Food Technol.* 36: 58–67.

Loughrin, J. 2006. Comparison of solid-phase microextraction and stir bar sorptive extraction for the quantification of malodors in wastewater. *J. Agric. Food Chem.* 54(9): 3237–41.

Lundie, S. and G. M. Peters. 2005. Life cycle assessment of food waste management options. *J. Clean. Prod.* 13(3): 275–86.

Manios T., E. Gaki, S. Banou, A. Klimathianou, N. Abramakis and Sakkas N. 2003. Closed wastewater cycle in a meat producing and processing industry. *Resour. Conserv. Recycl.* 38: 335–45.

Marculescua, C. and C. Stana. 2011. Poultry processing industry waste to energy conversion. *Energy Procedia.* 6: 550–7.

Marcotte, M., A. R. Taherian and Y. Karimi. 2008. Thermophysical properties of processed meat and poultry products. *J. Food Eng.* 88(3): 315–22.

Matsumura, E. M. and J. C. Mierzwa. 2008. Water conservation and reuse in poultry processing plant – A case study. *Resour. Conserv. Recycl.* 52(6): 835–42.

Maud, S. 2007. Sustainability of poultry production. *Agric. Ecosyst. Environ.* 120(2–4): 470–1.

Mead, G. C. 2004. Microbial hazards in production and processing. In G. C. Mead (Ed.), *Poultry Meat Processing and Quality*. CRC Press, New York, NY, pp. 232–57.

Mead, G. and M. Scott. 1997. Spread of an enteric 'marker' organism during evisceration of new york dressed poultry in a simulated kitchen environment. *Br. Poult. Sci.* 38(2): 195–8.

Merka, W. C. 2010. Chapter 18. Processing water and wastewater. In C. M. Owens, C. Z. Alvarado and A. R. Sams (Eds), *Poultry Meat Processing*, 2nd ed., pp. 371–80. CRC Press, Boca Raton, FL.

Merka, W. C. 1991. Environmental-issues for the poultry-industry – processing. *Poult. Sci.* 70(5): 1118–22.

Mohsen, M. and J. Jaber. 2003. Potential of industrial wastewater reuse. *Desalination* 152(1–3): 281–9.

Morawicki, R. 2010. Chapter 19. Coproducts and by-products from poultry processing. In C. M. Owens, C. Z. Alvarado and A. R. Sams (Eds), *Poultry Meat Processing*, 2nd ed., pp. 381–96. CRC Press, Boca Raton, FL.

Mountney, G. J. and C. R. Parkhurst. 1995. *Poultry Products Technology*, 3rd ed. Food Products Press, New York, NY., p. 446.

North Carolina Division of Environment and Natural Resources. 2010. Water and wastewater use in the food processing industry – meat and poultry processing. http://www.nwfpa.org/nwfpa.info/component/content/article/372-water-and-wastewater-use-in-the-food-processing-industry?start=2 (Accessed 21 July 2011).

Northcutt, J. K. and D. R. Jones. 2004. A survey of water use and common industry practices in commercial broiler processing facilities. *J. Appl. Poult. Res.* 13(1): 48–54.

O'Bryan, C. A., P. G. Crandall, M. L. Davis, G. Kostadini, K. E. Gibson, W. Q. Alali, D. Jaroni, S. C. Ricke and J. A. Marcy. 2014. Mobile poultry processing units: A safe and cost-effective poultry processing option for the small-scale farmer in the United States. *World's Poultry Sci. J.* 70: 787–802.

O'Bryan, C. A., E. J. Van Loo, S. C. Ricke and P. G. Crandall. 2012a. Chapter 2. Organic meat operations in the United States. In S. C. Ricke, E. J. Van Loo, M. G. Johnson and C. A. O'Bryan (Eds), *Organic Meat Production and Processing*, pp. 13–21. Wiley Scientific/IFT, New York, NY.

O'Bryan, C. A., K. E. Gibson, P. G. Crandall and S. C. Ricke. 2012b. Chapter 12. Slaughter options for organic meat products in the U.S. In S. C. Ricke, E. J. Van Loo, M. G. Johnson and C. A. O'Bryan (Eds), *Organic Meat Production and Processing*, pp. 201–9. Wiley Scientific/IFT, New York, NY.

O'Bryan, C. A., P. G. Crandall and S. C. Ricke. 2008. Organic poultry pathogen control from farm to fork. *Foodborne Pathogens and Disease* 5: 709–20.

Ollinger, M. 2011. Structural change in the meat and poultry industry and food safety regulations. *Agribusiness* 27(2): 244–57.

Oosterom, J., S. Notermans, H. Karman and G. Engels. 1983. Origin and prevalence of *Campylobacter jejuni* in poultry processing. *Journal of Food Protection* 46: 339–44.

Owens, C. M., C. Z. Alvarado and A. R. Sams. 2010. *Poultry Meat Processing*. CRC Press, Boca Raton, FL, p. 441.

Papinaho, P. A.1996. Physiological and processing factors affecting broiler *Musculus pectoralis* shear vales and tenderness. Ph.D. Thesis, Department of Food Technology, University of Helsinki. Helsinki, Finland.

Pelletier, N. 2008. Environmental performance in the US broiler poultry sector: Life cycle energy use and greenhouse gas, ozone depleting, acidifying and eutrophying emissions. *Agric. Syst.* 98(2): 67–73.

Pereira, R. N. and A. A. Vicente. 2010. Environmental impact of novel thermal and non-thermal technologies in food processing. *Food Res. Int.* 43(7): 1936–43.

Pittman, H. M., K. C. Boling and S. J. Mirus. 2012. Chapter 3. Regulatory issues in domestically raised and imported organic meats in the United States. In S. C. Ricke, E. J. Van Loo, M. G. Johnson and C. A. O'Bryan (Eds), *Organic Meat Production and Processing*, pp. 23–51. Wiley Scientific/IFT, New York, NY.

Plumber, H. S. and B. H. Kiepper. 2011. Impact of poultry processing by-products on wastewater generation, treatment, and discharges. *Proc. of the 2011 Georgia Water Resour. Conf.* 11–13 April 2011, University of Georgia. http://www.gawrc.org/2011paper_pdfs/6.5.3Plumber.pdf (Accessed 25 May 2011).

Porges, R. 1950. Wastes from poultry dressing establishments. *Sewage and Industrial Wastes* 22(4): 531–5.

Raj, A. B. M. 1998. Welfare during stunning and slaughter of poultry. *Poult. Sci.* 77: 1815–19.

Raj, A. B. M. and N. G. Gregory. 1991. Efficiency of bleeding of broilers after gaseous or electrical stunning. *Vet. Rec.* 128(6): 127–8.

Raj, A. B. M., T. C. Grey, A. R. Audsely and N. G. Gregory. 1990. Effect of electrical and gaseous stunning on the carcass and meat quality of broilers. *Br. Poult. Sci.* 31(4): 725–33.

Rasekh, J., A. M. Thaler, D. J. Engeljohn and N. H. Pihkala. 2005. Food safety and inspection service policy for control of poultry contaminated by digestive tract contents: A review. *J. Appl. Poult. Res.* 14: 603–11.

Ricke, S. C. 2010. Future prospects for advancing food - safety research in food animals. In S. C. Ricke and F. T. Jones (Eds), *Perspectives on Food Safety Issues of Food Animal Derived Foods*, pp. 335–50. University of Arkansas Press, Fayetteville, AR.

Ricke, S. C. 2003. Perspectives on the use of organic acids and short chain fatty acids as antimicrobials. *Poultry Sci.* 82: 632–9.

Ricke, S. C., E. J. Van Loo, M. G. Johnson and C. A. O'Bryan (Eds). 2012. *Organic Meat Production and Processing*. Wiley Scientific/IFT, New York, NY., p. 444.

Ricke, S. C., M. M. Kundinger, D. R. Miller and J. T. Keeton. 2005. Alternatives to antibiotics: Chemical and physical antimicrobial interventions and foodborne pathogen response. *Poultry Sci.* 84: 667–75.

Russell, S. 2008. Chemical residuals in the environment and on chicken carcasses associated with scalding chickens in an acidic, copper sulfate–based commercial sanitizer during poultry processing. *J. Food Protect.* 71 (1): 226–30.

Russell, S. M. 2001. Spoilage bacteria associated with poultry. In A. R. Sams (Ed.), *Poultry Meat Processing*, pp. 159–80. New York, CRC Press.

Sams, A. R. and C. Z. Alvarado. 2010. Chapter 1. Introduction to poultry meat processing. In C. M. Owens, C. Z. Alvarado and A. R. Sams (Eds), *Poultry Meat Processing*, 2nd ed., pp. 1–3. CRC Press, Boca Raton, FL.

Sams A. R. and S. R. McKee. 2010. Chapter 3. First processing: Slaughter through chilling. In C. M. Owens, C. Z. Alvarado and A. R. Sams (Eds), *Poultry Meat Processing*, 2nd ed., pp. 25–49. CRC Press, Boca Raton, FL.

Sams, A. R. 2001. Chapter 3. First processing: Slaughter through chilling. In A. R. Sams (Ed.), *Poultry Meat Processing*, 1st ed., pp. 19–33. CRC Press, Boca Raton, FL.

Sustainable Agricultural Research and Education (SARE). 2006. Profitable Poultry: Raising Birds on Pasture. http://www.sare.org/Learning-Center/Bulletins/National-SARE-Bulletins/Profitable-Poultry (Accessed 28 July 2011).

Sarkar, B., P. P. Chakrabarti, A. Vijaykumar and V. Kale. 2006. Wastewater treatment in dairy industries – possibility of reuse. *Desalination* 195: 141–52.

Seideman, S. C. T. R. Callaway, P. G. Crandall, S. C. Ricke and D. J. Nisbet. 2010. Organic beef production: Production and food safety issues. In S. C. Ricke and F. T. Jones (Eds), *Perspectives on Food Safety Issues of Food Animal Derived Foods*, pp. 307–21. University of Arkansas Press, Fayetteville, AR.

Shields, S. J. and A. B. M. Raj. 2010. A critical review of electrical water-bath stun systems for poultry slaughter and recent developments in alternative technologies. *J. Appl. Anim. Welf. Sci.* 13(4): 281–99.

Sirsat, S. A., A. Muthaiyan and S. C. Ricke. 2009. Antimicrobials for pathogen reduction in organic and natural poultry production. *J. Appl. Poultry Res.* 18: 379–88.

Skunca, D., I. Tomasevic and I. Djekic. 2015. Environmental performance of the poultry meat chain – LCA approach. *Procedia Food Science* 5: 258–61.

Sofos, J. N., G. Flick, G.-J. Nychas, C. A. O'Bryan, S. C. Ricke and P. G. Crandall. 2013. Chapter 6. Meat, poultry, and seafood. In M. P. Doyle and R. L. Buchanan (Eds), *Food Microbiology-Fundamentals and Frontiers*, 4th ed., pp. 111–67. American Society for Microbiology, Washington, DC.

Somsen, D., A. Capelle and J. Tramper. 2004. Production yield analysis in the poultry processing industry. *J. Food Eng.* 65(4): 479–87.

Sparrey, J. M., P. J., Kettlewell, M. E. R. Paice and W. C. Whetlor. 1993. Development of a constant current water bath stunner for poultry processing. *J. Agri. Eng. Rese.* 56: 267–74.

Stadelman, W. J., V. M. Olson, G. A. Shemwell and S. Pasch. 1988. *Egg and Poultry-Meat Processing*. Ellis Horwood, Ltd., Chichester, England, p. 211.

Sundrum, A. 2006. Protein supply in organic poultry and pig production. *Proc. 1st IFOAM Intl. Conf. Animal in Organic Production*, 23–5 August 2006, St. Paul, MN., pp. 195–9.

Uijttenboogaart, T. G. 1999. European perspective on poultry slaughter technology. *Poult. Sci.* 78(2): 295–7.

USDA. National Agricultural Statistics Service. 2011b. Poultry slaughter: 2010 summary. February 2011. http://usda.mannlib.cornell.edu/usda/current/PoulSlauSu/PoulSlauSu-02-25-2011_new_format.pdf (Accessed 26 May 2011).

USDA. Economic research service. 2010a. Energy use in the U.S. food system. Economic research report Number 94, March 2010. http://www.ers.usda.gov/Publications/err94/err94.pdf (Accessed 8 July 2011).

USDA. Economic research service. 2010b. Organic production. http://www.ers.usda.gov/Data/Organic/ (Accessed 28 July 2011).

USDA. U.S. Department of Agriculture–Economic Research Service. 2006. Organic poultry and eggs capture high price premiums and growing share of specialty markets. Outlook Report LDP-M-150-01. U.S. Department of Agriculture, Washington, DC.

Van Loo, E. J., W. Alali and S. C. Ricke. 2012a. Food safety and organic meats. *Annual Rev. Food Science and Technol.* 3: 205–25.

Van Loo, E. J., S. C. Ricke, C. A. O'Bryan and M. G. Johnson. 2012b. Chapter 1. Historical and current perspectives on organic meat production. In S. C. Ricke, E. J. Van Loo, M. G. Johnson and C. A. O'Bryan (Eds), *Organic Meat Production and Processing*, pp. 1–9. Wiley Scientific/IFT, New York, NY.

Van Loo, E. J., S. C. Ricke, C. A. O'Bryan and M. G. Johnson. 2012c. Chapter 24. The future of organic meats. In S. C. Ricke, E. J. Van Loo, M. G. Johnson and C. A. O'Bryan (Eds), *Organic Meat Production and Processing*, pp. 423–30. Wiley Scientific/IFT, New York, NY.

Veerkamp, C. H. 1989. Chilling, freezing and thawing. In G. C. Mead (Ed.), *Processing of Poultry*. Elsevier Science, Oxford, UK., p. 103.

Vizzier-Thaxton, Y., K. D. Christensen, M. W. Schilling, R. J. Buhr and J. P. Thaxton. 2010. A new humane method of stunning broilers using low atmospheric pressure. *J. Appl. Poult. Res.* 19(4): 341–8.

Waters, B. W., J. M. Tatum and Y.-C. Hung. 2014. Effect of chlorine-based sanitizers properties on corrosion of metals commonly found in food processing environment. *J. Food Eng* 121: 159–65.

Wei, C.-I., D. L. Cook and J. R. Kirk. 1985. Use of chlorine in the food industry. *Food Technology* 39: 107–15.

Wiedemann, S., E. McGahan and G. Poad. 2012. Using life cycle assessment to quantify the environmental impact of chicken meat production. Australian Government – Rural Industries Research and Development Corporation. RIRDC Publication No. 12/029, p. 73.

Whyte, P., K. McGill, C. Monahan and J. Collins. 2004. The effect of sampling time on the levels of micro-organisms recovered from broiler carcasses in a commercial slaughter plant. *Food Microbiol.* 21: 59–65.

Wilkins, L. J., N. G. Gregory and S. B. Wotton. 1999. Effectiveness of different electrical stunning regimens for turkeys and consequences for carcase quality. *British Poult. Sci.* 40: 478–84.

Williams, A. G., E. Audsley and D. L. Sandars. 2006. Determining the environmental burdens and resource use in the production of agricultural and horticultural commodities. Main Report. Defra Research Project IS0205. Cranfield University and Defra, Bedford, England.

Wilson, A. 1998. Wilson's Practical Meat Inspection. 6th ed. Blackwell Sci. Ltd., Malden, MA., p. 287.

Windhorst, H. W. 2009. Recent patterns of egg production and trade: A status report on regional basis. *World's Poultry Sci. J.* 65: 685–708.

Wojdalski, J., B. Dró d and A. Pow zka. 2009. Effectiveness of energy and water consumption in a poultry processing plant. *TEKA Kom. Mot. Energ. Roln. – OL PAN*. 9: 395–402.

Chapter 20

Waste management and emissions in poultry processing

D. Luján-Rhenals, University of Arkansas Fayetteville, USA and Universidad de Córdoba, Colombia; R. Morawicki, University of Arkansas Fayetteville, USA; E. J. Van Loo, University of Arkansas Fayetteville, USA and Ghent University, Belgium; and S. C. Ricke, University of Arkansas Fayetteville, USA

1 Introduction

Food production in developing countries must be doubled during the next 50 years to supply the needs of a growing population where, due to the supplementation of animal products in the diet, food consumption has increased from 9.9 to 11.7 kJ person^{-1} day $^{-1}$ (Verge et al., 2007). Environmental concerns are increasing, and future consumers will be confronted with ecological and ethical criteria as well as food safety considerations for food product choices (Andersson et al., 1994; Ricke, 2010; Ricke et al., 2012). It is, therefore, necessary to evaluate this environmental impact and the use of resources in the food industry system for sustainable consumption (Roy et al., 2009), considering that the food system's environmental impact differs among countries and regions due to the types of foods consumed and the efficiencies of the respective farm-to-fork systems (Sonesson et al., 2005). In addition, the movement towards organic systems and smaller-scale natural production systems such as free-range poultry will further complicate developing accurate assessments of contributions to the environmental footprint (Ricke et al., 2012).

The poultry industry is one the largest meat-producing industrial sectors in the United States, and the broiler industry is not only the largest but the most generally efficient in the world (Pelletier, 2008). In addition, for England and Wales, Williams et al. (2006) reported that on the livestock side, due to the low impact and resources used, poultry

http://dx.doi.org/10.19103/AS.2016.0010.23

production represents the most environmentally efficient activity, followed by swine and sheep (lamb), with beef being the least efficient. The main reasons for the high poultry production efficiency include a very low overhead of poultry breeding stock, very efficient feed conversion and poultry's considerable daily weight gain (Havenstein et al., 2003).

The concept 'poultry' means domesticated fowl raised for meat and egg production, with chicken being the primary commercial bird associated with large-scale production operations. During the past half century, the poultry industry has grown tremendously, leading to increases in environmental impacts due to increased localized concentration of manure excretion and litter associated with confined animal feed operations (Sekar et al., 2010). The USDA (2011) reported that in May 2011 (ready-to-cook weight), 3.78 billion pounds (1.72 billion kg) of slaughtered poultry was certified as wholesome, up 8% from the amount certified in May 2010. In April 2011, the total was 3.48 billion pounds (1.58 billion kg). The April revision represented an increase of 3.71 million pounds (1.69 billion kg) from the previous month's preliminary pounds wholesome certified.

In summary, poultry production and processing appears poised to continue to expand both domestically and globally in the near future. As this trend continues, it will be critical to assess and potentially adjust current practices to minimize perturbations to the environment as innovations are introduced. While extensive documentation has been conducted for live poultry operations and the impacts such operations have on the environment, the contributions of the processing side of conventional poultry production have been much less characterized and data is nearly non-existent for non-conventional poultry processing. The aim of this chapter is to provide an overview and discussion of what is known about the waste generation and emissions that occur during poultry processing both in conventional and in non-conventional operations.

2 Treatment of by-products and waste

Processing water and organic solid by-products are the primary wastes in the slaughter poultry processing industry including the small-scale processing facilities, organics and backyard flock operations (Morawicki, 2010; Williams, 2011; Dunkley et al., 2012; O'Bryan et al., 2012a,b; 2014). One million tonnes of organic solids are generated every year worldwide from slaughter poultry-processing plants including all the steps from the reception of the live birds, through slaughter and evisceration, to the simple waste processing that occurs at the terminal stages (World Bank Group, 2007). These materials represent considerable quantities of pollutants to the environment, but can be used as by-products to obtain other valuable materials. For instance, animal waste such as poultry heads and muscle tissue represent valuable energy material due to their high content of fats and proteins, hence could be used as a substrate for methane fermentation (Wymyslowski et al., 2010). The following sections discuss some of the approaches used in the poultry industry to treat processing waste.

2.1 Incineration

Incineration is considered one of the thermal destruction methods with the most effective potential of destruction for infectious agents (Ritter and Chirnside, 1995). In the incineration operation, ash is produced by burning carcasses or by-products at temperatures of 850°C, thus destroying all infective agents (Kastner and Phebus, 2004). Currently, there are three broad categories of incineration techniques: open-air burning, fixed-facility incineration and air-curtain incineration (Kastner and Phebus, 2004). Open-air burning can be applied in (a) open fields, (b) on combustible heaps called pyres and (c) other techniques that are unassisted by incineration equipment. All of these incineration methods usually require U.S. state permission to be applied (APHIS, 2003). Straw or hay, untreated timbers, kindling wood, coal and diesel fuel are the most common materials used for open-air burning (McDonald, 2001; USDA, 2005). The volume of ash generated by this technique is significantly high (NAO, 2002) and has the added problem of groundwater and soil contamination caused by the hydrocarbons (HCs) used as fuel (Crane, 1997). However, fixed-facility incinerations include (a) small on-farm incinerators, (b) small and large incineration facilities, (c) crematoria and (d) power plant incinerators. Unlike open-air burning, this technique is completely contained and highly controlled. Diesel, natural gas and propane are the most common fuels used for these systems (Rosenhaft, 1974). Finally, air-curtain incineration involves fans that force a mass of air as much as six times faster than open-air burning (Ford, 1994). It also generates less carbon monoxide and particulate matter emissions (Ford, 2003).

2.2 Rendering

Offal and blood represent a major problem as by-products in the slaughter poultry industry in the United States, where most of them are processed in rendering plants for animal feed supplements (Cai et al., 1995). Rendering can be used in the production of meat and bonemeal, which may be used in animal feed, as fertilizer or it can be used for anaerobic digestion processes or composting. Rendered products for animal feeds and pet foods represent economically important protein sources as well as a highly effective means to reduce greenhouse gases (GHGs) by recycling the by-product nitrogen and carbon sources as by-product protein rather than releasing it as carbon dioxide and other GHG from simple decomposition (Meeker and Meisinger, 2015). However, early concerns on uniform protein quality have abated as rendering equipment has improved, thus leading to digestibility improvements in the essential amino acids such as lysine, threonine, tryptophan and methionine (Meeker and Meisinger, 2015). Rendering has also been viewed as an effective means for eliminating food-borne pathogens in these proteins, which can be a contamination source as ingredients in pet food and animal diets (Maciorowski et al., 2004, 2006; Ge et al., 2013; Nemser et al., 2014; Meeker and Meisinger, 2015; Molina et al., 2016). Further feed processing physical treatments such as pelleting and addition of chemicals such as organic acids have been identified over the years as effective means to minimize food-borne pathogen contamination in feeds prior to feeding (Williams, 1981; Hinton and Hinton, 1988; Ricke, 2003, 2005; Maciorowski et al., 2004, 2006; Carrique-Mas et al., 2007; Wales et al. 2010; Jones, 2011).

In addition, the fat generated may be used for animal feed, in chemical industry products, or burned as fuel (Salminen and Rintala, 2002). Patil and Nag (2011) reported the use of these waste materials (viscera) for the production of polyunsaturated fatty acids, concentrated using a process with acetone and urea; this process could be applied with success at industrial levels. However, carcasses also represent an important environmental, economic and biological problem for this industry. Finally, a poor quality of poultry feed can occur especially during the summer season as high temperature accelerates the offal putrefaction process (Cai and Pancorbo, 1994). Poultry feathers correspond to approximately 5–8.5% of bird weight and they serve as an important source of pure keratin waste, which historically has not been used in the most efficient manner but, depending on processing approaches, can be utilized as a feed protein source, as discussed later in this chapter (Khardenavis et al., 2009; McGovern, 2000; Mokrejs et al., 2011). Gas and odour emissions are the main environmental concerns associated with the use of rendering in slaughter poultry plants, but scrubber-packed column systems are used to remove many of the malodorous compounds via contact of the gas with water-containing additives (Morawicki, 2010; Gwyther et al., 2011). However, as Morawicki (2010) points out, these scrubbers still represent an environmental cost themselves due to the use of fans and pumps along with the addition of chemicals and the treatment of the resulting wastewater.

2.3 Alkaline hydrolysis

Alkaline hydrolysis emerged as a more recent technology developed in the 1990s, which used sodium hydroxide or potassium hydroxide to hydrolyse biological material, such as carcasses, into an aqueous solution comprised mainly of peptides, amino acids, sugars and soaps (Kaye et al., 1998; Kastner and Phebus, 2004; Shafer et al., 2000, 2001). Carcasses are hydrolysed at 150°C for up to 6 hours at high pressure in a steel tank with an alkali concentration depending on the carcass weight (Kalambura et al., 2008). There are minimal gaseous emissions and odour problems associated with alkaline hydrolysis, but the effluent possesses a high pH and is rich in nutrients and thus can represent serious problems when discarded into the environment (Kastner and Phebus, 2004).

Considering that chicken feathers represent one of the consistent waste products of poultry processing (Shih, 1993), they can also be treated by a two-stage alkaline-enzymatic hydrolysis with high efficiency under mild reaction conditions (0.1 or 0.3% water KOH solution), leading to the production of keratin hydrolyzates, which have applications in packaging technology. Processing feathers exhibited 91% degradation after being treated in 0.3% KOH and were hydrolysed within 8 hours in the second stage at 70°C with a 5% dose of enzyme. This technology achieved an economic option and a high efficiency under mild conditions of the reaction (temperatures around 70°C and pH in the middle of the alkaline region) (Mokrejs et al., 2011). This is discussed in more detail in a later section.

2.4 Wastewater treatment

The accumulation of wastewater generated by uncollected blood, feathers, eviscerations and cleaning of the live haul area at slaughter plant is classified as the poultry processing wastewater (PPW) (Kiepper et al., 2008). Wastewater is generated in poultry processing with a chemical oxygen demand (COD) 1500 to 3500 mg L^{-1}, biological oxygen demand (BOD) 1000 to 2100 mg L^{-1}, total nitrogen 150 to 400 mg L^{-1} and total phosphorus 16 to 50 mg L^{-1}. The composition of wastewater (BOD and COD) primarily depends on how it is used, for example, washing and efficiency of blood recovery (Eremektar et al., 1999; Weakley et al., 1972).

Some poultry-processing plants, in addition to savings from sanitation, use bubbled accelerated floatation, ultra-filtration, ozone treatment and recycling for the clean-up and recycling of poultry chilling and scalding water (Carawan and Sheldon, 1989). With these techniques, chilling and scalding water use can be decreased by up to 80% (Carawan and Sheldon, 1989).

3 Environmental impact of processing plant emissions

3.1 General concepts

Food systems contribute appreciably to global GHG emissions. Even though all sectors in the supply chain contribute, overall, the agricultural sector is one of the largest GHG emitters, whereas meat and dairy products are among the most GHG-intensive food types (Garnett, 2011). Even though technological improvements are crucial, they alone will not be sufficient in reducing GHG emissions due to the combination of population growth and anticipated rises in per capita consumption of meat and dairy products that in all likelihood make this more difficult to achieve (Garnett, 2011).

Since the poultry and pig feed conversion efficiency is greater and CH_4 is less of an issue than ruminants, poultry and pigs require less land and produce fewer emissions. Therefore, monogastrics appear more profitable and 'efficient' to produce food products than ruminants. Currently, there is a rapid growth in production and per capita consumption of poultry and pig products. This trend will continue and it will positively contribute to the reduction of GHG (FAO, 2006; DERFRA, 2010).

Food accounts for 31% of the EU-25's total GHG impacts and an additional 9% comes from the hotel and restaurants sector (European Commission, 2006). Furthermore, studies in developed countries revealed that food consumption contributes between 15% and 28% to overall national emissions (Garnett, 2011). According to the working group I of the Intergovernmental Panel on Climate Change in Switzerland (IPCC, 2007), CO_2 is the most important anthropogenic GHG.

3.2 Emissions from refrigeration

Refrigeration is an essential operation in almost all processes related to food conservation, such as transportation of raw material and final product, storage for long periods of time and several intermediate operations for changing properties or improving quality (chilling, pasteurization, lyophilization and fermentation). Refrigeration not only produces greenhouse gases because of the energy used to operate the equipment, but also because of the intrinsic global warming potential (GWP) of the refrigerant gases (Garnett, 2007). Refrigerant's emissions, particularly the discharge of refrigerants containing chlorine, bromine and other halogens, are the main global environmental impacts generated from refrigeration systems and have direct effects on the stratospheric ozone equilibrium, such as ozone depletion following breakdown, direct action as GHGs and reduction in the efficiency with increases in energy use (Calm, 2002).

Fortunately, over time, the refrigerant releases from centrifugal chillers have been reduced (Calm, 2002; McCulloch et al., 2001). Hydrofluorocarbons (HFCs), the standard refrigerant fluid today, produce a high GWP; hence, a switch to other refrigerants such as HCs, which have a zero GWP can potentially lead to overall GHG reductions (Garnett, 2007). In the food industry, the refrigeration charge sizes usually range from 650 to 9100 kg, and the average lifetime is approximately 25 years. Ammonia, HCs, hydrochlorofluorocarbon (HCFC)-123 and HFC-134a are expected to be the most extensively used substitute refrigerants for new equipment (UNEP, 1999). Upon completion of the HCFC phase out, HFC-134a, R-404A and R-507A are expected to be the primary refrigerants used in this end use (EPA, 2006).

The shelf life of poultry products needs to be extended due to the increased demand of fresh poultry and a desire to transport these products to new and far-off markets (Jimenez et al., 1997; Ricke, 2010; Ricke et al., 2015). Nonetheless, even when stored at low temperatures, fresh chicken is a highly perishable product. Chicken and other poultry meats certainly carry microbial pathogen and non-pathogen loads as do many other raw foods (Snyder, 1998; Sofos et al., 2013; Handley et al., 2015). Hence, after slaughtering, the usual shelf life of chicken is less than five days (Ellis et al., 2006). Industrial refrigeration could represent up to 60% of the total cost in frozen food and 85% for cold storage poultry product. Improvements in industrial refrigeration systems have the potential to reduce energy consumption by 15–40% (FDF, 2007).

Several global conferences have established concerns resulting in international agreements related to the progressive greenhouse effects due to anthropogenic causes, as well as the strategies and actions be acted upon worldwide (U.N., 1987 – Montreal Protocol; U.N., 1997– Kyoto Protocol; World Resources Institute, 2015). In the United States, there has been a focus on industrial sources of GHGs, including those originating from food-processing facilities (EPA, 2008, 2009). Refrigeration and air conditioning are part of this issue with two main aspects: the direct environmental emission caused by the most common refrigerants used (HFC fluid family) and the direct effect caused by the energy consumption of these systems (Bovea et al., 2007). Overall, commercial refrigeration has the greatest emissions from refrigerants (CO_2eq), which represents 40% of the total refrigerant emissions (IPCC, 2005; EPA, 2010).

One of the most common refrigeration systems in the poultry industry is the chiller, which usually ranges in thermal capacity from approximately 10 kW (3 tonnes) to more than 30 MW (8500 tonnes), with a typically common range of 300 kW (85 tonnes) to 14 MW (4000 tonnes). Larger systems usually employ multiple chillers, with installations at 300 MW (85 000 tonnes) of cooling capacity (Calm, 2006).

There is no specific data reporting the GHG emissions from the refrigeration systems for the poultry industry, but an approach can be developed by knowing the plant capacities and the particular refrigerants used in each case. For instance, chlorofluorocarbon R-11 has been the most common refrigerant for large chillers employing centrifugal turbocompressors, based on both its efficiency and the cost advantages of low-pressure designs. This refrigerant has a GWP_{100} of 4680. However, chlorofluorocarbons are being replaced for environmental protection purposes primarily with HCFC R-123 and HFC R-134a, which have a GWP_{100} of 76 and 130, respectively. Furthermore, R-717 (NH_3) is extensively used in industrial refrigeration, especially in the food and beverage industries with a GWP_{100} less than 1 (Calm, 2006).

3.3 Odour control

Odour is defined by human olfactory perception of a mixture of chemical compounds in the environment. While not considered a GHG in the classical sense, odours from agricultural origins can certainly become a point of contention between municipalities and agricultural operations when in close proximity and they most likely draw negative publicity. Odours are emitted elementarily by all livestock production operations (Ullman et al., 2004). In modern production facilities, odours are generated primarily from the confinement buildings, from manure storage structures, from manure or storage effluent applied to cropland and from disposal of dead animals (Ullman et al., 2004; Nahm, 2003). The natural by-products of microbial degradation of uric acid and faeces produce most of the odorants, including ammonia in poultry facilities emitting malodours for the anaerobic metabolism of amino acids (Loughrin, 2006; Ullman et al., 2004).

In order to avoid the spread of possible zoonotic agents, on-farm depopulation is preferred instead of being a part of conventional slaughter at the processing plant. Hence, it is more effective to euthanize the chicken soon and on site to control the rapid dissemination of the highly contagious poultry diseases (Dawson et al., 2006). As a result of deficient or uncontrolled combustion during the depopulation in poultry facilities, compounds such as dioxins and furans are released into the air causing serious odour issues in the environment (Scudamore et al., 2002). Immediate rendering is required to reduce odour problems in and around the slaughter poultry plant due to the rapid putrefaction process of offals (Cai and Pancorbo, 1994).

For poultry processing, Russell (2008) demonstrated that there was no significant increase in the levels of the chemical components (sulphuric acid, ammonium sulphate and copper sulphate) in the air above the scalder when dosing the scalder with an acidic, copper sulphate-based commercial sanitizer. However, depending on the types of raw material and process equipment used, rendering plants generate a variety of odorous emissions. Aldehydes and small amounts of dialkyl disulphides occur predominantly in poultry rendering emissions. A variety of carbonyl components with traces of sulphurous and nitrogenous compounds represent the odours emitted mainly with dry rendering of dead stock, beef offal and slaughterhouse trimmings (Bethea et al., 1973). Krüger et al. (2009) found that the major classes of substances present in the condensed waste gas of feather and viscera meal manufacture were acids, aldehydes, amines, alcohols and sulphur compounds. Condensation followed by incineration, scrubbing or combinations of these methods are the most common odour control systems used in poultry-rendering plants. Incineration has been the most efficient method, especially for low-volume, high-concentration odours. Other methods with different levels of efficiency were also reported by Bethea et al. (1973), including a catalytic combustion method, which may not be practically viable due to the higher investment cost over a typical thermal incinerator. In addition, it has rapid catalyst poisoning with subsequent loss of activity and the inability to destroy rendering odours at temperatures lower than 1 200°F. Finally, ozone treatment approaches have been demonstrated to be an effective approach for reduction of odours arising from animal manure (Alkoaik, 2009).

4 Solid wastes

During the last decades, poultry slaughterhouses have increased not only their production but also the generation of organic solid wastes (Salminen and Rintala, 2002). In the poultry slaughter plants, most offal is produced during bleed out, scalding and evisceration (Plumber and Kiepper, 2011). The offals account, on average, for 28% of the live weight of a broiler chicken. Hence, a typical poultry slaughter plant processing 200000 broilers per day with a mean live weight of 5.0 pounds (2.3 kg) will produce 140 tonnes (217 MT) of offal (Boushy and Poel, 2000).

The size of the slaughter facilities is an important factor having an impact on the environment and food safety practices. Hence, small slaughter processors and backyard flock operations generate small quantities of solid wastes, and the treatment of these wastes has primarily focused on direct disposal or recycling (burial or composting)

(Williams, 2011). However, as markets such as organic pet food increase further processing to prepare some of the meat by-products for companion animal diets, this, in turn, may very well change approaches to disposal strategies (Dunkley et al., 2012).

Due to its environmental conservation, practical, economical and biosecurity advantages, on-farm disposal methods are preferred by the farming community. For instance, processes such as anaerobic digestion and composting have found predilection in the United States and Canada (Gwyther et al., 2011). Nevertheless, new methods must be developed and applied due to issues with the regulations; methods such as bioreduction and freezing appear promising as on-farm storage methods for livestock mortalities to reduce associated biosecurity risks (Gwyther et al., 2011). In the following sections, assessment of the main solid wastes generated in a poultry slaughter plant, such as blood, feathers, carcasses and offals, will be discussed.

4.1 Blood

According to USEPA (1975), blood is the greatest source of pollution in a poultry-processing plant. Blood represents approximately 8% of a broiler's live weight (Dukes et al., 1973). Approximately, 40–60% of a broiler's blood exits the carcass during the bleeding operation, taking place in a regular slaughter process (Wilson, 1998). Blood is one of the main causes of zoonotic disease originating from poultry. For instance, Johnson et al. (2010) reported that high exposure to poultry oncogenic viruses, coming from blood and other common offal sources, would increase risk of several types of cancers to poultry-processing workers.

Plumber and Kiepper (2011) reported that a bleed time treatment of 60 seconds produced, from the scalder wastewater, considerably greater amounts of COD (616 mg/L), total solids (TS; 510 mg/L), total suspended solids (TSS; 129 mg/L), total volatile solids (TVS; 425 mg/L) and total Kjeldahl nitrogen (TKN; 70 mg/L). This research did not reveal significant differences in TSS concentration between 60 and 120 seconds of bleeding. Increasing the bleed time to 120 seconds from 60 seconds resulted in mean per cent reductions of COD by 34%, TS by 33%, TSS by 34%, TVS by 36% and TKN by 29%.

4.2 Feathers

Poultry feather wastes cause serious environmental problems worldwide (Saucedo-Rivalcoba et al., 2011). Feathers represent approximately 5 to 8.5% of chicken weight (average of 2 kg grown chicken); the U.S. annually produces nearly 0.9 to 1.1 billion kg of feather wastes (Khardenavis et al., 2009; McGovern, 2000; Mokrejs et al., 2011). Worldwide feather production is estimated at 4 million tonnes per year (Salminen and Einola, 2003). Feathers are an important source of contamination with an average poultry slaughter plant producing nearly 29 500 kg of feathers/day. Additionally, they contribute to low profits per bird, so they must be processed rapidly at low cost (McGovern, 2000). Most feathers produced in the poultry industry are disposed of by incineration or landfills, which involve cost and can cause environmental problems. Hence, tremendous negative effects with high energy demand and CO_2 production would have to take place if keratin from poultry slaughter plants was to be disposed of via landfilling or combustion (Suzuki et al., 2006). Consequently, these methods generate ecological disadvantages in terms of the apparent energy loss and the production of a large amount of CO_2 (Marculescua and Stana, 2011).

When feathers mature they become highly keratinized to form keratin, a scleroprotein that is highly recalcitrant to proteolytic degradation (Leeson and Walsh, 2004). As a protein polymer, keratin is also highly chemically resistant due to the strong cross-linked nature of the polymers (Mokrejs et al., 2011). Approximately 90% of feather structure consists of keratin (Saucedo-Rivalcoba et al., 2011). Hence, a high content of protein is present in keratin wastes; the dry matter (DM) N content of feathers is 15% and for wool and horns is 15.8 and 16.6%, respectively (Graham et al., 1949; Fahnenstock and Steinbuchel, 2003), which could be processed to obtain animal feed after pre-treatments (Onifade et al., 1998; Dalev, 1994; El Boushy and van der Poel, 1990). However, feathers can also be used for bedding, decoration, manufacture of fibres, sporting equipment and various other uses (Fonkwe et al., 2001). A few studies have shown that this by-product can be used to produce biodiesel (Kondamudi et al., 2009).

Given the high protein content there has always been interest in examining the potential of feather meal as a dietary protein source for feeding livestock, but either chemical or enzymatic hydrolysis is required to improve digestibility, with autoclaving being most often applied (Papadopoulos, 1985). However, extending these fairly extreme conditions can reduce the amino acid content and nutritive quality (Papadopoulos, 1985). Consequently, combinations at milder conditions would appear to have advantages. For example, Mokrejs et al. (2011) reported that over 90% of feather degradation was achieved by treating chicken feathers with a two-stage alkaline-enzymatic hydrolysis. This technology offered a high-efficiency hydrolysis under mild conditions due to the limited NaOH concentration as long as it did not exceed 0.3% at temperatures less than 70°C and was followed by keratinase applied at concentrations between 1 and 5% at temperatures between 50°C and 70°C.

It has been known for some time that microorganisms do exist in the environment and can produce proteases with keratinolytic activity capable of hydrolyzing the keratin in feathers (Papadopoulos, 1985). Beginning in the late 1980s and early 1990s, microorganisms were isolated and characterized on the basis of their capability to degrade feathers. The feather-degrading bacteria, *Bacillus licheniformis*, was isolated from a poultry waste digester that contained feathers (Shih, 1993). The organism secreted a keratinase enzyme that hydrolysed the keratin in the feather and thus subsequently resulted in the formation of free amino acids (Williams and Shih, 1989; Williams et al., 1990, 1991; Lin et al., 1992; Shih, 1993; Ricke and Saengkerdsub, 2015). Once keratinase activities were better characterized, they were followed by several nutritional studies demonstrating the enzymatic generation of feather meal as a nutritionally viable protein source (Onifade et al., 1998; Kim and Patterson, 2000; Kim et al., 2002; Odetallah et al., 2003, 2005). Other *Bacillus* species capable of degrading feathers have also been isolated from soils after adding chicken feathers to the microbial culture media (Zaghloul et al., 1998; Riffel and Brandelli, 2002). Similarly, non-bacillus bacteria that can hydrolyse feathers have also been identified (Riffel and Brandelli, 2002; Khardenavis et al., 2009). For example, a *Flavobacterium* species was isolated from samples collected at a poultry-processing plant at several sites where feathers had been collected (Riffel and Brandelli, 2002). As more feather-degrading bacteria are identified and further characterized, it is anticipated that keratinase production will be optimized both in improvements on the cultivation and media composition and as molecular modification of the keratinolytic enzyme producing strains (Harde et al., 2011; Ricke and Saengkerdsub, 2015). Genetic modifications to improve hydrolytic activity and stability under adverse conditions, as well as production on a much greater industrial scale, would certainly be potential goals as well.

4.3 Carcasses (condemnations) and offals

One of the major problems faced by the poultry producers is the disposal of carcasses (Blake, 2004). Large volumes of by-products are generated during day-to-day commercial poultry production activities such as manure, farm mortalities and hatchery, and processing wastes require daily attention. Carcass losses by mortality on the farm must be faced on a daily basis and in the end the general balance will be equal to a considerable volume of dead animals. For instance, a flock of 50 000 broilers grown to 49 days of age that averages 0.1% daily mortality (4.9% total mortality) will produce approximately 2.2 tonnes of carcasses. A turkey flock of 30 000 birds averaging 0.5% weekly mortality (9% total mortality over an 18-week period) will produce approximately 12.6 tonnes (11 430.5 kg) of carcasses (Blake et al., 1990). These losses represent a tremendous amount of organic matter due to a fresh broiler carcass containing approximately 34.2% DM, consisting of 51.8% protein, 41.0% fat and 6.3% ash (Malone et al., 1987). Approximately, 70–75% of the live bird weight corresponds to the carcass and the amount of solid wastes to the environment depends of the previous bird health and the efficiency of the process (Williams, 2011). In India, for example, chicken viscera represent approximately 30% of the total wastes with an annual disposal of more than 1 800 000 tonnes of chicken viscera (Jamdar and Harikumar, 2008).

A few studies have shown that this by-product can be used to produce animal feed. For example, Adewolu et al. (2010) reported that a mixture comprising hydrolysed feather meal, chicken offal meal and maggot meal at a ratio of 4:3:2 served as a comparable alternative to fishmeal. This protein mixture could be used to replace up to 50% of the fishmeal component in diets for *C. gariepinus* fingerlings without causing adverse effects on growth. Van Lunen and Anderson (1991) reported that hydrolysed poultry offals could be included as part of the diet for finishing pigs at levels of up to 10% DM because of palatability problems, but did not affect carcass quality. Finally, Van Lunen et al. (1990) reported that changes in fatty acid profile of pork do not appear to affect sensorial qualities when hydrolysed poultry offal is incorporated in the diet at levels of up to 15% DM.

5 Future directions

From 1970 to 2005, poultry and egg production expanded more rapidly than beef and veal or pork on a global scale (Windhorst, 2006). However, the poultry industry's contribution to global warming by carbon emissions (0.03%) is not as high as other food industries such as cattle (2.4%), dairy cattle (1.0%) and swine (0.3%). Approximately 15% of the poultry GHGs are generated during postharvest operations, such as transport to the processing plant, slaughtering process and transport to the market, whereas 72% of GHGs are generated by feed production and breeding (IDD, 2001). According to Lesschen et al. (2011), beef has the highest GHG emissions with 22.6 kg CO_2-eq/kg, followed by pork, poultry and eggs with 3.5, 1.6 and 1.7 kg CO_2-eq/kg, respectively. This lower impact is, in part, due to the high efficiency of the poultry industry compared with other industries in the food sector.

It has been shown that when animal size increases, the footprint on the environment also increases due to high feed consumption, emissions and wastes (DERFRA, 2010;

FAO, 2006; Monteny et al., 2001). However, effects of the various food industries on the environment must be measured and analysed in more detail. For instance, the energy balance complexity of cattle feed consumption and emissions are, in part, based on the ingestion of grass. Grass is a much less expensive feed source than commercial animal feed products added to the diet formulation. Even though GHG emissions associated with pork and poultry are less per gram of protein than those for beef, it cannot be assumed that a simple change from red to white meat consumption is all that is required. The balance is more complicated than that since the amount of feed and its original sources have to be included as well (Millward and Garnett, 2010).

The most significant impact on the PPW stream is due to the operations of bleed out, scalding and evisceration (Plumber and Kiepper, 2011). The average water consumption reported in the current chapter is 33.6 L/bird for broiler chicken slaughter plants, which is far less than the water consumption for other food-processing plants such as red meat, swine or even turkey. Nevertheless, it would be even less if the poultry industry developed more initiatives to reduce their water consumption and the global environmental footprint. By making simple daily changes in hygiene practices, wastewater generated by the poultry slaughter industry could be reduced, for example not reusing the water that conveys waste and offal. Water could be replaced by mechanical or air-vacuum equipment-based approaches. As another approach, this water could be reused safely by installing appropriate filtration systems. By doing that, the flow pressure in the facility could be reduced and it in turn projects less energy consumption. Undoubtedly, the poultry industry needs to face the relevant challenges of high investment costs for water reuse treatment technology or seek new less expensive alternatives. This investment will be recovered in the middle or long term not only economically but environmentally.

In addition to growth in conventional poultry production, non-conventional production systems referred to as organic and natural, respectively, are also on the rise and may represent a different challenge to waste management and emissions control (Ricke et al., 2012). A livestock and agriculture product is considered organic when, among other requirements, agrochemicals (synthetic pesticides and fertilizers) are avoided during the grow-out and processing stages (Sundrum, 2006; Fanatico, 2008; Ricke et al., 2012; Van Loo et al., 2012). Even though organic production has been present in the U.S. for decades, this production system is considerably less developed than conventional poultry processing since the USDA established the National Organic Program in 2002 (Fanatico, 2008; Pittman et al., 2012). However, because of the increasing food demand, the demand in food safety regulations for large and even small meat-processing plants will also increase. Hence, small plants such as organic processors are more affected by the implementation of structural changes in terms of food safety, such as the hazard analysis and critical control points system, which can encourage them to easily exit the industry (Ollinger, 2011). Not surprisingly, much less is known about the environmental impact of these systems, but this will need to change if these non-conventional poultry operations continue to grow and expand their market share. In some way, this will be more of a challenge since a large part of the market appeal of these types of products is their perceived environmental and animal welfare friendly production practices. Consequently, public expectations to some extent would have greater requirements and be more demanding of these products than their conventionally produced counterparts.

6 Conclusions and recommendations

Poultry meat has a high overall efficiency of feed conversion in the non-organic sector compared with other meat products such as swine and cattle. Other differences exist as well. For example, methane production levels of poultry and swine are less of an issue than they are for ruminants (Jensen, 1996; Saengkerdsub and Ricke, 2014). While poultry do harbour methanogens, most of the methane production is believed to originate from the manure (Ricke et al., 2004; Saengkerdsub et al., 2007a,b; Saengkerdsub and Ricke, 2014; Dunkley and Dunkley, 2013). Although some of these outcomes at the farm may be beyond immediate control, a more comprehensive understanding of the environmental footprint from the farm to the processing plant may be helpful for designing and optimizing management strategies that take advantage of the vertical integration prevalent in the poultry industry. For example, over the years, there have been several proposed feed additives, both chemical and biological, that can limit methane production in farm animals (McAllister et al., 1996; Baker, 1999; Klieve and Hegarty, 1999; Machmüller et al., 2003; Saengkerdsub et al., 2006; Patra and Yu, 2015), but it is not known if such mitigation approaches would impact the processing environmental steps such as disposal of the viscera. Knowing this outcome could determine whether a chemical or biological feed additive would be the better choice from a waste disposal standpoint. Similar considerations may be warranted for pre-harvest control of food-borne pathogens during production of poultry and other farm animals as well. Consequently, it will be critical for poultry processors at the industrial and farm levels to have the knowledge, technology and economic motivation to better control their energy consumption, water use and waste disposal to further reduce the environmental footprint of this important industry.

Undoubtedly, it is necessary to maintain optimal hygienic conditions in processing plants in order to release a safe product for the consumers, as well as further improve their waste management programmes (Sofos et al., 2013; Handley et al., 2015). As Williams (2011) reported, the size of the slaughter facilities is an important factor in the effects on the environment and food safety practices. Hence, small slaughter processors and backyard flock operations should generate less quantities of solid wastes and the treatment of these wastes would be assumed to be primarily focused on direct disposal or recycling (burial or composting). However, very little has been documented thus far on these systems. In a recent study, O'Bryan et al. (2016) measured nitrogen and phosphorus environmental loads in the presence of alternative poultry-processing operations that were being used or were similar to those used for pasture-raised poultry production. The systems examined were on-farm processing units either as open-air or in enclosed sheds along with a mobile poultry-processing unit pilot plant that simulated a batch-processing system typical of these types of operations (Trimble et al., 2013; O'Bryan et al., 2016). Wastewater collected from both on-farm processing units and the mobile poultry-processing unit contained less TKN and total phosphorus than the levels reported for conventional processing (Kiepper, 2009; Plumber and Kiepper, 2011; O'Bryan et al., 2016). While such studies offer some initial assessments of the environmental impact, it remains difficult to draw general conclusions due to the high degree of variability in these small operations. To better establish a more comprehensive assessment of environmental impact will require more such studies across several geographical regions to develop general baseline values for impacts.

Emissions derived from poultry processing are less delineated at this time. Overall, as discussed earlier, the refrigerant releases from centrifugal chillers have been reduced during the past 20 years as a consequence of the Montreal Protocol (Calm, 2002; McCulloch et al., 2001). However, because there is no specific information regarding the GHG emissions coming from the refrigeration systems in the poultry industry, an approach could be calculated in each particular case by knowing the plant capacity and the particular refrigerants. Once this is established, known ranges could be estimated to account for geographical, climate and other environmental factors that may influence the level of refrigeration required on an individual processing plant basis.

Conventional poultry slaughter-processing plants represent complex systems with multiple outputs that can directly impact the environment, including the more obvious solid and liquid wastes and the less obvious impacts such as emissions. As estimations of these outputs become more precise, opportunities will certainly become apparent on the means to predict their respective impacts when alterations and adjustments in processing are made, such as introducing new interventions or changing line speeds. In addition, the ability to alter existing processing plants or construct new ones will offer design opportunities to improve operations and perhaps further reduce the environmental footprint. However, this will require economic incentives, in addition to regulatory mandates, to make substantial progress. Non-conventional processing does not necessarily generate the volumes of wastes and emissions seen with conventional processing but can be equally challenging for different reasons. In particular, mobile processing units that are on-farm sites and thus directly interfaced with the surrounding environment represent as of yet unknown waste contributions. As the popularity for the production of free-range chicken grows, the ensuing environmental footprint will need to be quantified so that recommendations can be made to producers on how best to minimize the environmental impact of their localized processing operations.

7 Where to look for further information

Meeker, D. L. and J. L. Meisinger. 2015. Rendered ingredients significantly influence sustainability, quality, and safety of pet food. *J. Anim. Sci.* 93: 835–47.

Morawicki, R. O. 2010. Chapter 19. Coproducts and by-products from poultry processing. In *Poultry Meat Processing*, C. M. Owens, C. Z. Alvarado and A. R. Sams (Eds.), 2nd edition. CRC Press, Boca Raton, FL, pp. 381–96.

Morawicki, R. O. 2012. *Handbook of Sustainability for the Food Sciences*. Wiley-Blackwell, Hoboken, NJ.

O'Bryan, C. A., P. G. Crandall, M. L. Davis, G. Kostadini, K. E. Gibson, W. Q. Alali, D. Jaroni, S. C. Ricke and J. A. Marcy. 2014. Mobile poultry processing units: A safe and cost-effective poultry processing option for the small-scale farmer in the United States. *World's J. Poult. Sci.* 70: 787–802.

Owens, C. M., C. Z. Alvarado and A. R. Sams (Eds.). 2010. *Poultry Meat Processing*. 2nd edition. CRC Press, Boca Raton, FL.

Ricke, S. C., E. J. Van Loo, M. G. Johnson and C. A. O'Bryan (Eds.). 2012. *Organic Meat Production and Processing*. Wiley Scientific/IFT, New York, NY, 444p.

Shih, J. C. H. 1993. Recent development in poultry waste digestion and feather utilization – A review. *Poult. Sci.* 72: 1617–20.

8 References

Adewolu, M. A., N. B. Ikenweiwe and S. M. Mulero. 2010. Evaluation of an animal protein mixture as a replacement for fishmeal in practical diets for fingerlings of *Clarias gariepinus* (Burchell, 1822). *Israeli J. Aquaculture-Bamidgeh* 62 (4): 237–44.

Alkoaik, F. A. 2009. Ozone treatment of animal manure for odor control. *Am. J. Environ.Sci.* 5 (6): 765–71.

Andersson, K., T. Ohlsson and P. Olsson. 1994. Life-cycle assessment (LCA) of food-products and production systems. *Trends Food Sci. Technol.* 5 (5): 134–8.

APHIS. 2003. Risk reduction strategies for potential BSE pathways involving downer cattle and dead stock of cattle and other species: Advance Notice of Proposed Rulemaking. *Fed. Reg.* 68 (13): 2703–11.

Baker, S. K. 1999. Rumen methanogens, and inhibition of methanogenesis. *Aust. J. Agric. Res.* 50: 1293–8.

Bethea, R. M., B. N. Murthy and D. F. Carey. 1973. Odor controls for rendering plants. *Environ. Sci. Tech.* 7 (6): 504–10.

Blake, J. P. 2004. Methods and technologies for handling mortality losses. *Worlds Poult. Sci. J.* 60 (4): 489–99.

Blake, J. P., M. E. Cook, C. C. Miller and D. Reynolds. 1990. Dry extrusion of poultry processing plant wastes and poultry farm mortalities. In *Proceedings of the 6th International symposium on Agricultural and Food Processing Wastes*, American Society of Agricultural Engineers, St. Joseph, MI, pp. 319–27.

Boushy, A. R. Y. and A. F. B. Van der Poel. 2000. *Handbook of Poultry Feed from Waste: Processing and use*. 2nd ed. Dordrecht: Kluwer Academic Publishers. The Netherlands, 409p.

Bovea, M. D., R. Cabello and D. Querol. 2007. Comparative life cycle assessment of commonly used refrigerants in commercial refrigeration systems. *Int. J. Life Cycle Assess.* 12 (5): 299–307.

Cai, T. and O. C. Pancorbo. 1994. Stabilization of poultry-processing by-products and wate and poultry carcasses through lactic acid fermentation. *J. Appl. Poult. Res.* 3: 17–25.

Cai, T., O. C. Pancorbo, W. C. Merka, J. E. Sander and H. M. Barnhart. 1995. Stabilization of poultry-processing by-products and poultry carcasses through direct chemical acidification. *Bioresour. Technol.* 52 (1): 69–77.

Calm, J. M. 2002. Emissions and environmental impacts from air-conditioning and refrigeration systems. *Int. J. Refrig. Rev. Int. Froid.* 25 (3): 293–305.

Calm, J. M. 2006. Comparative efficiencies and implications for greenhouse gas emissions of chiller refrigerants. *Int. J. Refrig. Rev. Int. Froid.* 29 (5): 833–41.

Carawan, R. E. and B. W. Sheldon. 1989. Systems for recycling water in poultry processing. North Carolina Cooperative Extension Service. Report No. 6/89-2M-TWK.

Carrique-Mas, J. J., S. Bedford and R. H. Davies. 2007. Organic acid and formaldehyde treatment of animal feeds to control *Salmonella*: Efficacy and masking during culture. *J. Appl. Microbiol.* 103: 88–96.

Crane, N. 1997. Animal disposal and the environment. *State Vet. J.* 7 (3): 3–5.

Dalev, P. 1994. Utilization of waste feathers from poultry slaughter for production of a protein-concentrate. *Bioresour. Technol.* 48 (3): 265–7.

Dawson, M. D., E. R. Benson, G. W. Malone, R. L. Alphin, I. Estevez and G. L. Van Wicklen. 2006. Evaluation of foam-based mass depopulation methodology for floor-reared meat-type poultry operations. *Appl. Eng. Agric.* 22 (5): 787–93.

DERFRA. Department for Environment Food and Rural Affairs. 2010. OECD FAO Agricultural Outlook2010 2019 Summary by Defra. Accessed 28 July 2011. http://www.defra.gov.uk/statistics/files/defra-stats-foodfarm-monthly-brief-1007-a.pdf.

Dukes, H. H., E. A. Hewitt, G. W. McNutt and S. A. Asdell. 1973. *The Physiology of Domestic Animals*. Ithaca, NY: Comstock Pub. Co. 695p.

Dunkley, C. S., D. Carter and K. Dunkley. 2012. Chapter 9. Organic meat by-products for affiliated food industries. In S. C. Ricke, E. J. Van Loo, M. G. Johnson and C. A. O'Bryan (Eds.), *Organic Meat Production and Processing*. New York, NY: Wiley Scientific/IFT, pp. 147–56.

Dunkley, C. S. and K. D. Dunkley. 2013. Greenhouse gas emissions from livestock and poultry. *Agric. Food Anal. Bacteriol.* 3: 17–29.

El Boushy, A. R. and A. F. B. van der Poel. 1990. Feather meal-a biological waste: Its processing and utilization as a feedstuff for poultry. *Biol. Wastes* 32: 39–74.

Ellis, M., K. Cooksey, P. Dawson, I. Han and P. Vergano. 2006. Quality of fresh chicken breasts using a combination of modified atmosphere packaging and chlorine dioxide sachets. *J. Food Prot.* 69 (8): 1991–6.

EPA. Environmental Protection Agency. 2006. Global mitigation of non-CO_2 greenhouse gases. Washington, DC. Accessed 12 July 2012. http://www.epa.gov/climatechange/economics/downloads/GlobalMitigationFullReport.pdf.

EPA. Environmental Protection Agency. 2008. Quantifying greenhouse gas emissions from key industrial sectors in the United States. Working draft. Sector strategies. Accessed 12 July 2011. http://www.epa.gov/sectors/pdf/greenhouse-report.pdf.

EPA. Environmental Protection Agency. 2009. Technical support document for food processing facilities: proposed rule for mandatory reporting of greenhouse gases. Climate change division. Accessed 22 June 2011. http://www.epa.gov/climatechange/emissions/downloads/tsd/TSD%20food%20processing_2_4_09.pdf.

EPA. Environmental Protection Agency. 2010. Transitioning to low-gwp alternatives in commercial refrigeration. Accessed 12 July 2011. http://www.epa.gov/ozone/downloads/EPA_HFC_ComRef.pdf.

Eremektar, G., E. U. Cokgor, S. Ovez, F. G. Babuna and D. Orhon. 1999. Biological treatability of poultry processing plant effluent – A case study. *Water Sci. Technol.* 40 (1): 323–9.

European Commission. 2006. Environmental Impact of Products (EIPRO): Analysis of the Life Cycle Environmental Impacts Related to the Total Final Consumption of the EU 25 (2006). European Commission Technical Report EUR 22284 EN.

Fahnestock, S. R. and A. Steinbuchel. 2003. *Biopolymers: Polyamides and Complex Proteinaceous Materials II*. Germany: Wiley-VCH, pp. 155–202.

Fanatico, A. 2008. Organic poultry production in the United States. National sustainable agriculture service. Accessed 25 May 2011. https://attra.ncat.org/attra-pub/PDF/organicpoultry.pdf.

FAO. 2006. Livestock's Long Shadow. Environmental issues and options. Rome.

FDF. Food and Drink Federation. 2007. Carbon trust networks project: food & drink industry refrigeration efficiency initiative. Accessed 12 July 2011. http://www.ior.org.uk/ior/images/pdf/general/REI-G5%20Site%20Guidance%20Topics%20-%20Final%20Jul-07.pdf.

Fonkwe, L. G., Singh, R. K. and J. H. Lee 2001. Utilization of poultry processing wastes. *J. Food Sci. Nutr.* 6 (4): 257–62.

Ford, G. 2003. *Disposal Technology Seminar on Air-curtain Incineration*. Kansas City, MO: Midwest Regional Carcass Disposal Conference.

Ford, W. B. 1994. Swine carcass disposal evaluation using air curtain incinerator system, model T 359 Foreign Animal Disease Report, 22–2; reprinted by Air Burners, LLC, Washington: USDA Animal and Plant Health Inspection Service. Accessed 5 July 2011. http://www.airburners.com/DATA-FILES_Tech/ab_swine_report.pdf.

Garnett, T. 2007. Food refrigeration: What is the contribution to greenhouse gas emissions and how might emissions be reduced? A working paper produced as part of the Food Climate Research Network. Center for Environmental Strategy. University of Surrey. United Kingdom.

Garnett, T. 2011. Where are the best opportunities for reducing greenhouse gas emissions in the food system (including the food chain)? *Food Policy*. 36: S23–32.

Ge, B., P. C. LaFon, P. J. Carter, S. D. McDermott, J. Abbott, A. Glenn, S. L. Ayers, S. L. Friedman, J. C. Paige, D. D. Wagner, S. Zhao, P. F. McDermott and M. A. Rasmussen 2013. Retrospective analysis of *Salmonella*, *Campylobacter*, *Escherichia coli*, and *Enterococcus* in animal feed ingredients. *Foodborne Pathogens Dis.* 10: 684–91.

Graham, C. E., H. K. Waitkoff and S. W. Hier. 1949. The amino acid content of some scleroproteins. *J. Biol. Chem.* 177 (2): 529–32.

Gwyther, C. L., A. P. Williams, P. N. Golyshin, G. Edwards-Jones and D. L. Jones. 2011. The environmental and biosecurity characteristics of livestock carcass disposal methods: A review. *Waste Manage.* 31 (4): 767–78.

Handley, J. A., S. H. Park, Z. Shi, T. M. Dawoud, Y. M. Kwon and S. C. Ricke. 2015. Chapter 6. *Salmonella* and the potential role for microbial process indicators on chicken carcasses. In S. C. Ricke, J. R. Donaldson and C. A. Phillips (Eds.), *Food Safety: Emerging Issues, Technologies and Systems.* Oxford, UK: Elsevier, pp. 81–104.

Harde, S. M., I. B. Bajaj and R. S. Singha. 2011. Optimization of fermentative production of keratinase from *Bacillus subtilis* NCIM 2724. *Agric. Food Anal. Bacteriol.* 1: 54–65.

Havenstein, G. B., P. R. Ferket and M. A. Qureshi. 2003. Growth, livability, and feed conversion of 1957 versus 2001 broilers when fed representative 1957 and 2001 broiler diets. *Poult. Sci.* 82: 1500–8.

Hinton, M. and A. H. Hinton. 1988. Control of salmonella infections in broiler chickens by the acid treatment of their feed. *Vet. Rec.* 123: 416–21.

IPCC. 2005. *Intergovernmental Panel on Climate Change. Safeguarding the Ozone Layer and the Global Climate System.* London, UK: Cambridge University Press.

IPCC. 2007. Climate change 2007: Synthesis report. Contribution of working groups I, II and III to the fourth assessment report of the Intergovernmental Panel on Climate Change. Core Writing Team. In R. K. Pachauri and A. Reisinger (Eds.). IPCC, Geneva, Switzerland.

Jamdar, S. N. and P. Harikumar. 2008. Radiation decontamination of poultry viscera. *Radiat. Phys. Chem.* 77 (4): 467–72.

Jensen, B. B. 1996. Methanogenesis in monogastric animals. *Environ. Monit. Assess.* 42: 99–112.

Jimenez, S. M., M. S. Salsi, M. C. Tiburzi, R. C. Rafaghelli, M. A. Tessi and V. R. Coutaz. 1997. Spoilage microflora in fresh chicken breast stored at 4 degrees C: Influence of packaging methods. *J. Appl. Microbiol.* 83 (5): 613–18.

Johnson, E. S., H. Detan and K.-M. Lo. 2010. Cancer mortality in poultry slaughtering/processing plant workers belonging to a union pension fund. *Environ. Res.* 110: 588–94.

Jones, F. T. 2011. A review of practical *Salmonella* control measures in animal feed. *J. Appl. Poult. Res.* 20: 102–13.

Kalambura, S., T. Kricka, V. Jurisic and Z. Janjecic. 2008. Alkaline hydrolysis of animal waste as pretreatment in production of fermented fertilizers. *Cereal Res. Commun.* 36: 179–82.

Kastner, J. and R. Phebus. 2004. Carcass disposal: A comprehensive review. National Agricultural Biosecurity Center Consortium, USDA APHIS Cooperative Agreement Project, Carcass Disposal Working Group. Kansas State University, Manhattan, KS.

Kaye, G. I., P. B. Weber, A. Evans and R. A. Venezia. 1998. Efficacy of alkaline hydrolysis as an alternative method for treatment and disposal of infectious animal waste. *Contemp. Topics Lab. Anim. Sci.* 37 (3): 43–6.

Khardenavis, A. A., A. Kapley and H. J. Purohit. 2009. Processing of poultry feathers by alkaline keratin hydrolyzing enzyme from *Serratia* sp HPC 1383. *Waste Manage* 29 (4): 1409–15.

Kiepper, B. H. 2009. Effects of tertiary microsieving on the composition of poultry processing wastewater. *J. Appl. Poult. Res.* 18: 716–24.

Kiepper, B. H., W. C. Merka and D. L. Fletcher. 2008. Proximate composition of poultry processing wastewater particulate matter from broiler slaughter plants. *Poult. Sci.* 87 (8): 1633–6.

Kim, W. K., E. S. Lorenz and P. H. Patterson. 2002. Effect of enzymatic and chemical treatments on feather solubility and digestibility. *Poult. Sci.* 81: 95–8.

Kim, W. K. and P. H. Patterson. 2000. Nutritional value of enzyme- or sodium hydroxide-treated feathers from dead hens. *Poult. Sci.* 79: 528–34.

Klieve, A. V. and R. S. Hegarty. 1999. Opportunities for biological control of ruminal methanogenesis. *Aust. J. Agric. Res.* 50: 1315–19.

Kondamudi, N., J. Strull, M. Misra and S. K. Mohapatra. 2009. A green process for producing biodiesel from feather meal. *J. Agric. Food Chem.* 57 (14): 6163–6.

Krüger, R. L., R. M. Dallago, I. do Nascimento Filho and M. Di Luccio. 2009. Study of odor compounds in gaseous effluents generated during production of poultry feather and viscera meal using headspace solid phase microextraction. *Environ. Monit. Assess.* 158 (1–4): 355–63.

Leeson, S. and T. Walsh. 2004. Feathering in commercial poultry I. Feather growth and composition. *World's Poult. Sci. J.* 60: 42–51.

Lesschen, J. P., M. van den Berg, H. J. Westhoek, H. P. Witzke and O. Oenema. 2011. Greenhouse gas emission profiles of European livestock sectors. *Anim. Feed Sci. Tech.* 166–7: 16–28.

Lin, X., C.-G. Lee, E. S. Casale and J. C. H. Shih. 1992. Purification and characterization of a keratinase from a feather-degrading *Bacillus licheniformis* strain. *Appl. Environ. Microbiol.* 58: 3271–5.

Loughrin, J. 2006. Comparison of solid-phase microextraction and stir bar sorptive extraction for the quantification of malodors in wastewater. *J. Agric. Food Chem.* 54 (9): 3237–41.

Machmüller, A., C. R. Soliva and M. Kreuzer. 2003. Methane-suppressing effect of myristic acid in sheep as affected by dietary calcium and forage proportion. *Br. J. Nutr.* 90: 529–40.

Maciorowski, K. G., P. Herrera, M. M. Kundinger and S. C. Ricke. 2006. Animal production and contamination by foodborne *Salmonella*. *J. Consumer Prot. Food Saf.* 1: 197–209.

Maciorowski, K. G., F. T. Jones, S. D. Pillai and S. C. Ricke. 2004. Incidence and control of food borne *Salmonella* spp. in poultry feeds- A review. *World's Poult. Sci. J.* 60: 446–57.

Malone, G. W., W. W. Saylor, M. G. Ariza, K. M. Lomax and C. R. Kaifer. 1987. Acid preservation and utilization of poultry carcasses resulting from mortality losses. In Progress Through Research and Extension 1987, Report 11. University of Delaware College of Agricultural Sciences, Newark, DE, pp. 13–16.

Marculescua, C. and C. Stana. 2011. Poultry processing industry waste to energy conversion. *Energy Procedia*. 6: 550–7.

McAllister, T. A., Okine, E. K., Mathison, G. W. and K.-J. Cheng. 1996. Dietary, environmental. and microbiological aspects of methane production in ruminants. *Can. J. Anim. Sci.* 76: 231–43.

McCulloch, A., P. Ashford and P. M. Midgley. 2001. Historic emissions of fluorotrichloromethane (CFC-11) based on a market survey. *Atmos. Environ.* 35 (26): 4387–97.

McDonald, C. 2001. Report to Dumfries and Galloway Council: Air monitoring of carcass pyre at whithorn. Glasgow: Glasgow scientific service. Accessed 5 July 2011. http://www.fmd-enviroimpact.scieh.scot.nhs.uk/Papers/FMD%20Whithorn.pdf.

McGovern, V. 2000. Recycling poultry feathers: More bang for the cluck. *Environ. Health Perspect.* 108 (8): A366–9.

Meeker, D. L. and J. L. Meisinger. 2015. Rendered ingredients significantly influence sustainability, quality, and safety of pet food. *J. Anim. Sci.* 93: 835–47.

Millward, D. J. and T. Garnett. 2010. Food and the planet: Nutritional dilemmas of greenhouse gas emission reductions through reduced intakes of meat and dairy foods. *Proc. Nutr. Soc.* 69 (1): 103–18.

Mokrejs, P., P. Svoboda, J. Hrncirik, D. Janacova and V. Vasek. 2011. Processing poultry feathers into keratin hydrolysate through alkaline-enzymatic hydrolysis. *Waste Manage. Res.* 29 (3): 260–7.

Molina, A., F. Granados-Chinchilla, M. Jiménez, M. T. Acuña-Calvo, M. Alfaro and G. Chavarría. 2016. Vigilance for *Salmonella* in feedstuffs available in Costa Rica: Prevalence, serotyping and tetracycline resistance of isolates obtained from 2009 to 2014. *Foodborne Pathogens Dis.* 13: 119–27.

Monteny, G. J., C. M. Groenestein and M. A. Hilhorst. 2001. Interactions and coupling between emissions of methane and nitrous oxide from animal husbandry. *Nutr. Cycl. Agroecosyst.* 60 (1–3): 123–32.

Morawicki, R. O. 2010. Chapter 19. Coproducts and by-products from poultry processing. In C. M. Owens, C. Z. Alvarado and A. R. Sams (Eds.), *Poultry Meat Processing*. 2nd ed. Boca Raton, FL: CRC Press, pp. 381–96.

Nahm, K. H. 2003. Current pollution and odor control technologies for poultry production. *Avian Poult. Biol. Rev.* 14: 151–74.

NAO. (2002). The 2001 outbreak of foot and mouth disease. London: UK National Audit Office. Accessed 5 July 2011. http://www.nao.org.uk/publications/0102/the_2001_outbreak_of_foot_and.aspx.

Nemser, S. M., T. Doran, M. Grabenstein, T. McConnell, T. McGrath, R. Pamboukian, A. C. Smith, M. Achen, G. Danzeisen, S. Kim, Y. Liu, S. Robeson, G. Rosario, K. M. Wilson and R. Reimschuessel. 2014. Investigation of *Listeria*, *Salmonella*, and toxigenic *Escherichia coli* in various pet foods. *Foodborne Pathogens Dis.* 11: 706–9.

O'Bryan, C. A., P. Crandall, D. Jaroni, S. C. Ricke and K. E. Gibson. 2016. Assessment of nitrogen and phosphorus loads present in environments impacted by alternative poultry processing operations utilized in pasture-raised poultry production. *Renew. Agric. Food Syst.* doi:10.1017/S1742170515000514. (In press)

O'Bryan, C. A., P. G. Crandall, M. L. Davis, G. Kostadini, K. E. Gibson, W. Q. Alali, D. Jaroni, S. C. Ricke and J. A. Marcy. 2014. Mobile poultry processing units: A safe and cost-effective poultry processing option for the small-scale farmer in the United States. *World's Poult. Sci. J.* 70: 787–802.

O'Bryan, C. A., E. J. Van Loo, S. C. Ricke and P. G. Crandall. 2012a. Chapter 2. Organic meat operations in the United States. In S. C. Ricke, E. J. Van Loo, M. G. Johnson and C. A. O'Bryan (Eds.), *Organic Meat Production and Processing*. New York, NY: Wiley Scientific/IFT, pp. 13–21.

O'Bryan, C. A., K. E. Gibson, P. G. Crandall and S. C. Ricke. 2012b. Chapter 12. Slaughter options for organic meat products in the U.S. In S. C. Ricke, E. J. Van Loo, M. G. Johnson and C. A. O'Bryan (Eds.), *Organic Meat Production and Processing*. New York, NY: Wiley Scientific/IFT, pp. 201–9.

Odetallah, N. H., J. J. Wang, J. D. Garlich and J. C. H. Shih. 2003. Keratinase in starter diets improves growth of broiler chicks. *Poult. Sci.* 82: 664–70.

Odetallah, N. H., J. J. Wang, J. D. Garlich and J. C. H. Shih. 2005. Versazyme supplementation of broiler diets improves market growth performance. *Poult. Sci.* 84: 858–64.

Ollinger, M. 2011. Structural change in the meat and poultry industry and food safety regulations. *Agribusiness* 27 (2): 244–57.

Onifade, A. A., Al-Sane, N. A., Al-Musallam, A. A. and S. Al-Zarban. 1998. A review: Potentials for biotechnological applications of keratin-degrading microorganisms and their enzymes for nutritional improvement of feathers and other keratins as livestock feed resource. *Bioresour. Technol.* 66: 1–11.

Papadopoulos, M. C. 1985. Processed chicken feathers as feedstuff for poultry and swine: A review. *Agric. Wastes* 14: 275–90.

Patil, D. and A. Nag. 2011. Production of PUFA concentrates from poultry and fish processing waste. *J. Am. Oil Chem. Soc.* 88 (4): 589–93.

Patra, A. K. and Z. Yu. 2015. Effects of adaptation of *in vitro* rumen culture to garlic oil, nitrate, and saponin and their combinations on methanogenesis, fermentation, and abundances and diversity of microbial populations. *Front. Microbiol.* 22. http://dx.doi.org/10.3389/fmicb.2015.01434.

Pelletier, N. 2008. Environmental performance in the US broiler poultry sector: Life cycle energy use and greenhouse gas, ozone depleting, acidifying and eutrophying emissions. *Agric. Syst.* 98 (2): 67–73.

Pittman, H. M., K. C. Boling and S. J. Mirus. 2012. Chapter 3. Regulatory issues in domestically raised and imported organic meats in the United States. In S. C. Ricke, E. J. Van Loo, M. G. Johnson and C. A. O'Bryan (Eds.), *Organic Meat Production and Processing*. New York, NY: Wiley Scientific/IFT, pp. 23–51.

Plumber, H. S. and B. H. Kiepper. 2011. Impact of poultry processing by-products on wastewater generation, treatment, and discharges. *Proc. of the 2011 Georgia Water Resources Conference*. 11–13 April 2011, University of Georgia. Accessed 25 May 2011. http://www.gawrc.org/2011paper_pdfs/6.5.3Plumber.pdf.

Ricke, S. C. 2003. Perspectives on the use of organic acids and short chain fatty acids as antimicrobials. *Poult. Sci.* 82: 632–9.

Ricke, S. C. 2005. Chapter 7 – Ensuring the safety of poultry feed. In G. C. Mead (Ed.), *Food Safety Control in the Poultry Industry*, Cambridge, UK: Woodhead Publishing, pp. 174–94.

Ricke, S. C. 2010. Future prospects for advancing food – safety research in food animals. In S. C. Ricke and F. T. Jones (Eds.), *Perspectives on Food Safety Issues of Food Animal Derived Foods*. Fayetteville, AR: University of Arkansas Press, pp. 335–50.

Ricke, S. C., T. M. Dawoud and Y. M. Kwon. 2015. Chapter 4. Application of molecular methods for traceability of foodborne pathogens in food safety systems. In S. C. Ricke, J. R. Donaldson and C. A. Phillips (Eds.), *Food Safety: Emerging Issues, Technologies and Systems*. Oxford, UK: Elsevier, pp. 37–63.

Ricke, S. C. and S. Saengkerdsub. 2015. Chapter 19. *Bacillus* probiotics and biologicals for improving animal and human health: Current applications and future prospects In V. Ravishankar Rai and A Jamuna Bai (Eds.), *Beneficial Microbes in Fermented and Functional Foods*. Boca Raton, FL: CRC Press/Taylor & Francis Group, pp. 341–60.

Ricke, S. C., E. J. Van Loo, M. G. Johnson and C. A. O'Bryan (Eds.). 2012. *Organic Meat Production and Processing*. New York, NY: Wiley Scientific/IFT, 444p.

Ricke, S. C., C. L. Woodward, Y. M. Kwon, L. F. Kubena and D. J. Nisbet. 2004. Chapter 11 – Limiting avian gastrointestinal tract *Salmonella* colonization by cecal anaerobic bacteria and a potential role for methanogens. In R. C. Beier, S. D. Pillai, T. D. Phillips and R. L. Ziprin (Eds.), *Pre-Harvest and Post-Harvest Food Safety: Contemporary Issues and Future Directions*. Ames, IA: Blackwell Publishing Professional, pp. 141–50.

Riffel, A. and A. Brandelli. 2002. Isolation and characterization of a feather-degrading bacterium from the poultry processing industry. *J. Ind. Microbiol. Biotechnol.* 29: 255–8.

Ritter, W. F. and A. E. M. Chirnside. 1995. Impact of dead bird disposal pits on groundwater quality on the Delmarva peninsula. *Bioresour. Technol.* 53 (2): 105–11.

Rosenhaft, M. E. 1974. Evaluation of requirements for pathological incineration facilities. *Lab. Anim. Sci.* 24 (6): 905–9.

Roy, P., D. Nei, T. Orikasa, Q. Xu, H. Okadome, N. Nakamura and T. Shiina. 2009. A review of life cycle assessment (LCA) on some food products. *J. Food Eng.* 90 (1): 1–10.

Russell, S. 2008. Chemical residuals in the environment and on chicken carcasses associated with scalding chickens in an acidic, copper sulfate – based commercial sanitizer during poultry processing. *J. Food Protect.* 71 (1): 226–30.

Saengkerdsub, S., R. C. Anderson, H. H. Wilkinson, W.-K. Kim, D. J. Nisbet and S. C. Ricke. 2007a. Identification and quantification of methanogenic archaea in adult chicken ceca. *Appl. Environ. Microbiol.* 73: 353–6.

Saengkerdsub, S., P. Herrera, C. L. Woodward, R. C. Anderson, D. J. Nisbet and S. C. Ricke. 2007b. Detection of methane and quantification of methanogenic archaea in faeces from young broiler chickens using real-time PCR. *Letts. Appl. Microbiol.* 45: 629–34.

Saengkerdsub, S., W.-K. Kim, R. C. Anderson, C. L. Woodward, D. J. Nisbet and S. C. Ricke. 2006. Effects of nitrocompounds and feedstuffs on *in vitro* methane production in chicken cecal contents and rumen fluid. *Anaerobe* 12: 85–92.

Saengkerdsub, S. and S. C. Ricke. 2014. Ecology and characteristics of methanogenic archaea in animals and humans. *Crit. Revs. Microbiol.* 40: 97–116.

Salminen, E. and J. Einola. 2003. The methane production of poultry slaughtering residues and effect of pretreatments on the methane productivity of poultry feathers. *Environ. Tech.* 24: 1079–86.

Salminen, E. and J. Rintala. 2002. Anaerobic digestion of organic solid poultry slaughterhouse waste – A review. *Bioresour. Technol.* 83: 13–26.

Saucedo-Rivalcoba, V., A. L. Martinez-Hernandez, G. Martinez-Barrera, C. Velasco-Santos and V. M. Castano. 2011. (Chicken feathers keratin)/polyurethane membranes. *Appl. Phys. A-Mater. Sci. Process.* 104 (1): 219–28.

Scudamore, J. M., G. M. Trevelyan, M. V. Tas, E. M. Varley and G. A. W. Hickman. 2002. Carcass disposal: Lessons from Great Britain following the foot and mouth disease outbreaks of 2001. *Rev. Sci. Tech. Off. Int. Epizoot.* 21 (3): 775–87.

Sekar, S., S. Karthikeyan and P. Iyappan. 2010. Trends in patenting and commercial utilization of poultry farm excreta. *World's Poult. Sci. J.* 66 (3): 533–71.

Shafer, D. J., R. P. Burgess, K. A. Conrad, J. F. Prochaska and J. B. Carey. 2001. Characterization of alkaline hydroxide-preserved whole poultry as a dry byproduct meal. *Poult. Sci.* 80 (11): 1543–8.

Shafer, D. J., J. B. Carey, R. P. Burgess, K. A. Conrad and J. F. Prochaska. 2000. Chemical preservation of whole broiler carcasses utilizing alkaline hydroxides. *Poult. Sci.* 79 (11): 1517–23.

Shih, J. C. H. 1993. Recent development in poultry waste digestion and feather utilization – A review. *Poult. Sci.* 72: 1617–20.

Snyder, O. P. 1998. Menu management and purchasing. In O. P. Snyder (Ed.), *Food Safety Through Quality Assurance Management*. Saint Paul, MN: Hospitality Institute of Technology and Management, pp. 11–13.

Sofos, J. N., G. Flick, G.-J. Nychas, C. A. O'Bryan, S. C. Ricke and P. G. Crandall. 2013. Chapter 6. Meat, Poultry, and Seafood. In M. P. Doyle and R. L. Buchanan (Eds.), *Food Microbiology-Fundamentals and Frontiers*, 4th ed. Washington, DC: American Society for Microbiology, pp. 111–67.

Sonesson, U., B. Mattsson, T. Nybrant and T. Ohlsson. 2005. Industrial processing versus home cooking: An environmental comparison between three ways to prepare a meal. *Ambio* 34 (4–5): 414–21.

Sundrum, A. 2006. Protein supply in organic poultry and pig production. In *Proceedings of the 1st International Conference on Animals in Organic Production*, 23–25 August 2006. St. Paul, MN, pp. 195–9.

Suzuki, Y., Y. Tsujimoto, H. Matsui and K. Watanabe. 2006. Decomposition of extremely hard-to-degrade animal proteins by thermophilic bacteria. *J. Biosci. Bioeng.* 102 (2): 73–81.

Trimble, L. M., W. Q. Alali, K. E. Gibson, S. C. Ricke, P. Crandall, D. Jaroni and M. Y. Habteselassie. 2013. Prevalence and concentration of *Salmonella* and *Campylobacter* in the processing environment of smallscale pastured broiler farms. *Poult. Sci.* 34: 3060–6.

Ullman, J. L., S. Mukhtar, R. E. Lacey and J. B. Carey. 2004. A review of literature concerning odors, ammonia, and dust from broiler production facilities: 4. Remedial management practices. *J. Appl. Poult. Res.* 13 (3): 521–31.

U.N. United Nations. 1997. Kyoto protocol to the United Nations frame work convention on climate change. New York, NY.

U.N. United Nations. 1987 Montreal protocol on substances that deplete the ozone layer. New York, NY.

UNEP. United Nations Environment Programme. 1999. Report of the TEAP HFC and PFC Task Force. Accessed 12 July 2011. http://ozone.unep.org/teap/Reports/Other_Task_Force/HFCPFC.pdf.

USDA, National Agriculture Statistics Service. 2011. Poultry slaughter (June 2011). Accessed 26 May 2011. http://www.nass.usda.gov/Publications/Todays_Reports/reports/psla0611.pdf.

USDA. 2005. Operational guidelines: disposal, APHIS. National Animal Health Emergency Management System Guidelines Riverdale, Maryland: United States. Accessed 5 July 2011. http://www.aphis.usda.gov/emergency_response/tools/onsite/htdocs/images/nahems_disposal.pdf.

USEPA. 1975. Development document for proposed effluent limitations guidelines and new source performance standards for the poultry segment of the meat product and rendering process point source category. U.S. Environmental Protection Agency, Washington, DC.

Van Loo, E. J., W. Alali and S. C. Ricke. 2012. Food safety and organic meats. *Annu. Rev. Food Sci. Technol.* 3: 205–25.

Van Lunen, T. A. and D. M. Anderson. 1991. The utilization of acid-preserved poultry offal by growing-finishing pigs. *Can. J. Anim. Sci.* 71 (3): 935–8.

Van Lunen, T. A., R. L. Wilson, L. M. Poste and G. Butler. 1990. The effect of dietary poultry offal hydrolysate on the fatty acid composition and meat quality of pork. *Can. J. Anim. Sci.* 70 (4): 1041–52.

Verge, X. P. C., C. De Kimpe and R. L. Desjardins. 2007. Agricultural production, greenhouse gas emissions and mitigation potential. *Agric. For. Meteorol.* 142 (2–4): 255–69.

Wales, A. D., V. M. Allen and R. H. Davies. 2010. Chemical treatment of animal feed and water for the control of *Salmonella*. *Foodborne Pathogens Dis.* 7: 3–15.

Weakley, F., W. Roth, D. Hamm and C. L. Mehltret. 1972. Crosslinked protein glue containing chicken blood for interior-type plywood. *Poult. Sci.* 51 (2): 378–81.

Williams, A. G., E. Audsley and D. L Sandars. 2006. Determining the environmental burdens and resource use in the production of agricultural and horticultural commodities. Main Report. Defra Research Project IS0205. Cranfield University and Defra, Bedford, England.

Williams, C. M. 2011. Poultry waste management in developing countries. Food and Agriculture Organization of the United Nations: Poultry Development Review. Accessed 25 May 2011. http://www.fao.org/docrep/013/al716e/al716e00.pdf.

Williams, C. M., C. G. Lee, J. D. Garlich and J. C. H. Shih. 1991. Evaluation of a bacterial feather fermentation product, feather-lysate, as a feed protein. *Poult. Sci.* 70: 85–94.

Williams, C. M., C. S. Richter, J. M. Mackenzie, Jr. and J. C. H. Shih. 1990. Isolation, identification, and characterization of a feather-degrading bacterium. *Appl. Environ. Microbiol.* 56: 1509–15.

Williams, C. M. and J. C. H. Shih. 1989. Enumeration of some microbial groups in thermophilic poultry waste digesters and enrichment of a feather-degrading culture. *J. Appl. Microbiol.* 67: 25–35.

Williams, J. E. 1981. Salmonellas in poultry feeds – A worldwide review. Part III. Methods in control and elimination. *World's Poult. Sci. J.* 37: 97–105.

Wilson, A. 1998. *Wilson's Practical Meat Inspection.* 6th ed. Malden, MA: Blackwell Sci. Ltd, 287p.

Windhorst, H.-W. 2006. Changes in poultry production and trade worldwide. *World's Poult. Sci. J.* 62: 585–602.

World Bank Group. 2007. Environmental, health, and safety guidelines: Poultry processing. Accessed 25 May 2011. www.ifc.org/ifcext/sustainability.nsf/content/environmental guidelines.

World Resources Institute. 2015. Greenhouse gas protocol, stage 2 guidance. Available at: http://www.ghgprotocol.org/scope_2_guidance.

Wymyslowski, M., M. Luczak, A. Zawadzka, M. Imbierowicz and A. Chacuk. 2010. Methane fermentation of poultry slaughterhouse waste. *Pol. J. Chem. Technol.* 12 (3): 15–18.

Zaghloul, T. I., M. Al-Bahra and H. Al-Azmeh. 1998. Isolation, identification, and keratinolytic activity of several feather-degrading bacterial isolates. *Appl. Biochem. Biotechnol.* 70–2: 207–13.

Chapter 21

Organic systems for raising poultry

R. Michael Hulet, Penn State University, USA

1 Introduction

The organic production system is an important rearing system that is compatible with sustainable animal production (Castellini et al., 2006). The US Congress passed the Organic Food Production Act in 1990 (Oberholtzer et al., 2006); organic poultry production system standards were subsequently established within the National Organic Program (NOP) by the United States Department of Agriculture (USDA) in 2002. The USDA determined the goals of organic poultry production and these were aligned with 'cultural, biological and mechanical practices that support the cycling of on-farm resources, promote ecological balance, and conserve biodiversity' (http://www.ams.usda.gov/sites/default/files/media/NOP-2015StrategicPlan.pdf). These standards were similar to those established by the International Federation of Organic Agriculture Movement (IFOAM) and were overseen by the joint Food and Agricultural Organization/World Health Organization (FAO/WHO), and these standards were established in 1963 (www.ams.usda.gov/nop). The guidelines were established to promote a sustainable method of rearing poultry in an environment that allows natural behaviour to be expressed and eliminates pesticides, herbicides and other chemicals from feed, growing environment and carcass processing. While the definition concentrates on the physical factors of rearing, greater focus on the humane aspects such as outside access, sunlight, bird density and enrichment have dramatically changed the housing systems for meat and egg production.

http://dx.doi.org/10.19103/AS.2016.0010.24

Since its inception, the organic poultry market has experienced 20% annual growth (van der Sluis, 2007). As a result, there has been a change from rearing small numbers of birds to large commercial enterprises, which has led to the development of large-scale housing and processing systems to meet the NOP requirements. Organic production has also increased greatly in other regions such as Canada, the European Union and Australia and accounted for up to 46% of Australia's total egg production. In the United States, the leading states for organic poultry meat production are California, North Carolina, New York and Pennsylvania (Dimitri and Greene, 2002; van der Sluis, 2007). The change in production of US-certified organic poultry increased from 1997 to 2005 from nearly 800 000 to over 13 million animals (van der Sluis, 2007). A geographic summary of relative marketing features for US chicken eggs and meat showed that the Northwest region had 50% organic broiler meat featured, while the Southwest region had 15% organic egg and 38% organic chicken meat items. Other US regions featured organic eggs 39, 31, 25 and 0% of total egg and meat offerings, respectively, for Midwest, South, Northeast and Southeast regions (AMS – USDA, 2016). The important components of the organic production systems to be discussed are 'housing and environment, nutrition, health, food safety and processing/marketing' (Fanatico, 2008; Fanatico et al., 2009).

2 Bird housing for organic egg and meat production

Housing of birds for organic egg or meat production should be designed to maintain the comfort and health of the bird while still providing outdoor access, fresh air, sunlight, areas to dust-bathe and scratch. Outdoor access through pop-holes or doors allows the birds' access to the outside when the environmental conditions are equal to or better than the internal house environment. Windows that allow natural sunlight are a requirement for organic houses (Riddle, 2013). Because of the high brooding temperature requirements for broiler chicks and turkey poults, they are not allowed outdoor access until after four weeks of age, while pullets are not allowed outdoor access until the initiation of lay (approximately 18 weeks of age). Other housing condition factors include bird density, proper heating and ventilation, litter management, natural lighting and exercise and behaviour activities (Fanatico, 2008).

While not specified by the NOP, bird density is usually reduced for organic production when compared to commercial non-organic stocking rates; this is done in an effort to reduce microbiological loads and behavioural stress and strain on the birds, thereby reducing possible housing causes of morbidity and mortality. For turkeys and broilers, brooding densities are nearly doubled for organically reared in an effort to improve the starting environment. Housing for organic pullets have also employed reduced stocking densities when compared to non-organic commercial birds and are not reared on slatted floors; rather they are reared on littered floors (Berg, 2001). All litter sources must be organic and many producers clean out the litter after each flock to decrease chances of viral or bacterial disease challenges.

As previously mentioned, because of the birds' need to maintain its body temperature, no outdoor access is required for chicks or poults during the brooding period. Equipment necessary to provide heating and ventilation for organic birds is not different or limited to that used for non-organic poultry production. Heating equipment should keep the birds comfortable. With proper ventilation, house ammonia levels should be less than 25 ppm, while litter is kept dry in order to prevent foot pad dermatitis (NOSB, 2011).

Litter amendments are not used in organic programmes and that is why the reduced bird density and adequate ventilation are important strategies to reduce ammonia emissions and maintain a dry litter (Cengiz et al., 2011). Many organic producers prefer facilities that house fewer numbers of birds and use natural lighting and ventilation. Artificial lighting is allowed for organic production, but many local-certifying agencies encourage placement of windows to increase light intensity and bird activity (Fanatico, 2008).

3 Nutrient management for organic poultry farming

Nutrient management for organic production has to be done in a way that supports the cycling of on-farm resources and promotes ecological balance. Because of the efforts to decrease microbiological load, many producers clean out the litter after each flock, while others compost the litter in-house before starting a new flock (Ritz et al., 2004; McCrea et al., 2008). A complete management plan to ecologically handle the nutrients in a sustainable manner is a must. The idea of recycling nutrients is a basic tenet of the organic production system. Responsible storage, composting, delivery and spreading of poultry waste on soil at the correct agronomical rates should be part of the system. A nutrient management plan would also include properly taking care of bird mortalities either by composting, burial, rendering or by incineration. Proper disposal of mortality is also important for biosecurity and is a method of disease control (Blake and Donald, 1992).

Sustainability of the poultry production is evaluated by looking at an analytical assessment of energy flows and emissions to see how inputs and outputs contribute to the economical use of resources. The current production (2010), instead of the egg production of 50 years ago, was applied to a life cycle assessment (LCA) to see how our production methods have changed our use of energy resources. This technology, as established by the International Organization for Standardization (ISO, 2006), was used to quantify changes in environmental impacts. Pelletier et al. (2014) found that our current environmental footprint is 65% less for each kilogram of eggs produced than what we had in 1960. The reasons for this improvement in efficiency were improvement in egg production, lower emissions of ammonia and differences in feed composition (a major portion of the cost of production). Pelletier (2008) also performed an LCA for the US broiler industry and found that the feed accounted for 80% of the supply chain energy use, 82% of greenhouse gas emissions and around 97% of ozone-depleting, acidifying and eutrophying emissions for the broiler industry, while on-farm energy inputs due to ventilation, heating and lights contributed only 9%. Therefore, more concentration on improving LCA should be directed towards the feed supply chain portion rather than efficiency of egg and meat production.

4 Nutrition for organically raised poultry

One of the critical differences between non-organic and organic production is that in the latter, dietary ingredients must not be genetically modified, can only be mechanically processed and must be grown in an organically approved manner (Fanatico et al., 2009).

Corn and soybean meal are typically the main dietary energy and protein sources, respectively. As the birds' dietary protein requirements are based on their amino acid needs, the goal is to formulate diets using the minimum possible protein content, and then supplementing with the limiting synthetic essential amino acids up to the birds' requirement. However, only a limited amount of synthetic methionine (the most limiting amino acid in a corn–soybean meal diet) is currently permitted for use in an organic diet (Burley et al., 2015). Thus, the inability to supplement organic diets with adequate amounts of synthetic methionine limits the birds' performance or requires higher dietary protein levels. Alternatively, using protein supplements to meet the birds' methionine requirement results in excesses of other amino acids, leading to increased nitrogen excretion which is an environmental issue, and wet litter, which can impair bird health through foot pad problems and increased atmospheric ammonia (Riddle, 2013). Many ingredients such as Brazil nuts, spray dried egg whites, egg blend, fly larvae meal, naked oats, hulless barley, dehulled sunflower seed meal and fishmeal have been used as methionine-rich supplements but with limited success (Fanatico, 2010; Burley et al., 2013, 2015). Antibiotics, growth-promoting drugs or animal by-products are also nutrient supplements that are not allowed in organic diets (Van Overbeke et al., 2006).

Organic feed must be manufactured in a feed mill approved only for organic feed. This means that the mill must not have any conventional or genetically modified feedstuffs as well as herbicides or insecticides on its premises (Fanatico, 2008). Another limiting factor for the organic feed manufacturer is the local availability of feed corn and soy. Currently, many of the organic feed ingredients have to be transported long distances from as far away as Asia or South America, which greatly adds to the expense and availability of the diets. Organic diets are currently nearly twice the price of conventional diets for a similar dietary formulation (van der Sluis, 2007).

5 Managing bird health in organic flocks

Since organically reared birds are produced in a sustainable environment that does not rely on antibiotics, pesticides, herbicides and chemical tools to sustain production and maximize growth and performance, other management methods such as adequate growing space, litter depth and good nutrition are important for growth and development. Despite all efforts, birds reared in a NOP manner can be affected by virus, bacterial and coccidia similar to non-organically reared birds. *Clostridium perfringens* is another organism that can cause significant health problems in organic flocks in the form of a disease called necrotic enteritis (Shojadoost et al., 2012). This organism is ubiquitous in the environment; thus without antibiotic control, it can and does cause significant loses in organic flocks (Van De Weerd et al., 2009).

Furthermore, it was found in other studies (Berg, 2001) that organic birds had increased problems with endo- and ecto-parasites. Mites, lice and leg parasites are problems in organic layers and are hard to control with current available methods. Allowing birds to dust-bathe with diatomaceous earth, applying linseed oil externally in the house and dietary use of diatomaceous earth for worms have all had limited success without the use of pesticides (Fanatico, 2008). Other hock and leg issues in broilers and turkeys have been reported with organic production systems that are typically not observed in conventional systems (Van De Weerd et al., 2009).

The tools to help maintain health in organic flocks consist of vaccines, probiotics, prebiotics and essential oils. These products tend to reduce the intensity of the disease effects on the birds. However, if birds have a serious disease, then they must be treated, but they can no longer be marketed as organic products (Novak and Troche, 2006).

Probiotics, which are beneficial microbes, are fed to birds in an attempt to colonize the gut, thereby out-competing pathogenic microbes for available nutrients. Prebiotics are non-digestible ingredients that shift the gut environment in favour of the beneficial microbes. Examples of prebiotics are *Lactobacillus*, fructo- and mannen-oligosaccharides (Pedroso, et al., 2013). Vaccinations have been an essential part of the health programme to prevent disease in organic flocks. Control of diseases such as coccidiosis, Marek's disease, Newcastle, infectious laryngotracheitis and infectious bronchitis would be impossible without the use of vaccinations (Berg, 2002).

As previously mentioned, necrotic enteritis causes significant losses in flocks of organic and antibiotic-free flocks and is due to exposure to *Clostridium perfringens*. Mortalities from this and other pathogenic organisms can be significantly higher than for non-organic flocks. For example, where a non-organic broiler flock would have an average mortality of 4%, organic flocks of broilers, turkeys or layers can generate mortalities 2–2.5 times higher (Fanatico, 2008). Unfortunately, many times the flocks will remain untreated because of the lack of marketability as an organic flock or due to the lack of options for processing the flock after treatment. Either way the economic significance for the producer is great. For this reason, biosecurity and sanitation of feed and water are of utmost importance. While birds are allowed outside, care to prevent wild birds, and especially waterfowl, from mixing with the domestic birds is very important to prevent the possible transfer of disease-causing organisms (Berg, 2002).

6 Food safety in organic poultry production

Much of the sustainable nature of organic and pasture poultry production has been based on the perceived improvement in food quality and safety, that is, the organic products and the ingredients that they are fed have less contamination from herbicides, pesticides, chemically processed feed ingredients, growth-promoting antibiotics and from the use of non-genetically modified cereal grains and show improvements in carcass cleanliness, that is, decreases in bacteria such as *Salmonella*. Let us evaluate the evidence and see if this is correct (Bailey and Cosby, 2005).

Van Loo et al. (2012) remarked that 'consumers are not aware that USDA organic standards are based only on methods of production and processing rather than product quality or safety'. Further, the general perception is that birds grown on a range or in organically controlled conditions are 'healthier' or 'safer' as a food item. In a comparison between organic and non-organically grown broilers, Heuer et al. (2001) found that 100% of the organic birds had *Campylobacter*, while only 36.7% of birds grown in a conventional manner had *Campylobacter* isolated from their carcass. However, a study by Sapkota et al. (2011) found a significant reduction in the presence of multi-drug resistant *Enterococcus* in broiler houses that transitioned to organic production versus those that remained in non-organic production. A study by Bailey and Cosby (2005) that examined the incidence of *Salmonella* in organic and free-range poultry flocks reported 31% for the organic free-range birds and 25% for birds grown in an all-natural (dietary animal protein) antibiotic-free manner. These values are far greater than that reported by USDA Food Safety and Inspection Service (2003) rates for non-organically reared broilers that ranged from 9% to 12%. In a recent article, Alali et al. (2010) concluded that in a survey for the prevalence of *Salmonella*, the organic broiler farms had significantly less (4.3%, 13/300) incidence when compared to non-organic flocks (28.8%, 115/400). Multi-drug resistance was also found greater for the non-organic broiler flocks. However, when sources of *Salmonella* were investigated, the feed source for many farms was highly contaminated and appeared to be a significant factor for the prevalence of contamination on the farms (5.3% for organic flocks and 27.5% for conventional broiler flocks). In an epidemiological survey by the CDC, they found a *Salmonella* incidence of 1.62% for the general population and poultry was identified as an important vehicle for its transmission. Other critical factors listed for contamination by poultry products were seasonal effects, hatchery, breeder flock, feed, water, insects and rodents.

7 Processing and marketing

The last stage of production overseen and certified by the organic boards is processing and product labelling. In order to legally process organic poultry, the processing plant must be approved by the NOP. This applies to using approved detergents, sanitizers, pest control methods and preventing the comingling of organic and non-organic carcasses (Fanatico, 2008). For plants that process both organic and non-organic flocks, the former will usually be run through the plant first in the morning with an adequate break time, so that the non-organic birds are not comingled with organic birds in the chill tank. Labelling of birds as organic must also be approved by the USDA prior to use. Producers slaughtering less than $5000 worth of product per year are exempt from the label certification, but they cannot use the USDA organic label (Fanatico, 2008).

8 Summary

The market demand continues to increase for organic poultry that are raised in an environmentally sound, sustainable and healthy manner. Investigation areas that would be beneficial in supporting a sustainable organic production of eggs and broiler meat is identification of organic ingredients that could provide required levels of non-essential amino acids, specifically methionine and lysine. Another research area is establishing levels of probiotics, prebiotics, essential oils and organic acids that are readily available, effective and economical to meet the need of this marketing niche. Management and production strategies for poultry production and food safety continue to evolve and are consistent with the NOP goals of 'promoting cultural, biological, and mechanical practices that support the recycling of on-farm resources, create an ecological balance, and conserve biodiversity'.

9 Where to look for further information

Further information and updates for rules and regulation regarding organic poultry production can be found in the following websites:

- www.oie.int
- https://www.ams.usda.gov/rules-regulations/organic/nosb

10 References

Alali, W. Q., S. Thakur, R. D. Berghams, M. P. Martin and W. A. Gebreyesm, 2010. Prevalence and distribution of Salmonella in organic and conventional broiler poultry farms. *Foodborne Pathog. Dis*. 10: 1078–89.

Anonymous, 2004. Progress report on Salmonella testing of raw meat and poultry products. 1998–2003. U.S. Department of Agriculture Food Safety and Inspection Service. August. http://www.fsis.usda.gov/OPHS/haccp/salm6year.htm.

Agricultural Marketing Service, 2016. Weekly USDA Certified Organic Poultry and Eggs. USDA Livestock, Poultry & Grain Market News. http://search.ams.usda.gov/MNDMS/2013/07/PY20130701WOrganic.pdf.

Bailey, J. S., and D. E. Cosby, 2005. Research Note: Salmonella prevalence in free-range and certified organic chickens. *J. Food Prot*. 68 (11): 2451–3.

Berg, C., 2001. Health and welfare in organic poultry production. *Acta Vet. Scan* 95 (Suppl 1): 37–45.

Berg, C., 2002. Health and welfare in organic poultry production. Proceedings of the 14th Internordic Symposium of the Nordic Committee for Veterinary Scientific Cooperation (NKVet). *Acts Vet. Scan* 43 (Suppl 1): S37.

Blake, J. P., and J. O. Donald, 1992. Alternatives for the disposal of poultry carcasses. *Poult. Sci*. 71 (7): 1130–5.

Burley, H. K., P. H. Patterson and K. E. Anderson, 2015. Alternative ingredients for providing adequate methionine in organic poultry diets in the US with limited synthetic amino acid use. *World's Poult. Sci. J*. 71 (3): 493–504.

Burley, H., P. H. Patterson and R. M. Hulet, 2013. Alternative ingredients to synthetic methionine for organic broiler diets. *Poult Sci*. 92 (Suppl.): M65 (Abstract).

Castellini, C., S. Bastianoni, C. Granai, A. D. Besco and M. Brandt, 2006. Sustainability of poultry production using the energy approach; Comparison of conventional and organic rearing systems. *Agric. Ecosyst. Environ*. 114 (2–4): 343–50.

Cengiz, O, J. B. Hess and S. F. Bilgili, 2011. Effect of bedding type and transient wetness on footpad dermatitis in broiler chickens. *J. Appl. Poult. Sci*. 20 (4): 554–60.

Dimitri, C., and C. Greene, 2002. Recent growth patterns in the U.S. organic foods Market. Economic Research Service/USDA, AIB-797, pp. 1–42. http://www.ers.usda.gov/publications/aib-agricultural-information-bulletin/aib777.aspx (accessed December 2015).

Fanatico, A. C., 2008. Organic Poultry Production. National Sustainable Agriculture Service, pp. 1–16. https://attra.ncat.org/attra-pub/summaries/summary.php?pub=336 (accessed December 2015).

Fanatico, A. C., 2010. Organic Poultry Production: Providing Adequate Methionine. National Sustainable Agriculture Service, pp. 1–20. https://attra.ncat.org/attra-pub/summaries/summary.php?pub=336 (accessed December 2015).

Fanatico, A. C., C. M. Owens and J. L. Emmert, 2009. Organic poultry production in the U.S. *Broilers. J. Appl. Poult. Res*. 18 (2): 355–66.

Heuer, O. E., K. Pederson, J. S. Anderson and M. Madsen, 2001. Prevalence and antimicrobial susceptibility of thermophilic *Campylobacter* in organic and conventional broiler flocks. *Lett. Appl. Microbiol*. 33: 269–74.

ISO, 2006. Life cycle assessment principles and framework. International Organization for standardization. Geneva, Switzerland.

McCrea, B. A., K. S. Macklin, R. A. Norton, J. B. Hess and S. F. Bilgili, 2008. Recovery and genetic diversity of *Escherishia coli* isolates from deep litter shallow litter and surgical shoe covers. *J. Appl. Poult. Res*. 17 (2): 237–42.

National Organic Standards Board (NOSB) to the National Organic Program (NOP), 2011. Agricultural Marketing Service – USDA. https://www.ams.usda.gov/sites/default/files/media/NOP%20Livestock%20Final%20Rec%20Animal%20Welfare%20and%20Stocking%20Rates.pdf.

Novak, C., and C. Troche, 2006. Use of Bio-Mos® to Control Salmonella and Campylobacter in Organic Poultry. https://www.researchgate.net/publication/254365676_Use_of_Bio-MosR_to_Control_

Salmonella_and_Campylobacter_in_Organic_Poultry?enrichId=rgreq-a0cf0727-b767-4689-a5ad-7a290ff5dfb9&enrichSource=Y292ZXJQYWdlOzI1NDM2NTY3NjtBUzoyNDcyMTkzNTgwMDcyOTZAMTQzNTk1MzA5NjIzOQ%3D%3D&el=1_x_2.

Oberholtzer, L., C. Greene and E. Lopez, 2006. Organic poultry and eggs capture high price premiums and growing share of specialty markets. Outlook Report from the Economic Research Service. http://www.ers.usda.gov/media/866670/ldpm15001_002.pdf.

Pedroso, A. A., A. L. Hurley-Bacon, A. S. Zedek, T. W. Kwan, A. P. Jordan, G. Avellaneda, C. L. Hofacre, B. B. Oakley, S. R. Collett, J. J. Maurer and M. D. Lee, 2013. Can probiotics improve the environmental microbiome and resistome of commercial poultry production? *Int. J. Environ. Res. Public Health* 10: 4534–59.

Pelletier, N. 2008. Environmental performance in the U.S. broiler poultry sector: Life cycle energy use and greenhouse gas, ozone depleting, acidifying and eutrophying emissions. *Agric. Syst.* 98: 67–73.

Pelletier, N., M. Ibarburu and H. Xin, 2014. Comparison of the environmental footprint of the egg industry in the United States in 1960 and 2010. *Poult. Sci.* 93 (2): 241–55.

Riddle, J., 2013. Requirements for organic poultry production – extension. 27 August. http://articles.extension.org/pages/69041/requirements-for-organic-poultry-production.

Ritz, C. W., B. D. Fairchild and M. P. Lacy, 2004. Implications of ammonia production and emissions from commercial poultry facilities: A review. *J. Appl. Poult. Res.* 13 (4): 684–92.

Sapkota, A. R., R. M. Hulet, G. Zhang, P. McDermott, E. Kinney, K. J. Schwab and S. W. Joseph, 2011. Lower prevalence of antibiotic-resistant *Enterococci* on US conventional poultry farms that transitioned to organic. http://dx.doi.org/10.1289/ehp.1003350 (accessed December 2015).

Shojadoost, B., A. R. Vince and J. F. Prescott, 2012. The successful experimental induction of necrotic enteritis in chickens by *Clostridium perfringens*: A critical review. *Vet. Res.* 43 (26 October): 74.

Van De Weerd, H. A., R. Keatinge and S. Roderick, 2009. A review of key health-related welfare issues in organic poultry production. *World's Poult. Sci. J.* 65 (4): 649–84.

van der Sluis, Weibe, 2007. More organic poultry meat and eggs in the US. *World Poult.* 23 (3): 26–8.

Van Loo, C. T., W. Alai and S. C. Ricke, 2012. Food safety and organic meats. *Ann. Rev. Food Sci. Technol.* 3: 203–25

Van Overbeke, I., L. Duchateau, L. De Zutter, G. Albers and R. Ducatelle, 2006. A comparison survey of organic and conventional broiler chickens for infectious agents affecting health and food safety. *Avian Dis.* 50 (2): 196–200.

Chapter 22

Helping smallholders to improve poultry production

Robert Pym, University of Queensland, Australia; and Robyn Alders, University of Sydney, Australia

1 Introduction

The world's human population as of 2016 is estimated at about 7.4 billion, but is predicted to increase to more than nine billion by 2050 (UN, 2015a). Presently, almost 85% of people live in developing countries and this proportion is predicted to increase due to essentially stabilized population numbers in developed countries compared to an average population increase of about 1.7% p.a. in developing countries (PRB, 2015). Accompanying the rapid population increase in developing countries is an even greater increase in the demand for animal products as a result of increasing affluence and product availability. It is estimated that the livestock sub-sector presently accounts for about 30% of the GDP in developing countries and is growing rapidly (FAO, 2014).

http://dx.doi.org/10.19103/AS.2016.0010.25

In rural communities in developing countries, livestock are fundamental to the livelihoods of families and are often the sole asset of resource poor farmers. In such communities, the majority of households keep a small flock of semi-scavenging poultry, which in many cases contributes in a meaningful way to household nutrition and food security, poverty alleviation and the empowerment of women as the principal poultry keepers (Gueye, 2005; Alders and Pym, 2009; de Bruyn et al., 2015). Whilst such flocks numerically make up the large majority of poultry kept in rural regions of most developing countries (Pym et al., 2006), more intensive and/or larger-scale production systems are also present. The latter flocks, often kept by the more entrepreneurial and wealthier members of the community, also make a valuable contribution to livelihoods and to the fabric of society in rural communities. Details of the different production systems and of their contribution to livelihoods are provided in the following section.

Many factors have an impact on the efficiency and sustainability of the more extensive production systems involving poultry, making them extremely complex. Such factors include the genetic makeup of the birds, other livestock and cropping, the climate, available shelter, available scavenging feed resources, availability and level of supplementary feeding, access to water, population density, endemic diseases, predation, security, level of management, gender issues, the keeper's financial resources, access to microfinance and credit, access to markets, etc. (Alders and Pym, 2009; Kryger et al., 2010). Even with the more intensive forms of smallholder poultry production, the system is complex and solutions to problems generally far from simple. It is widely recognized that any attempt to improve the well-being of smallholders through livestock production requires a farming systems approach, wherein the entire system is considered and farmers are closely involved in the process of identifying problems and in developing appropriate and achievable solutions (FAO, 2014). For the most part, effective interventions need to be simple and affordable.

Thus, in any consideration of how research might benefit poultry flock-owning smallholder families in developing countries, it is critical to focus on the most effective ways in which improvements can be identified and, very importantly, on how to facilitate adoption of appropriate interventions by the smallholders (Rota et al., 2014). To achieve this, there needs to be on-the-ground direct contact between members of the extended research team and the smallholder farmers within the region/area identified, which suggests the need for at least moderate-scale development project work, as discussed in the following section. In most situations this should include involvement of and support from the country's agricultural ministry, regional agricultural extension services and the village leadership and community (Bagnol, 2014).

2 Types of smallholder poultry production systems

Poultry species kept by smallholders in developing countries include chickens, ducks, turkeys, guinea fowl, geese, quail, pigeons and pheasants. As in the larger-scale commercial industry, chickens make up more than 90% of the total in most countries and regions, and are by far the most important poultry species (Table 1). The term 'poultry' is thus often synonymous with 'chickens' and for the purposes of this discussion, this will be assumed unless other species are referred to specifically.

Smallholder poultry production systems in developing countries range from small- to medium-size scavenging flocks through semi-intensive operations to small-scale intensive

Table 1 Distribution of poultry species populations by region in 2011

	Chickens (%)	Ducks (%)	Geese and Guinea fowl (%)	Turkeys (%)	Other poultry (%)
Africa	96.03	1.10	0.85	1.21	0.81
Americas	93.95	0.45	0.01	5.58	0.00
Asia	88.07	8.99	2.70	0.10	0.14
Europe	91.30	2.65	0.89	5.03	0.13
Oceania	96.45	1.60	0.07	1.88	0.00
World	90.55	5.53	1.67	2.09	0.15

Source: FAOSTAT (2012).

systems utilizing commercial broilers or layers (Besbes et al., 2012). Over the years there has been a variety of classification systems used to describe the different smallholder poultry production systems encountered in developing countries. For the purposes of this discussion, the classification system proposed in FAO (2014) has been adopted, as shown in Table 2.

Under this classification system, there are four recognized systems, namely (i) small-extensive scavenging, (ii) extensive scavenging, (iii) semi-intensive and (iv) small-scale intensive (Thieme et al., 2014) which developed from earlier classification systems. Because of access to land and availability of scavengable food, the first two categories are mostly restricted to rural areas, whereas the latter two categories can be found in rural, peri-urban and urban areas. In the biosecurity-based 'sector' classification system introduced by FAO (FAO, 2008a), sector 4 would more or less equate to presently classified systems (i) and (ii).

The term 'family poultry' (Gueye, 2000) is now widely used to describe smallholder poultry production which includes all of the above four systems, as all are practised by individual families as a means of obtaining food security, income and gainful employment (Besbes et al., 2012). Because of the wide variation in the elements of the four systems, it is important that the production system in question is clearly defined when discussing issues, problems and solutions relating to family poultry.

2.1 Small- to medium-scale extensive scavenging systems (i) and (ii)

In the rural regions of most developing counties, the large majority of families keep a small household scavenging poultry flock of between 1 and 50 birds; mostly indigenous breed chickens, but sometimes other avian species including ducks, turkeys and guinea fowl (Aini, 1990; Gueye, 1998). These 'village' poultry constitute about 80% of poultry stocks in the lower-income developing countries (Gueye, 2003; Pym et al., 2006), thus contributing considerably to livelihoods of rural families.

Poultry keeping by villagers is very traditional and has its roots in ancient cultures. Not only do the birds supply the family with meat and eggs for home consumption, but they are a resource that can be sold to provide cash to purchase household items and

Table 2 Characteristics of the four family poultry production systems

Criteria	Small-extensive scavenging	Extensive scavenging	Semi-intensive	Small-scale intensive
Production/ farming system	Mixed, poultry and crops, often landless	Mixed, livestock and crops	Usually poultry only	Poultry only
Other livestock raised	Rarely	Usually	Sometimes	No
Flock size	1–5 adult birds	5–50 adult birds	50–200 adult birds	>200 broilers >100 layers
Poultry breeds	Local	Local or cross-bred	Commercial, cross-bred or local	Commercial
Source of new chicks	Natural incubation	Natural incubation	Commercial day-old chicks or natural incubation	Commercial day-old chicks or pullets
Feed source	Scavenging; almost no supplementation	Scavenging; occasional supplementation	Scavenging; regular supplementation	Commercial balanced ration
Poultry housing	Seldom; usually made from local materials or kept in the house	Sometimes; usually made from local materials	Yes; conventional materials; houses of variable quality	Yes; conventional materials; good-quality houses
Access to veterinary services and veterinary pharmaceuticals	Rarely	Sometimes	Yes	Yes
Mortality	Very High, >70%	Very High >70%	Medium to High 20% to >50%	Low to Medium <20%
Access to reliable electricity supply	No	No	Yes	Yes
Existence of conventional cold chain	No	Rarely	Yes	Yes
Access to urban markets	Rarely	No, or indirect	Yes	Yes
Products	Live birds, meat	Live birds, meat, eggs	Live birds, meat, eggs	Live birds, meat, eggs
Time devoted each day to poultry management	<30 minutes	<1 hour	>1 hour	>1 hour

Source: Thieme et al. (2014).

other food or pay for the educational needs of the children (Gueye, 2003; Kryger et al., 2010). Additionally, chickens and eggs are widely used in many traditional and religious ceremonies, can be given as gifts to honoured guests and the birds are active in pest control and produce manure for use as a fertilizer (Alders and Pym, 2009). A particularly desirable feature of smallholder poultry keeping in comparison with other livestock keeping/farming is that women are generally responsible for looking after the poultry flock. This empowers women and translates into better nutrition and greater educational opportunities for the children, particularly where productivity is improved through poultry production improvement programmes (Gueye, 2005; Bagnol, 2009).

The scavenging system means that feed costs are eliminated or dramatically reduced in comparison with intensive forms of production. The impact of this on productivity depends on the nature of the scavenging feed resource base (SFRB) and the level of supplemental feeding (Roberts and Gunaratne, 1992). The latter, however, is usually quite limited and typically restricted to household food scraps and small amounts of grain or grain by-products (Sonaiya, 1995). The SFRB includes plant seeds and fruits, grain, earthworms, snails, frogs, insects etc., and is influenced by geographic location, local vegetation and microclimate, physical restrictions on the scavenging area, the poultry and other scavenging animal population density, as well as season. In most cases, the SFRB is limited, which restricts the overall effective scavenging poultry population (Sonaiya, 2004, 2014).

The requirement to scavenge means that the birds are not confined in enclosures and are free to range and forage, at least in the daytime. In many cases housing for night-time protection can be very rudimentary or even absent, in which case the birds roost at night in trees or on buildings (Gueye, 1998). Shelters, if provided, range from in or beneath the family's home or in the houses/barns of other livestock, through rudimentary poultry houses built from locally sourced materials, to more sophisticated housing providing shelter from the elements and reasonable security from predators (Alders, 2014a). In nearly all cases, the cost and sophistication of housing is very much lower than used in the more intensive production systems (Glatz and Pym, 2013). Night-time predation and theft are often important sources of loss in young chicks and adult birds respectively, so secure night-time housing is an important element in reducing losses (Pym and Alders, 2012).

There are a number of reasons why indigenous breed poultry rather than commercial broilers or layers are used almost exclusively in scavenging production systems (see Table 2): Indigenous breeds have retained the capacity to go broody and thus are capable of hatching their own eggs and reproducing the flock, something that commercial broilers and layers have not (Gueye, 1998); whilst productivity under confined rearing and high level feeding is much greater in commercial breeds of broilers and layers than in indigenous breeds, under village scavenging conditions, commercial birds perform very poorly due to much higher nutrient requirements and lack of adaptation to the typically harsh environment (Besbes, 2009); due to natural selection, indigenous breed birds can run fast and fly (to roost in trees) and thus escape from predators, something that commercial breeds (especially broilers) are unable to (Gueye, 2000); as a result of natural selection, indigenous breeds are for the most part more high-temperature tolerant (Ndiweni, 2010) and more resistant to a number of diseases typically encountered in the village environment, than commercial genotypes (Besbes et al., 2012); and the meat and eggs from indigenous breed birds are generally preferred to that from commercial broilers and layers, and buyers will often pay a premium for the former, particularly when calculated on a weight basis (Sonaiya, 2009).

The typical annual egg production of indigenous breed hens ranges from about 40 to 100 with an average around 60, compared to commercial layers at about 300 (Pym et al., 2006). This very large difference, however, needs to be seen in the context of the different physical and nutritional environments under which the birds are reared, and to take account of the fact that the village hen spends a very significant proportion of the year (65–75%) hatching the eggs and rearing the chicks, during which time she is out of lay (Pym and Alders, 2012). Having said this, even when these birds are placed in laying cages and given *ad libitum* access to good-quality layer diets, their laying performance is much lower than commercial layers (Sorensen, 2010). This has particular ramifications for farmers contemplating intensive egg production by using indigenous breeds to capitalize on consumer preference for the eggs from indigenous hens; the premium paid for their eggs must be very considerable to compensate for their much lower level of production.

Typical growth rates of indigenous birds are also much lower than that of commercial broilers, even under ideal management and feeding conditions (Besbes, 2009). Under village conditions, male birds that are surplus to breeding and cock fighting requirements are typically sold at between 15 and 20 weeks of age weighing between 800 and 1200 g (Pym et al., 2006). The preference for indigenous poultry meat expressed not only by rural communities but also by a significant proportion of urban and peri-urban dwellers (Sonaiya et al., 1999) is related to the older, firmer and more flavoursome meat from the indigenous birds, which is more suitable for traditional forms of cooking than broiler meat (Aini, 1990).

The village hen is for the most part a very effective incubator and brooder. She will typically lay a clutch of between 10 and 14 eggs and sit on these to hatch between 8 and 12 chicks, with a typical hatchability of 75 to 85% (Sonaiya et al., 1999). Whilst it is commonplace to see a hen scavenging in the village environment with up to 10 or 12 baby chicks not long after they have hatched, it is rare to see a hen with more than 4 or 5 chicks once they have reached about 6 weeks of age (Pym and Alders, 2012). This typical attrition rate of between 50 and 80% is due to the effects of malnutrition, disease, predation and climatic exposure (Farrell et al., 2000) and is undoubtedly the greatest single factor limiting the impact of poultry keeping on household food security and income generation.

Mortality in older birds is also often quite high, due principally to the effects of disease with a lesser impact of malnutrition and predation. It is generally recognized that in the absence of effective vaccination programmes, Newcastle disease (ND) frequently accounts for a high proportion of the disease-related deaths, but other diseases such as avian influenza, fowl pox and fowl cholera are also involved in significant losses at times (Alders, 2013).

2.2 Semi-intensive and small-scale intensive production systems (iii) and (iv)

As noted in Table 2, these systems differ from the scavenging systems with the requirement to provide feed and housing and the use of 'improved' breed (commercial) or cross-bred genotypes. The semi-intensive system is an intermediate step between scavenging and intensive systems, with access to foraging but with most of the energy and protein requirements of the birds being met through the supplied feed (Ravindran and Blair, 1993). The feed supplied is often a mixture of local ingredients and commercially prepared feed, and is frequently not particularly well balanced to meet the nutritional requirements of the birds (Ravindran, 2013). This system also frequently utilizes 'crossbred' birds which include some infusion from both commercial and indigenous breeds to capitalize on the desirable characteristics of each (Fotsa et al., 2014).

The small-scale intensive system typically utilizes commercial genotypes and commercially prepared 'balanced' diets. The broiler or layer enterprises essentially mirror larger-scale commercial operations, but with limited control of the environment, lower levels of biosecurity and no automation. Feed is often sourced from small local feed mills and the quality is frequently sub-standard compared to that from the larger urban-based feed mills (Ravindran, 2013). Added to this, the cost and quality of day-old chicks is often higher and lower respectively than those available to the larger operators, which can impact significantly on profitability particularly where the smallholder does not enjoy the benefits of a niche market (Besbes et al., 2012).

Whilst smallholders producing birds under these two systems usually have access to veterinary diagnostic services and vaccines as required, such access is often limited and relatively expensive unless it is government subsidized (Alders and Pym, 2009).

There are significant obstacles to transition from small-scale scavenging production to intensive production. The many millions of families in the former group are essentially poultry *keepers*, not poultry *producers*, and they are for the most part very poor and risk-averse. There is undeniably a place for the more entrepreneurial of these to move to more intensive production, but if more than a small proportion were to make this move, the consequences could be quite devastating in terms of oversupply of meat and eggs and the resultant loss in profitability. Through poultry improvement programmes employing simple, affordable and effective interventions, small-scale extensive poultry production can impact meaningfully on household income, food security and nutrition, the empowerment of women as the traditional poultry keepers and improved educational opportunities for children (Pym and Alders, 2012). The increased returns from the enterprise provide an opportunity for the keeper to move to other income-generating activities, including farming of other livestock.

At the FAO-sponsored and organized International Poultry Conference *Poultry in the 21st Century: Avian Influenza and Beyond* held in Bangkok in November 2007 (FAO, 2008b), the question was posed 'Will there be smallholder poultry development in 2030?' Delegates were unanimous in answering this in the affirmative (McLeod et al., 2009). It would seem that smallholder poultry production will be with us for a very long time, which argues strongly for global efforts to maximize the efficiency and profitability of this activity with such potential impact upon the well-being of rural poor communities in developing countries.

3 Development objectives for family poultry production

Because it is a fundamental part of the fabric of society in poor rural communities in developing countries, small-scale family poultry production has impacted meaningfully on the United Nations' Millennium Development Goals (MDGs) (UN, 2008) and now has the capacity to contribute to a number of the UN Sustainable Development Goals (SDGs; Table 3) (UN, 2015b). The overall development objectives of family poultry production are to improve food security (MDG 1; SDGs 1 and 2), contribute to gender equity and empowerment of women (MDG 3; SDG 5), assist the well-being of the rural population (MDGs 2, 4, 5, 6 and 8; SDGs 3, 4, 8 and 12) and contribute to environmental sustainability (MDG 7; SDG 15) in

Table 3 The United Nations Sustainable Development Goals 2016–30 impacted by family poultry development

Goal 1	End poverty in all its forms everywhere
Goal 2	End hunger, achieve food security and improved nutrition and promote sustainable agriculture
Goal 3	Ensure healthy lives and promote well-being for all at all ages
Goal 4	Ensure inclusive and equitable quality education and promote lifelong learning opportunities for all
Goal 5	Achieve gender equality and empower all women and girls
Goal 8	Promote sustained, inclusive and sustainable economic growth, full and productive employment and decent work for all
Goal 12	Ensure sustainable consumption and production patterns
Goal 15	Protect, restore and promote sustainable use of terrestrial ecosystems, sustainably manage forests, combat desertification, halt and reverse land degradation and halt biodiversity loss

Source: UN (2015b).

developing countries (Alders and Pym, 2009; de Bruyn et al., 2015). This can be achieved through the following specific development objectives (Besbes et al., 2012):

- Improve family poultry production and productivity in a sustainable manner
- Adapt family poultry production to the changing environment
- Maintain the diversity of local bird types
- Improve access to markets
- Build the capacity of stakeholders in family poultry production
- Raise awareness and influence livestock policy to promote family poultry production.

Achieving the above development objectives involves the development and implementation of strategies which comprehensively incorporate the fundamental elements of one or more of the objectives, namely (i) bird health and disease, (ii) breeding and genetics, (iii) management and nutrition, (iv) marketing and economics, (v) training and education and vi) strengthening the policy environment. The following section discusses the importance of each of these and the opportunities to impact on the development objectives.

4 Preventing disease in poultry

Health is an essential component of family poultry production with the production system(s) involved dictating the relevant poultry and public health issues (Alders et al., 2014a). Reducing the impact of disease on poultry production is achieved through prevention and control and the overall principles are similar irrespective of the production system concerned. Disease incorporates conditions due to infectious and transmissible pathogens and non-communicable disease (NCD) such as ascites syndrome and lameness.

4.1 Lack of robust data on the prevalence of key infectious diseases

Worldwide, vast numbers of family poultry birds continue to die annually from preventable diseases, such as ND (Alders, 2012). Despite significant investment in the control of HPAI H5N1, the development of cost-efficient and robust poultry disease surveillance, prevention and control systems remains a significant challenge (Gardner and Alders, 2014) with under-reporting of poultry diseases a chronic problem in many countries, both 'developed' and 'developing' (Cattoli et al., 2010; Ogundipe et al., 1989). This lack of data makes it difficult to estimate the true impact of poultry disease on communities and therefore its impact on food security and household poverty. For example, basic estimates of the direct economic impact of ND in Tanzania range from around 22 million US dollars (USD; Harun et al., 2009) to over USD 34 million (Minga and Nkini, 1986) annually.

In the south-eastern region of Africa, fowl pox has emerged as an important problem in village chickens following the control of ND, while in some parts of Asia, fowl cholera is widespread. Duck plague is a serious constraint in Southeast Asia. Infectious bursal disease (also known as Gumboro disease) has proved to be a major problem for small-scale intensive poultry units. Diseases related to poor nutrition, for example, vitamin A deficiency, may have a seasonal appearance in areas where the SFRB is limiting. Internal and external parasitism is also widespread (Alders, 2012).

Sensitive disease surveillance systems that provide the earliest warnings possible are those that are capable of detecting diseases that have clinical signs compatible with priority diseases. For example, in the case of HPAI, a zoonotic disease of global importance, the list of differential diagnoses includes: ND, acute fowl cholera (FC), infectious laryngotracheitis (ILT) and infectious bursal disease (IBD, Gumboro disease) (Alders et al., 2007; OIE, 2012). Each of these diseases also has a significant economic impact on poultry production in addition to presenting a clinical picture similar to HPAI (Alders, 2012). Further research and development activities are required to support the development of cost-efficient and sensitive surveillance systems that work for both producers and birds and for the wider human society. Where poultry disease surveillance and diagnosis is incomplete, participatory epidemiology can be employed to identify diseases or disease syndromes of importance, which can be confirmed by laboratory diagnosis. Participatory epidemiology (Catley et al., 2012) provides a cost-effective complement to classical approaches to disease surveillance and diagnosis.

4.2 Options for the sustainable prevention of infectious disease in poultry

Prevention can be achieved through bioexclusion, vaccination and genetic resistance:

- *Bioexclusion* – investing in adequate biosecurity practices remains difficult for family poultry producers, especially small-scale intensive producers with low profit margins

(Alders, 2012; FAO, 2008a). A lack of understanding of living conditions and resource-limitations facing family poultry producers can lead to miscommunication between producers and animal health officers with biosecurity and other husbandry measures for family poultry rarely developed in collaboration with the producers themselves (Hickler, 2007).

- *Vaccination* – an important component of commercial flock management strategies in many countries, as it is the most cost-effective means of preventing and controlling the spread of economically important diseases such as Marek's disease, ND and infectious bursal disease (Zander et al., 1997). Vaccination of village poultry in rural areas against killer diseases such as ND has become an increasingly viable option with the development of thermotolerant vaccines (Alexander et al., 2004). Vaccination can support programmes for eradicating diseases such as highly pathogenic avian influenza (HPAI) when effectively implemented in conjunction with other control methods such as quarantine and movement control, culling, improved producer and trader knowledge; improved sanitation and husbandry; improved biosecurity (both bioexclusion and biocontainment) and improved food safety (Capua and Alexander, 2006). It is important to combine vaccination with other control methods because the vaccines currently available prevent clinical disease but do not prevent infection (Ahlers et al., 2009; Alders et al., 2007). Research continues into cost-efficient options for delivering vaccine to the village level (such as the development of a low-dose fast-dissolving vaccine tablet that enhances the shelf life of ND vaccines; Lal et al., 2014) and non-veterinary factors that improve vaccination success rates (Alders, 2014b); and/or
- *Genetically resistant species and breeds* – genetic selection for disease resistance has had past success, for example, the development of chicken strains resistant to Marek's disease. However, over the past two decades, the commercial chicken industry has emphasized the selection of production traits over disease resistance. A 2008 study reported that commercial pure lines of chicken, both broiler (meat) and layer (egg) lines, are missing significant genetic diversity found in non-commercial chickens. Future food production challenges are unpredictable and likely will include new diseases or more virulent recurring diseases, environmental changes necessitating alternatives. Therefore, a healthy genetic reservoir in food-producing animals remains as crucial as ever. Muir et al. (2008) suggested that non-commercial flocks, including those found in many underdeveloped and developing countries, could potentially represent the reservoir opportunity for alleles 'missing' from commercial pure line stocks. This interest in the genetic heterogeneity of indigenous chickens now also includes programmes to enhance genetic resistance through selective breeding for disease control in low-input poultry production systems (ILRI, 2010).

4.3 Achieving efficient control of poultry diseases

Control is based on biocontainment to stop or reduce to a minimum the transmission of the causative agent. The efficient and effective control of a number of infectious poultry diseases require the quarantining, vaccination where appropriate and slaughter of affected flocks. This approach has been successful where prior agreements have been made between producers and government agencies and both parties are in accord and trust the other party to uphold their side of the agreement. In cases where quarantine and slaughter have been used by animal health authorities without prior agreement and adequate compensation of producers, the long-term consequences for disease control have been seriously impacted as producers no longer report diseases and handle outbreaks by salvaging what value they can from infected flocks (i.e. consumption and sale). This situation simply aids the further spread of the infectious agent (Alders, 2012).

4.4 Non-communicable disease and health

The selection of fast-growing broilers has been accompanied by a range of NCDs, such as ascites syndrome and lameness, which have also led to health problems in intensively raised broilers at the smallholder level (Gupta, 2011). Concurrently, the composition of the broiler carcasses has changed with changing ration formulation. This impacts on the quality of food entering the human food system at a time when obesity and other NCDs have become major public health issues in both developing and developed countries (Doku and Neupane, 2015). Wang et al. (2009) report a substantial increase in the amount of non-essential fats and a loss of essential fats derived from contemporary animal husbandry, including poultry meat, with modern organic and broiler chickens sold for human consumption providing more energy from fat than protein.

4.5 Food safety

Over the past 50 years, poultry production has contributed significantly to food security while at the same time receiving increased attention from the public health community because of its links to zoonotic disease. Three zoonotic diseases of major interest are salmonellosis, highly pathogenic avian influenza (HPAI; subtype H5N1) and campylobacteriosis. Each of these diseases has been more problematic in intensive production systems. Prevention by purchasing birds from flocks certified free of key diseases and training on, and implementation of, appropriate biosecurity is crucial irrespective of the production system.

Millions of poultry were killed or slaughtered to control the spread of the HPAI H5N1 virus (Alders et al., 2013). These measures severely impacted the livelihoods of many households, especially women, in some countries (FAO, 2009). The HPAI H5N1 pandemic also highlighted the readiness of vulnerable households to slaughter and consume unhealthy birds or carcasses of poultry that have died of infectious disease because of food insecurity, a practice that pre-dates the HPAI H5N1 pandemic (Alders et al., 2014a). Improving overall health and production levels of birds and incomes of farmers will significantly reduce such practices.

5 Breeding and genetics

Whilst genetics has the capacity to impact profoundly upon productivity, as seen in the development of specialized commercial meat and egg strains of chickens, such performance can be achieved only where nutritional and management inputs permit the expression of the birds' genetic potential. As noted above, there are considerable constraints to the provision of such inputs under extensive small-scale family poultry farming. As noted in Table 2, different genetic stock are likely to be used in the four production systems, according to the innate characteristics of the birds and to their need for food and other inputs.

Whilst selection of the appropriate genotype for the production system in question is a fundamental requirement, any genetic improvement of stock is a significant undertaking requiring good management, accurate recording and in most cases, considerable inputs, either by the farmer or by the source of the breeding stock (Pym, 2013). The literature is full of reports of failed genetic upgrading and cockerel replacement programmes in rural areas of developing countries (see Sorensen, 2010). Furthermore, where breeding programmes are directed towards the production of birds more suited to the climatic and nutritional constraints of the village poultry environment, the effectiveness of such programmes has often been impacted due to significant between-bird environmental variation resulting in difficulty in identifying whether differences between birds are due to genetic rather than environmental factors (Besbes, 2009; Pym, 2013).

Where genetic improvement is contemplated, the chosen strategy will differ not only according to production system but also from region to region depending on the local market and needs for eggs and meat, and on attitudes towards other traits such as feather colour and other characteristics that may have a meaning in religious and ritual contexts (Alders and Pym, 2009).

General-purpose indigenous breed birds are ubiquitous in the rural regions of nearly all developing countries (Gueye, 2003). By way of quite extreme contrast with specialized commercial strains of broilers and layers, these birds have, for the most part, considerably lower genetic potential for meat and egg production, but are able to survive, reproduce and produce meat and eggs under the often harsh semi-scavenging village environment (Fotsa et al., 2014). There is, however, significant variation in productivity between the various indigenous breeds and ecotypes in different regions within and between countries, and indeed in the climatic and nutritional environments typically experienced by (or provided to) the birds in the different regions (Sorensen, 2010). Before a development programme commences, the performance of the available ecotypes under local conditions needs to be determined.

Attempts by farmers at genetic 'upgrading' of flocks through the introduction of bigger and faster growing males are widespread (Sorensen, 2010; Fotsa et al., 2014). Whilst this generally results in an increased growth potential in the progeny, this may not be achieved due to nutritional and other limitations, and, depending upon the chosen genotype, there may be other effects upon broodiness and reproductive efficiency and performance of the female progeny (Pym, 2013). There may be further negative consequences similar to those encountered with cross-breeding.

In some government and NGO poultry improvement programmes and in some larger-scale extensive and semi-intensive production systems, local indigenous and commercial genotypes have been crossed in attempts to provide birds that are tolerant to local conditions while also capable of reasonable performance (Sorensen, 2010; Okeno, 2013). This involves the need for maintaining separate parent lines/breeds for the generation of the F1 cross-bred progeny. In nearly all cross-breeding programmes, the cross-bred bird exhibits considerably better egg production and/or growth rate than the indigenous breed parent, but where the progeny are intended for use under extensive production systems, problems can be encountered with:

- loss of broodiness in hens, making them incapable of reproducing naturally;
- the need for additional inputs (particularly feed) to achieve the birds' genetic potential for production;

- a change in appearance and 'type', which may affect the birds' acceptability to farmers and the consumers of poultry eggs and meat.

Notwithstanding these limitations, there is an undeniable case for genetic improvement through cross-breeding for birds intended for large-scale extensive and semi-intensive production systems involving significant supplementary feeding (Fotsa et al., 2014).

Under small- and larger-scale intensive production systems, due to the requirement for high productivity, the only economically viable genotypes to use are the genetically 'improved' commercial broiler or layer genotypes. There is a case for using commercial genotypes that may be better adapted to the local environment by virtue of some single-gene heat adaptive capability (e.g. naked neck), if such are available (Cahaner, 2008). Given the cost and complexity of breeding programmes at this level and the use of three- or four-way cross commercial stock, however, there would seem to be little point in contemplating further genetic improvement in, or genetic modification of, commercial birds.

One of the greatest limitations to profitability under extensive production systems is the high mortality rate of the birds (Sonaiya et al., 1999; Gueye, 2003). Notwithstanding its importance, however, given the complexity and low heritability of livability, it is difficult to have a meaningful impact on this through artificial selection. However, under the semi-scavenging village system, there is considerable natural selection for survival rate which has resulted in indigenous 'breeds' being less susceptible than commercial strains to a range of diseases (Besbes, 2009; Pym, 2013). The most effective way of reducing mortality in indigenous birds under semi-scavenging conditions is through health and general management procedures, as described in the previous and following sections.

Whilst it is very important to improve egg production, there is a negative relationship between broodiness/mothering ability and egg production. There is much greater opportunity for hens that do not go broody to lay more eggs than those that do. Broodiness is, however, essential where artificial incubation is not an option and chick production is regarded as important.

Meat production is the prime focus for small-and larger-scale extensive family poultry production in most developing countries (Gueye, 2003; Sonaiya, 2009). A high reproduction rate, and hence good egg production, is an essential element. Thus, any attempts at genetic improvement in extensive poultry production should focus on improving both egg and meat production within the constraints of the physical and nutritional environment. Possible approaches to this are provided by Fotsa et al. (2014).

6 Husbandry and nutrition

Notwithstanding the generally low genetic potential of indigenous village chickens for meat and egg production, in many cases they do not achieve their genetic potential due to limitation to their management and/or nutrition (Besbes, 2009; Sorensen, 2010). Such limitation is largely due to perceived more pressing needs for the normally very limited available funds in the household and to long-term cultural attitudes that chickens are part of the village environment, will always be so and that intervention by way of changing their management, is not cost-effective (Sonaiya, 2009). Prior to the advent of effective ND vaccination, this attitude was not unreasonable due to the periodic decimation of chicken populations in ND outbreaks (Spradbrow, 1993). With effective ND control (Alders et al., 2010) comes the opportunity to improve the efficiency and profitability of small-scale village poultry production through improvements to nutrition, management and housing.

As noted above, one of the main causes of low productivity and profitability in semi-scavenging village flocks is the very high rate of attrition in the chicks over the first six or so weeks post hatch, due to predation, disease, malnutrition and climatic exposure (Farrell et al., 2000). One of the consequences of the high mortality rate is that a large proportion of village households in developing countries allow the hen to sit on nearly all of the eggs laid in a clutch in order to produce chicks. The hen is normally an effective incubator and typically 75–80% of the eggs hatch to produce chicks, most of which die before six weeks of age due to the above-mentioned reasons (Farrell et al., 2000; Pym and Alders, 2012). This means that frequently, very few eggs are available for home consumption or sale, which has an impact on household nutrition (particularly that of the children) and income.

The fairly obvious management and nutritional solution to this problem is to reduce the attrition rate by protecting the chicks from predators and climatic stress and provide appropriate food for them to an age where they can scavenge effectively in relative safety. This approach was successfully tested in a field study in Myanmar between 2003 and 2008, and is described in some detail in the following section under Case study 1.

Access to and quality of feed is a critical element in semi-intensive and intensive production systems and also in the more extensive production systems where bird numbers outstrip the capacity of the SFRB, combined with the typically limited supplementary household food scraps and broken grain provided to the birds, to provide adequate nutrition (Ravindran and Blair, 1993; Sonaiya, 2004). A particular problem in rural areas of developing countries is the availability of well-compounded feeds to meet the nutritional requirements of the different classes of stock, and/or the availability of quality feedstuff ingredients to use in the preparation of such feeds (Sonaiya, 2004; Ravindran, 2013). There is considerable competition with the human population for quality feed ingredients, which means that poultry feeds are often based on poor quality or alternative ingredients. The main challenges to the use of effective nutritionally balanced poultry diets are: (i) the widespread use of poor quality feedstuffs, the nutrient composition of which is often unknown, (ii) very limited information on the nutrient composition of many of the alternative feedstuffs used, (iii) limited understanding by local feed millers of the nutrient requirements of the different classes of stock and (iv) adulteration of diets with non-nutritive materials like sand and sawdust (Sonaiya, 1995, 2004; Ravindran, 2013).

Because of competition with humans for grains, it is often not advisable to develop a wholly grain-based animal feed system. The recommended approach is to identify and use locally available feed resources to formulate diets that are as balanced as possible (Branckaert et al., 2000). This requires information on their nutrient composition, which is often missing or difficult to obtain. By-products of processing local crops (brans, oil and seed cakes) can be used as both energy and protein sources (Hutagalung, 1981), but cannot be used to form a balanced diet on their own. Again, information on nutrient composition of these is a prerequisite.

Given the opportunity to do so, birds have a remarkable capacity to balance their diets for energy, protein (amino acids), minerals and vitamins (Emmans, 1977, 1991; Gous and Swatson, 2000), which is why the scavenging system is so widespread and effective. In most circumstances, however, there is a deficit of both energy and protein in the SFRB (Sonaiya, 2004). The former can be corrected through supplementation with energy sources such as grains, cassava, coco yam, arrow root and coconut residues, and the protein deficit can be addressed through provision of such materials as termites, maggots and earthworms. Sonaiya (2014) provides details of how these materials can be sourced and provided to the birds.

Once bird numbers are allowed to increase beyond the capacity of the SFRB to feed them, apart from the additional requirement for feed, there are considerations of availability of finance to purchase the required feed, as well as access to markets and other outlets for sale/ barter of the birds produced. Housing may also be a requirement if land area limits holding the flock under a semi-scavenging environment. An alternative approach to producing more chickens for sale or for home consumption as a result of a reduction in chick attrition rate through management procedures as described in the following Case study 1, is to set fewer eggs under the hens and either use the eggs so saved for home consumption or sale. As noted by Pym and Alders (2012), the impact of the former approach on the nutrition of children in these often nutrient-deficit communities has the potential to be quite profound.

With the more intensive forms of management, provided basic bird physiological requirements are met, there is no reason why simple forms of housing and management cannot result in reasonable productivity (Glatz and Pym, 2013). The basic requirements for poultry housing are space, ventilation, light and protection (Alders, 2014a). This means avoidance of overcrowding, protection from predators and climatic extremes, ability to shelter from the sun and rain, access to clean (and preferably cool) water and access to compounded feed or feedstuffs which reasonably meet the birds' nutrient requirements. In the interests of sustainability, structures for housing and shelter should preferably be constructed from simple locally available materials. Such structures will vary considerably from region to region depending upon the environment, available materials and cultural considerations (Alders, 2014a).

7 Marketing and economics

In any consideration of sustainability of small-scale family poultry production, the ability of smallholders to obtain a fair return for their birds and/or eggs is paramount. This applies particularly to the more intensive forms of production where the farmer is normally required to pay for both the birds (as chicks or layer pullets) and the feed used in the enterprise. Thus, together with considerations of markets and marketing, there is the issue of credit available to smallholder families, particularly to allow them to undertake the more intensive forms of poultry production.

7.1 Access to credit and microfinance

It is estimated that about 90% of people in developing countries lack access to financial services from the banks and other larger institutions, either for credit or for savings (Saleque, 2014). Starting in the 1980s, thousands of microfinance NGOs were established in Asia, Africa and Latin America to provide small loans to poor people. Whilst a significant proportion of these were not sustainable, others went on to demonstrate that microfinance loans to poor people are an effective way to help them increase income and assets, manage risk and work their way out of poverty (Saleque, 2014).

A number of poultry projects in Bangladesh demonstrated the positive impact of microfinance on the livelihoods of the rural poor (Saleque, 2007). These projects demonstrated convincingly the capacity of family poultry to increase food security and alleviate poverty in the poorest of households. As a consequence of these and other studies, microfinance institutions around the developing world now recommend family poultry farming as a viable income-generating activity (Saleque, 2014). One of the elements contributing to the success of microfinance in family poultry farming is that women, as the principal poultry keepers and borrowers, have demonstrated themselves to be excellent borrowers with generally prompt and reliable repayment of loans (Bagnol, 2014).

As noted in the following Case study 1, there may well be a need for microfinance to fund the purchase of materials required for agreed interventions to improve production and profitability of small-scale extensive village poultry production (e.g. the coops, creep feeders and starter diet for the chicks which comprise the management intervention of the above study aimed at reducing attrition rate in chicks). The study in question demonstrated a good benefit–cost ratio for the intervention, but each situation needs to be assessed individually with inputs from all players. All improvements in production come at a cost and the challenge is to assess the risks and benefits and to provide the necessary support where there is evidence that the risks are manageable and that benefits will be significantly greater than the costs. Part of the challenge is to accommodate the understandably very risk-averse nature of many smallholders. These issues will be addressed in the following section on training and extension.

7.2 Marketing and value chain development

One characterizing feature of small-scale extensive poultry production in developing countries is that the birds used and produced for sale are typically indigenous breeds and ecotypes, with meat and egg characteristics very different respectively from commercial broilers and layers. Also in most countries, as noted earlier, there is a considerable market for these products due to a preference by a significant proportion of the population based on flavour and texture and other cultural considerations (Alders and Pym, 2009). This generally has resulted in premium prices paid for both meat and eggs from indigenous birds, particularly so when calculated on a weight rather than a per bird or egg basis.

One of the constraints to smallholders capitalizing on these high prices for birds and eggs produced often with relatively little inputs is that normally few birds or eggs are available at any one time from each household and that the households are typically quite remote from potential consumers in the larger population centres. As a consequence of this, birds and eggs surplus to home consumption or local sale or barter are typically sold to middle-men traders who take the produce to local or regional markets. The price paid by the middlemen is often an issue for the farmers, but the middlemen themselves face considerable collection and transport costs, particularly so in rural areas (Hinrichs et al., 2014).

A good understanding of the value chain (i.e. the groups of people and processes through which a commodity is supplied to the final consumer) is an essential feature of any determination of how returns can be improved or new markets established. Hinrichs et al. (2014) provided a listing of steps which address the constraints to improving the family poultry value chain with the aim of achieving greater returns for smallholders. These include assessing consumer preferences, mapping and characterizing people involved, assessing existing relationship structures, identifying potential equity issues and developing assessment tools to map priorities for public policy research and investment into supply chains. They make the point that a rigorous and multidisciplinary value chain analysis plays an important role in ensuring the sustainability of development projects.

Because of the nature of production, the majority of trade is in live birds, even to the end consumer (Sonaiya, 2009). Whilst this ensures that the birds purchased are indeed indigenous birds and that the meat will be fresh, there is a significant risk for disease transmission between birds during the various stages of marketing. This has particular ramifications where HPAI is endemic.

8 Education and training

8.1 Why training and extension is vital

Increased access to education and training for producers will yield benefits beyond simple increases in production (Alders, 2012). Improved education of women, in particular, will lead to better health of their families as well as their poultry. With increased education of women, child survival improves and the number of children born decreases (UN, 2011). Given the current predictions for increases in the human population over the coming decades, allowing women access to education and control over their reproductive health would appear to be beneficial. Improved family poultry production has been demonstrated to increase the number and percentage of school-age children attending school in participating households (Sonaiya et al., 2002).

Improving the education of producers (men and women) will, in turn, increase the efficiency of their poultry enterprises (Alders, 2012). Curricula on small-scale family poultry production can enrich the material available to primary and secondary school teachers and those involved in adult literacy programmes. Poultry modules within Farmer Field Schools and Junior Farmer Field and Life Schools contribute to improved skills of farmers and information sharing between them (IRPC, 2006).

Those involved with small-scale family poultry production from the producers to Ministers of Agriculture need access to reliable information (Alders, 2012). Ensuring that accurate and timely market information is available to poultry producers is also critical to enable them to make sound decisions relating to the purchase of inputs and the sale of their products. The use of digital technology (i.e. mobile phones and the internet) is benefiting producers across the globe (Bradfield-Moody and Nogrady, 2010). Digital technology can increasingly link family poultry producers with markets and government and private sector support services.

Lack of access to information and education continues to result in households and producers who are unfamiliar with the germ theory of disease and the science behind good nutrition and poultry husbandry. Recommendations or directives that caution households against the raising of family poultry fly in the face of centuries of experience of the positive contribution of poultry to family well-being (Alders, 2012). The result is a lack of compliance with biosecurity and other recommendations.

8.2 What are the characteristics of effective extension and training?

The preparation of effective and cost-efficient extension and training material is a continuous process from initial planning, drafting and pre-testing to evaluation of materials released for use (Alders et al., 2007; Sasidhar et al., 2014). Comprehensive extension packages should be developed for use with all available communication options, in particular, radio, newspapers, group meetings, field days, drama, newspapers, group meetings and school lessons (Alders, 2001). Where literacy levels are low, more attention should be given to gender-sensitive audiovisual and non-formal means of extension and training communication that are tailored (by using local languages and locally appropriate examples) to those who will apply the information on their own farms (Alders et al., 2007). Adequate time and resources must be invested in the development and evaluation of the extension material. The effectiveness of the extension and training material is critical in situations where farmers are to pay for inputs (Alders and Spradbrow, 2001).

8.3 Who should implement the extension and training activities?

Extension and training activities will have most impact if they are an integral part of the poultry production system as lessons learnt and can be incorporated into subsequent activities. In the case of smallholder intensive poultry production, close links with the private sector which is responsible for the provision of birds, feed and other inputs will be invaluable.

For semi-intensive and extensive production systems, NGOs and government agencies are frequently involved with the provision of extension and training activities. Ensuring that entities with ongoing linkages with producers are involved with the training and extension activities will likely ensure that future support activities will be informed by interactions to date.

9 Strengthening policy environments

Sustainable and appropriate support to family poultry production is dependent on a facilitating policy environment. Enabling family poultry production to effectively contribute to sustainable food and nutrition security requires improved functional linkages between a range of public and private agencies.

9.1 Key elements of a facilitating policy environment

Pica-Ciamarra and Otte (2014) suggest that family poultry policies will be likely to succeed if they adhere to three major high-order policy principles that are applicable in all agro-ecological conditions and policy contexts, that is, (i) appropriate targeting, (ii) incentives and (iii) public goods and smart subsidies. These policies underpin six major policy domains, namely, sourcing of birds, poultry health, poultry nutrition, basic infrastructure and/or equipment, marketing and research.

9.2 Building multisectoral collaboration in support of food and nutrition security

Food and nutrition security is a global priority requiring a multisectoral approach. Despite increases in agricultural production over the past two decades, undernutrition rates in children have not diminished significantly in many developing countries (Masset et al., 2012). The context to these poor achievements is that gender inequality and cultural issues have been inadequately addressed in most research linking agriculture and nutrition (Hawkes et al., 2012). Undernutrition is a result of complex causes and to date little research has examined nutrition-specific, health-based approaches in collaboration with food system and livelihood-based interventions. Crucially, supplying women of childbearing age and their children with sufficient calories is important but it is not enough to optimize epigenetic programming; the proper balance of micronutrients is also essential for both short- and long-term health (Kaput, 2010).

Efforts among human health-related multilateral agencies have historically focused on issues such as the promotion of infant and young child feeding, and micronutrient fortification and supplementation, through Ministries of Health (Bagnol et al., 2016). In contrast, agriculture-related multilateral agencies have supported increased production of agricultural commodities. The long-term sustainability of these interventions is being questioned as many of the rural poor are not able to access fortified foods and increased agricultural production has tended to emphasize energy-rich and nutrient-poor staples such as hybrid maize (Idikut et al., 2009). There is a need for sustainable solutions that will bring the two sectors together, that is, to improve human nutrition through improved household income and dietary diversification. Local initiatives, such as enhancing traditional livestock-crop systems, can provide a sustainable solution to the ongoing demographic challenges to developing countries which are driving the need for more food, improved livelihood opportunities and reduced migration to urban centres (Bagnol et al., 2016). The Australian Centre for International Agricultural Research (ACIAR)-funded project presented in Case study 2 in this chapter provides an example of a transdisciplinary and multisectoral approach to developing a sustainable food-based approach to food and nutrition security.

Launched in 2010, Scaling Up Nutrition (SUN) is working to improve both direct nutrition interventions and to integrate nutrition into broader national health and development efforts (UN, 2011). SUN unites a broad range of stakeholders from governments, civil society, the United Nations, donors, businesses and researchers in a collective effort to improve nutrition. Family poultry activities in countries involved with the SUN initiatives would be well advised to become part of the national SUN activities. The Sustainable Development Goals, launched in 2015, are a response to the complex development challenges facing our world today and demand multisectoral, multidimensional approaches (Cabellero, 2014).

10 Case study 1: improving small-scale poultry production in Myanmar

In this study, RP was the poultry production specialist member of the research team in the ACIAR-supported project '*Control of Newcastle disease and identification of major constraints to village chicken production systems in Myanmar*' (2003–8).

Following a participatory rural appraisal study of major constraints to village chicken production in the Yangon Division of Myanmar, a field research project was conducted in the region to evaluate interventions designed to overcome these constraints.

The initial participatory rural appraisal involved collection of data on village chicken production, health, husbandry and marketing from 296 households across 10 villages in the Yangon Division in November 2003 (Henning et al., 2007). Average flock size in the study area was about 30 birds made up of 12 chicks, 12 growers, 4 hens and 2 cocks. Farmers listed disease as the predominant cause of mortality in growers, whereas predation and exposure to unfavourable environmental conditions were considered the main causes of loss in chicks. The main predators identified were birds, rats, snakes, dogs and cats. From this study, two main strategies were identified as putative interventions aimed at reducing losses and improving returns from village chicken production, namely, (i) vaccination against ND and (ii) confinement rearing of the chicks with the hen combined with supplemental creep feeding of the chick with chick starter diet (Henning et al., 2006).

The field research project involved a total of 124 households across nine villages in the Yangon Division of Myanmar (Henning et al., 2009). Households were randomly allocated to one of the three treatments: T1 – ND vaccinated (62 households), T2 – no treatment (31 households) and T3 – improved chick rearing management (31 households). Differences in household numbers allocated to the treatments were based on calculations of mortality incident risk at the bird level. ND vaccination used the thermotolerant I-2 ND vaccine, administered by eye-drop and produced by the Myanmar Livestock Breeding and Veterinary Department, and the improved chick management intervention involved confining the chicks with the hen in a bamboo coop for one week with access to chick starter diet in a creep feeder. In weeks 2 and 3, the chicks were confined at night with the hen and with creep feeding but allowed to scavenge during the day, and in weeks 4–6, chicks were confined alone at night but scavenged freely during the daytime. During the entire period the hen was provided with cracked rice by the farmer, a common practice in Myanmar. The creep feeders were specially designed to fit under the coops and to allow entrance of the chicks but prevent access to the feed by the hen.

Concomitant with the randomized control trial, there was a nested double-blinded controlled trial at bird level to compare serological titres between I-2 vaccinated and placebo-treated birds both between and within households (Henning et al., 2008).

Vaccination against ND reduced overall mortality assumed to be caused by disease, in both growers and chicks, and odds of having protective ND titres two weeks after vaccination were up to 125 times higher in vaccinated birds and 47 times higher in control birds in contact with I-2 ND vaccinates compared to birds without I-2 contact. Crude mortality incidence was lower and income from sale of birds higher ($2.50 per month) in households that applied the improved chick management treatment (Henning et al., 2009).

In a cost–benefit analysis of the treatments by using deterministic and stochastic modelling to determine net benefits, the net present values for ND vaccination and chick management treatments were about $31 and $168 respectively, but the benefit–cost ratio for ND vaccination was much higher (28.8) compared to 4.7 for chick management) due to the considerably higher cost of the latter treatment (Henning et al., 2013). The benefit–cost ratio of the ND treatment was influenced by the provision of subsidized I-2 ND vaccine in Myanmar at the time of the study. Notwithstanding this, ND vaccination is unquestionably an essential element in any attempt to improve the profitability of small-scale family poultry production in developing countries (Alders et al., 2010).

The two treatments chosen for the study are complementary and focus on addressing two different sources of loss. Due to constraints in design space, it was regrettably not possible to include the two treatments concurrently. The chick management treatment was focused on reducing the attrition rate in chicks, due largely to predation, malnutrition and climatic exposure (Farrell et al., 2000), whereas the ND vaccination treatment was focused on reducing losses due to ND in chicks, growers and adult birds. The impact of the latter treatment can change quite dramatically depending upon whether the birds are exposed to an outbreak of pathogenic ND. The impact of the chick management treatment is likely to have a much more consistent positive impact on productivity and profitability due to the ever-present problems of predation, malnutrition and climatic exposure. This was observed in the study. In the absence of such management interventions, mortality rate in young chicks is high irrespective of whether ND is a problem (Pym and Alders, 2012).

One area where further study is indicated is in defining the appropriate specifications and optimal timing of creep feeding for the young chicks. In our study we initially used a locally produced chick starter diet, but it was unpalatable to the chicks, which resulted in our replacing it with an imported starter diet (at significantly increased cost).

Based on this and other studies, the combination of these two interventions (or variations thereof) would appear to have the potential to impact very positively on the efficiency and profitability of small-scale semi-scavenging village poultry production.

11 Case study 2: contribution of village chickens to diets in Tanzania and Zambia

In this study, RA is the principal investigator in the ACIAR-supported project '*Strengthening food and nutrition security through family poultry and crop integration in Tanzania and Zambia*' which commenced in 2014 (Alders et al., 2014b).

Optimal growth and health requires not only sufficient energy but also the appropriate balance of essential micronutrients for each stage of development (Neumann et al., 2003). The World Bank states that the economic costs of poor nutrition are very high, perpetuating the cycle of poverty and malnutrition through three main routes: (i) direct losses in productivity from poor physical status and losses caused by disease linked with malnutrition; (ii) indirect losses from poor cognitive development and (iii) losses in schooling; and losses caused by increased health care costs (World Bank, 2006). Food and nutrition security is a global international development priority and one that can be achieved in parallel with improved village chicken health and welfare. For example, in Tanzania, approximately 42% of children and 45% of children in Zambia suffer from stunting (growth and neurodevelopmental failure) due to malnutrition.

Our project aims to enhance traditional village chicken-crop systems as a sustainable solution to the ongoing nutritional challenges in Africa. Rural communities that rely on rainfed crops often go through severe hunger periods just prior to the major harvesting season when their stored grains have been exhausted. These significant peaks and troughs in household food availability are reduced when there is diversity in family farming activities (Alders et al., 2015). By improving village poultry health and welfare, families have greater access to poultry meat and eggs which are a source of high-quality protein, bioavailable micronutrients and income (de Bruyn et al., 2015). Poultry manure can also contribute to increased soil fertility for the production of indigenous vegetables at the household level, further diversifying the range of foods consumed.

Village poultry are frequently the only livestock under the control of women in developing countries in Africa and Southeast Asia (Alders and Pym, 2009). By improving the health of village poultry we can make major contributions to ensuring that households, at all times, have physical and economic access to adequate amounts of nutritious, safe and culturally appropriate foods (Alders et al., 2014b). Village chicken health activities commence with the control of ND through 4-monthly vaccination campaigns against ND implemented by community vaccinators via eye-drop administration on a fee-for-service basis (Alders, 2009; Dias et al., 2001).

Evidence shows that empowered women who can make decisions on how household incomes are allocated, spend more money on nutritious food, health care and education (Quisumbing et al., 1995). Women's work can lead to increased income, which may be spent on food, resulting in improved nutrition outcomes. According to the Food and Agricultural Organization of the United Nations (FAO), the International Fund for Agricultural Development (IFAD) and the World Bank, an increase in women's income of $10 achieves the same improvements to children's nutrition and health as an increase in a man's income of $110 (World Bank, FAO and IFAD, 2008).

Much of our work on village poultry has been supported by the Australia Government (including several research projects funded by ACIAR and the FAO. The model of cross-sectoral collaboration that we have used, incorporating producers and traders, as well as government and local and international research institutions, has been critical to the success of our village poultry programme (Alders, 2001). It has allowed us to tailor our approach for each community to identify the strategies that work best for the long term and that can gradually be handed over to the villagers to maintain.

Our work also has a strong focus on capacity building by incorporating graduate students into our research projects. Many of our students benefit from Australian Government funding (either through the Australian Postgraduate Award scheme or Australia Awards), and this funding support is vital. We work in direct collaboration with partner frontline ministries and regional economic communities as well as regional and international research organizations and donors.

12 Acknowledgements

In relation to these two case studies, we are grateful for funding provided by the Australian Government, especially ACIAR, in support of research into the sustainable improvement of smallholder poultry production in Africa and SE Asia.

13 Future trends

Future research in support of family poultry and their owners will cover the spectrum of research tools and methodologies available to us as we seek to deliver sustainable food and nutrition security in the face of climate change, weather variability and global economic uncertainty.

The complex nature of the mixed farming systems and value chains incorporating family poultry production will increasingly benefit from inter- and transdisciplinary research (Alders et al., 2014b; Bagnol et al., 2016). Nutrition-sensitive food value chains (Hawkes and Ruel, 2012) will be vital to achieving sustainable food and nutrition security as will adopting an Ecohealth (Rapport and Singh, 2006) or One Health (AVMA, 2008) framework. EcoHealth can be defined as the recognition that 'health and well-being are the result of complex and dynamic interactions between determinants, and between people, social and economic conditions, and ecosystems' (Charron, 2012), whereas One Health has been described as 'an integrative effort of multiple disciplines working locally, nationally, and globally to attain optimal health for people, animals, and the environment' (AVMA, 2008).

Ongoing and new research areas seen as important in achieving sustainable improvement in smallholder well-being through family poultry production include:

- social science research into smallholder poultry value chains. One of the greatest challenges to improvement in smallholder well-being through family poultry production is, in fact, sustainability of the procedures/interventions which have been demonstrated to increase profitability and household food security after the development project ceases. A closely related challenge is, where appropriate, determining the most effective way of extending the application of such interventions to other regions. Approaches provided in the above *training and extension* and *strengthening the policy environment* sections are crucial in facilitating sustainable development;
- social science research into nutrition-sensitive family poultry value chains;
- basic physical science research into sustainable options for disinfecting fomites. With increasing linkages between large commercial and smallholder production systems, new pathogens can now circulate more quickly than ever before. Developing options for preventing disease transmission between production units and systems is of paramount importance;
- molecular biology in support of vaccines that are easy to administer, provoke immunity of longer duration and against multiple pathogens;
- selective breeding, including molecular approaches, for improved disease resistance and for performance under suboptimal and high-temperature conditions. Allied to this is the development of mechanisms by which such birds can be made available at reasonable cost to smallholders;
- with the dramatic reduction in arable land globally due to population growth and climate change, in-country agronomic research into environmentally and ecologically appropriate crops for animal feeding;
- research into the nutrient composition and feeding value of available alternative feedstuffs in the different regions and their potential use in both intensive and extensive poultry production; and
- given the often poor utilization of nutrients in available feedstuffs due to feed formulation problems in many regional feed mills, social science research into education and training of personnel, and the development of mechanisms for feedstuff analysis for or by these feed mills.

14 Where to look for further information

Because of the broad nature of the topic and the multidisciplinary nature of studies aimed at assisting smallholders, many of the references provided fall into the category of reports (or papers developed from reports) from medium- to large-scale development projects, review articles and FAO and other aid agency monographs. Together these provide a comprehensive coverage of past and current studies focused on small-scale family poultry production in developing countries. A listing of the key references, societies and associations, journals and conferences, projects, and research organizations follows:

Review articles:

General: Alders and Pym (2009), Pym and Alders (2012), Thieme et al. (2014); *Poverty alleviation and food security*: Gueye (2003), de Bruyn et al. (2015); *Newcastle disease*: Spradbrow (1993), Alders et al. (2010); *Highly Pathogenic Avian Influenza*: Alders et al. (2013); *Nutrition*: Sonaiya (2004), Ravindran (2013); *Breeding and genetics*: Sorensen (2010), Pym (2013), Fotsa et al. (2014).

Seminal articles and monographs:

Ahlers, et al. (2009). *Improving Village Chicken Production.* Monograph 139. Australian Centre for International Agricultural Research. Canberra.

FAO. (2008). *Poultry in the 21st Century: Avian Influenza and Beyond.* International Poultry Conference, Bangkok, November 2007. http://www.fao.org/docrep/011/i0323e/i0323e00.htm.

FAO. (2013). *Poultry Development Review:* Poultry production in developing countries, FAO, Rome. www.fao.org/3/a-i3531e.pdf.

FAO. (2014a). *Family Poultry Development – Issues, Opportunities and Constraints.* FAO Animal Production and Health Working Paper. No. 12. Rome, Italy. http://www.fao.org/3/a-i3595e.pdf.

FAO. (2014b). *Decision Tools for Family Poultry Development.* FAO Animal Production and Health Guidelines No 16. Rome, Italy, 104p.

Sandilands, V. and Hocking, P. M. (eds) (2012). *Alternative Systems for Poultry: Health Welfare and Productivity.* Poultry Science Symposium No 30. Wallingford, CAB International, 359p.

Sorensen, P. (2010). *Chicken Genetic Resources used in Smallholder Production Systems and Opportunities for their Development.* FAO Smallholder Poultry Production Paper, No 5. FAO, Rome. www.fao.org/docrep/013/al675e/al675e00.pdf.

Key societies, associations and organizations:

FAO – The Food and Agricultural Organization of the United Nations Organization, Rome, Italy. www.fao.org.

IAEA – International Atomic Energy Agency. https://www.iaea.org.

IFAD – International Fund for Agricultural Development. Rome, Italy. www.ifad.org.

International Rural Poultry Centre of the Kyeema Foundation. Brisbane, Australia. https://kyeemafoundation.org.

World Veterinary Poultry Association. http://www.wvpa.net/.

WPSA – The World's Poultry Science Association. www.wpsa.com.

Key journals and conferences:

Family Poultry Communications – the official biannual electronic publication of the International network for Family Poultry Development – Jean-Claude Fotsa ed. www.infpd.net.

World's Poultry Science Journal – official journal of the World's Poultry Science Association. www.wpsa.com.

World's Poultry Congresses – Istanbul, 2004; Brisbane, 2008; Salvador, 2012; Beijing, 2016; Paris, 2020. www.wpsa.com.

World Veterinary Poultry Association Congresses – Beijing, 2007; Marrakesh, 2009, Cancun, 2011, Nantes, 2013; Cape Town 2015; Edinburgh, 2017; Bangkok, 2019. www.wvpa.net.

Major international research projects:

'African Chicken Genetic Gains' (commenced 2014). ILRI.

'Feed the Future Innovation Lab for Genomics to Improve Poultry' (commenced 2013). United States Agency for International Development (USAID).

'Improving Vaccination of Poultry against Newcastle Disease' (Mozambique) Implementation period: June 2013 to June 2018. FAO Mozambique.

'Supporting Food Security And Capacity Building In African Union Member States Through The Sustainable Control Of Newcastle Disease In Village Chickens' (commenced 2013). Australian Government Department of Foreign Affairs and Trade.

'Strengthening food and nutrition security through family poultry and crop integration in Tanzania and Zambia' (commenced 2014). ACIAR.

'Timor-Leste Village Poultry Health and Biosecurity Program' (commenced 2014). Australian Department of Foreign Affairs and the Australian Department of Agriculture and Water Resources.

Research centres focused on smallholder poultry production:

International Livestock Research Institute, Nairobi and Addis Ababa www.ilri.org.
International Rural Poultry Centre of the Kyeema Foundation, Brisbane, https://kyeemafoundation.org.
Faculty of Veterinary Science, University of Sydney, Australia, http://sydney.edu.au/vetscience/.
School of Veterinary Science, University of Queensland, www.uq.edu.au.

15 References

Ahlers, C., Alders, R. G., Bagnol, B., Cambaza, A. B., Mgomezulu, R., Msami, H., Pym, R., Wegener, P., Wethli, E. and Young, M. (2009). *Improving Village Chicken Production: A Manual for Field Workers and Trainers*. Monograph, 139. Canberra: Australian Centre for International Agricultural Research. http://aciar.gov.au/publication/mn139.

Aini, I. (1990). Indigenous chicken production in South-East Asia. *World's Poultry Science Journal*, 46: 51–7.

Alders, R. G. (2001). Sustainable control of Newcastle disease in rural areas. In R. G. Alders and P. B. Spradbrow (eds), SADC Planning Workshop on Newcastle Disease Control in Village Chickens. Proceedings of an International Workshop, Maputo, Mozambique, 6–9 March 2000. ACIAR Proceedings No. 103, pp. 80–7. http://aciar.gov.au/publication/pr103.

Alders, R. G. (2009). The Southern Africa Newcastle Disease Control Project: its history, approach and lessons learnt. In R. G. Alders, P. B. Spradbrow and M. P. Young (eds), Village chickens, poverty alleviation and the sustainable control of Newcastle disease. Proceedings of an International

Conference held in Dar es Salaam, Tanzania, 5–7 October 2005. ACIAR Proceedings No. 131, pp. 62–6. http://aciar.gov.au/publication/pr131.

Alders, R. G. (2012). Challenges and opportunities for small-scale family poultry production in developing countries. XXIV World's Poultry Congress, Salvador, Brazil, 5–9 August 2012. *World's Poultry Science Journal* 68(Suppl. 1): 153. https://www.researchgate.net/publication/258832842_Challenges_and_opportunities_for_small-scale_family_poultry_production_in_developing_countries.

Alders, R. G. (2013). Healthy family poultry, healthy communities: sustainable food security and safety in the 21st century. Proceedings of the XVIII World Veterinary Poultry Association Congress, Nantes, France, 19–23 August 2013, pp. 35–41.

Alders, R. G. (2014a). Identifying appropriate interventions- housing and other infrastructure. In *Decision Tools for Family Poultry Development.* FAO Animal Production and Health Guidelines No. 16. Rome, Italy, pp. 33–7. http://www.fao.org/3/a-i3542e.pdf.

Alders, R. G. (2014b). Making Newcastle disease vaccines available at village level. *Veterinary Record* 174: 502–3.

Alders, R. G., Ankers, P. and Watkins, E. (2014a). Health, public health and biosecurity. In *Decision Tools for Family Poultry Development.* FAO Animal Production and Health Guidelines No. 16. Rome, Italy, pp. 29–33. http://www.fao.org/3/a-i3542e.pdf.

Alders, R. G., Aongola, A., Bagnol, B., de Bruyn, J., Kimboka, S., Kock, R., Li, M., Maulaga, W., McConchie, R., Mor, S., Msami, H., Mulenga, F., Mwala, M., Mwale, S., Rushton, J., Simpson, J., Victor, R., Yongolo, C. and Young, M. (2014b). Using a One Health approach to promote food and nutrition security in Tanzania and Zambia. *Planet@Risk* (Special Issue on One Health) 2(3): 187–90. https://planet-risk.org/index.php/pr/article/viewFile/67/210.

Alders, R. G., Aongola, A., Bagnol, B., de Bruyn, J., Darnton-Hill, I., Jong, J., Kimboka, S., Li, M., Lumbwe, H., Mor, S., Maulaga, W., Mulenga, F., Rukambile, E. and Wong, J. (2015). Village chickens and their contributions to balanced diverse diets throughout the seasons. World Veterinary Poultry Association Congress. Cape Town, 7–11 September 2015, e-Booklet, p. 115.

Alders, R. G., Awuni, J., Bagnol, B., Farrell, P. and de Haan, N. (2013). Impact of Avian Influenza on village poultry production globally. *EcoHealth* 11(1): 63–72.

Alders, R. G., Bagnol, B. and Young, M. (2010). Technically sound and sustainable Newcastle disease control in village chickens: lessons learnt over fifteen years. *World's Poultry Science Journal* 66: 433–40.

Alders, R. G., Bagnol, B., Young, M. P., Ahlers, C., Brum, E. and Rushton, J. (2007). Challenges and constraints to immunization in developing countries. In B. Dodet and the Scientific & Technical Department of the OIE (eds), *Vaccination: A Tool for the Control of Avian Influenza.* Dev Biol (Basel). Basel, Karger, 2007, vol. 130, pp. 73–82.

Alders, R. G. and Pym, R. A. E. (2009). Village poultry: still important to millions, eight thousand years after domestication. *World's Poultry Science Journal* 65: 181–90.

Alders, R. G. and Spradbrow, P. B. (2001). *Controlling Newcastle Disease in Village Chickens: A Field Manual.* Canberra: Australian Centre for International Agricultural Research. Monograph 82. 112p. http://aciar.gov.au/publication/mn082.

Alexander, D. J., Bell, J. G. and Alders, R. G. (2004). Technology review: Newcastle disease with special emphasis on its effect on village chickens. FAO Animal Production and Health Paper No. 161. FAO, Rome, 63p. http://www.fao.org/docrep/006/y5162e/y5162e00.HTM.

American Veterinary Medical Association (AVMA). (2008). One Health: A new professional imperative. One Health Initiative Task Force: Final report. Available: https://www.avma.org/KB/Resources/Reports/Pages/One-Health.aspx.

Bagnol, B. (2009). Gender issues in small-scale family poultry production: Experiences with ND and HPAI control. *World's Poultry Science Journal* 65: 231–40.

Bagnol, B. (2014). Conducting participatory monitoring and evaluation. In *Decision Tools for Family Poultry Development.* FAO Animal Production and Health Guidelines No. 16. Rome, Italy, pp. 81–5. http://www.fao.org/3/a-i3542e.pdf.

Bagnol, B., Clarke, E., Li, M., Maulaga, W., Lumbwe, H., McConchie, R., de Bruyn, J. and Alders, R. G. (2016). Transdisciplinary project communication and knowledge sharing experiences in Tanzania and Zambia through a One Health lens. *Frontiers in Public Health* 4: 10. doi: 10.3389/fpubh.2016.00010.

Besbes, B. (2009). Genotype evaluation and breeding of poultry for performance under sub-optimal village conditions. *World's Poultry Science Journal* 65: 260–71.

Besbes, B., Thieme, O., Rota, A., Gueye, E. F. and Alders, R. G. (2012). Technology and programmes for sustainable improvement of village poultry production. In V. Sandilands and P. M. Hocking (eds), *Alternative Systems for Poultry: Health, Welfare and Productivity*, 110–27. Wallingford: CAB International.

Bradfield-Moody, J. and Nogrady, B. (2010). *The Sixth Wave: How to Succeed in a Resource-limited World*. Sydney: Vintage Books.

Branckaert, R. D. S., Gavira, L., Jallade, J. and Seiders, R. W. (2000). *Transfer in Technology in Poultry Production for Developing Countries*. Rome: FAO Sustainable development Dimensions.

Cabellero, P. (2014). Sustainable Development Challenges Post 2015 Need a Multisectoral Focus. Voices: Perspectives on Development. World Bank, Washington DC. Available: http://blogs.worldbank.org/voices/sustainable-development-challenges-post-2015-need-multisectoral-focus.

Cahaner, A. (2008). Breeding fast-growing, high-yielding broilers for hot conditions. In N. J. Daghir (ed.), *Poultry Production in Hot Climates*, 2nd ed, 30–47. Wallingford: CAB International.

Capua, I. and Alexander, D. J. (2006). The challenge of avian influenza to the veterinary community. *Avian Pathology* 35: 189–205.

Catley, A., Alders, R. G. and Woods, J. (2012). Participatory epidemiology: approaches, methods, experiences. *The Veterinary Journal* 191: 151–60.

Cattoli, G., Fusaro, A., Monne, I., Molia, S., Le Menach, A., Maregeya, B., Nchare, A., Bangana, I., Garb Maina, A., N'goran Koffi, J. N., Thiam, H., Bezeid, O. E. M. A., Salviato, A., Nisi, R., Terregino, C., Capua, I. (2010) Emergence of a new genetic lineage of Newcastle disease virus in West and Central Africa – implications for diagnosis and control. *Veterinary Microbiology* 142(3–4): 168–76.

Charron, D. F. (2012). *Ecohealth: Origins and Approach. Ecohealth Research in Practice: Innovative Applications of an Ecosystem Approach to Health*. Edited by Charron DF. New York: Springer, pp. 1–32. http://www.idrc.ca/EN/Resources/Publications/Pages/IDRCBookDetails.aspx?PublicationID=1051.

de Bruyn, J., Wong, J., Bagnol, B., Pengelly, B. and Alders, R. (2015). Family poultry and food and nutrition security. *CAB Reviews* 10(13): 1–9.

Dias, P. T., Alders, R. G., Fringe, R. and Mata, B. V. (2001). Laboratory and field trials with Thermostable Live Newcastle Disease Vaccines in Mozambique. In R. G. Alders and P. B. Spradbrow. SADC Planning Workshop on Newcastle Disease Control in Village Chickens. Proceedings of an International Workshop, Maputo, Mozambique, 6–9 March 2000. ACIAR Proceedings No. 103, pp. 91–6. http://aciar.gov.au/publication/pr103.

Doku, D. T. and Neupane, S. (2015). Double burden of malnutrition: Increasing overweight and obesity and stall underweight trends among Ghanaian women. *BMC Public Health* 15: 670. doi: 10.1186/s12889-015-2033-6.

Emmans, G. C. (1977). The nutrient intake of laying hens given a choice of diets, in relation to their production requirements. *British Poultry Science* 18: 227–36.

Emmans, G. C. (1991). Diet selection by animals: theory and experimental design. *Proceedings of the Nutrition Society* 50: 59–64.

FAO. (2008a). Biosecurity for highly pathogenic avian influenza. FAO Animal Production and health Paper, 165. FAO, Rome. http://www.fao.org/docrep/011/i0359e/i0359e00.htm.

FAO. (2008b). *Poultry in the 21st Century: Avian Influenza and Beyond*. International Poultry Conference, Bangkok, November 2007. http://www.fao.org/docrep/011/i0323e/i0323e00.htm.

FAO. (2009). Highly pathogenic avian influenza: a rapid assessment of its socio-economic impact on vulnerable households in Egypt. G. Limon & N. Haan. FAO, Rome. http://www.fao.org/docrep/013/al686e/al686e00.pdf.

FAO. (2014). *Decision Tools for Family Poultry Development*. FAO Animal Production and Health Guidelines No 16. Rome, Italy. http://www.fao.org/3/a-i3542e.pdf.

Farrell, D. J., Bagust, T. J., Pym, R. A. E. and Sheldon, B. L. (2000). Strategies for improving the production of scavenging chickens. *Asian-Australian Journal of Animal Science* 13(Suppl.): 79–85.

Fotsa, J. C., Sorensen, P. and Pym, R. A. (2014). Identifying appropriate interventions- Breeding and reproduction. In *Decision Tools for Family Poultry Development*. FAO Animal Production and Health Guidelines No. 16. Rome, Italy, pp. 18–25.

Gardner, E. and Alders, R. G. (2014). Livestock Risks and Opportunities: Newcastle Disease and Avian Influenza in Africa. *Planet@Risk* (Special Issue on One Health) 2(4): 208–11. https://planet-risk.org/index.php/pr/article/view/99/211.

Glatz, P. and Pym, R. (2013). Poultry housing and management in developing countries. *Poultry Development Review*. FAO, Rome, pp. 24–44. www.fao.org/3/a-i3531e.pdf.

Gous, R. M. and Swatson, H. K. (2000). Mixture experiments: a severe test of the ability of a broiler chicken to make the right choice. *British Poultry Science* 41: 136–40.

Gueye, E. F. (1998). Village egg and fowl meat production in Africa. *World's Poultry Science Journal* 54: 73–86.

Gueye, E. F. (2000). The role of family poultry in poverty alleviation, food security and the promotion of gender equality in rural Africa. *Outlook on Agriculture* 29: 129–36.

Gueye, E. F. (2003). Poverty alleviation, food security and the well-being of the human population through family poultry in low-income food-deficit countries. *Journal of Food, Agriculture and Environment* 1: 2–21.

Gueye, E. F. (2005). Gender aspects in family poultry management systems in developing countries. *World's Poultry Science Journal* 61: 39–46.

Gupta, A. R. (2011). Ascites syndrome in poultry: a review. *World's Poultry Science Journal* 67(03): 457–68.

Harun, M., Alders, R. G., Sprowles, L., Bagnol, B., Cambaza, A. B., Msami, H., Mgomezulu, R. (2009). Southern Africa Newcastle Disease Control Project impact studies: baseline and participatory rural appraisal results. In R. G. Alders, P. B. Spradbrow and M.P. Young (eds), Village chickens, poverty alleviation and the sustainable control of Newcastle disease. Proceedings of an international conference held in Dar es Salaam, Tanzania, 5–7 October 2005. ACIAR Proceedings No. 131, pp. 96–101. http://aciar.gov.au/publication/pr131.

Hawkes, C. and Ruel, M. T. (2012). Value Chains for Nutrition. In *Reshaping Agriculture for Nutrition and Health*. Washington: International Food Policy Research Institute, pp. 73–81. http://www10.iadb.org/intal/intalcdi/PE/2012/09841.pdf.

Hawkes, C., Turner R. and Waage, J. (2012). Current and Planned Research on Agriculture for Improved Nutrition: A Mapping and a Gap Analysis. Aberdeen: Leverhulme Centre for Integrative Research on Agriculture and Health (LCIRAH) and Centre for Sustainable International Development, University of Aberdeen. http://www.lidc.org.uk/_assets/DfID%20report%20(LCIRAH%20mapping%20and%20gap%20analysis)%2001%20Aug%202012.pdf.

Henning, J., Khin, A., Hla, T. and Meers, J. (2006). Husbandry and trade of indigenous chickens in Myanmar: results of a participatory rural appraisal in the Yangon and Mandalay divisions. *Tropical Animal Health and Production* 38: 611–18.

Henning, J., Morton, J., Hla, T. and Meers, J. (2008). Mortality rates adjusted for unobserved deaths and associations with Newcastle disease virus serology among unvaccinated village chickens in Myanmar. *Preventive Veterinary Medicine* 85: 241–52.

Henning, J., Pym, R., Hla, T., Kyaw, N and Meers, J. (2007). Village chicken production in Myanmar: purpose, magnitude and major constraints. *World's Poultry Science Journal* 63: 308–22.

Henning, J., Morton, J., Pym, R., Hla, T. and Meers, J. (2009). Evaluation of strategies to improve village chicken production: controlled field trials to assess effects of Newcastle disease vaccination and altered chick rearing in Myanmar. *Preventive Veterinary Medicine* 90: 17–30.

Henning, J., Morton, J., Pym, R., Hla, T. and Meers, J. (2013). Economic analysis of interventions to improve village chicken production in Myanmar. *Preventive Veterinary Medicine* 110: 525–40.

Hickler, B. (2007). Bridging the Gap between HPAI 'awareness' and practice in Cambodia. Food and Agriculture Organization of the United Nations, Bangkok. http://www.fao.org/docs/eims/upload//241483/ai301e00.pdf.

Hinrichs, J., Ifft, J. and Heft-Neal, S. (2014). Identifying appropriate interventions- Marketing and value chain development. In *Decision Tools for Family Poultry Development*. FAO Animal Production and Health Guidelines No. 16. Rome, Italy, pp. 37–41. http://www.fao.org/3/a-i3542e.pdf.

Hutagalung, R. I. (1981). The use of tree crops and their by-products for intensive animal production. In A. J Smith and R. G. Gunn (eds), Intensive animal production in the developing countries. Occasional Publication. *British Society of Animal Production* 4: 151–84.

Idikut, L., Atalay, A. I., Kara, S. N. and Kamalak, A. (2009). Effect of Hybrid on Starch, Protein and Yields of Maize Grain. *Journal of Animal and Veterinary Advances* 8(10): 1945–7.

ILRI. (2010). Reducing the impact of infectious diseases on village poultry production in Ethiopia. International Livestock Research Institute, Nairobi. Available: http://www.ilri.org/ReducingtheImpactofInfectiousDiseases.

IRPC. (2006). Improvement of village chicken production by Junior Farmers & People living with HIV/AIDS (Component of FAO Project GCP/MOZ/079/BEL). Trip Report by the International Rural Poultry Centre, KYEEMA Foundation, Maputo, Mozambique. October 2006.

Kaput, J. (2010). Using genetics to tackle malnutrition. SciDevNet. http://www.scidev.net/en/opinions/using-genetics-to-tackle-malnutrition.html.

Kryger, K. N., Thomsen, K. A., Whyte, M. A. and Dissing, M. (2010). Smallholder poultry production: Livelihoods, food security and socio-cultural significance. FAO Smallholder Poultry Production Paper No 4. FAO, Rome.

Lal, M., Zhu, C., McClurkan, C., Koelle, D., Miller, P., Afonso, C., Donadeu, M., Dungu, B. and Chen, D. (2014). Development of a low-dose fast-dissolving tablet formulation of Newcastle disease vaccine for low-cost backyard poultry immunisation. *Veterinary Record*. doi: 10.1136/vr.101926.

Masset, E., Haddad, L., Cornelius, A. and Isaza-Castro, J. (2012). Effectiveness of agricultural interventions that aim to improve nutritional status of children: systematic review. *British Medical Journal* 344: d8222. doi: 10.1136/bmj.d8222.

McLeod, A., Thieme, O. and Mack, S. (2009). Structural changes in the poultry sector: will there be smallholder poultry development in 2030? *World's Poultry Science Journal* 65: 191–9.

Minga, U. M. and Nkini, R. P. (1986). Diseases as a major constraint to chicken production in Tanzania. In *Proceedings of the 13th Scientific Conference of the TSAP*, Arusha, pp. 284–92.

Muir, W. M., Wong, G. K., Zhang, Y., Wang, J., Groenend, M. A. M., Crooijmans, R. P. M. A., Megensd, H.-K., Zhang, H., Okimoto, R., Vereijkeng, A., Jungerius, A., Albers, G. A. A., Taylor Lawley, C., Delanyi, M. E., MacEachern, E. and Cheng, H. H. (2008). Genome-wide assessment of worldwide chicken SNP genetic diversity indicates significant absence of rare alleles in commercial breeds. *PNAS* 105(45): 17312–17.

Ndiweni, N. J. (2010). Prudent poultry farming as a source of livelihood and food security in a changing climate: the case of Zhombe communal lands, Zimbabwe. *International Journal of Scientific and Research Publications* 3(10): 1–5.

Neumann, C. G., Bwibo, N. O., Murphy, S. P., Sigman, M., Whaley, S., Allen, L. H., Guthrie, D., Weiss, R. E. and Demment, M. W. (2003). Animal source foods improve dietary quality, micronutrient status, growth and cognitive function in Kenyan school children: background, study design and baseline findings. *Journal of Nutrition* 133(11, Suppl. 2): 3941S–9S.

Ogundipe, G. A. T., Oluokun, S. B. and Esuruoso, G. O. (1989). The development and efficiency of the Animal Health Information System in Nigeria. *Preventive Veterinary Medicine* 7: 121–35.

OIE. (2012). Highly Pathogenic Avian Influenza. OIE Technical Disease Cards. http://www.oie.int/fileadmin/Home/eng/Animal_Health_in_the_World/docs/pdf/AVIAN_INFLUENZA_FINAL.pdf.

Okeno, T. O., Kahi, A. K. and Peters, K. J. (2013). Evaluation of breeding objectives for purebred and crossbred selection schemes for adoption in indigenous chicken breeding programmes. *British Poultry Science* 54: 62–75.

Pica-Ciamarra, U. and Otte, J. (2014). Creating an enabling policy environment. In *Decision Tools for Family Poultry Development*. FAO Animal Production and Health Guidelines No. 16. Rome, Italy, pp. 57–62. http://www.fao.org/3/a-i3542e.pdf.

Population Reference Bureau (PRB). (2015). http://www.prb.org/Publications/Lesson-Plans/HumanPopulation/PopulationGrowth.aspx.

Pym, R. A. E. (2013). Poultry genetics and breeding in developing countries. *Poultry Development Review*. FAO, Rome, pp. 80–94. www.fao.org/3/a-i3531e.pdf.

Pym, R. A. E. and Alders, R. G. (2012). Introduction to village and backyard poultry production. In V. Sandilands and P. M. Hocking (eds), *Alternative systems for poultry: Health, welfare and productivity*, 97–109. Wallingford: CAB International.

Pym, R. A. E., Guerne Bleich, E. and Hoffmann, I. (2006). The relative contribution of indigenous chicken breeds to poultry meat and egg production and consumption in the developing countries of Africa and Asia. *Proceedings XII European Poultry Conference*, 10–14 September 2006, EPC Verona, Italy, CDROM.

Quisumbing, A. R., Brown, L. R., Feldstein, H. S., Haddad, L. and Pena, C. (1995). *Women: The Key to Food Security*. Food Policy Statement No. 21, August 1995. Washington, DC: International Food Policy Research Institute.

Rapport, D. J. and Singh, A. (2006). An EcoHealth-based framework for state of environment reporting. *Ecological Indicators* 6(2): 409–28.

Ravindran, V. (2013). Poultry feed availability and nutrition in developing countries: Alternative feedstuffs for use in poultry feed formulations. *Poultry Development Review*. FAO, Rome. http//www.fao.org/docrep/013/al706e/al706e00.pdf.

Ravindran, V. and Blair, R. (1993). Feed resources for poultry production in Asia and the Pacific. iii) Animal protein sources. *World's Poultry Science Journal* 49: 219–35.

Roberts, J. A. and Gunaratne, S. P. (1992). The scavenging feed resource base for village chickens in a developing country. *Proceedings of the XIX World's Poultry Congress*, Amsterdam.

Rota, A., Thieme, O., De' Besi, G. and Gilchrist, P. (2014). Designing successful projects. In *Decision Tools for Family Poultry Development*. FAO Animal Production and Health Guidelines No. 16. Rome, Italy, pp. 63–80. http://www.fao.org/3/a-i3542e.pdf.

Saleque, M. A. (2007). Effective way to integrate small farmers in the value chain: experience of BRAC. *Southeast Asian Regional Conference on Agricultural Value Chain Financing*, jointly organized by APO, NPC, FAO and APRACA, Malaysia.

Saleque, M. A. (2014). Identifying appropriate interventions- Microfinance and access to credit. In *Decision Tools for Family Poultry Development*. FAO Animal Production and Health Guidelines No. 16. Rome, Italy, pp. 41–6. http://www.fao.org/3/a-i3542e.pdf.

Sasidhar, P. V. K., Hadrill, D., Bagnol. B. and Alders, R. (2014). Training and extension. In *Decision Tools for Family Poultry Development*. FAO Animal Production and Health Guidelines No. 16. Rome, Italy, pp. 50–7. http://www.fao.org/3/a-i3542e.pdf.

Sonaiya, E. B. (1995). Feed resources for smallholder rural poultry in Nigeria. *World Animal Review* 82: 25–33.

Sonaiya, E. B. (2004). Direct assessment of nutrient resources in free-range and scavenging systems. *World's Poultry Science Journal* 60: 523–35.

Sonaiya, E. B. (2009). Some technical and socioeconomic factors affecting productivity and profitability of smallholder family poultry. *World's Poultry Science Journal* 65: 204–9.

Sonaiya, E. B. (2014). Identifying appropriate interventions- Feeds and feeding. In *Decision Tools for Family Poultry Development*. FAO Animal Production and Health Guidelines No. 16. Rome, Italy, pp. 26–9. http://www.fao.org/3/a-i3542e.pdf.

Sonaiya, E. B., Branckaert, R. D. S. and Gueye, E. F. (1999). Research and development options for family poultry. *First INFPD/FAO Electronic Conference on Family Poultry*, 7 December 1988 to 5

March 1999. Introductory paper. http://www.fao.org/ag/aginfo/themes/en/infpd/econf_scope.html.

Sonaiya, E. B., Permin, A., Pedersen, G., Riise, J. C. (2002). *Using Poultry as a Tool for Poverty Alleviation*. Copenhagen: Network for Smallholder Poultry Development.

Sorensen, P. (2010). Chicken genetic resources used in smallholder production systems and opportunities for their development. FAO Smallholder Poultry Production Paper, No 5. FAO, Rome. http://www.fao.org/3/a-al675e.pdf.

Spradbrow, P. B. (1993). Newcastle disease in village chickens. *Poultry Science Review* 5: 57–96.

Thieme, O., Sonaiya, F., Rota, A., Guèye, E. F., Dolberg, F. and Alders, R. (2014). Defining family poultry production systems and their contribution to livelihoods. In *Decision Tools for Family Poultry Development*. FAO Animal Production and Health Guidelines No. 16. Rome, Italy, pp. 3–8. http://www.fao.org/3/a-i3542e.pdf.

UN. (2008). *The Millennium Development Goals Report 2008*. New York City, NY: United Nations.

UN. (2011). Improving maternal and child nutrition during the 1,000 days between pregnancy and age two, and the Scaling Up Nutrition (SUN) movement. Available: http://un-foodsecurity.org/sites/default/files/UN_SUN_FactSheet.pdf.

UN. (2015a). *World Population Prospects: The 2015 Revision*. United Nations, Department of Economic and Social Affairs, Population Division. http://esa.un.org/unpd/wpp/.

UN. (2015b). *Transforming our World: The 2030 Agenda for Sustainable Development*. New York City, NY: United Nations.

Wang, Y, Lehane, C., Ghebremeskel, K. and Crawford, M. A. (2009). Modern organic and broiler chickens sold for human consumption provide more energy from fat than protein. *Public Health Nutrition* 13(3): 400–8.

World Bank. (2006). *Repositioning Nutrition as Central to Development A Strategy for Large-Scale Action*. Washington DC, USA: The World Bank.

World Bank, FAO and IFAD. (2008). *Gender in Agriculture Sourcebook*. Washington DC, USA: The World Bank.

Zander, D. V., Bermudez, A. J. and Mallinson, E. T. (1997). Principles of disease prevention: diagnosis and control. In B. W. Calnek, H. J. Barnes, C. W. Beard, L. R. McDougald, Y. M. Saif (eds), *Diseases of Poultry*, 10th ed., 3–45. London: Mosby-Wolfe.

Index

www.ingramcontent.com/pod-product-compliance
Lightning Source LLC
LaVergne TN
LVHW020628300926
849161LV00004B/640